Siam Physics Congress (SPC 2021)

Journal of Physics: Conference Series Volume 2145

Online
24 – 25 May 2021

ISBN: 978-1-7138-5116-5
ISSN: 1742-6588

Printed from e-media with permission by:

Curran Associates, Inc.
57 Morehouse Lane
Red Hook, NY 12571

Some format issues inherent in the e-media version may also appear in this print version.

Institute of Physics
Dirac House, Temple Back
Bristol BS1 6BE UK

Phone: 44 1 17 929 7481
Fax: 44 1 17 920 0979

techtracking@iop.org

Additional copies of this publication are available from:

Curran Associates, Inc.
57 Morehouse Lane
Red Hook, NY 12571 USA
Phone: 845-758-0400
Fax: 845-758-2633
Email: curran@proceedings.com
Web: www.proceedings.com

TABLE OF CONTENTS

The 16th SIAM PHYSICS CONGRESS (SPC2021)

B Chatthong[1,a], C Buranachai[1,b], P Kalasuwan[1,c], S Rakkapao[1,d], C Putson[1,e], B Soonthornthum[2,3,f,g] and S Aukkaravittayapun[2,3,f,h]

[1] Division of Physical Science, Faculty of Science, Prince of Songkla University, Songkhla, Thailand

[2] Thai Physics Society, Chiang Mai, Thailand

[3] National Astronomical Research Institute, Chiang Mai, Thailand

Email: [a]boonyarit.ch@psu.ac.th, [b]chittanon.b@psu.ac.th, [c]pruet.k@psu.ac.th, [d]suttida.r@psu.ac.th, [e]chatchai.p@psu.ac.th, [f]thaiphysoc@gmail.com, [g]boonrucksar@narit.or.th, [h]suparerk.a@narit.or.th

The 16th Siam Physics Congress (SPC2021) was successfully held 100% online during May 24–25, 2021. Initially, the Thai Physics Society and Prince of Songkla University had planned to co-host the SPC2021 in Songkhla province in Southern part of Thailand via hybrid form. Unfortunately, due to the rise of coronavirus (COVID-19) pandemic cases in Thailand, the organizing committee had considered that physical conference would no longer be possible, and made 100% online decision. This is to keep up to the highest standard of health regulation and self-practice (from WHO) in avoiding the extremely serious pandemic. Consequently, the Thai Physics Society and Prince of Songkla University had facilitated and hosted the SPC2021 as national teleconference during May 24–25, 2021.

This year, we had 185 research abstracts and papers submitted for both oral and poster online presentations. There were 1 plenary speaker, 25 invited speakers, 76 oral presentations, 83 poster presentations, and 99 full papers submitted for possible publication in all 21 tracks(topics). The scientific program was announced on the SPC2021 website (https://spc2021.sci.psu.ac.th/) and also distributed via email to all participants. The SPC2021 oral sessions were held via Zoom video conference. Participants can join online sessions either by using their individual Zoom account. The Zoom's chat channel as well as professional management such as co-hosting the Zoom room from the host-side were employed to alleviate some glitches (e.g. delayed or unclear signals due to slow internet connection from the client-side). In addition, the chat channel was also used for exchanging contact details for further detailed discussion about the presentation.

The online opening ceremony and plenary talk was live-broadcasted via Faculty of Science, Prince of Songkla University official Facebook page and YouTube account. After the online opening ceremony, a plenary talk entitled "Exploring the Universe with neutrinos and gravitational waves" by Professor Takaaki Kajita, 2015 Nobel Laureate in Physics from Institute for Cosmic Ray Research, The University of Tokyo, Tokyo, Japan, was given to the audience of approximately 200 participants. The oral presentations were arranged into 3 parallel Zoom rooms in total of 16 sessions. Each invited speaker was allowed 30-minute presentation while regular presenter had 15 minutes including questions and answers. The online audience-size of each room varied between 30 and 60 people. The poster presentations were uploaded onto the SPC2021 Indico website (https://indico.cern.ch/event/973315/), where only registered participants could access. In addition, all participants can login, access all abstracts as well as presentation materials, and submit the full paper.

Journal of Physics: Conference Series **2145** (2022) 011001 doi:10.1088/1742-6596/2145/1/011001

We are also very grateful to the organizing committee for all their contribution in making this conference successful even during the COVID-19 crisis. The authors and reviewers had been working very hard to maintain the highest academic standard of the presented articles in this volume. It is very much hoped that these articles may serve as references across broad topics in physical science research communities (both to Thailand and the world). We also feel thankful and would like to acknowledge our sponsors for their supports.

Track

1. Astronomy, Astrophysics and Cosmology
2. Accelerators and Synchrotron Radiations
3. High Energy and Particle Physics
4. Nuclear and Radiation Physics
5. Plasma Fusion and Technology
6. Ion and plasma Physics
7. Atomic Physics, Quantum Physics, Molecular and Chemical Physics
8. Quantum Technology
9. Biological Physics, Biosensors and Chemical Sensors
10. Condensed Matter Physics
11. Surface, Interface and Thin Films
12. Magnetic and Semiconductor Physics
13. Material Physics
14. Nanoscale Physics and Nanotechnology
15. Energy Materials and Physics
16. Environmental Physics, Atmospheric Physics, Geophysics and Renewable Energy
17. Optics, Ultrafast Phenomena, and Photonics
18. Instrumentation, Metrology and Standards
19. Physics Innovation
20. Physics Education
21. Statistical and Theoretical Physics

Organizing Committee

- The Dean of Faculty of Science, Prince of Songkla University

- Associate Dean for Academic Affairs and Education, Faculty of Science, Prince of Songkla University
- The President of Thai Physics Society
- Head of Division of Physical Science, Faculty of Science, Prince of Songkla University
- Head of Secretary and Finance Committee
- Head of Scientific Committee
- Head of Bachelor Independent Project Competition Committee
- Head of Registration, Ceremony and Reception Committee
- Head of Audio Visual Aid Committee
- Head of Public Relations, Evaluation and Emergency Committee
- Head of Food, Beverage and Catering Committee
- Head of Sponsorship and Fundraising Committee
- Prof. Dr. SUKIT LIMPIJUMNONG
- Asst. Prof. Dr. TANAKORN OSOTCHAN
- Dr. SUPARERK AUKKARAVITTAYAPUN
- Asst. Prof. Dr. SIRAMAS KOMONJINDA
- Asst. Prof. Dr. SUPAB CHOOPUN
- Assoc. Prof. Dr. YONGYUT LAOSIRITAWORN
- Dr. SARAN POSHYACHINDA
- Assoc. Prof. Dr. DHEERAWAN BOONYAWAN
- Asst. Prof. Dr. KAJORNYOD YOODEE
- Asst. Prof. Dr. JARIN KANCHANAWARIN
- Assoc. Prof. Dr. JIRAROJ T-THIENPRASERT
- Asst. Prof. RATCHANA CHINPITAK
- Dr. CHAIYAPONG RUANGSUWAN
- Assoc. Prof. Dr. SIRICHOK JUNGTHAWAN
- Asst. Prof. Dr. THIRANEE KHUMLUMLERT
- Mr. CHALOEMCHON WANNATHONG
- Asst. Prof. Dr. PHICHET KITTARA
- Asst. Prof. Dr. SOJIPHONG CHATRAPHORN
- Asst. Prof. Dr. PRASONG KESSARATIKOON
- Asst. Prof. Dr. SARAWUT THOUNTOM

Scientific Committee

- Asst. Prof. Dr. CHITTANON BURANACHAI
- Asst. Prof. Dr. BOONYARIT CHATTHONG
- Dr. PRUET KALASUWAN
- Assoc. Prof. Dr. TEPARKSORN PENGPAN
- Assoc. Prof. Dr. CHATCHAI PUTSON
- Assoc. Prof. Dr. YUTTHANA TIRAWANICHAKUL
- Assoc. Prof. Dr. PANOTE THAVARUNGKUL
- Assoc. Prof. Dr. NANTAKAN MUENSIT
- Assoc. Prof. Dr. SAWASDEE YORDKAYHUN
- Asst. Prof. Dr. SUTTIDA RAKKAPAO
- Asst. Prof. Dr. PAPHAVEE VAN DOMMELEN

- Asst. Prof. Dr. CHALONGRAT DAENGNGAM
- Asst. Prof. Dr. Helmut JOSEF DUERRAST
- Dr. NARAGORN KAEWKAO
- Dr. KOMRIT WATTANAVATEE
- Dr. PRAOPIM LIMSAKUL
- MR. WANICBUT WATTANAMATIPHOT
- Mr. NATJADEE DORLOH
- MR. ISARAPAB CHUMRUKSA
- MR. PANUMAS ONMARK

Bachelor Independent Project Competition Committee
- Asst. Prof. Dr. JARIN KANCHANAWARIN
- Assoc. Prof. Dr. DHEERAWAN BOONYAWAN
- Asst. Prof. Dr. SOJIPHONG CHATRAPHONG
- Assoc. Prof. Dr. SIRICHOK JUNGTHAWAN
- Asst. Prof. Dr. CHITTANON BURANACHAI
- Asst. Prof. Dr. KRITSANU TIVAKORNSASITHORN

Registration, Ceremony and Reception Committee
- Assoc. Prof. Dr. CHATCHAI PUTSON
- Asst. Prof. Dr. BOONYARIT CHATTHONG
- Asst. Prof. Dr. SUTTIDA RAKKAPAO
- Dr. SUREERAT HOMHUAN
- Dr. PRAOPIM LIMSAKUL
- Dr. RADCHAGRIT SUPAKULOPAS
- MR. WANTAPON NGAMDEE
- MISS RATCHANEEWAN SIRI
- MISS VITCHUTDA SUKNUI
- MISS NUTCHANAT SUCHATPONG
- MR. WISSARUT PUAYPUNG
- MRS. JITTIMA PHOSENA
- MRS. ANNA CHATTHONG
- MR. TOM HENDRIK FREIMUTH
- MISS SUNISA MEESEN
- MR. ISARAPAB CHUMRUKSA
- MR. PANUMAS ONMARK
- MR. WANICBUT WATTANAMATIPHOT
- MISS CHANYA INTAKARN
- MR. TANAWAT THIANGTRONG
- MR. TEERAYUT MALAKUL
- MR. ATHASIT RONGSAWAD
- MR. WANCHAI WANNO
- MR. CHAYANON PROMMOON

Audio Visual Aid Committee
- DR. KOMRIT WATTANAVATEE
- Asst. Prof. Dr. KAMHAENG WATTANASEN

Journal of Physics: Conference Series **2145** (2022) 011001 doi:10.1088/1742-6596/2145/1/011001

- Asst. Prof. Dr. SUKSAWAT SIRIJARUKUL
- Asst. Prof. Dr. KRISADA RAWIRASWATTANA
- MR. WANTAPON NGAMDEE
- MR. WISSARUT PUAYPUNG
- MISS NUTCHANAT SUCHATPONG
- MR. KHAMPANART PHANNARAI
- MR. ATHASIT RONGSAWAD
- MR. CHALAM PROMPAT
- MR. ADISORN VANGPREECHA
- MR. SANTI BUAKING
- MR. TANAWAT THANGTRONG
- MR. PRAPHUN INTACHAI
- MR. JUMRAS NASUWAN
- MR. THUNYARODE INSUVANNO

Public Relations, Evaluation and Emergency Committee

- Asst. Prof. Dr. BOONYARIT CHATTHONG
- MR. WISSARUT PUAYPUNG
- MISS VITCHUTDA SUKNUI
- MR. ISARAPAB CHUMRUKSA
- MR. PANUMAS ONMARK
- MR. WANICBUT WATTANAMATIPHOT
- MISS CHANYA INTAKARN

Food, Beverage and Catering Committee

- MISS WANIDA SUMATHAKULAWAT
- Dr. KOMRIT WATTANAVATEE
- MR. WISSARUT PUAYPUNG
- MISS NUTCHANAT SUCHATPONG
- MISS VITCHUTDA SUKNUI
- MISS RATCHANEEWAN SIRI
- MISS VARAPORN KARANSANTI
- MRS. SUPRANEE SAMNAKPONG

Sponsorship and Fundraising Committee

- Asst. Prof. Dr. PAPHAVEE VAN DOMMELEN
- Assoc. Prof. Dr. YUTTHANA TIRAWANICHAKUL
- Asst. Prof. Dr. CHALONGRAT DAENGNGAM
- Dr. PRUET KALASUWAN
- MISS VITCHUTDA SUKNUI

Plenary Speaker

- Prof. Dr. Takaaki Kajita, Institute for Cosmic Ray Research, University of Tokyo, Tokyo, Japan

Organizing Institutions

- Prince of Songkla University, Thailand
- Thai Physics Society

Sponsors

Editorial Materials

It gives us great pleasure to announce that the *Siam Physics Congress 2021 (SPC2021)* held on 24–25 May 2021 through online hubs at https://spc2021.sci.psu.ac.th and https://indico.cern.ch/event/973315 was a resounding success. Even with the COVID-19 situation that prohibited the conventional face-to-face meeting, there were still 258 participants from Thailand and other countries registered the conference. In the SPC2021, unique opportunities had been provided for experts, academics, and educators to exchange knowledge and build friendships. The meeting consisted of plenary lecture, invited speaker sessions, and a number of contributed oral and poster presentations. Resulting from the success of SPC2021, there are qualified full research articles from presentations, which have been reviewed and compiled in IOP's *Journal of Physics: Conference Series.* It is our sincere hope that these articles will contribute to advancing our understanding of Physics knowledge and applications and serve as initiators for even greater achievements in the field. Lastly but most importantly, on behalf of the organizers (Prince of Songkla University and Thai Physics Society), we would like to express our gratitude for the support received from The Council of Scientific and Technological Association of Thailand, National Astronomical Research Institute of Thailand (NARIT), Synchrotron Light Research Institute (Public Organization), Institute for the Promotion of Teaching Science and Technology (IPST), Thailand Center of Excellence in Physics (ThEP), National Research Council of Thailand (NRCT), Ryts Instruments. Co. Ltd., Thailand Institute of Nuclear Technology (TINT), Society of Exploration Geophysicists (SEG), Nuclear Society of Thailand (NST) and CP All Public Company Limited, whose generous contributions made the SPC2021 a successful and noteworthy meeting.

The SPC2021 Editorial Team

Peer review declaration

All papers published in this volume of Journal of Physics: Conference Series have been peer reviewed through processes administered by the Editors. Reviews were conducted by expert referees to the professional and scientific standards expected of a proceedings journal published by IOP Publishing.

- **Type of peer review:** Single-blind
- **Describe criteria used by Reviewers when accepting/declining papers. Was there the opportunity to resubmit articles after revisions?:**
  The criteria for reviewers consist of:
  1) Importance of the topic
  2) Sufficient originality and interest to merit publication
  3) General layout of the paper
  4) Clarity of expression
  5) Standard use of English
  6) Use of figures and tables
  7) Validity of arguments
  8) Sufficient, correct and appropriate procedures and methods
  9) Adequacy of references
  10) Overall impression.
  The authors have opportunity to resubmit the revision as many times as required by the reviewers.
- **Conference submission management system:** indico.cern.ch
  All the papers were submitted online through the conference site:
  https://spc2021.sci.psu.ac.th and https://indico.cern.ch/event/973315/
- **Number of submissions received:** 99
- **Number of submissions sent for review:** 99
- **Number of submissions accepted:** 80
- **Acceptance Rate (Number of Submissions Accepted / Number of Submissions Received X 100):** 80.81%
- **Average number of reviews per paper:** 4.27
- **Total number of reviewers involved:** 86
- **Any additional info on review process:** A plagiarism software was used to check for plagiarism. The similarity index must be less then 50% unless the submission gets rejected.

Contact person for queries:

Dr. Yongyut Laosiritaworn (Editor in Chief)
Chiang Mai University, Thailand
yongyut.laosiri@cmu.ac.th

Dr. Boonyarit Chatthong (Coeditor in Chief)
Prince of Songkla University
boonyarit.ch@psu.ac.th

An investigation of Aharonov-Bohm effect towards the potential use for the gravitational wave detection

C Deesamer[1], N Wanwieng[2], P Chainakun[1,3,*] and A Watcharangkool[4]

[1]School of Physics, Institute of Science, Suranaree University of Technology, Nakhon Ratchasima 30000, Thailand
[2]Department of Physics and Materials Science, Faculty of Science, Chiang Mai University, Chiang Mai 50200, Thailand
[3]Centre of Excellence in High Energy Physics and Astrophysics, Suranaree University of Technology, Nakhon Ratchasima 30000, Thailand
[4]National Astronomical Research Institute of Thailand, Chiang Mai 50200, Thailand

*E-mail: pchainakun@g.sut.ac.th

Abstract. We investigate an alternative way to detect the gravitational wave using the concept of Aharonov-Bohm experiment in curved space-time. Our system consists of an electron beam which is split into two beams passing opposite sides of the solenoid and producing interference patterns. The change in interference patterns can be observed if the system is perturbed by the gravitational wave, and can be used to trace back to the nature of the gravitational wave. This system is described by the cylindrical coordinate in Minkowski space-time where we set the incoming wave propagating in the z-direction, perpendicular to the solenoid's cross-section. We found that the perturbation on the cross-section area due to gravitational strength is not strong enough to significantly change the phase shift. Contrarily, by changing the magnetic field generated by the current inside the solenoid, the results suggest that the significant phase shift could potentially be detected if the gravitational wave is allowed to propagate in the direction that is perpendicular to z-direction.

1. Introduction

The gravitational wave has been robustly detected for the first time in the stellar-mass binary black hole merging, GW150914, by the Laser Interferometer Gravitational-Wave Observatory (LIGO) in 2015 [1]. The measurement method of LIGO is based on the laser interferometry technique to measure the strain of the displacement of two perpendicular arms connecting to reflection mirrors, relating to the phase difference. Here, we investigate the gravitational wave detection in an alternative way using the Aharonov-Bohm experiment which provides the magnetic vector potential that can produce a specific interference pattern of the electron beams [2] once perturbed by the gravitational wave. Due to its nature, the wave passes through the earth with strength (h) weaker than 10^{-20} [3] and currently be detected with the minimum strain $\sim 10^{-21}$ that, when being treated as a weak perturbation from linearized theory [4], can slightly affect the space-time continuum [5]. Hence, the change in the interference pattern would be insignificant and difficult to be detected if the perturbation is beyond the first order. By setting the magnetic field in the Aharonov-Bohm experiment to be uniform, the change of magnetic

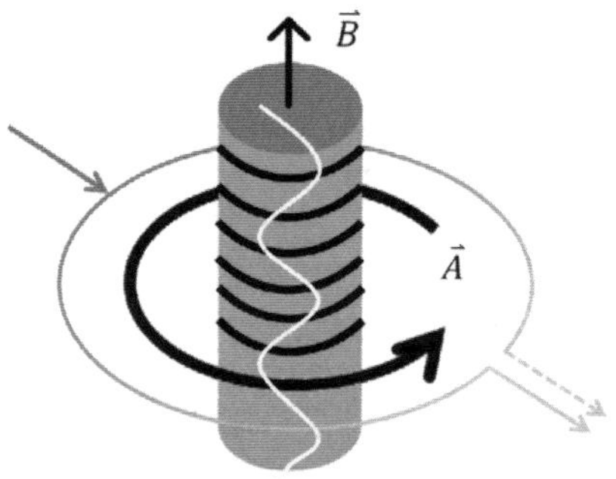

Figure 1. The schematic of Aharonov-Bohm experiment with gravitational wave. See text for more details.

flux and corresponding phase shift can be calculated from the change in the cross-section area of the solenoid. We then test the coupling between the gravitational wave and electromagnetic field from the variation of the current density using Maxwell's equation in curved space-time [6]. The metric tensor carrying the information of the weak perturbation contains the variation term that leads us to evaluate the potential use of this experimental set-up to detect the gravitational wave.

2. Experimental model

In the Aharonov-Bohm experiment, the incoming electron beam traveling in the x-y plane can be split into two beams passing opposite sides of the solenoid (figure 1). The solenoid is assumed to have infinite length where the magnetic field is zero outside the solenoid. Therefore, the magnetic vector potential $\mathbf{A}$ is only proportional to the magnetic field $\mathbf{B}$ inside the solenoid circuit, hence $\nabla \times \mathbf{A} = \mathbf{B}$. The magnetic flux inside the solenoid circuit then can be written as

$$\Phi = \int \mathbf{B} \cdot d\mathbf{S} = \oint \mathbf{A} \cdot d\mathbf{l} \,, \tag{1}$$

where $\mathbf{S}$ is the cross-section area of the solenoid and $\mathbf{l}$ is the path around the solenoid. Without gravitational wave, the phase shift produced by the Aharonov-Bohm effect is suggested to be [2]

$$\Delta\phi = \frac{q}{\hbar}\int \mathbf{B} \cdot d\mathbf{S} = \frac{q}{\hbar}\oint \mathbf{A} \cdot d\mathbf{l} = \frac{q}{\hbar}\Phi \,. \tag{2}$$

Since the gravitational wave perturbation influences the ripples in the geometry of space-time, we employ a uniform magnetic field and observe the change in the path length and the cross-section area due to the perturbation by the gravitational wave. This leads to the change in the current density that, finally, allows us to quantify the change in phase shift when the gravitational wave is present. In this case, we do not concern the interaction between each particle, i.e., weak or strong force.

We employ the metric tensor g consisting of the Minkowski metric η in local flat space-time and a weak perturbation h for the wave propagating in the vacuum in z-direction using transverse-traceless gauge [4], having the polarization h_+ and $h_\times$ on the x-y plane [7]. Then, the length L from point P to Q can be described by

$$L = \int_P^Q \sqrt{g_{\mu\nu}dx^\mu dx^\nu} = \int_P^Q \sqrt{(\eta_{\mu\nu} + h_{\mu\nu})dx^\mu dx^\nu} \,. \tag{3}$$

In cartesian coordinate, the metric tensor for the wave propagating in z-direction is

$$g_{\mu\nu} = \begin{bmatrix} -1 & 0 & 0 & 0 \\ 0 & 1 & 0 & 0 \\ 0 & 0 & 1 & 0 \\ 0 & 0 & 0 & 1 \end{bmatrix} + \begin{bmatrix} 0 & 0 & 0 & 0 \\ 0 & h_{+} & h_{\times} & 0 \\ 0 & h_{\times} & -h_{+} & 0 \\ 0 & 0 & 0 & 0 \end{bmatrix} \cos\left[\omega(t - z/c)\right] . \tag{4}$$

In cylindrical coordinate, using tensor transformation, the gauge is still transverse-traceless. The metric tensor then becomes

$$g_{\mu\nu} = \begin{bmatrix} -1 & 0 & 0 & 0 \\ 0 & 1 & 0 & 0 \\ 0 & 0 & r^2 & 0 \\ 0 & 0 & 0 & 1 \end{bmatrix} + \begin{bmatrix} 0 & 0 & 0 & 0 \\ 0 & H_{+} & rH_{\times} & 0 \\ 0 & rH_{\times} & -r^2H_{+} & 0 \\ 0 & 0 & 0 & 0 \end{bmatrix} , \tag{5}$$

where

$$\begin{aligned} H_{+}(\phi, t) &= M \sin(A + 2\phi) \cos\left[\omega(t - z/c)\right] , \\ H_{\times}(\phi, t) &= M \cos(A + 2\phi) \cos\left[\omega(t - z/c)\right] , \\ M &= \sqrt{h_{+}^2 + h_{\times}^2} . \end{aligned}$$

Furthermore, we define the initial length in x- and y-axis as L_{x0} and L_{y0}, respectively. When the effect of perturbation due to the gravitational wave is included (figure 2), we calculate the cross-section area at a particular value of z and t.

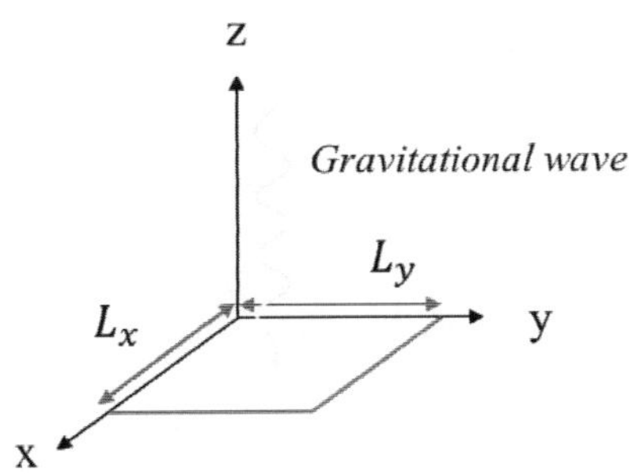

Figure 2. The schematic of the change in the cross-section area in Aharonov-Bohm experiment due to the gravitational wave perturbation.

The variation of current density j is evaluated via the derivative of metric tensor in case of a uniform magnetic field. According to Maxwell's equations in curved space-time [6], we can write

$$j^{\nu} = \frac{1}{\sqrt{-\det(g)}} \partial_{\mu}[\sqrt{-\det(g)} g^{\mu\rho} g^{\nu\sigma} F_{\rho\sigma}] \tag{6}$$

$$= \frac{1}{r} \partial_{\mu}[r g^{\mu\rho} g^{\nu\sigma} F_{\rho\sigma}] , \tag{7}$$

where $F_{\rho\sigma}$ is the electromagnetic strength tensor in cylindrical coordinate. For the infinite length solenoid, we do not consider the point charge and leave the current alone, so the electric field is negligible due to the uniform magnetic field configuration.

3. Result and discussion

To calculate the change in the cross-section area shown in figure 2, we multiply the proper length from two axes given in the cartesian coordinate (x,y)

$$L_x = L_{x0}\{1 + h_+ \cos\left[\omega(t - z/c)\right]\}^{\frac{1}{2}} \,, \tag{8}$$

$$L_y = L_{y0}\{1 - h_+ \cos\left[\omega(t - z/c)\right]\}^{\frac{1}{2}} \,, \tag{9}$$

to estimate the cross-section area, from equations (8) and (9),

$$S = L_x L_y = L_{x0} L_{y0} + O(h^2) \,. \tag{10}$$

The result shows that the difference between perturbed and unperturbed cross-section is in the order of h^2, which is negligible in the linearized theory. However, more precise calculations can be done by taking into account the cross polarization in the surface integral, but the results are not significantly different from our estimation here.

Furthermore, in cylindrical coordinate, we set the solenoid's radius r to be constant. Then, Maxwell's equation for fields which are constant in time with the gravitation effect can be expressed in terms of

$$j_h^r = \frac{1}{r^2}\partial_\phi\left[(1 - H_\times^2)B_z\right] - \partial_z\left[(1 - H_+)B_\phi\right] + \frac{1}{r}\partial_z(H_\times B_r) \,, \tag{11}$$

$$j_h^\phi = -\frac{1}{r}\partial_r\left[\frac{1}{r}(1 - H_\times^2)B_z\right] + \frac{1}{r^2}\partial_z\left[(1 + H_+)B_r\right] + \frac{1}{r}\partial_z(H_\times B_\phi) \,, \tag{12}$$

$$j_h^z = \frac{1}{r}\partial_r\left[(1 - H_+)rB_\phi\right] + \frac{1}{r}\partial_r(H_\times B_r) - \frac{1}{r^2}\partial_\phi\left[(1 + H_+)B_r\right] + \frac{1}{r}\partial_\phi(H_\times B_\phi) \,, \tag{13}$$

while the variation of current density in each dimension without non-gravitational effect is

$$j^{r*} = -\frac{1}{r^2}\partial_\phi(H_\times^2 B_z) - \partial_z(H_+ B_\phi) + \frac{1}{r}\partial_z(H_\times B_r) \,, \tag{14}$$

$$j^{\phi *} = \frac{1}{r}\partial_r(\frac{1}{r}H_\times^2 B_z) + \frac{1}{r^2}\partial_z(H_+ B_r) + \frac{1}{r}\partial_z(H_\times B_\phi) \,, \tag{15}$$

$$j^{z*} = -\frac{1}{r}\partial_r(H_+ r B_\phi) + \frac{1}{r}\partial_r(H_\times B_r) - \frac{1}{r^2}\partial_\phi(H_+ B_r) + \frac{1}{r}\partial_\phi(H_\times B_\phi) \,. \tag{16}$$

Note that we discard the term that shows the second order of H^2 in $g^{\mu\rho}g^{\nu\sigma}$. As a result, all terms of each current density equation show no presence of B_z, hence the variation due to the perturbation in this way is approximately be zero.

On the other hand, by changing the orientation of the solenoid so that the wave propagation is tangent to the cross-section area, i.e. fixes x-axis and rotates 90°degree from the original z-direction, the perturbation is occurred only in one direction of the solenoid's cross-section as shown in figure 3. The perturbed area then becomes

$$S = \int\!\!\int \sqrt{g_{xx}g_{zz}}\, dx dz = \int\!\!\int \sqrt{1 + O(h)}\, dx dz \;\approx\; L_{x_0} L_{z_0}[1 + O(h)]^{\frac{1}{2}} \,. \tag{17}$$

Consequently, some of the metric tensor multiplication terms can be in the order of h which, in turn, allow the variation of the current density. Under this configuration, the significant phase shift should potentially be detected, which will be investigated further in the future.

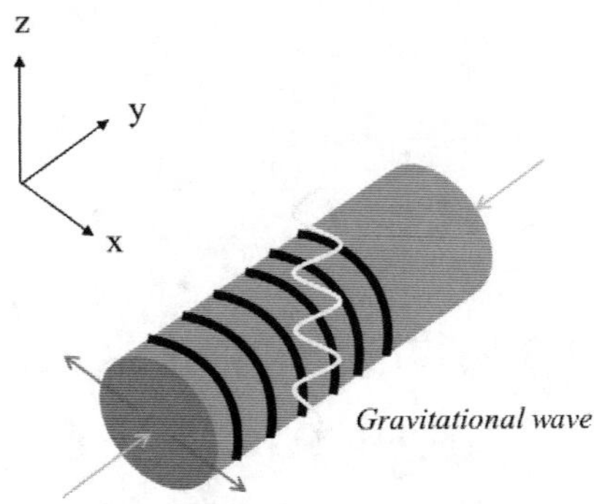

Figure 3. The schematic of the solenoid, rotating 90° from the original z-direction.

4. Conclusion

Our calculations show that the change in phase shift induced by the perturbation on the cross-section area due to the gravitational strength is not significant, especially if the gravitational wave is travelling along the z-direction. This system, instead, could be used to detect gravitational wave travelling in the tangent direction to the solenoid cross-section area.

Acknoledgement

This research was supported by the Program Management Unit for Human Resources & Institution Development, Research and Innovation [grant number B05F630108]

Reference

[1] Abbott B P *et al.* 2016 *Phys. Rev. Lett.* **116**(6) 061102
[2] Aharonov Y and Bohm D 1959 *Phys. Rev.* **115** 485
[3] Blair D 1991 *The Detection of Gravitational Waves* (Cambridge : University Press)
[4] Carroll S M 2019 *Spacetime and Geometry* (Cambridge : University Press)
[5] Einstein A 1916 *Ann. Phys.* **354** 769
[6] Hall G S 1984 *Gen. Relativ. Gravit.* **16** 495
[7] Hawking S W and Israel W 1987 *Three Hundred Years of Gravitation* (Cambridge : University Press)

Galaxy evolution in different environments along redshift within the local universe $z < 0.8$

Ponlawat Yoifoi[1,2,*] **and Wichean Kriwattanawong**[1]

[1] Department of Physics and Materials Science, Faculty of Science, Chiang Mai University, Chiang Mai, 50200, Thailand

[2] Graduate School, Chiang Mai University, Chiang Mai, 50200, Thailand

*E-mail: ponlawat_yoifoi@cmu.ac.th

Abstract. This study presents the evolution of the galaxies in different matter density along redshift within the local universe. A sample of 702,352 galaxies was collected from the Sloan Digital Sky Survey (SDSS). Under the limitation of the spectroscopic data, the appropriate photometric redshift was used to represent the spectroscopic redshift in the range of $0.0 \leq z \leq 0.8$. Number density of galaxies, galaxy's colors, and star formation activities are considered to describe the evolution of galaxies. In summary, the number density is not clearly different although the Dec and RA of the sky areas are disparate, but it steeply declines along the redshift direction. Considering the number density together with galaxies' Hα emission line from spectroscopic data, we find that both equivalent of hydrogen alpha and Hα flux tend to decrease along the redshift, similar to the decreasing trend of the number density. Furthermore, the galaxy color trend is found to be redder as a function of the redshift for the magnitude range of $-19 \leq M_g \leq -17$. It implies that the overview of the star formation activity of the fainter galaxies at the lower redshift tend to show higher than the ones at higher redshift.

1. Introduction

Many sky surveys have observed galaxies in several sky regions, depending on the objective of the mission (e.g., The APM galaxy survey [1] and The 2dF Galaxy survey [2]). The APM survey found that the distribution of galaxies showed the presence of some structures connected together in the southern sky area. However, the distribution in the two-dimensional map cannot indicate the entire structure with the combinations of galaxy groups or clusters. Therefore, the three-dimensional galaxy distribution in the 2dF redshift survey is more reasonable way to clearly present the connected structures in the universe. The survey shows that some sky areas are crowded by the numerous galaxies alternating with empty space. For the locations with dense environment in the local universe, galaxy-galaxy and galaxy-cluster interactions can induce star formation activity in the galaxies. The star formation can be traced by Hα emission measurements and corresponding directly to the galaxy color indices [3]. This work aims to study the dependence of star formation activity on the locations having different matter density along the redshift, mainly using the data from the Sloan Digital Sky Survey (SDSS). Sample selection is described in section 2. Section 3 explains the Hα emission and galaxy color analysis. Finally, section 4 contains conclusions of the main results.

2. Sample selection

The SDSS is a large survey, that observed astronomical objects, covering one-third of the entire sky, mostly in the northern galactic and some parts of southern galactic hemisphere. In this work, we collected photometric data from the SDSS Data Released 15 (DR15). In addition, we used spectroscopic data from the Max Planck Institute for Astrophysics and the Johns Hopkins University (MPA-JHU) that measured the galaxies' emission lines, using spectroscopic data in the SDSS Data Released 7 (DR7) [4]. The difference between the two DRs is the observed sky area. For the SDSS DR7, the survey covers the sky areas, RA from 8^h to 16^h and declination (δ) from 0° to 70°. While the SDSS DR15 covers the observed sky areas of all previous versions and additional parts in southern galactic sky. Therefore, our sample data are limited by the sky areas of the observations in SDSS DR7. We selected the two sample areas as shown in table 1. However, the spectroscopic redshift data from each selected area are only ~6000 galaxies. In order to increase more statistical results, it is necessarily to use photometric redshift for the sample galaxies that have no available spectroscopic redshift. Thus, we need to find the variability of the photometric redshift comparing with the accurate spectroscopic redshift data. The standard error of the difference between photometric and spectroscopic redshift was calculated and plotted in figure 1. It is clearly seen that the photometric redshift can appropriately represent the spectroscopic redshift in the range of $z < 0.8$. As a consequence, the galaxies, which have the photometric redshift $z_{photo} < 0.8$, will be counted as our sample. Therefore, we obtained 702,352 sample galaxies, that can provide much better statistical results than the sample used only spectroscopic redshift.

Table 1. The selected sky areas that used in this research.

No.	Centered position with 1° width	
	α	δ
1	10h00m00s (149.5° - 150.5°)	0° - 23° (-0.5° - 23.5°)
2	14h00m00s (209.5° - 210.5°)	0° - 23° (-0.5° - 23.5°)

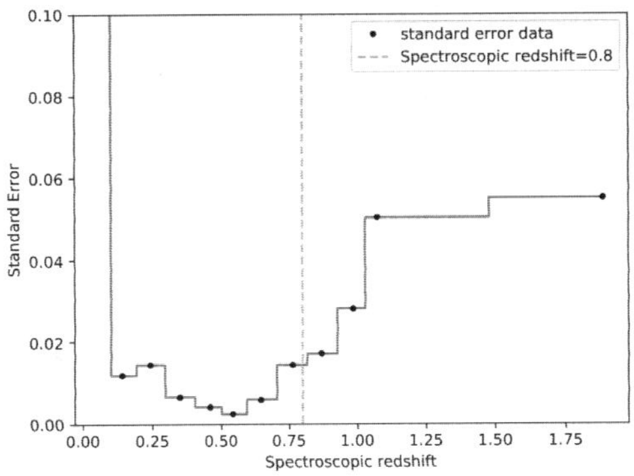

Figure 1. The standard error of the difference between photometric and spectroscopic redshift. The *orange dashed* line shows the selected redshift limit.

The SDSS data were corrected for the Galactic extinction but not for the K-correction. Thus, The K-correction model [5,6] was applied to correct photometric magnitudes for our sample. The sample was subdivided into 48 parts of sky areas with 1×1 deg^2. Besides, the sample was subdivided into 13 redshift intervals with a fixed distance of 180 Mpc. In this work, we applied the Hubble constant $H_0 = 67.66$ km s^{-1} Mpc^{-1} [7] to calculate the distance of the sample galaxies. As a result, we have 624 subsamples, which are statistically sufficient to compare the variation of galaxy parameters distributed in different matter density within $z < 0.8$.

3. Results and discussions

3.1. Number density

The number densities of 624 subsamples were calculated for each volume of a three-dimensional spherical sector [8]. The number densities in some sky areas are plotted against the redshift, as shown in figure 2. The number density in other sky areas show a similar trend as shown in figure-2. The trends of number density gradually drop as a function of redshift in both RA directions. However, there is a "bump" near the redshift $z \sim 0.4$ for all subsamples. With considering the number densities together with the spatial distribution of galaxies in figure 3, they are clearly seen that the number of galaxies is densely populated near the redshift ~0.4 and lightly for higher redshift.

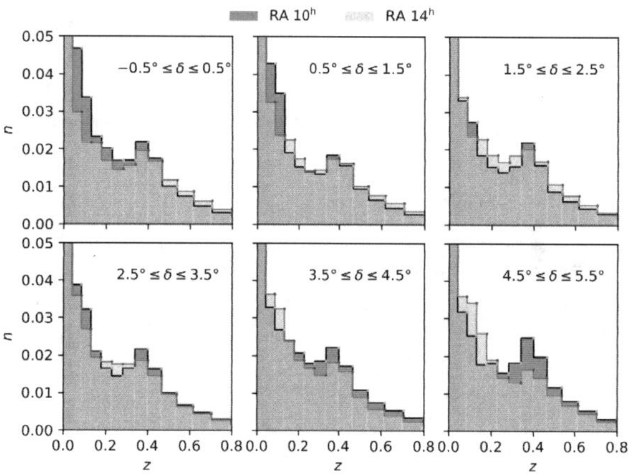

Figure 2. The number density against the redshift bin for both RAs with the declination range from -0.5° to 5.5°.

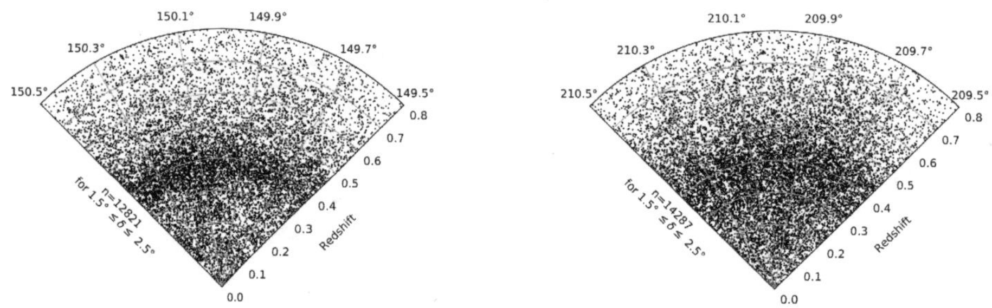

Figure 3. The spatial distribution of sample galaxies in some 1°×1° sky areas for both RAs.

3.2. The Hα emission

The galaxies' emission line from the MPA-JHU spectroscopic data, used in this work, is Hα line. This emission line is typically used to study the star formation activity in the galaxy (e.g., [3,9]). The two important parameters of the Hα emission in this work are the equivalent width of hydrogen alpha (EW(Hα)) and Hα flux density. However, the available MPA-JHU data for the Hα line of our sample are only ~5000 galaxies, i.e., 0.7% of the photometric data. As a consequence, we combined subsamples in all sky areas together for each redshift bin. The average

$EW(\mathrm{H}\alpha)$ and average Hα flux density were then calculated to find the mean characteristic along the redshift. Figures 4 and 5 clearly show that both Hα parameters tend to decrease as a function of redshift. These decreasing trends imply that the star formation activity of galaxies increases toward the very low redshift. Furthermore, there are high fluctuations in both Hα emission densities at the redshift more than 0.3. However, these variations are considered to be negligible when comparing with the values at the very low redshift.

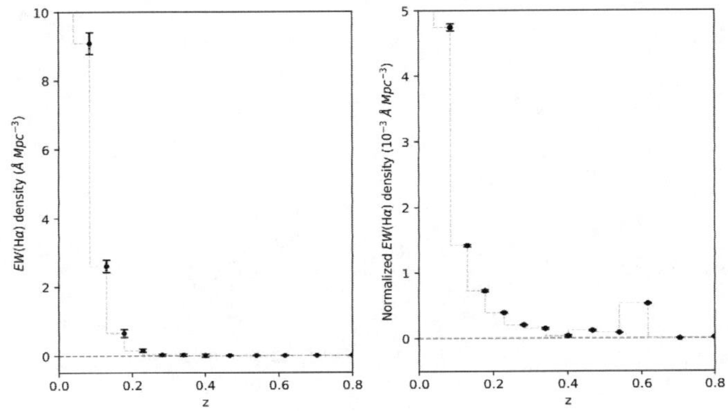

Figure 4. The $EW(\mathrm{H}\alpha)$ density along the redshift. *Left:* The average $EW(\mathrm{H}\alpha)$ density *Right:* The normalised $EW(\mathrm{H}\alpha)$ density by the number of galaxies in those redshift bins.

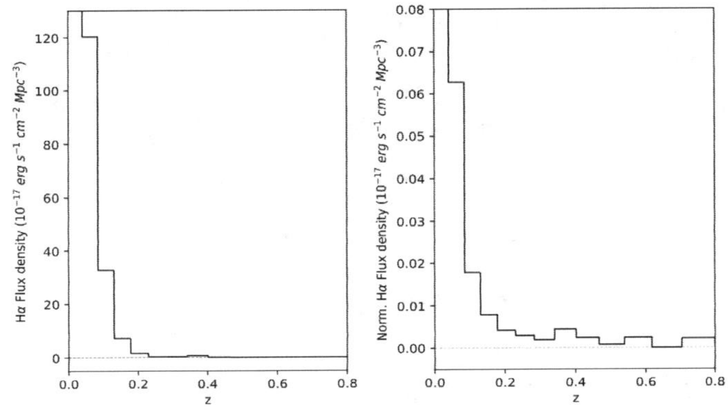

Figure 5. The Hα flux density along the redshift. *Left:* The average Hα flux density *Right:* The normalised Hα flux density by the number of galaxies in those redshift bins.

3.3. The galaxy's color

The color of galaxy is a parameter, widely used to study the stellar content in the galaxy [10]. The three color indices $g-r$, $g-i$, and $r-i$, were calculated and plotted against the redshift, as shown in figures 6 and 7. The M_g band absolute magnitude of the sample is in the range of $-21 \leq M_g \leq -17$. For both RAs, the $g-i$ and $r-i$ colors show upward trends for fainter sample (-19 $\leq M_g \leq$ -17). While a trend of the $g-r$ color trend is nearly constant along the redshift, similar to the trends of the three color indices for the brighter sample (-21 $\leq M_g \leq$ -19). This means that the fainter sample galaxies tend to be bluer as we look toward the lower redshift, consistent to the higher star formation activity at the lower redshift as mentioned above.

4. Conclusions

In this work, the SDSS DR15 and the Hα emission data from the MPA-JHU project were analyzed to investigate the evolution of galaxy parameters within redshift $z \sim 0.8$. Four selected parameters, which are the number density, $EW(\mathrm{H}\alpha)$, Hα flux density, and the color index, were considered to indicate the difference between the two sky areas and along the redshift. The photometric results imply that there exist different matter densities along the redshift. It is clear that Hα emission and the number density show the decreasing trends as a function of redshift. While the $g - i$ and $r - i$ color indices seem to be redder along the redshift for the fainter galaxies (-19 $\leq M_g \leq$ -17). The results show that the fainter sample galaxies at lower redshift tend to have higher star formation activity than the ones at higher redshift.

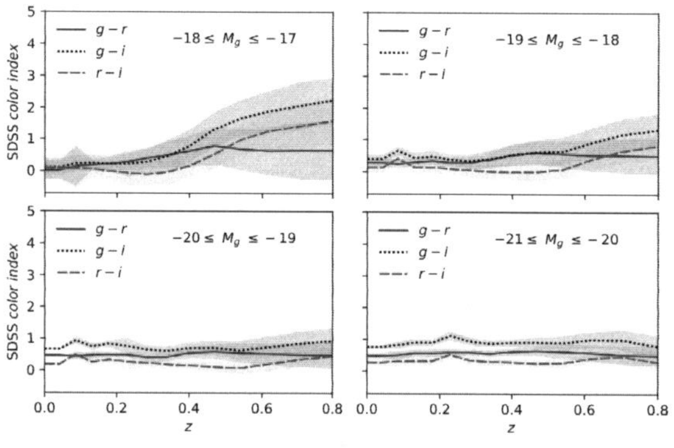

Figure 6. The evolution of average SDSS colors along the redshift at RA 10^{h} for each M_g magnitude bin from -21 to -17. *Shaded region:* 1-σ range of uncertainty in SDSS colors.

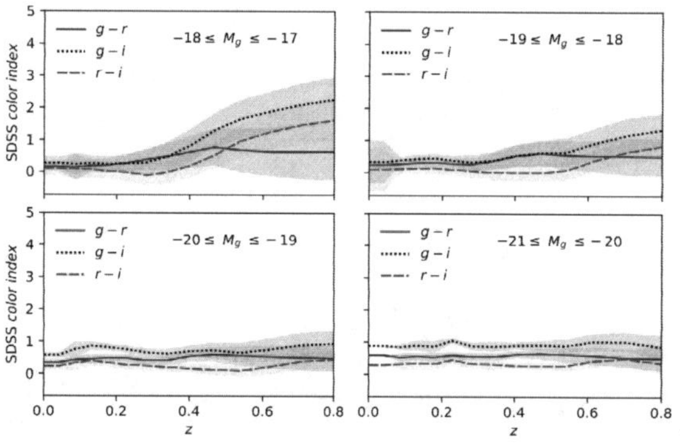

Figure 7. The evolution of average SDSS colors along the redshift at RA 14^{h} for each M_g magnitude bin from -21 to -17. *Shaded region:* 1-σ range of uncertainty in SDSS colors.

Acknowledgments

This work has made use of the database of the SDSS. Funding for the SDSS IV has been provided by the Alfred P. Sloan Foundation, the U.S. Department of Energy Office of Science, and the Participating Institutions. The anonymous reviewers are thanked for many helpful comments regarding the content and presentation of this paper.

References

[1] Dalton G B, Maddox S J, Sutherland W J and Efstathiou G 1997 *Mon. Not. R. Astron. Soc.* **289** 263
[2] Peacock J A *et al* 2001 *Nature* **410** 169
[3] Kennicutt Jr R C 1983 *Astrophys. J.* **272** 54
[4] Brinchmann J, Charlot S, White S D M, Tremonti C, Kauffmann G, Heckman T and Brinkmann J 2004 *Mon. Not. R. Astron. Soc.* **351** 1151
[5] Chilingarian I V, Melchior A L and Zolotukhin I Y 2010 *Mon. Not. R. Astron. Soc.* **405** 1409
[6] Chilingarian I V and Zolotukhin I Y 2012 *Mon. Not. R. Astron. Soc.* **419** 1727
[7] Planck Collaboration *et al* 2020 *Astron. Astrophys.* **641** A6
[8] Ly C *et al* 2011 *Astrophys. J.* **726** 109
[9] Kriwattanawong W, Moss C, James P A and Carter D 2011 *Astron. Astrophys.* **527** A101
[10] Baldry I K *et al* 2004 *Astrophys. J.* **600** 681

Fitting electron spectrum from AMS-02 by pulsar electrons

Kritaporn Butsaracom[1,2], Brandon Khan Cantlay[1] and Maneenate Wechakama[1,2,*]

[1]Department of Physics, Faculty of Science, Kasetsart University, Bangkok, 10900, Thailand
[2]National Astronomical Research Institute of Thailand (Public Organization), Chiangmai, 50180, Thailand

[*]E-mail: `fscimnw@ku.ac.th`

Abstract. In this work, we aim to explain the latest data of cosmic-ray electrons from AMS-02 by an electron background model and pulsar electrons. We consider an electron background model which includes primary and secondary electrons. We assume that pulsars are major sources of the electron excess. Since electrons easily lose their energy through the interstellar radiation field and the magnetic field via inverse Compton scattering and synchrotron radiation, respectively, they propagate in a short length. We adopt nearby pulsar data in the distance of 1 kpc from the Australia Telescope National Facility (ATNF) pulsar catalogue. By using a Green's function of an electron propagation model, we then fit pulsar parameters (i.e. the spectral index, the fraction of the total spin-down energy and the cutoff energy) for several cases of a single pulsar. With a combination of the electron background model, several cases of pulsar spectrum are able to explain the electron excess.

1. Introduction

An excess of electron spectrum has been measured by remarkable detectors such as AMS-02 [1], PAMELA [2], Fermi-LAT [3], H.E.S.S.[4], CALET [5] and DAMPE [6]. There are interpretations that try to explain origins of the electron excess by supernova remnants (SNRs) [7], pulsars [8] and dark matter annihilation or decay which requires the existence of new particles [9].

In this work, we try to explain the electron excess based on the latest data of electron fluxes from Alpha Magnetic Spectrometer-02 (AMS-02) mounted on the international space station [1] by electron spectrum from pulsars. We consider nearby pulsars that produce a burst-like source as the source of the electron excess instead of SNRs, which are assumed to be a stable continuous source of cosmic-ray particles. The electron data are comprehensive in between a few GeV to about 1 TeV. In our model, we consider an electron background model that consists of primary electrons from SNRs and secondary electrons from the inelastic collisions between cosmic-ray nuclei and the interstellar medium. The production of electrons from pulsars is considered as the source of the electron excess. The propagation of the pulsar electrons is determined by a solution of a Green's function of a diffusion model. We adopt nearby pulsar data in the distance of 1 kpc from the ATNF pulsar catalogue. We use the solution of the Green's function of an electron propagation model and the electron background model to find the fitting pulsar parameters, i.e. the spectral index, the fraction of the total spin-down energy and the cutoff energy, that provide the explanation for the electron excess of the AMS-02 data.

2. Method and data

In this work, we are interested in the cosmic-ray electron data from AMS-02 [1] that have revealed an electron excess against the electron background. We aim to explain the electron excess by using the pulsar electron spectrum which is explained by a Green's function of an electron propagation model. In this section, we will explain the model of the electron background and the model of the pulsar electron spectrum that we use in this work. Finally, the details for the pulsar data that we adopt from the ATNF pulsar catalogue and the method of the data fitting will also be expressed here.

2.1. The electron background

We consider that background electrons consist of primary electrons from SNRs and secondary electrons from the inelastic collisions between cosmic-ray nuclei and the interstellar medium [10]. The total flux of background electrons is given by

$$\Phi_{e^-} = \Phi_{e^-_{\rm pri}} + 0.6\Phi_{e^-_{\rm sec}}. \tag{1}$$

Due to the neutron production process, there are only 60 percents of secondary electrons left. The fluxes of primary and secondary electrons can be described by broken power-law functions,

$$\Phi_{e^-_{\rm pri}} = C_1 E^{-\gamma_1}\left[1+\left(\frac{E}{E_{br1}}\right)^{\gamma_2}\right]^{-1} \exp\left(\frac{-E}{E_c}\right) \tag{2}$$

and

$$\Phi_{e^-_{\rm sec}} = C_2 E^{-\gamma_3}\left[1+\left(\frac{E}{E_{br2}}\right)^{\gamma_4}\right]^{-1}, \tag{3}$$

where C_1 and C_2 are the normalization factor of the primary and secondary electrons, respectively, in the units of $\mathrm{GeV^{-1}m^{-2}s^{-1}sr^{-1}}$. γ_i is the spectral index, E_{br} is the break of energy in GeV and the energy cutoff, E_c, occurs only for the primary electrons. We adopt the best-fit parameters of the backgrounds from [10].

2.2. Model of electron spectrum from pulsars

The electrons propagation can be determined by a diffusion equation which is a function of diffusion parameters, loss rate and source term [11]. By using a Green's function, the solution of the electrons spectrum in a function of energy E, time t and pulsar distance $\boldsymbol{r}$ can be given by

$$\psi(E,t,\boldsymbol{r}) = \frac{1}{2}\frac{Q_0}{\pi^{3/2}r^3_{\rm diff}}\left(1-\frac{E}{E_{\rm max}}\right)^{\gamma-2}\left(\frac{E}{\rm GeV}\right)^{-\gamma}\exp\left(-\frac{E}{(1-E/E_{\rm max})E_c}\right)\exp\left(-\frac{r^2}{r^2_{\rm diff}}\right), \tag{4}$$

where $E_{\rm max} = (b_0 t)^{-1}$ is the maximum energy of electrons after the propagation and γ is the spectral index. The electrons can lose their energy through inverse Compton scattering via interstellar radiation field, i.e. low energy photons of cosmic microwave background, star light and dust emission, and synchrotron radiation via magnetic field which provide the loss rate $b_0 = 1.4 \times 10^{-16}\ \mathrm{GeV^{-1}s^{-1}}$. The diffusion length $r_{\rm diff}$ is explained in [11]. E_c is the cutoff energy.

Q_0 is the normalization factor related to the total injection energy $E_{\rm tot}$ which is expressed by

$$E_{\rm tot} = \int_{E_{\rm min}}^{\infty} EQ\left(E,t,\boldsymbol{r}\right) dE d\boldsymbol{r} dt, \tag{5}$$

where $E_{\rm min} = 0.1$ GeV and $Q\left(E,t,\boldsymbol{r}\right)$ is the source term of electrons provided by pulsars. The pulsar electron spectrum is described in a power-law energy spectrum in a term of delta fuctions,

$$Q\left(E,t,\boldsymbol{r}\right) = Q_0 \left(\frac{E}{\rm GeV}\right)^{-\gamma} \exp\left(-\frac{E}{E_c}\right) \delta\left(\boldsymbol{r}\right) \delta\left(t\right). \tag{6}$$

Finally, $E_{\rm tot}$ can be written in the term of a fraction of the total spin-down energy η, the spin-down luminosity of the pulsar $\dot{\varepsilon}$, the age of pulsar t and the typical luminosity decay time τ_0 which is equal to 10 kyr, as given by

$$E_{\rm tot} = \eta \dot{\varepsilon} t \left(1 + t/\tau_0\right). \tag{7}$$

2.3. The pulsar catalogue and the fitting method
We use data of nearby pulsars in the distance of 1 kpc which are observed by the ATNF pulsar catalogue. The catalogue provides the distance $\boldsymbol{r}$, the age t and the spin-down luminosity $\dot{\varepsilon}$ for individual pulsars. The catalogue reveals that the age of the nearby pulsars is in the order of 10^4 to 10^{10} years, and the spin-down luminosity is in the order of 10^{28} to 10^{35} $\rm ergs^{-1}$.

By using the electron background model and the parameters of pulsars provided by the ATNF catalogue, we are able to constrain the other parameters, i.e. the spectral index γ, the fraction of the total spin-down energy η and the cutoff energy E_c by fitting the modeled electron spectrum with the measured electron spectrum from AMS-02.

We provide the fitting parameters for each pulsar. We use the χ^2 to test the best-fit for the pulsar spectrum. The ratio of χ^2 with degrees of freedom is given by

$$\frac{\chi^2}{\rm df} = \frac{\sum \frac{(\text{Model}_i - \text{Observed}_i)^2}{\sigma_i^2}}{51 - 3}, \tag{8}$$

where σ_i represents the uncertainty of the observed data and the degrees of freedom is expressed by 51 observations and 3 fitted parameters.

3. Results
We have selected 65 possible pulsars in the distance of 1 kpc and adopted parameters such as the distance $\boldsymbol{r}$, the age t and the spin-down luminosity $\dot{\varepsilon}$ of each pulsar from the ATNF catalogue. We have adopted the best-fit parameters for the electron background model from [10]. We then constrain parameters such as the spectral index γ, the fraction of the total spin-down energy η and the cutoff energy E_c in our model for each pulsar to find the best fitting compared with the measured electron spectrum from AMS-02.

From 65 pulsars, there are 8 pulsars whose spectrum can explain the AMS-02 electron data well. The results of electron spectrum from the 8 pulsars, i.e. J0954-5430, J1825-0935, J0659+1414, J1732-3131, J1731-4744, J1809-2332, J2337+6151 and J1740-3015, are presented in figure 1 together with χ^2/df. The best-fit parameters of each pulsar are shown in table 1. They are young pulsars with ages less than 3×10^5 years and the spin-down luminosity $\dot{\varepsilon}$ are more than 4×10^{33} erg/s. Due to the high energy loss and low total injection energy, older pulsars and pulsars with the spin-down luminosity $\dot{\varepsilon}$ less than 4×10^{33} erg/s are unable to provide the electron spectrum at energy higher than 100 GeV where the electron excess starts to rise against the background. Surprisingly, the spectral index γ for the 8 pulsars is all equal to 1.50 where the fraction of the total spin-down energy η and the energy cutoff E_c are varied.

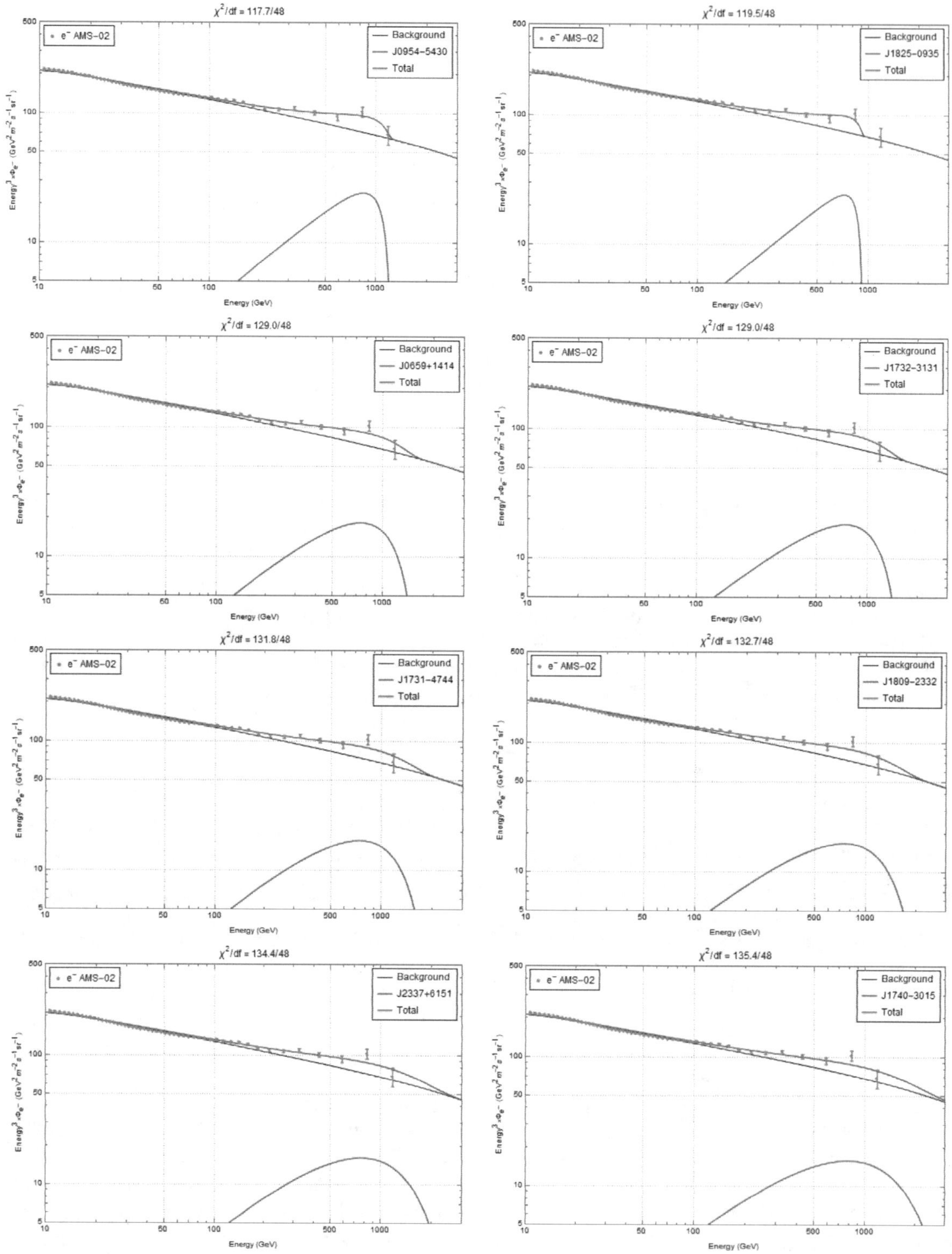

Figure 1. Cosmic-ray electron flux from AMS-02 (red dots) compares with modeled electron flux (blue line) which is a combination of background electron flux (black line) and electron flux from a single pulsar (brown line) for 8 pulsars that provide electron spectrum cover the electron excess.

Table 1. The best-fit parameters of 8 pulsars.

Parameters	r (kpc)	t (years)	$\dot{\varepsilon}$ (erg/s)	γ	η	E_c (GeV)
J0954-5430	0.433	1.71×10^5	1.60×10^{34}	1.50	0.693	3834
J1825-0935	0.300	2.33×10^5	4.50×10^{33}	1.50	1.92	5000
J0659+1414	0.288	1.11×10^5	3.80×10^{34}	1.50	0.415	1463
J1732-3131	0.640	1.11×10^5	1.50×10^{35}	1.50	0.105	1463
J1731-4744	0.700	8.04×10^4	1.10×10^{34}	1.50	1.64	1156
J1809-2332	0.880	6.76×10^4	4.30×10^{35}	1.50	0.0445	1066
J2337+6151	0.700	4.06×10^4	6.30×10^{34}	1.50	0.349	920
J1740-3015	0.400	2.06×10^4	8.20×10^{34}	1.50	0.308	838

4. Conclusions

We have used an electron background model and a model of pulsar spectrum to explain the latest data of cosmic-ray electrons from AMS-02. We have considered an electron background model which includes primary and secondary electrons and we have adapted the background model from [10]. We have used a pulsar spectrum model which is a solution of a diffusion equation in a function of diffusion parameters, loss rate and source term. We have adopted nearby pulsar data in the distance of 1 kpc from the Australia Telescope National Facility (ATNF) pulsar catalogue which provides parameters such as the distance r, the age t and the spin-down luminosity $\dot{\varepsilon}$ of each pulsar. We have constrained parameters of pulsars in the pulsar spectrum equation, i.e. the spectral index γ, the fraction of the total spin-down energy η and the cutoff energy E_c, to find the best-fit parameters to explain the AMS-02 data.

We have found that the pulsar spectrum from 8 pulsars, i.e. J0954-5430, J1825-0935, J0659+1414, J1732-3131, J1731-4744, J1809-2332, J2337+6151 and J1740-3015, combine with the electron background fit well with the cosmic-ray electron data from AMS-02. The pulsars are younger than 3×10^5 years and the spin-down luminosity $\dot{\varepsilon}$ are more than 4×10^{33} erg/s otherwise, they are unable to provide the electron spectrum that can explain the electron excess at energy higher than 100 GeV. In the future, we aim to extend our model for double and triple pulsars with the purpose to find a better explanation of the cosmic-ray electron sources.

Acknowledgments

We would like to thanks the members of the Astroparticle Physics group at Kasetsart University for very useful suggestions. M. Wechakama would like to acknowledge that this research project is supported by Kasetsart University Research and Development Institute (KURDI).

References

[1] Aguilar M *et al* (AMS Collaboration) 2019 *Phys. Rev. Lett.* **112** 101101
[2] Adriani O *et al* 2011 *Phys. Rev. Lett.* **106** 201101
[3] Ackermann M *et al* (Fermi LAT Collaboration) 2012 *Phys. Rev. Lett.* **108** 011103
[4] Aharonian F *et al* (H.E.S.S. Collaboration) 2008 *Phys. Rev. Lett.* **101** 261104
[5] Adriani O *et al* (CALET Collaboration) 2018 *Phys. Rev. Lett.* **120** 261102
[6] Ambrosi G *et al* (DAMPE Collaborationn) 2017 *Nature* **552** 63
[7] Di Mauro M *et al* 2014 *J. Cosmol. Astropart. Phys.* **2014** 006
[8] Lin S J, Yuan Q and Bi X J 2015 *Phys. Rev.* D **91** 063508
[9] Cirelli M, Kadastik M, Raidal M and Strumia A 2009 *Nucl. Phys.* B **813** 1
[10] Zu L *et al* 2018 *Phys. Rev.* D **98** 063010
[11] Ding Y C *et al* 2021 *Phys. Rev.* D **103** 115010

Isochrone fitting of ACS survey globular clusters using the PAdova and TRieste Stellar Evolution Code (PARSEC)

N Chehlaeh

Department of Science, Faculty of Science and Technology, Prince of Songkla University, Pattani, 94000, Thailand

E-mail: nareemas.c@psu.ac.th

Abstract. We present new isochrone fits to color magnitude diagrams (CMDs) of five globular clusters (GCs) including NGC 1261, NGC 1851, NGC 2298, NGC 3201, and NGC 4590. We used archival data obtained from the Advanced Camera for Survey (ACS) on board the Hubble Space Telescope (HST). The data of these five GCs were collected in F606W (V) and F814W (I) filters. In this study, the isochrone fitting to GC CMDs was analyzed using the PAdova and TRieste Stellar Evolution Code (PARSEC), which is the fundamental tool for age and distance estimation and modelling the evolution of stellar clusters and other galaxies. The main purpose is to estimate the fundamental physical properties of the GC samples using the PARSEC code and compare with results from published articles. The fundamental physical parameters determined in the study are age, metallicity, reddening, and distance modulus. The theoretical isochrone fits properly with the shape of CMD at the turn-off point that can be used to estimate the age and metallicity of clusters. We found that the age of these five GCs; NGC 1261, NGC 1851, NGC 2298, NGC 3201, and NGC 4590 are 12.6±1.0 Gyr, 12.0±1.0 Gyr, 12.7±1.0 Gyr, 12.0±1.0 Gyr, and 13.0±1.0 Gyr, respectively. Among the analyzed clusters, the results show that NGC 4590 is the oldest GC and has lowest metallicity value compare with other cluster samples. Studies of the properties and distribution of GCs play an important role to understand formation and evolution of the Milky Way.

1. Introduction

Globular clusters (GCs) are spherical compact group of stars mostly distributed within the Milky Way's Halo. They are approximately the same age, metallicity, distance, and chemical composition. GCs play an important role to reveal the information about structure of the Galaxy and assist us to understand the Galaxy. Theoretical stellar evolution models are extremely useful tools for the study of GCs, and the comparison of GC photometry to models is the traditional method to test stellar evolution models. Isochrone comparisons to GC photometry allows for the determination of cluster parameters [1]. Studying the ages of the Galactic Globular Clusters (GGCs) provide the relative ages and their behaviour with Galactocentric distance, and chemical constituent. Moreover, the absolute ages of the oldest GCs provide a lower limit on the Universe's age [2].

Targets in our study are NGC 1261, NGC 1851, NGC 2298, NGC 3201, and NGC 4590. These five GCs were selected from the list of Salaris and Weiss [3], who presented relationship between age and metallicity for 55 GGCs aim to describe the mechanisms of Galaxy formation. The celestial and Galactic coordinates for the cluster samples are presented in table 1.

1.1. NGC 1261
NGC 1261 is a medium metallicity highly concentrated GC, which located in constellation Horologium [4]. It is located at a distance of 17.2 kpc [5] and has metallicity of $[Fe/H] = -1.27$ dex [6]. It has low interstellar reddening and less contamination by field stars because of its high galactic latitude [7]. First observation and photographic study were made by Alcaino and Contreras [8]. The first CCD photometric observations were published by Bolte and Marleau [9], aimed to estimate the cluster's age. After space technology had been developed in the 20th century, the photometric data of 74 GGCs from the HST F439W and F555W bands were carried out [10], and 18 probable blue straggler candidates in NGC 1261 were discovered [11].

1.2. NGC 1851
NGC 1851 is an interesting high density GGC, which contains different types of stars with different values of chemical abundance [12]. It is located at a distance of 12.1 kpc with an intermediate metallicity of $[Fe/H] = -1.18$. A further advantage to photometrically studying this cluster is its very low reddening of $E(B-V) = 0.02$ [13], so there is no concern about the photometric effects of a large variable reddening. In 2017, the first UV CMDs of the cluster were published by Subramaniam *et al* using UVIT/AstroSat telescope [14].

1.3. NGC 2298
NGC 2298 is a cluster of intermediate concentration and low metallicity ($[Fe/H] = -1.85$) [13]. The first photographic study was presented by Alcaino [15] with 60 stars and calibrated from a photoelectric sequence of 14 stars. The study only hinted at the horizontal-branch position. In 1986, all 48 *BVRI* CCD frames were carried out by Alcaino and Liller [16], who presented photometric analysis for 117 stars. The results are deducing turnoff points for all diagrams ($V_{TO} = 19.5$, $B-V = 0.6$) and an age of 17.0 ± 1.5 Gyr.

1.4. NGC 3201
NGC 3201 is a low concentration cluster, which is known to have a retrograde Galactic orbit with distance of 5.2 kpc. Recent studies include many publications [17–19]. The effects of differential reddening across the face of NGC 3201 are minimal, because of its low-latitude position in Galactic coordinate. Gonzalez and Wallerstein reported a range in of as much $E(B-V) = 0.21$ [20]. Moreover, they presented spectroscopic data for 18 giants in the cluster. A recent study in 2018, Monty *et al* estimated metallicity of $[Fe/H] = -1.6$ dex [19].

1.5. NGC 4590
NGC 4590 or M68 is one of the most metal-poor GGCs. It contains 50 known variable stars. The first photometric analysis in *BVIR* was performed in 1990 by Alcaino *et al* [21]. The comparison between four-color CMDs and the theoretical isochrones with a cluster age of 13 Gyr. They estimated values for the interstellar reddening of 0.02 mag with what would be expected assuming $E(B-V) = 0.07$ mag. In 2015, 42 RR Lyrae (RRL) variables in the cluster include 12 doublemode RRL (RR01) stars were discovered to present the relationship between mass and metallicity of field and cluster RRL stars [22].

In this paper, we present new isochrone fits to CMDs of ACS survey GCs from HST. Our main purpose of the study is to estimate the fundamental physical properties of the GC samples using PARSEC code. The data and analysis are described in section 2. The results and discussion are present in section 3. Finally, a brief summary is given in section 4.

2. Data and analysis

We obtained the archival data of GC samples from the Advanced Camera for Survey (ACS) on board the HST. The main goal of the survey was to observe the central areas of 65 GCs in order to generate enormous star catalogues in homogeneous set for several fields of scientific studies. Project number of the GCs ACS Survey is GO-10775 (PI: Ata Sarajedini, University of Florida). All data were taken in two filters including F606W (V) and F814W (I). The total exposure times for each cluster and each filter are presented in table 1. The exposures for each cluster were taken in 4-5 deep exposures for each of the F606W and F814W filters. In this study, we used calibrated magnitude values of F606W and F606W-F814W for our CMDs analysis.

Table 1. The celestial and Galactic coordinates for the cluster samples in the study. Cluster coordinates are taken from the Milky Way GC McMaster Catalog [23] and summary of cluster observations were takem from publication of Anderson *et al* [24].

Cluster	R.A.(J2000) (hh:mm:ss)	Dec.(J2000) (dd:mm:ss)	l (deg)	b (deg)	Date (mm/dd/yy)	F606W (V)	F814W (I)
NGC 1261	03:12:16.21	-55:12:58.4	270.54	-52.12	03/10/2006	5×350s	5×360s
NGC 1851	05:14:06.76	-40:02:47.6	244.51	-35.03	05/01/2006	5×350s	5×350s
NGC 2298	06:48:59.41	-36:00:19.1	245.63	-16.00	06/12/2006	5×350s	5×350s
NGC 3201	10:17:36.82	-46:24:44.9	277.23	+08.64	03/14/2006	4×100s	4×100s
NGC 4590	12:39:27.98	-26:44:38.6	299.63	+36.05	03/07/2006	4×130s	4×150s

Data reduction and calibration were done using the ACS pipeline [24] before publishing the on-line archival data. Firstly, data reduction was performed using the ACS pipeline, this step includes bias subtracted and flat fielded. Secondly, The detection and identification of all stars in the frame were performed using the automated IRAF daofind routine on combined frame in the F606W band [25]. Next step, stellar photometry based on the point-spread function fitting (PSF photometry) were performed using the DOLPHOT software package. Consequently, instrumental magnitudes were calibrated into the VEGAMAG system using the photometric zero points for the ACS VEGAMAG magnitude system [26]. The procedure also included and combined the same band images.

Magnitude values in F606W and F814W bands from archival data were used to plot CMDs for our GCs samples. The stars in the CMDs of all cluster samples are very crowed, especially for high magnitude stars. In order to avoid the effects of the crowding and scattering stars in CMDs, we only selected the stars with magnitude errors of less than 0.2 in both filters following Barker and Paust [1]. We fitted the theoretical isochrone to the data using the Padova isochrones library or PARSEC code [27]. It consists of isochrone tracks with metallicity from $0.0005 \leq Z \leq 0.07$, and initial mass values from $0.1M_{\odot} \leq M \leq 12M_{\odot}$. There are several photometric systems in the library, including HST-WFC, which is used in this study. The major physics concepts in the code are initial chemical composition, nuclear reaction rates, stellar atmosphere, opacities, microscopic diffusion, stellar extinction, initial mass function, circumstellar dust and so on [27]. The PARSEC model in 1.2S version was specially used in this study. This is the most updated version, which includes an improvement for very low mass main sequence (MS) stars [1]. The best fits of isochrones to the CMDs of each cluster sample are presented in figure 1. The white solid lines represent the best fit isochrones to the data by adopting the specific values of metallicity and age.

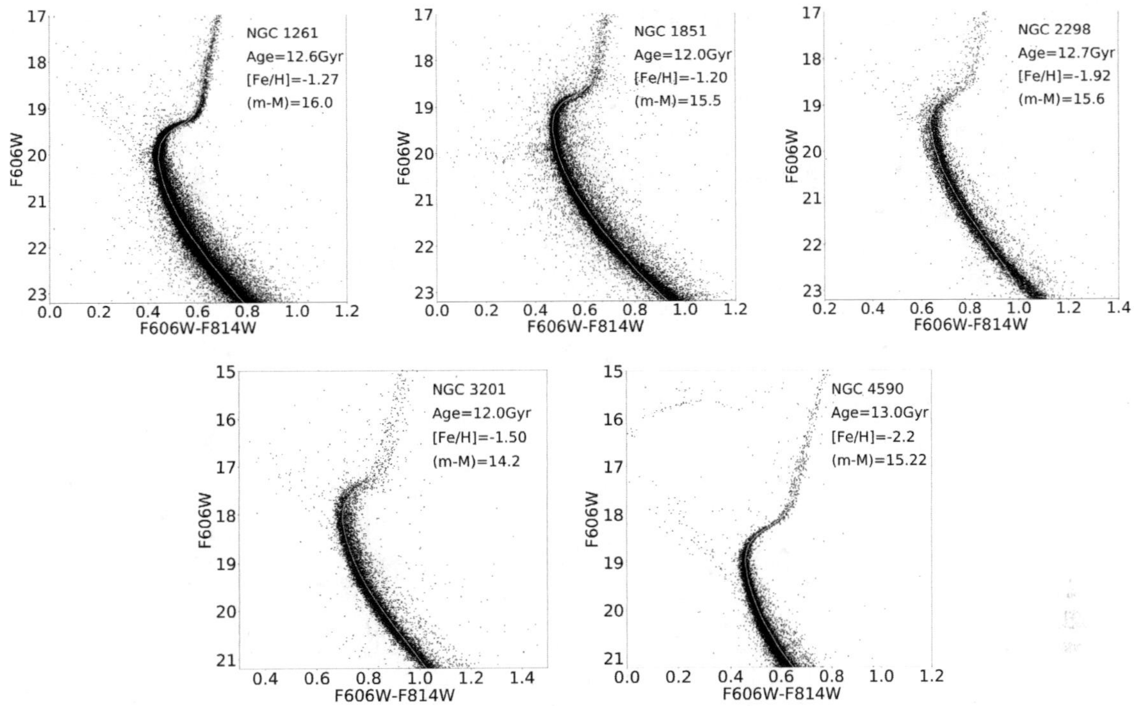

Figure 1. The PARSEC isochrones fit to CMDs of NGC 1261, NGC 1851, NGC 2298, NGC 3201, and NGC 4590, in F606W and F814W filters. The PARSEC isochrones have good fit to the CMDs with our age and metallicity values.

3. Result and discussion

After crowded stars are excluded from the CMDs of the cluster samples, each CMD presents very well defined pattern of turn-off point (TO) and red giant branch (RGB), which are the most outstanding structure in general GC CMDs. It represents that the stars in the cluster are mainly old. Interestingly, the cluster samples, except NGC 1851, obviously show blue straggler sequence in their CMDs. A blue straggler (BS) is a type of massive star that is brighter and bluer than stars in the TO. It lies along over the MS [28] and has evolved into white dwarf (WD) long ago [29].

From isochrone fitting analysis, we also obtained a reddening and distance modulus of the cluster samples as present in table 2. The table shows summary of the fundamental properties for our GC samples determined in this study compare with those from previous published articles. We found that CMD of NGC 1261 has very good fit with metallicity value of $[Fe/H] = -1.27$ and age of 12.6±1.0 Gyr. Our final results have good agreement with those values from Rani *et al* [4]. The CMD of NGC 1851 has good fit with metallicity value of $[Fe/H] = -1.20$ and age of 12.0±1.0 Gyr.

In this study, we obtained the new values of both age and metallicity, which have significant differences with the results from previous works. We found that CMD of NGC 2298 has very good fit with metallicity value of $[Fe/H] = -1.92$ and age of 12.7±1.0 Gyr. The final results have good agreement with those reported by Haris [23] and we obtained value of distance modulus of $(m-M)_V = 15.60$, which is consistent with that from Alcaino [16] and Harris [23]. The CMD of NGC 3201 has good fit with metallicity value of $[Fe/H] = -1.5$ and age of 12.0±1.0 Gyr. The age value of NGC 3201 is in good agreement with that from Dotter *et al* [17] and Monty *et al* [19]. The CMD of NGC 4590 has very good fit with metallicity value of $[Fe/H] = -2.2$. Our final results are approximately the same as those values reported by Sollima *et al*, Harris, and Kains *et al* [23, 28], and [22].

Table 2. Fundamental parameters of GC samples from this study with comparison values from previous works.

Cluster	Age(Gyr)	[Fe/H]	(m-M)	E(B-V)	Reference
NGC 1261	14-15	-1.27	16.05±0.25	0.04	[9]
	8-18	-	16.00±0.23	0.07	[7]
	-	-1.27	-	-	[30]
	-	-1.27	16.09	0.01	[23]
	12.6±1.0	-	-	-	[31]
	12.6	-1.27	-	-	[4]
	12.6±1.0	-1.27	16.00	0.01	This study
NGC 1851	9.1±1.1	-1.4	15.50	0.034	[32]
	-	-1.18	15.47	0.02	[23]
	12.0±1.0	-1.20	15.50	0.04	This study
NGC 2298	17.0±1.5	-1.77	15.6	-	[16]
	12.9±1.4	-1.85	-	-	[3]
	-	-1.92	15.6	0.14	[23]
	12.7±0.7	-	-	-	[33]
	13.1±0.6	-	-	0.17±0.2	[19]
	12.7±1.0	-1.92	15.60	0.23	This study
NGC 3201	12.3	-	-	-0.22±0.03	[34]
	11.48 ± 1.27	-	-	-0.24±0.02	[18]
	12.0 ± 0.75	-	-	-	[17]
	-	-1.59	14.2	0.24	[23]
	12.0 ± 0.7	-1.6	-	0.24 ± 0.02	[19]
	12.0±1.0	-1.50	14.20	0.28	This study
NGC 4590	13±2	-2.03	15.25	0.07	[21]
	11.2±0.9	-2.11	-	-	[3]
	-	-2.23	15.210	0.05	[28]
	11.2	-2.23	15.21	0.05	[23]
	13.0	-2.2	-	-	[22]
	13.0±1.0	-2.2	15.22	0.04	This study

4. Summary

In this research, we present new isochrone fitting on HST photometry archival data of five GCs including NGC 1261, NGC 1851, NGC 2298, NGC 3201, and NGC 4590. All data were collected from HST ACS Wide Field Channel (WFC) images in the F606W and F814W bands, on board the HST (PI: A. Sarajedini). Isochrones were generated from the PARSEC stellar evolution model, which are the fundamental tool for age and distance estimation and modelling the evolution of stellar clusters and other galaxies. There are several codes to use for GCs age and metallicity estimation including PARSEC, MIST, DSEP, ASteCa codes. Among the analyzed clusters using PARSEC code, we found that CMDs of our cluster samples are very good fit with metallicity value of $-2.20 \leq [Fe/H] \leq -1.20$ and $12.0 \leq Age \leq 13.0$ Gyr. The final results show that NGC 4590 is the oldest GC and has lowest metallicity value compare with other cluster samples.

Acknowledgments

The archived data based on observations made with the NASA/ESA Hubble Space Telescope, obtained from the data archive at the Space Telescope Science Institute. This research has made use of the SIMBAD database, operated at CDS, Strasbourg, France.

References

[1] Barker H and Paust N E Q 2018 *Publ. Astron. Soc. Pac.* **130** 034204
[2] Sarajedini A 2009 *The Ages of Stars* vol 258 ed E E Mamajek *et al* (Cambridge: Cambridge University Press) p 221
[3] Salaris M and Weiss A 2002 *Astron. Astrophys.* **388** 492
[4] Rani S, Pandey G, Subramaniam A, Sahu S and Rao N 2020 *J. Astrophys. Astron.* **41** 35
[5] Arellano F A, Bustos I H, Calderón J H and Ahumada J A 2019 *Rev. Mex. Astron. Astrofis.* **55** 337
[6] Carretta E, Bragaglia A, Gratton R G, Recio B A, Lucatello S, D'Orazi V and Cassisi S 2010 *Astron. Astrophys.* **516** A55
[7] Alcaino G, Liller W, Alvarado F and Wenderoth E 1992 *Astron. J.* **104** 1850
[8] Alcaino G and Contreras C 1971 *Astron. Astrophys.* **11** 14
[9] Bolte M and Marleau F 1989 *Publ. Astron. Soc. Pac.* **101** 1088
[10] Piotto G *et al* 2002 *Astron. Astrophys.* **391** 945 (*Preprint* astro-ph/0207124)
[11] Kravtsov V, Alcaíno G, Marconi G and Alvarado F 2010 *Astron. Astrophys.* **516** A23
[12] Navin C, Martell S and Zucker D 2015 *Mon. Not. R. Astron. Soc.* **453** 531
[13] Harris W E 1996 *Astron. J.* **112** 1487
[14] Subramaniam A, Sahu S, Postma J E, Côté P, Hutchings J B, Darukhanawalla N, Chung C and Tandon S N 2017 *Astron. J.* **154** 233
[15] Alcaino G 1974 *Astron. Astrophys. Suppl.* **13** 55
[16] Alcaino G and Liller W 1986 *Astron. Astrophys.* **161** 61
[17] Dotter A *et al* 2010 *Astrophys. J.* **708** 698
[18] Bono G, Stetson P B, VandenBerg D A, Calamida A, Dall'Ora M, Iannicola G and Amico P 2010 *Astrophys. J. Lett.* **708** L74
[19] Monty S *et al* 2018 *Astrophys. J.* **865** 160
[20] Gonzalez G and Wallerstein G 1998 *Astron. J.* **116** 765
[21] Alcaino G, Liller W, Alvarado F and Wenderoth E 1990 *Astron. Astrophys. Suppl.* **83** 269
[22] Kains N *et al* 2015 *Astron. Astrophys.* **578** A128
[23] Harris W E 2010 *arXiv e-prints* arXiv:1012.3224 (*Preprint* astro-ph.GA/1012.3224)
[24] Anderson J *et al* 2008 *Astron. J.* **135** 2055
[25] de Marchi G and Pulone L 2007 *Astron. Astrophys.* **467** 107
[26] Sirianni M *et al* 2005 *Publ. Astron. Soc. Pac.* **117** 1049
[27] Bressan A, Marigo P, Girardi L, Salasnich B, Cero C D, Rubele S and Nanni A 2012 *Mon. Not. R. Astron. Soc.* **427** 127
[28] Sollima A, Lanzoni B, Beccari G, Ferraro F R and Fusi Pecci F 2008 *Astron. Astrophys.* **481** 701 (*Preprint* astro-ph/0801.4511)
[29] Ferraro F R, Beccari G, Dalessandro E, Lanzoni B, Sills A, Rood R T, Pecci F F, Karakas A I, Miocchi P and Bovinelli S 2009 *Nature* **462** 1028 (*Preprint* astro-ph.GA/1001.1096)
[30] Carretta E, Bragaglia A, Gratton R, D'Orazi V and Lucatello S 2009 *Astron. Astrophys.* **508** 695 (*Preprint* astro-ph.GA/0910.0675)
[31] Kharchenko N V, Piskunov A E, Schilbach E, Röser S and Scholz R D 2013 *Astron. Astrophys.* **558** A53
[32] Brasseur C M, Stetson P B, VandenBerg D A, Casagrande L, Bono G and Dall'Ora M 2010 *Astron. J.* **140** 1672
[33] Roediger J C, Courteau S, Graves G and Schiavon R P 2014 *Astrophys. J., Suppl. Ser.* **210** 10 (*Preprint* astro-ph.GA/1310.3275)
[34] Covino S and Ortolani S 1997 *Astron. Astrophys.* **318** 40

Light curve analysis of the eclipsing binary system V781 Tau

N Lamlert[1] and W Maithong[2*]

[1]Department of Physics and General Science, Faculty of Science and Technology, Phetchaburi Rajabhat University, Phetchaburi, 76000, Thailand
[2]Department of Physics and General Science, Faculty of Science and Technology, Chiang Mai Rajabhat University, Chiang Mai, 50300, Thailand

*E-mail : wiraporn_mai@cmru.ac.th

Abstract. V781 Tau is one of W UMa eclipsing binary systems whose orbital period is 0.34 days. The 0.7-meter telescope with CCD photometric system in B and V filters was conducted at the Regional Observatory for the Public, Chachoengsao, Thailand during December 2018, UT. The Wilson-Devinney Technique was used for calculating the physical properties of V781 Tau. The results showed the inclination of their orbital is 66.140°± 0.14. The effective temperature of the primary and secondary star is 6,060 and 5,881 K, respectively and the degree of contact is 4.38 %

1. Introduction

V781 Tau is one of W UMa eclipsing binary systems, which R.A. 05h 50m 13.1s and Dec 26°57'43". The orbital period of V781 Tau is 0.3449098 days [1]. It was discovered by Harris in 1979 [2]. Lu *et al.* studied by using the cross-correlation technique for measured for radial velocities V0 = +24.4 ± 1.3 kms^{-1}. The estimated distance of V 781 was 70 pc [3]. Liu and Yang found that the period change is $\delta p/p = -5.0\times10^{-11}$ and the minimum shrinking velocity is 6.77×10^{-5} cms^{-1}[4]. Zwitter *et al.* using GAIA performance the properties of binary systems. The result showed the spectrum type of V781 was G0 and the period change same to Liu and Yang. Determined V781 Tau is a W-subtype medium contact binary star, which the rate of orbital period change $-6.01(\pm2.28)\times10^{-8}$ d/yr [4,5].

In this work, the physical properties of V781 Tau were analyzed from the photometric observed data by Wilson-Devinney Technique to construct the V781 Tau binary star model. There were compared with the result from the past and collected in the binary star database for the future.

2. Observation and data analysis

V781 Tau was observed in the clear night sky on 18 December 2018, UT at the Regional Observatory for the Public, Chachoengsao, Thailand. The 0.7 - meter reflecting telescope and CCD with 30 seconds exposure time each in the blue (*B*) and standard visual (*V*) filters of the *UBV* system were used. The data reduction and the photometry were analyzed by MaxIn DL6 program. The Comparison and the Check stars are TYC 1870-582-1 and BD +26 972, respectively. The observational light curve of V781 Tau from 667 images in *B* and 657 images in *V* wavelength bandwidth is shown in figure 1. The Wilson-Devinney Technique was used to calculate the physical properties values.

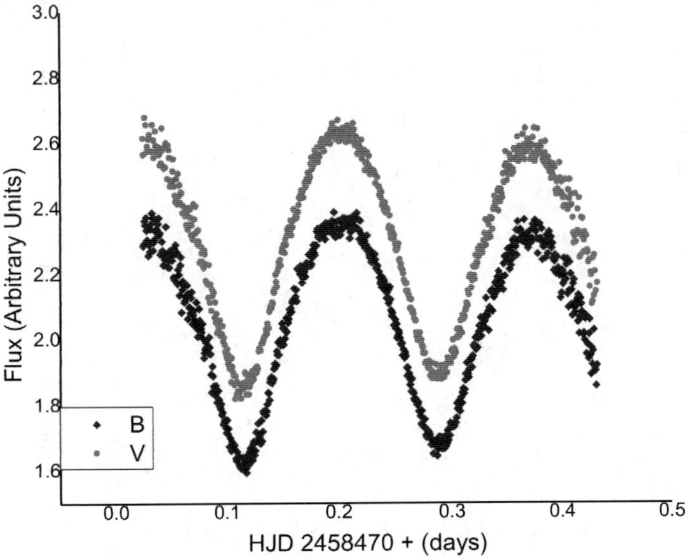

Figure 1. The light curve of V781 Tau.

3. Results and discussion

The synthetic light curve between phase and relative light of V781 Tau in *B* and *V* filters, as shown in figure 2, was constructed from photometric data computation by using Wilson-Devinney software which was developed by Kang Y.W. [6]. The red line shows the calculated data and the dots are observed data. We operated the light curve in mode 1 of the differential correction (DC) in Wilson–Devinney (WD) code [7]. The initial parameters were used from Li K. that publish in 2016 [8]. The best solution of the V781 Tau in this study, as shown in table 1., is the subscripts 1 and 2 refer to the primary (hotter) and the secondary (cooler) components, respectively.

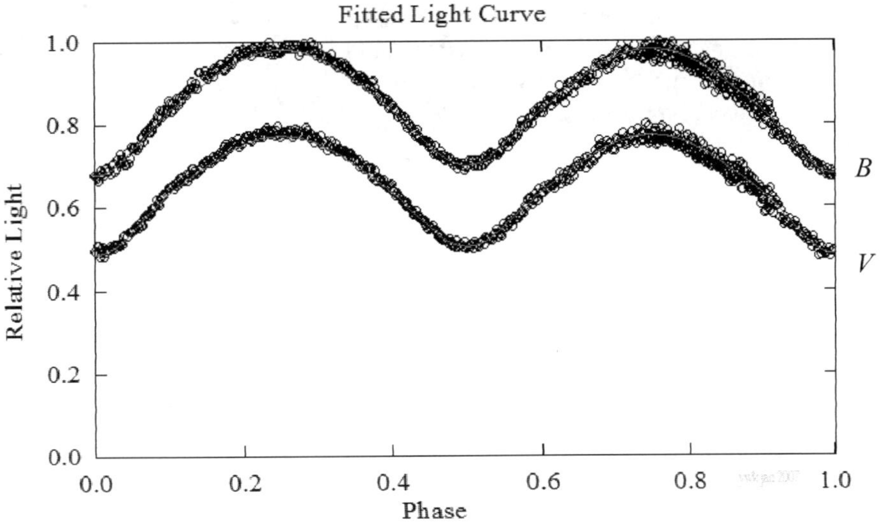

Figure 2. Synthetic light curve of the V781 Tau.

Table 1. Parameters for simulated of the V781 Tau Model.

Parameter	The Best Solution
i	$66.14^0 \pm 0.14$
g_1	0.32
g_2	0.32
$\Omega_1 = \Omega_2$	5.123377 ± 0.011423
Ω_{in}	5.239784
Ω_{out}	2.580423
T_1 (K)	6060 ± 152
T_2 (K)	5881 ± 136
A_1	0.50
A_2	0.50
q	1.991664 ± 0.007192
$L_1/(L_1+L_2)_V$	0.38296 ± 0.03222
$L_1/(L_1+L_2)_R$	0.37744 ± 0.02727
Degree of Contact (%)	4.38

From table 1, The inclination of V781 Tau is 66.14°±0.14 nearly Cereda *et al.*, J. Yakut *et al.* and Kallrath *et al.*, [2,9,10]. They calculated the value of 68°, 65.9° and 62.3°, respectively. The gravity-darkening coefficients and bolometric albedo coefficients of the two components were fixed at $g_{1,2}$ = 0.32 and $A_{1,2}$ = 0.50 appropriate for stars with convective envelope [4]. Surface temperature $T_{1,2}$ were 6,060 K and 5,881 K., that corresponding to G0 spectra in Li *et al.*, Kallrath *et al.* and Donato *et al.* The mass ratio (q) approximate to 1.99 mean the mass of the first star less than the secondary star [8,10,11]. There are not significantly different from Yakut *et al.* and Li *et al.* [2,8]. The degree of contact of V781 Tau is 4.38 percent that is smaller than the result from Li *et al.* [8], the value of contact degree of f = 21.6%, might be the effect of mass transfer. The plotted geometric structure models are 0.00, 0.25, 0.50, and 0.75 are shown in figure 3.

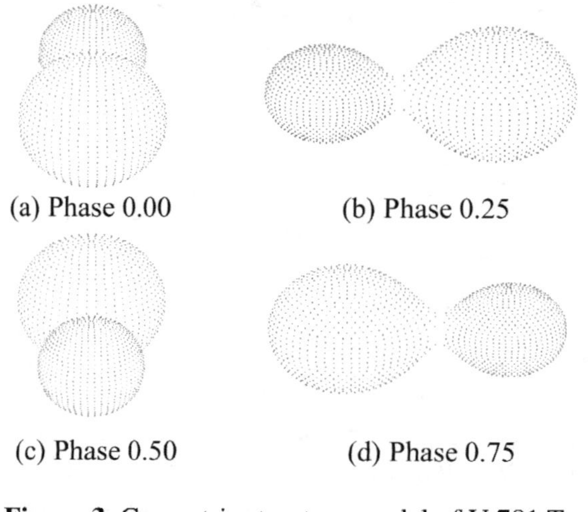

(a) Phase 0.00 (b) Phase 0.25

(c) Phase 0.50 (d) Phase 0.75

Figure 3. Geometric structure model of V 781 Tau.

4. Conclusion
The W UMa eclipsing binary system V781 Tau was observed on 18 December 2018, UT at the Regional Observatory for the Public, Chachoengsao, Thailand. The 0.7 - meter reflecting telescope and CCD with the blue (*B*) and standard visual (*V*) filters of the *UBV* system were used. The Wilson-Devinney technique was used for investigated photometric data at the Faculty of Science and Technology, Chiang Mai Rajabhat University, Chiang Mai, and Phetchaburi Rajabhat University, Phetchaburi, Thailand. We found that the inclination of their orbital is 66.14°±0.14. The effective temperature of the primary and secondary stars is 6,060 and 5,881 K, respectively. The mass ratio is 1.99 and the degree of contact is 4.38 %.

Acknowledgment
The author would like to acknowledge the National Astronomical Research Institute of Thailand (Public Organization) for this observation.

References
[1] Kreiner J M 2004 An atlas of O-C diagrams of eclipsing binary stars *Acta Astron.* **54** 207–10
[2] Yakut K, Ulas B, Kalomeni B and Gülmen O 2005 Basic physical properties of the low temperature contact binary system V781 Tau and the near-contact binary system V836 Cyg *Mon. Not. R. Astron. Soc.* **363**(4) 1272–8
[3] Lu W 1992 Radial-velocity observations and absolute dimensions of eclipsing binaries - V781 Tau *Astron. J.* **105** 646–51
[4] Liu Q and Yang Y 2000 A period study of the W UMa type contact binary V781 Tauri *Astron. Astrophys. Suppl. Ser.* **142** 31–4
[5] Zwitter T, Munari U, Marrese P M, Prša A, Milone E F, Boschi F, Tomov T and Siviero A 2003 Evaluating GAIA performances on eclipsing binaries II: Orbits and stellar parameters for V781 Tau UV Leo and GK Dra *Astron. Astrophys.* **404** 333–40
[6] Kang Y W 2010 New light curve analysis for large numbers of eclipsing binaries I. Detached and semi-detached binaries *J. Astron. Space Sci.* **27**(2) 75–80
[7] Wilson R E and Devinney E J 1971 Realization of accurate close-binary light curves: Application to MR Cygni *Astrophys. J.* **166** 605–19
[8] Li K, Gao D Y, Hu S, Guo D, Jiang Y and Chen X 2016 The active W UMa type binary star V781 Tau revisited *Astrophys. Space Sci.* **361**(63) 1–16
[9] Cereda L, Misto A, Niarchos P G and Poretti E 1988 BV photometry of the eclipsing W Ursae Majoris systems V781 Tauri and GR Virginis *Astron. Astrophys. Suppl. Ser.* **76** 255–61
[10] Kallrath J, Milone E F, Breinhorst R A, Wilson R E, Schnell A and Purgathofer A 2006 V781 Tauri a W Ursae Majoris binary with decreasing period *Astron. Astrophys.* **452** 959–67
[11] Donato L, Lepardo A, Santini V, Tomov T, Munari U and Zwitter T 2003 V781 Tau: Improved evidence for an orbital period change *Inf. Bull. Var. Stars* **5391** 1–3

Ganymede's magnetic footprint brightness and location in respond to main emission

T Promfu[1,2,3*], S Wannawichian[2,3], J D Nichols[4] and J T Clarke[5] and K Haewsantati[2,3,6,7]

[1]Ph.D. Program in Applied Physics, Department of Physics and Materials Science, Faculty of Science, Chiang Mai University, 50200, Thailand
[2]Department of Physics and Materials Science, Faculty of Science, Chiang Mai University, Chiang Mai 50200, Thailand
[3]National Astronomical Research Institute of Thailand (Public Organization), Maerim Chiang Mai, 50180, Thailand
[4]Department of Physics and Astronomy, University of Leicester, Leicester, LE1 7RH, UK
[5]Center for Space Physics, Boston University, Boston, MA 02215, USA
[6]Ph.D. program in Physics, Department of Physics and Materials Science, Faculty of Science, Chiang Mai University, Chiang Mai 50200, Thailand
[7]LPAP, STAR Institute, Université de Liège, Liège, 4000, Belgium

*E-mail: tatphicha_p@cmu.ac.th

Abstract. Jupiter's aurora features have been observed by the Hubble space telescope (HST) for over two decades. One of the auroral features, Ganymede's magnetic footprint, appears close to the main emission and is sometimes embedded in the main emission. The latter case causes difficulty in identifying Ganymede's magnetic footprint from in the main emission. The FUV aurora images were taken by Advanced Camera for Surveys (ACS) onboard the HST. The fluctuations of Ganymede's footprint brightness over time will be analyzed. Moreover, the correlation between the brightness and locations of the main emission and Ganymede's magnetic footprint will be analyzed to characterize the connection between ionospheric phenomena and the magnetospheric dynamics. Since the main emission is very bright in comparison with the footprint, therefore, the variation of the main emission can affect the Ganymede's magnetic footprint. Furthermore, the expansion of the main emission is consistent with the location shift of Ganymede's magnetic footprint in equatorward direction. The brightness and location of the main emission can be influenced by the plasma variation in Jupiter's magnetosphere which is affected partly by the volcanic eruption on Io and solar wind dynamic pressure. The variation of Ganymede magnetic footprint's brightness and location in respond to the main emission could be an important indicator of the magnetospheric variation under the influences of internal and external factors.

1. Introduction

Jupiter's aurora is mainly divided into three parts which are main emission, polar emission, and satellites' magnetic footprints (figure 1). This work focuses on the location shifts of main emission and Ganymede's magnetic footprint. Jupiter's magnetosphere contains plasma which is mainly from a volcanic eruption on Io. Charged particles inside the magnetosphere travel along Jupiter's magnetic field lines into Jupiter's ionosphere. These charged particles interact with Jupiter's atmospheric particles resulting in atmospheric particles emitting the auroral emissions.

The satellite magnetic footprints are connected to the magnetohydrodynamic interactions near the satellites which are embedded in Jupiter's magnetosphere. Due to the interactions, charged particles near the satellites are carried from the vicinity of the satellites along the magnetic flux tubes toward Jupiter's ionosphere causing the auroral magnetic footprints.

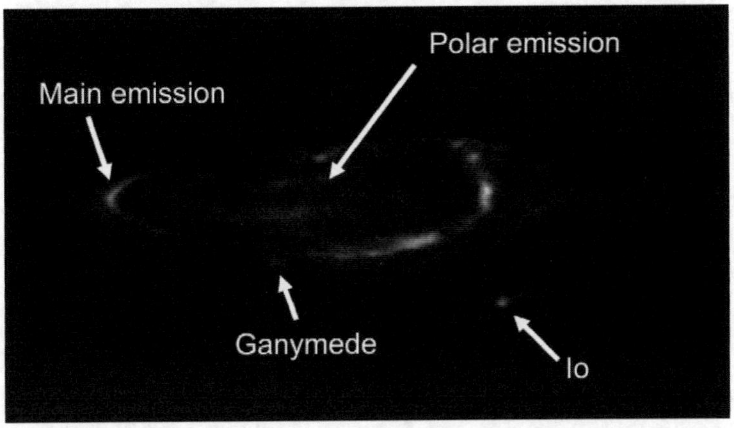

Figure 1. Jupiter's auroral components in northern hemisphere detected on day of year (DOY) 052 or 21 February 2007 by ACS instrument onboard Hubble Space Telescope.

From previous work [1, 2], Galileo observation revealed a region of high energy electron at distance between 20 - $30R_J$ ($1R_J$ = ~71492 km) in Jupiter's magnetosphere. On the other hand, Ganymede orbits at distance about $15R_J$ from Jupiter in Jupiter's middle magnetosphere. Figure 2 shows the stretching of the magnetic field lines in the middle magnetosphere which is influenced by strong variation of plasma. As a result, the magnetic field stretching affects Ganymede magnetic footprint locations in Jupiter's ionosphere to shift in poleward or equatorward directions depending on the plasma condition near the satellite. Ganymede's magnetic footprint typically appears close to the main emission. Therefore, the simultaneous shifts of the main emission and Ganymede's magnetic footprint locations could be used to identify the variation of plasma and magnetic field structure inside Jupiter's magnetosphere.

This work presents the location shifts of the main emission and Ganymede's magnetic footprint from the center of main emission. The center of main emission in northern hemisphere is at 74° latitude and 185° longitude [3]. From previous work [4], the expansion of main emission can be influenced by the plasma variation in Jupiter's magnetosphere which is affected partly by volcanic eruption on Io. Moreover, the structure of the main emission is affected by the solar wind dynamic pressure [5]. Therefore, in this work, the influence of both volcanic materials and solar wind dynamic pressure on the locations of main emission and Ganymede's magnetic footprint will be studied. The structure of the main emission was analyzed with the average path [5] to investigate the effect of solar wind property. The fluctuations of solar wind property could affect Jupiter's magnetic field structure. In the case of compression, the magnetic field lines can be compressed closer to Jupiter resulting in the shift of the main emission in poleward direction. This result could be implied to the shift of Ganymede magnetic footprint as well.

2. Data analysis

Jupiter's FUV images in this study were obtained by Hubble Space Telescope (HST) in 2007, using Solar blind camera (SBC) from the Advanced Camera for Survey (ACS) (115 -170 nm). The reduction process and auroral brightness conversion are based on pipeline routine from

Boston University [6, 7]. The brightness of Ganymed's magnetic footprint is presented in kilo-Rayleighs (1 kR = 10^9 photons cm^{-2} s^{-1}). The polar projection and location of the main emission were analyzed with (Interactive Data Language) IDL and Python software.

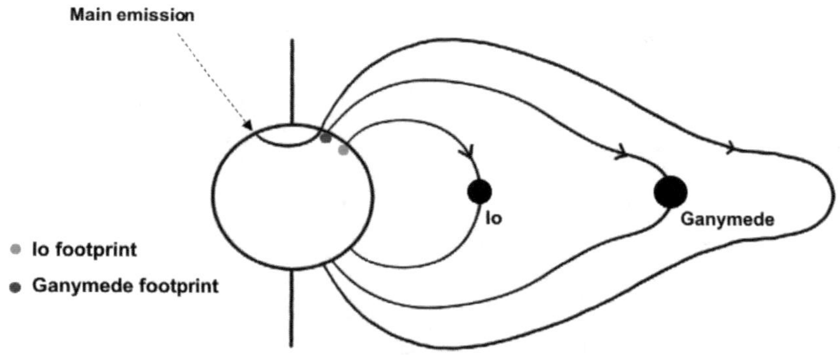

Figure 2. Sketch of the cross section through the inner and middle magnetosphere of Jupiter. Jupiter's magnetic field lines are represented by the solid lines with arrows. Io orbits in the inner magnetosphere (~$6R_J$), while Ganymede orbits in the middle magnetosphere (~$15R_J$). Adapted from figure 1 in Cowley and Bunce [8].

3. Result and discussion

Polar projections of Jupiter's aurora (figure 3) show the structures of main emission and the locations of Ganymede's magnetic footprint from the average path [5] (red dash line). The main emission structure is close to the average path at longitude beyond ~180°. However, at longitude ~150°-160°, the main emissions observed on DOY 052 and DOY 059 shift from the average path with similar distance (figure 3a and 3b). In the bottom panels, the main emission observed on DOY 068 shifts more further toward the equator in comparison with the data on DOY 053. Moreover, the shifted of Ganymede's magnetic footprint in equatorward direction is also detected on DOY 068.

The polar projection showed the similar feature on DOY 052 and DOY 059 (figure 3a and 3b). The solar wind propagation model showed that the solar wind dynamic pressure decreased slightly from DOY 052 to DOY 059 [6]. The New Horizons spacecraft observations revealed that, during February 2007, there were continuous volcanic eruptions on Io [9]. As a result, Jupiter's main emission and Ganymede's magnetic footprint located at similar locations while comparing data from DOY 059 with DOY 052.

The main emission had a noticeable change at longitude ~150°-160° on DOY 053 from DOY 068 (figure 3c and 3d). Between these two events, the main emission expanded in equatorward direction with maximum shift of ~1.6° at longitude ~150° on DOY 068 in comparison to the shift observed on DOY 053. Based on the propagating solar wind model [6], the solar wind dynamic pressure increased on DOY 053. The high solar wind dynamic pressure should correspond with the location shift of the main emission in polweard direction on DOY 053 in comparison to the main emission location in DOY 068. For the comparison, the expansion of the main emission is consistent with the location shift of Ganymede's magnetic footprint. Table 1 shows that the location of Ganymede's magnetic footprint shifted by ~0.6° in equatorward direction at similar longitude on DOY 068 in comparison to the footprint observed on DOY 053. The equatorward shift on DOY 068 could correspond with the increase of plasma sheet density in Jupiter's magnetosphere due to accumulated volcanic materials from Io [9]. The increase in plasma density causes the stretching of the magnetic field lines away from Jupiter. Thus, the

Journal of Physics: Conference Series **2145** (2022) 012006 doi:10.1088/1742-6596/2145/1/012006

locations of main emission and Ganymede's footprint are shifted in equatorward direction on DOY 068.

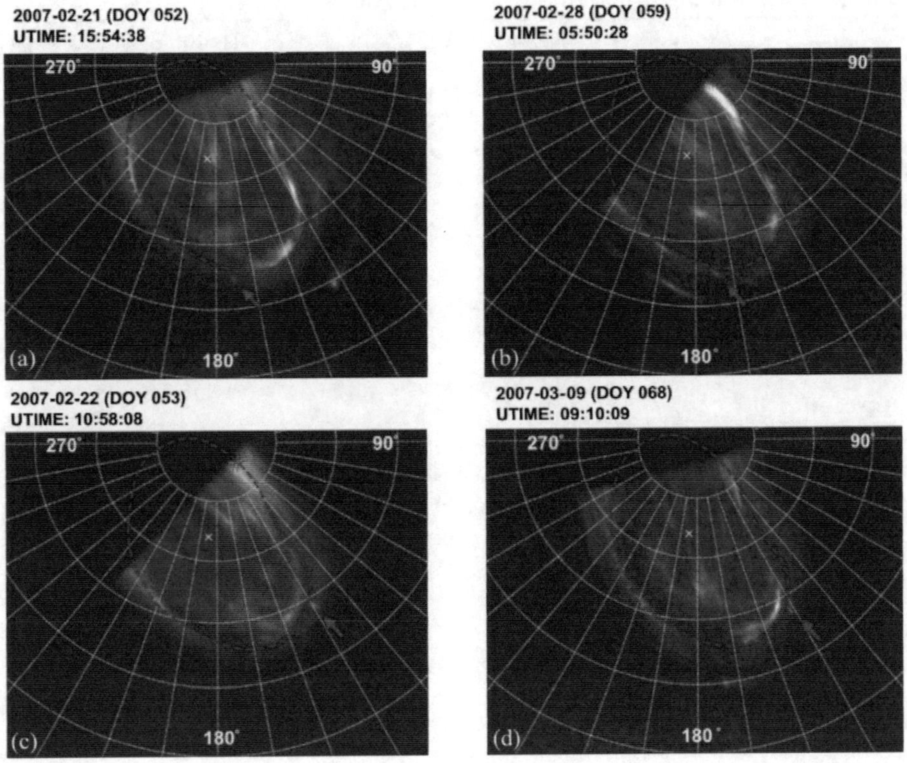

Figure 3. Polar projections show the variations of Jupiter's auroral structures, observed in 2007 between DOY 052 and 059 (a and b) and between DOY 053 and 068 (c and d). The statistical position of main emission is represented by red dashed line [5]. The center of Jupiter's main emission is represented by yellow cross. The red arrows point at Ganymede's magnetic footprint.

Table 1 shows that Ganymede's magnetic footprint brightness on DOY 052 is ~71.6 kR greater than the brightness on DOY 059. However, Ganymede's magnetic footprint brightness on DOY 053 is approximately the same as the footprint brightness observed on DOY 068. It must be noted that Ganymede's magnetic footprint appears close to the main emission. Since the main emission is very bright in comparison with the Ganymede's magnetic footprint. Thus, the variation of location and brightness of the main emission could affect the analysis of Ganymede's magnetic footprint brightness. Moreover, the brightness of Ganymede's magnetic footprint is related to the plasma density around the moon. The plasma density in Jupiter's magnetosphere increase mainly from Io volcanic materials. Therefore, the detail interpretation of Ganymede's magnetic footprint brightness will be studied in corresponding with the location shift and the brightness variation of the main emission in the future.

4. Conclusion

Jupiter's main emission could be affected by both internal (Io volcanic materials) and external (solar wind dynamic pressure) factors. During the solar wind compression, Jupiter's magnetopause is compressed resulting in the poleward shift of the main emission and Ganymede's magnetic footprint. The increasing of plasma from volcanic eruption on Io affects the structure of

Table 1. The comparison of main emission (ME) and Ganymede's magnetic footprint (GFP) locations from the center of Jupiter's aurora [3] observed in 2007 between DOY 052 and 059 and between DOY 053 and 068.

DOY	GFP	ME location(~150°-160°)	Brightness of GFP(kR)
052-059	Similar location	Similar location	71.6(052), 38.0(059)
053-068	~0.6° equatorward	~0.5°-1.6° equatorward	80.6(053), 71.6(068)

Jupiter's magnetic field, especially in the inner magnetosphere. While other factors are needed to be considered, both volcanic materials from Io and solar wind dynamic pressure strongly influence the structure of magnetosphere, causing the magnetic field and plasma fluctuations. Consequently, the modulation of main emission and Ganymede's magnetic footprint locations were detected and should be investigated in detail.

Acknowledgments
Aurora images obtained by NASA/ESA Hubble Space Telescope which was supported by grant HST-GO-10862.01-A from STScI to Boston University. This research was supported by Department of Physics and Materials Science, Faculty of Science, Chiang Mai University and National Astronomical Research Institute of Thailand (NARIT). Additional support was from Thailand Science Research and Innovation grant RTA6280002.

References
[1] Frank L A and Paterson W R 2002 *J. Geophys. Res.* **107** 1478
[2] Bagenal F, Dowling T E and McKinnon W B 2006 *Jupiter, The Planet, Satellites and Magnetosphere* Cambridge Planetary Science (Cambridge University Press)
[3] Bonfond B, Saur J, Groden D, Badman S V, Bisikalo D, Shematovich V, Gerard J C and Radioti A 2017 *J. Geophys. Res.* **122** 7985–96
[4] Bonfond B, Grodent D, Gerard J C, Stallard T, Clarke J T, Yoneda M, Radioti A and Gustin J 2012 *Geophys. Res. Lett* **39** L01105
[5] Nichols J D *et al* 2017 *Geophys. Res. Lett.* **44** 7643–52
[6] Clarke J T *et al* 2009 *J. Geophys. Res.* **114** A05210
[7] Wannawichian S, Clarke J T and Nichols J D 2010 *J. Geophys. Res.* **115** A02206
[8] Cowley S W H and Bunce E J 2001 *Planet. Space Sci.* **49** 1067–88
[9] Spencer J R *et al* 2007 *Science* **318** 240–3

Constraining the annihilation of dark matter via cosmic-ray positrons and electrons

Suwitchaya Setthahirun[1] and Maneenate Wechakama[1,2,*]

[1]Department of Physics, Faculty of Science, Kasetsart University, Bangkok, 10900, Thailand
[2]National Astronomical Research Institute of Thailand (Public Organization), Chiangmai, 50180, Thailand

*E-mail: fscimnw@ku.ac.th

Abstract. We aim to constrain the properties of dark matter particles by several measurements of positrons and electrons from cosmic-rays. We assume that collisions of dark matter particles and dark matter anti-particles can produce positrons and electrons. The electron-positron propagation is explained by a diffusion-loss equation including loss rates, diffusion, as well as source function. We use data of cosmic-ray positrons and electrons detected by PAMELA, H.E.S.S., AMS-02 and Fermi-LAT. We compare the observational data with the electron and positron spectrum from five annihilation channels in our model to derive constraining factors regarding the cross-section of the annihilation of dark matter. The tightest constraint is provided by cosmic-ray positrons of AMS-02 for the electron channel. Dark matter with mass below a few GeV gets excluded by the cosmic-ray positrons of AMS-02 for the electron, muon and tau channels.

1. Introduction

Dark matter properties can be examined by indirect detection based on the assumption that dark matter particles and dark matter anti-particles can collide and annihilate into positrons and electrons [1]. Several experiments have shown that electron and positron data exceeded the predicted astrophysical backgrounds in the solar neighbourhood [2]. This led to various attempts to detect their abnormalities which can be assumed to be the result of the annihilation of dark matter. Data regarding cosmic-ray electrons and positrons have been detected by several experiments; the Alpha Magnetic Spectrometer (AMS-02) [3], Fermi-Large Area Telescope (Fermi-LAT) [4–6], the Payload for Antimatter Matter Exploration and Light-nuclei Astrophysics (PAMELA) [7,8]) as well as the High Energy Stereoscopic System (H.E.S.S.) [9].

Within this paper we aim to derive constraining factors of the cross-section of the annihilation of dark matter and masses of dark matter particles by using data from several experiments that detected cosmic-ray positrons and electrons. We examine positrons and electrons produced from quark and lepton channels of the annihilation of dark matter. Based on the result from a diffusion-loss equation, we aim to explain the spectrum of positrons and electrons taking loss rates, diffusion, and source term into consideration. The upper limits of the cross-section of the annihilation of dark matter are derived by comparing the observational electron and positron data to our model spectrum.

2. Method and data

2.1. Model predictions

Our model assumes that dark matter particles and dark matter anti-particles can annihilate into positrons and electrons through quark and lepton channels. The energy of positrons and electrons diminishes as a result of traveling through the Galaxy due to ionization with hydrogen atoms, inverse Compton scattering by photons with low energy and synchrotron radiation through magnetic fields. Assuming spherical symmetry; positron and electron propagation at r from the centre of the Galaxy can be expressed as a diffusion-loss equation as shown in [10] where the spectrum of electrons as well as positrons is given by

$$\begin{aligned}\frac{\mathrm{d}n}{\mathrm{d}\gamma}(r,\gamma) &= \frac{1}{b(\gamma)}\frac{\exp\left(-\frac{r^2}{2\Delta\lambda^2}\right)}{(2\pi r^2\Delta\lambda^2)^{1/2}} \times \left\{ \int_{\gamma}^{\infty} \mathrm{d}\gamma_0 \int_0^{\infty} \mathrm{d}r_0 \; r_0 \; \exp\left(-\frac{r_0^2}{2\Delta\lambda^2}\right)\right. \\ &\left.\times \left[\exp\left(\frac{rr_0}{\Delta\lambda^2}\right) - \exp\left(-\frac{rr_0}{\Delta\lambda^2}\right)\right] Q(r_0,\gamma_0)\right\}, \end{aligned} \tag{1}$$

where γ is the Lorentz factor related to the energy $E = \gamma m_e c^2$ of the positrons and electrons, m_e represents the electron mass while at rest, c is the light speed constant, γ_0 and r_0 represent their initial energy and initial location where they are created, respectively. $\Delta\lambda^2 = \lambda^2(\gamma) - \lambda^2(\gamma_s)$, where λ is the positron and electron's diffusion length in a function of the coefficient of diffusion $K(\gamma)$ and energy loss-rate $b(\gamma)$, as given by

$$\lambda^2(\gamma) = \int_{\gamma}^{\infty} \frac{2K(\gamma)}{b(\gamma)}\mathrm{d}\gamma. \tag{2}$$

We adopt the model of the energy loss rate and the coefficient of diffusion from [10] and [11], respectively.

The source term that represents the rate of the positron and electron production from the annihilation of dark matter is given by

$$Q(r,\gamma) = \frac{1}{2}\left[\frac{\rho_{\mathrm{dm}}(r)}{m_{\mathrm{dm}}}\right]^2 \langle\sigma v\rangle_{e^\pm}\frac{\mathrm{d}N_{e^\pm}}{\mathrm{d}\gamma}\delta(\gamma-\gamma_0), \tag{3}$$

where m_{dm} is the mass of dark matter particle, $\langle\sigma v\rangle_{e^\pm}$ is the average annihilation cross-section multiplied by the relative particle velocity of dark matter and its anti-particle equivalents. Given our assumption that particles of dark matter are Majorana particles, i.e. self-conjugate particles, we add a factor 1/2 to avoid double counting. We adopt the final state injection spectrum of positrons and electrons, $\frac{\mathrm{d}N_{e^\pm}}{\mathrm{d}\gamma}$, for each annihilating channel from PPPC4DMID including electroweak corrections [12, 13].

The question regarding profiling the density of dark matter distributed throughout the Milky Way is still not settled. Either the shape of the profile can be very steep at the inner regions due to the adiabatic contraction from gas and stars [14], or it can be shallower as suggested in, e.g., [15]. In our model, we use the most common universal profile for dark matter halos identified in N-body simulations by [16] a profile developed by Frenk, Navarro and White with a spherically-symmetric halo for the profile of dark matter density,

$$\rho_{\mathrm{dm}}(r) = \frac{\rho_{\mathrm{s}}}{\left(\frac{r}{r_{\mathrm{s}}}\right)\left(1+\frac{r}{r_{\mathrm{s}}}\right)^2}, \tag{4}$$

where r_{s} and ρ_{s} are the characteristic density and radius of the halo of the Milky Way, respectively. The characteristic density $\rho_s c^2$ and radius r_s are equal to 0.349 GeV cm^{-3} and 16.7 kpc respectively as shown in [17].

2.2. Electron and positron data

Data from several remarkable experiments such as AMS-02, PAMELA, H.E.S.S. as well as Fermi-LAT is used with the intention of constraining the properties of dark matter particles. The data from AMS-02 contain the flux for electrons within the energy span between 0.5 GeV to 700 GeV as well as flux for positrons within the energy span between 0.5 GeV to 500 GeV [3]. We further utilize the data about positrons and electrons from cosmic-rays provided by PAMELA, where the electron data are represented within the energy span between 1 GeV to 625 GeV [7] and the positron data are represented in the energy span between 1 GeV to 300 GeV, [8]. The observed spectra of positrons as well as electrons from cosmic-rays that were detected within Fermi-LAT in the energy span between 7 GeV to 1 TeV [5] are also used as well as the electron plus positron data with the energy span between 7 GeV to 2 TeV [6]. Finally, we use the high energy electron spectrum with energy above 600 GeV detected by H.E.S.S. [9]. All aforementioned data are shown in figure 1.

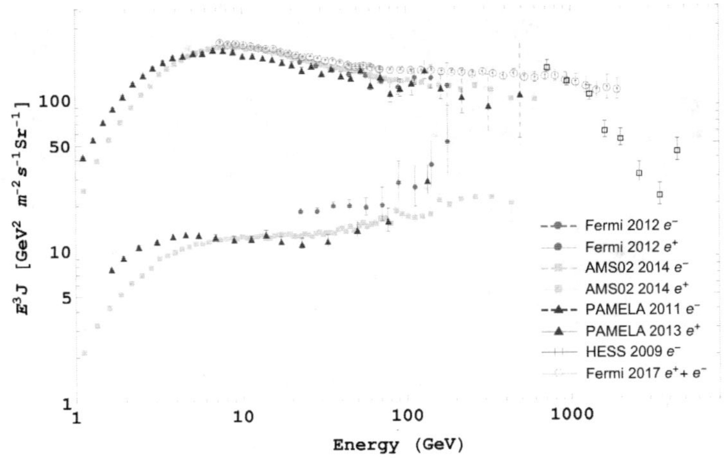

Figure 1. Spectrum of electron plus positron, only electron and only positron as measured by H.E.S.S., PAMELA, AMS-02 as well as Fermi-LAT.

3. Upper limits of the cross-section of the annihilation of dark matter

Our focus here is the property investigation regarding particles of dark matter, i.e. masses and the cross-section of annihilation. By comparing positron and electron data in the solar neighbourhood with electron and positron spectrum in our model, we can derive the upper limits of the cross-section of the annihilation of dark matter multiplied by the relative particle velocity of dark matter and dark matter anti-particles, $\langle\sigma v\rangle_{e^\pm}$, for the mass of dark matter $m_{\rm dm}$ in the range of GeV to PeV. In order to calculate these, we have to exclude the integration term $\int_\gamma^\infty d\gamma_0$ from equation (1). Therefore, the electron and positron spectrum created by dark matter particles with a mass $m_{\rm dm} \sim \gamma_0 m_{\rm e}$ is given by

$$\frac{{\rm d}n}{{\rm d}\gamma}(r,\gamma,\gamma_0) = \frac{1}{b(\gamma)}\frac{\exp\left(-\frac{r^2}{2\Delta\lambda^2}\right)}{(2\pi r^2\Delta\lambda^2)^{1/2}} \times \left\{ \int_0^\infty {\rm d}r_0\, r_0 \exp\left(-\frac{r_0^2}{2\Delta\lambda^2}\right) \times \left[\exp\left(\frac{rr_0}{\Delta\lambda^2}\right) - \exp\left(-\frac{rr_0}{\Delta\lambda^2}\right)\right] Q(r_0,\gamma_0)\right\}, \quad (5)$$

where $r = 8.5$ kpc for the case of the solar neighbourhood. We then calculate the electron and positron spectrum for five different channels of annihilation, i.e. muon, electron, tau,

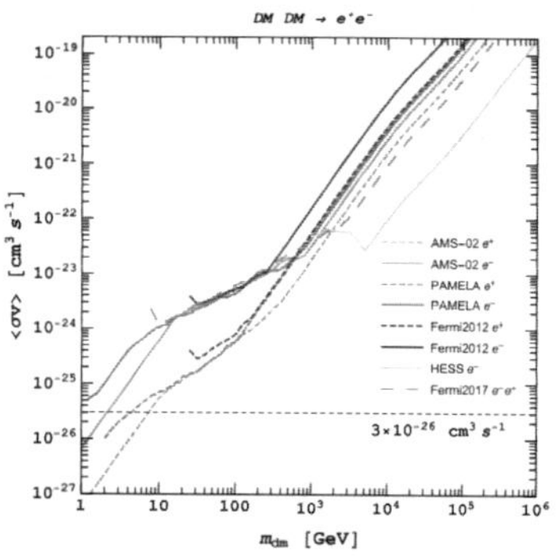

(a) electron channel

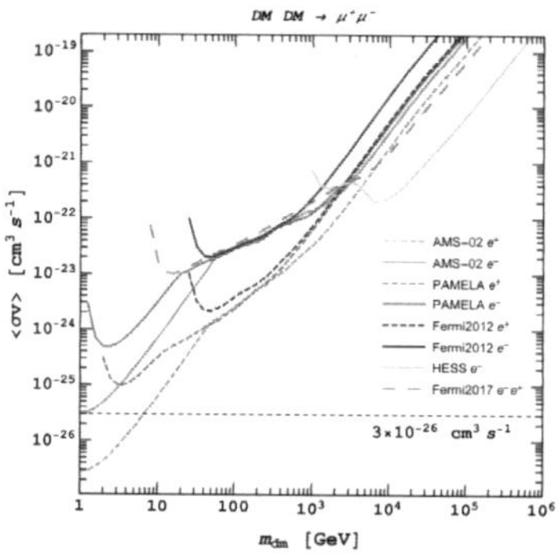

(b) muon channel

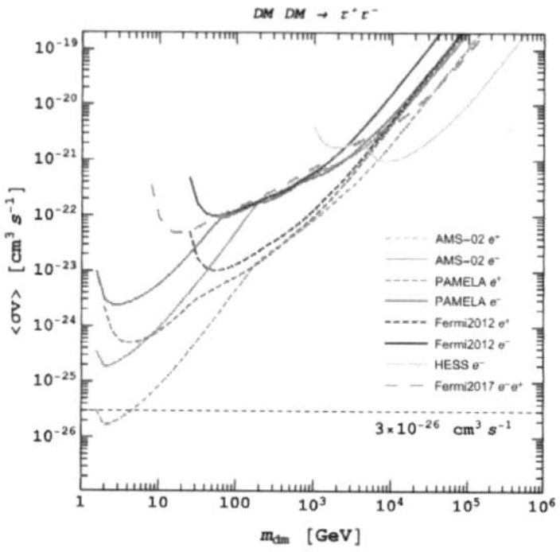

(c) tau channel

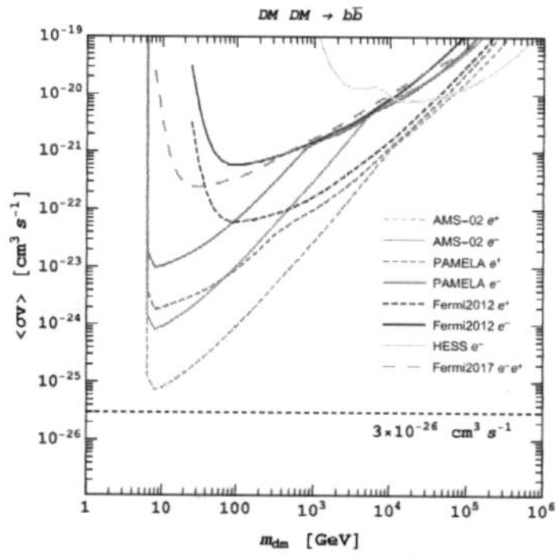

(d) bottom channel

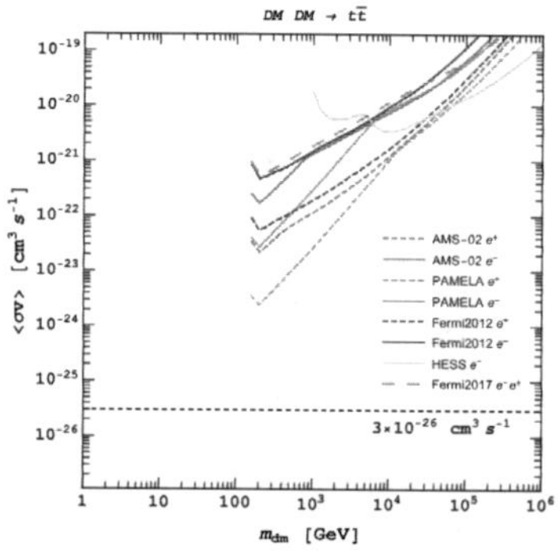

(e) top channel

Figure 2. Upper limits of the cross-section of the annihilation of dark matter for electron, muon, tau, bottom and top channels derived from the various cosmic-ray electron and positron measurements.

bottom and top channels, which we adopt the final injection spectrum of positrons and electrons for each channel, $\frac{dN_{e\pm}}{d\gamma}$ from PPPC4DMID including electroweak corrections [12, 13], as we have mentioned in the previous section. Finally, the upper limits of the cross-section of the annihilation of dark matter $\langle\sigma v\rangle_{e^\pm}$ for each mass of dark matter $m_{\rm dm}$ and for each channel of annihilation is provided by constraining the parameter $\langle\sigma v\rangle_{e^\pm}$ to imposed that the model spectrum does not exceed the measured spectrum by AMS-02, Fermi-LAT, PAMELA and H.E.S.S.. We derive the upper limits within the mass range of 1 GeV to 1 PeV. The upper limits of the cross-section of the annihilation of dark matter for each annihilating channel are shown in figure 2 where the thermal relic cross-section, $\langle\sigma v\rangle_{e^\pm} = 3\times10^{-26}$ cm s^{-1}, is represented in the dashed-black line.

As shown in figure 2, our results indicate that for the electron, muon and tau channels, the cross-section of annihilation is lower than the thermal relic cross-section for masses of dark matter lighter than a few GeV. Therefore, the expected value for a thermal relic cross-section, described as $\langle\sigma v\rangle_{e^\pm} = 3\times10^{-26}$ cm s^{-1}, can be ruled out for these candidates of dark matter that have this lighter mass. For masses of dark matter that are lighter than $\sim$ 1 TeV, the lowest upper limit is provided by the positron data by AMS-02 whereas the electron spectrum by H.E.S.S. provides the lowest limits for dark matter masses larger than $\sim$ 1 TeV. The electron channel provides the tightest constraints for all measurements which give the upper limits for masses of dark matter between 10 GeV to 1 TeV about 1 to 2 order of magnitude higher than the thermal relic cross-section. As for masses of dark matter that are larger than $\sim$ 1 TeV, the annihilation cross-sections are substantially higher compared to the thermal cross-section for all measurements and all annihilation channels, as also mentioned in [18–20].

4. Conclusions

Thus we have arrived at the upper limits of the cross-section of the annihilation of dark matter into positrons and electrons by using the cosmic-ray positron and electron data measured by H.E.S.S., AMS-02, Fermi-LAT and PAMELA. In our model, the propagation of the positrons and electrons is determined by a diffusion-loss equation which depends on diffusion, source term and energy loss rate. Consideration has also been given to self-conjugate dark matter particles with NFW profile in the Milky Way galaxy. The positrons and electrons are produced from the annihilation of dark matter through electron, muon, tau, bottom and top channels. After the production, the positrons and electrons travel through the Galaxy and lose their energy by inverse Compton scattering, synchrotron radiation and ionization. We have adopted parameters of the energy loss rate and the diffusion coefficient from the literature. The upper limits of the cross-section of the annihilation of dark matter can thereby be derived from the comparison of the predicted electron and positron spectrum for each annihilation channel with the observational data of positrons and electrons from cosmic-rays. The upper limits of the cross-section of the annihilation of dark matter for each dark matter mass are provided by imposing that the predicted values do not exceed the observational data.

The tightest limits on the cross-section of the annihilation of dark matter are derived from the positron data by AMS-02. Masses of dark matter smaller than a few GeV are excluded for the electron, muon and tau channels. However, the upper limits for the larger masses of dark matter particles are substantially higher compared to the thermal cross-section. The positrons and electrons produced from astrophysical backgrounds should be subtracted from the data before deriving the upper limits to get the better constraints.

Acknowledgments

S. Setthahirunin gratefully acknowledges that this research is supported in part by the Graduate Program Scholarship from the Graduate School, Kasetsart University. M. Wechakama would like to acknowledge that this research project is supported by Kasetsart University Research and Development Institute (KURDI).

References

[1] Bertone G and Hooper D 2018 *Rev. Mod. Phys.* **90** 045002
[2] Cirelli M 2012 *Pramana* **79** 1021
[3] Aguilar M *et al* 2014 *Phys. Rev. Lett.* **113** 121102
[4] Ackermann M *et al* 2010 *Phys. Rev.* D **82** 092004
[5] Ackermann M *et al* 2012 *Phys. Rev. Lett.* **108** 011103
[6] Abdollahi S *et al* 2017 *Phys. Rev.* D **95** 082007
[7] Adriani O *et al* 2011 *Phys. Rev. Lett.* **106** 201101
[8] Adriani O *et al* 2013 *Phys. Rev. Lett.* **111** 081102
[9] Aharonian F*et al* 2008 *Phys. Rev. Lett.* **101** 261104
[10] Wechakama M and Ascasibar Y 2011 *Mon. Not. R. Astron. Soc.* **413** 1991
[11] Donato F *et al* 2004 *Phys. Rev.* D **69** 063501
[12] Cirelli M *et al* 2011 *J. Cosmol. Astropart. Phys.* **1103** 051
[13] Ciafaloni P *et al* 2011 *J. Cosmol. Astropart. Phys.* **1103** 019
[14] Blumenthal G R *et al* 1986 *Astrophys. J.* **301** 27
[15] Oh S *et al* 2010 *Astrophys. J.* **142** 24
[16] Navarro J F, Frenk C S and White S D M 1997 *Astrophys. J.* **490** 493
[17] Wechakama M and Ascasibar Y 2014 *Mon. Not. R. Astron. Soc.* **439** 566
[18] Duangchan C and Wechakama M 2021 *J. Phys.: Conf. Ser.* **1719** 012012
[19] Wechakama M and Khan Cantlay B 2019 *J. Phys.: Conf. Ser.* **1380** 012144
[20] Khan Cantlay B and Wechakama M 2019 *J. Phys.: Conf. Ser.* **1380** 012071

The pointing model of 4.5-m small radio telescope at NARIT

N Namkham*, **N Chanwedchasart, P Jaroenjittichai and N Prasert**

National Astronomical Research Institute of Thailand (Public Organization), Chiang Mai, 50180, Thailand

*E-mail: nakornping@narit.or.th

Abstract. The efficiency of a radio telescope decisively depends on its pointing accuracy. Telescope's pointing model (PM) contains repeatable errors due to the antenna control system's imperfections, which can be corrected during observation. The 4.5m Small Radio Telescope (SRT's) has been developed for education and experiments at Astropark, National Astronomical Research Institute of Thailand (NARIT), Chiang Mai (18°N 51' 5" and 98°E 57' 27"). We have implemented a 10-cm optical camera system installed on the SRT's antenna structure to measure the offset of individual pointing covering all sky direction, which then are modeled, and the telescope's PM is obtained. Here, we report preliminary results of SRT's PM, where we obtain for each epoch -551.116 and -3811.549 arcsec for Azimuth encoder offset, and 1217.105 and -3343.866 arcsec for Elevation encoder offset. More accurate results can be obtained with better sky coverage observation.

1. Introduction

Pointing model is one of the methods to find the efficiency of utility by reporting in term of the encoder offset. Thus this method will be applied to the Thai National Radio Telescope (TNRT) after completely installed. Make sure that this method will be worked properly, we have applied the same manner with 4.5m Thai Small Radio Telescope (SRT's). This telescope is installed at Astro-Park, Chiang Mai, Thailand and used for educational purposes. Its located at 18°N 51' 5" and 98°E 57' 27" and above the mean sea level of 340 meters. The uncertainty of SRT's pointing is measured in term of the error of the stars by pointing them in both of radio frequency and visible light. The small optical telescope have been installed on the main surface of SRT's which track and point to the same direction with SRT's. We report preliminary results of SRT's pointing model in this report.

2. Pointing model

The pointing model is used to correct for miss-alignments from the radio telescope [1] in both of azimuth (Az) and elevation (El) which azimuth angle is the angle formed between a north celestial pole (determine at the north pole have azimuth is 0 degree) and a line from the observer to a point of interest projected along the horizon (maximum at 360 degree that is the north pole) and elevation is the angle of altitude of the object that maximum at 90 degree. The algorithm of pointing model which used for SRT's and TNRT's 40 meters is similar with ALMA [2] but add gravity term into the model [3]. This pointing model algorithm for azimuth error is given by

$$\delta Az = P_1 + P_2 \sec El + P_3 \tan El - P_4 \cos Az \tan El + P_5 \sin Az \tan El \tag{1}$$

But, this algorithm should be correct for cosine of elevation (cross-elevation), which is given by $\delta Az_{corr} = \delta Az \cos El$. The pointing model algorithm for elevation error is given by

$$\delta El = P_4 \sin Az + P_5 \cos Az + P_7 + P_8 \sin El + P_9 \cos El \tag{2}$$

where the individual pointing coefficients are defined in table 1.

Table 1. Explanation of the pointing coefficients whose pointing errors are described by equation (1) and (2).

Parameter	Physical meaning
P_1	Azimuth encoder offset, includes positioning error for receivers in the Nasmyth focus
P_2	Collimation error, includes positioning error for receivers in the Nasmyth focus
P_3	Lack of orthogonality between the azimuth and elevation axis, includes positioning error for receivers in the Nasmyth focus
P_4	Tilt of azimuth axis along a E-W direction
P_5	Tilt of azimuth axis along a N-S direction
P_6	Declination error of source, this parameter can be ignored as long as their positions have well-determined
P_7	Elevation encoder offset, includes positioning error for receivers in the Nasmyth focus
P_8	Gravitational effect, includes positioning error for receivers in the Nasmyth focus. Combines with P_9 and has no simple meaning
P_9	Gravitational effect, includes positioning error for receivers in the Nasmyth focus. Combines with P_8 and has no simple meaning

3. Planning observation and observable

We planned to observe the pointing model for 5 days along 1 month. But, we were able to observe only 2 out of 5 days (see the detail in table 2).

Table 2. Observation plan and what is happened in each day

Date	Remarks
15th January 2021	Complete data collection, but not cover the sky
21st January 2021	Cancelled due to cloudy weather
22nd January 2021	Cancelled due to cloudy weather
11th February 2021	Cancelled due to the technical problem
16th February 2021	Complete data collection, but not cover the sky

We determined the order of the observation in each object which depend on the position and time until we could observes whole sky. But, We could not observe the whole sky because the object disappeared out of optical telescope frame in some sky positions. We show positions of the observation plan and observable position in figure 1 and 2.

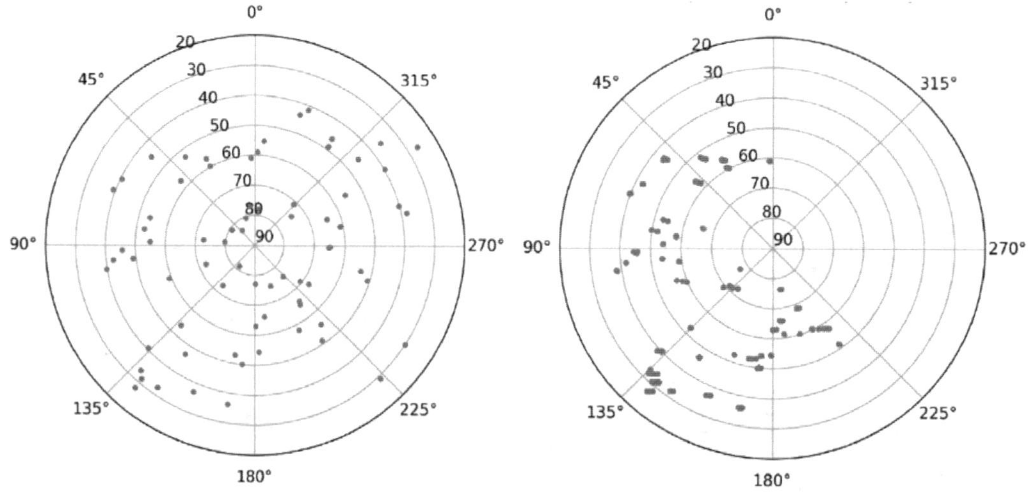

(a) Planning observation (b) Observable

Figure 1. (a) Planning observation on 15th January 2021 which coverage position. (b) Observable position at 15th January 2021 which observed in some area.

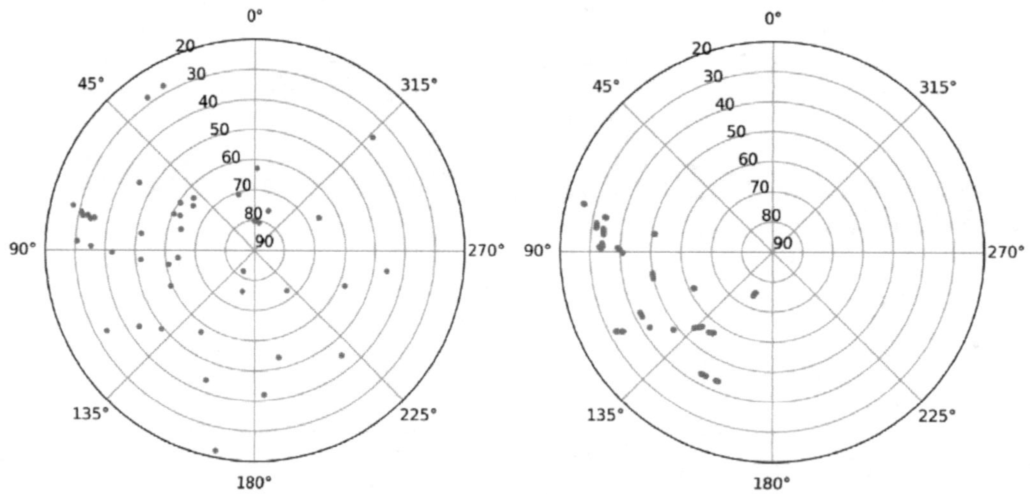

(a) Planning observation (b) Observable

Figure 2. (a) Planning observation on 16th February 2021 which coverage position. (b) Observable position at 16th February 2021 which observed in some area.

4. Results and discussion

When the observations were completed, we solved the pointing parameter by using equation (1) and (2) as shown the result in table 3. For SRT's, we focus on parameter P_1 and P_7 as only 2 parameters, which can be modified to the SRT's encoder. We get the result of both P_1 are -551.116 and -3811.549 arcsecond, respectively. And we get the result of both P_7 are 1217.105 and -3343.866 arcsecond, respectively. Neither plus nor minus signs have any physical meaning. On 15th January 2021, we were able to observe the azimuth angle in range of 0 to 220 degrees, which estimate cover 60 percentage of the whole sky. On 16th February 2021, we were able to observe the azimuth angle in range of 60 to 160 degrees, which estimate cover 27 percentage of the whole sky.

The pointing model parameters (P1 to P9) depend on a number of observable points (Regression). If there are a large number of observable points, We will get more accuracy results. In figure 1(b) or 15th January 2021, could observe as a wider area or greater coverage position of the whole sky than figure 2(b) or 16th.It is a reason why the pointing model parameters of 15th January 2021 is better (get lower values) than 16th February 2021.

Table 3. The pointing model parameters in each epoch which get from SRT's observation.

Parameter	15th January 2021	16th February 2021
P_1	-551.116	-3811.549
P_2	90.346	6570.450
P_3	614.102	-5184.783
P_4	-662.982	-988.760
P_5	-653.078	1919.906
P_7	1217.105	-3343.866
P_8	-1345.131	5076.639
P_9	-156.435	3931.304

5. Conclusion

We used a pointing model to find the efficiency of SRT's pointing. We planned to observe the pointing model for 5 days along 1 month. But, we were able to observe only 2 out of 5 days due to technical and weather issues. We get the pointing model parameters P_1 in each epoch are -551.116 and -3811.549 arcsecond, respectively. We get the pointing parameters P_7 in each epoch are 1217.105 and -3343.866 arcsecond, respectively. We get an encoder offset in both of azimuth and elevation angle on 15th January 2021 less than on 16th February 2021. If we need to get more accuracy results, we need to observe the whole sky. We will apply this result to enter a correction value to SRT's encoder in both of azimuth and elevation angle. The next observations should give a better result.

References

[1] de Vicente P and Barcia A 2007 Deconstructing a pointing model for the 40M OAN radiotelescope *Tech. Rep. IT-OAN 2007–26* pp 1–25
[2] Mangum J G 2001 *A Telescope Pointing Algorithm for ALMA ALMA Memo 366* (Charlottesville: National Radio Astronomy Observatory)
[3] Greve A, Panis J F and Thum C 1996 *Astron. Astrophys. Suppl. Ser.* **115** 379

Detectability of exomoons by examining the signals from a model of transiting exoplanets with moons

P Cherdwongsung[1*], S Awiphan[2†], P Kittara[3‡], K Matan[3] and N Nakharutai[4]

[1]Master's Degree Program in Physics, Department of Physics, Faculty of Science, Mahidol University, Bangkok, 10400, Thailand
[2]National Astronomical Research Institute of Thailand (Public Organization), Chiang Mai, 50180, Thailand
[3]Department of Physics, Faculty of Science, Mahidol University, Bangkok, 10400, Thailand
[4]Department of Statistics, Faculty of Science, Chiang Mai University, Chiang Mai, 50200, Thailand

E-mail: *`prangsutip.chr@student.mahidol.ac.th`, †`supachai@narit.or.th`, ‡`phichet.kit@mahidol.ac.th`

Abstract. Exomoons are natural satellites of exoplanets. Nowadays, none has been confirmed. However, a number of detection techniques have been proposed, including Transit Timing Variations (TTV) and Transit Duration Variations (TDV) techniques. From a recent study, fitting observed transit with the traditional photocentric fitting model shows unique features around the primary and secondary exomoon transits in TDV and transit depth signals, which might reduce the detectability. The aim of this work is to retrieve the variation of TTV, TDV and transit depth signals of exomoon systems with the photocentric fitting model. One year star-planet-moon transit light curves are simulated with `LUNA` algorithm and fit with `TransitFit`. The results show that neglecting the TDV and transit depth data with phase around exomoon's primary and secondary transits improve the exomoon detectability by a factor of ten and the systems with large moon orbital semi-major axis with nearly edge-on orbit around low mass stars can be detected.

1. Introduction

In the Solar system, more than 200 moons have been discovered. However, to date, none of moon orbiting exoplanet, an exomoon, has been discovered. Only two exomoon candidates: MOA-2011-BLG-262Lb [1] and Kepler-1625 b I [2], have been proposed. A number of exomoon detection techniques have been proposed, including the transit [3], microlensing [4], transit timing variation [5, 6] and transit duration variation [7] techniques.

The transit method is one of the most promising methods for searching exoplanets. More than 2,000 exoplanets have been discovered with the technique. The shape of transit light curves provides physical information of the exoplanetary system, including the mid-transit time and transit duration. In a planetary system with a satellite, the gravitation interactions of an exomoon can alter the exoplanet's transit time, called transit timing variation (TTV) effect. Moreover, the transverse velocity of the planet also varies with the moon position which causes the change in transit duration, called transit duration variation effects. The root mean square amplitude of TTV and TDV sinusoidal signal can be written as,

$$\delta_{TTV} \simeq \frac{1}{\sqrt{2}} \frac{a_m M_m \sqrt{a_p}}{(M_p + M_m)\sqrt{G(M_s + M_p + M_m)}} , \tag{1}$$

and

$$\delta_{TDV} \simeq \left[\frac{a_w a_p \cos^2 i}{(R_s + R_p)^2 - a_p^2 \cos^2 i} \pm \frac{2\pi a_w}{P_m} \frac{1}{v_{B\perp}} \right] \cdot \frac{\overline{\tau}}{\sqrt{2}} , \tag{2}$$

respectively, where a_m and a_w are the semi-major axes moon's and planet's orbit around the planet-moon barycentre, respectively, a_p is the planet orbital semi-major axis, M_s is the stellar mass, M_p is the planetary mass, M_m is the moon mass, i is the orbital inclination angle of both planet and moon orbits with the coplanar orbit is assumed, R_s is the star radius, R_p is the planet radius, $v_{B\perp}$ is the projected velocity of the planet-moon barycentre across the face of the star during transit, P_m is the moon orbital period, $\overline{\tau}$ is the transit duration, and G is the universal gravitational constant. Therefore, the TTV and TDV signals depend on eight parameters: M_s, M_p, M_m, R_s, R_p, a_p, a_m and i.

However, fitting transiting exoplanet with moon light curves using the photocentric fitting model, the TTV, TDV and transit depth signals are perturbed due to exomoon transits at the moon phase $\phi_m = 0.25, 0.75$ or around the phases when the moon transits inside the planetary disc at primary exomoon transit and secondary exomoon transit. The exomoon transit induces larger variation in TDV and transit depth signals than the TTV signal (figure 1).

In this paper, the TDV and transit depth perturbations effects at the primary and secondary moon's transits on the detectabilities are investigated. The transiting exoplanet with exomoon light curves are simulated using `LUNA` algorithm [8]. The light curves are fitted with `TransitFit`, which is a photocentric fitting model [9]. The moon phases are computed from the cyclic variation of TTV signals. The sinusoidal variation of the TDV and transit depth data with cutting and non-cutting the moon phases around the moon's transit models are compared.

2. Generating transiting exoplanet with moon light curves

The transiting exoplanet with exomoon light curves were generated by `LUNA` algorithm [8]. One year light curves are simulated. The number of light curves depend on their planetary orbital period. The orbital periods are set to be less than 18 days in order to have the number of transits large enough (> 20 transits) to perform the statistic test. In each light curve, 20,000 data points are simulated throughout the orbital period.

A Neptune-mass exomoon orbiting a Jupiter-mass exoplanet is focused in this work. The moon's massses and radii are followed mass-radius relationship of Neptune-mass exoplanet [10]. The moon radii are limited to be four times smaller than their planetary host radii. The planet-moon separations are constrained mainly by the Hill's stability. The moon periods that are orbital resonance with their host planets orbital period are neglected. The moons are orbiting in the same orbital plane of their host planet orbit ($i_m = i_p$). The systems whose the moon do not transit the host star ($b_m > 1$) is neglected. Both planets and their satellites are assume to have circular orbital ($e_p = 0$ and $e_m = 0$). The host stars are main sequence stars followed mass-radius relation by Eker Z *et al* [11] with TESS limb darkening coefficients [12]. The TESS bandpass covers the wavelength from 600 - 1,000 nm and is centered on the traditional Cousins I-band. The summarized of simulated systems is shown in table 1.

3. Generating TTV, TDV and transit depth signals

The simulated light curves are fitted with `TransitFit` package [9]. The simulated light curves are shorten to 1,000 data points for transit event. First of all, all simulated light curves are fitted in order to obtain the planetary orbital period and inclination of the systems. The period and

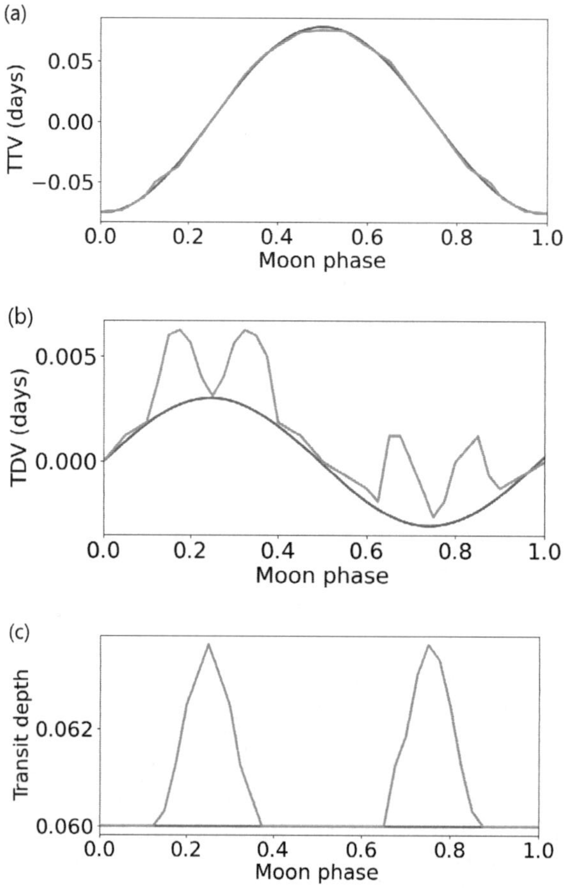

Figure 1. The theoretical moon phase evolution of TTVs (a), TDVs (b) and transit depth (c) signals (Blue) with photocentric fitting model (Orange). The model TDV and transit depth signal show a large variation around exomoon's transit phases.

inclination from the fitting are set to be fixed value for the second fitting where each transit is fitted individually (figure 2). In the second fitting, mid-transit time (t_0), planetary semi-major axis (a_p/R_*), planet-star radius ratio (R_p/R_s) and limb darkening coefficients of each transit are obtained. The TTV, TDV and transit depth are calculated from fitted mid-transit times, planetary semi-major axes and planet-star radius ratios, respectively.

4. TTV, TDV and transit depth analyses

Since the TTV signal is less affected from the perturbing of moon transit as shown in figure 1, the moon periods are derived from the highest peak of TTV's periodgram obtained from Lomb-Scargle periodogram package [13]. The systems with moon periods that have false alarm probability < 0.1 is selected. Only 83% of the simulated systems pass this criterion. The obtained period is used to calculate the moon phase. An example of phase-folded of TTV signal is shown in figure 3.

The example of phase-folded of TDV and transit depth signals are shown in figure 4 and 5. The signals are fitted with sine function. The goodness of fitting is defined by the chi-square value, χ^2. However, both TDV and transit depth signals show anomalies signals around primary

Journal of Physics: Conference Series **2145** (2022) 012009 doi:10.1088/1742-6596/2145/1/012009

Table 1. The input parameters for generating Neptune-mass exomoon orbiting a Jupiter-mass exomoon around main sequence stars light curves with `LUNA` algorithm.

Parameters		**Value**			
Planetary mass	(M_p)	150.0	-	5000.0	$M_\oplus$
Planetary radius	(R_p)	12.225	-	14.264	$R_\oplus$
Moon mass	(M_m)	10.0	-	50.0	$M_\oplus$
Moon radius	(R_m)	3.137	-	3.560	$R_\oplus$
Planetary semi-major axis	(a_p)	0.1	-	1.0	AU
Moon semi-major axis	(a_m)	0.00857	-	0.193	AU
Planet inclination	(i_p)	78.707	-	90.0	degree
Stellar mass	(M_s)	0.179	-	2.80	$M_\odot$
Stellar radius	(R_s)	0.175	-	4.209	$R_\odot$

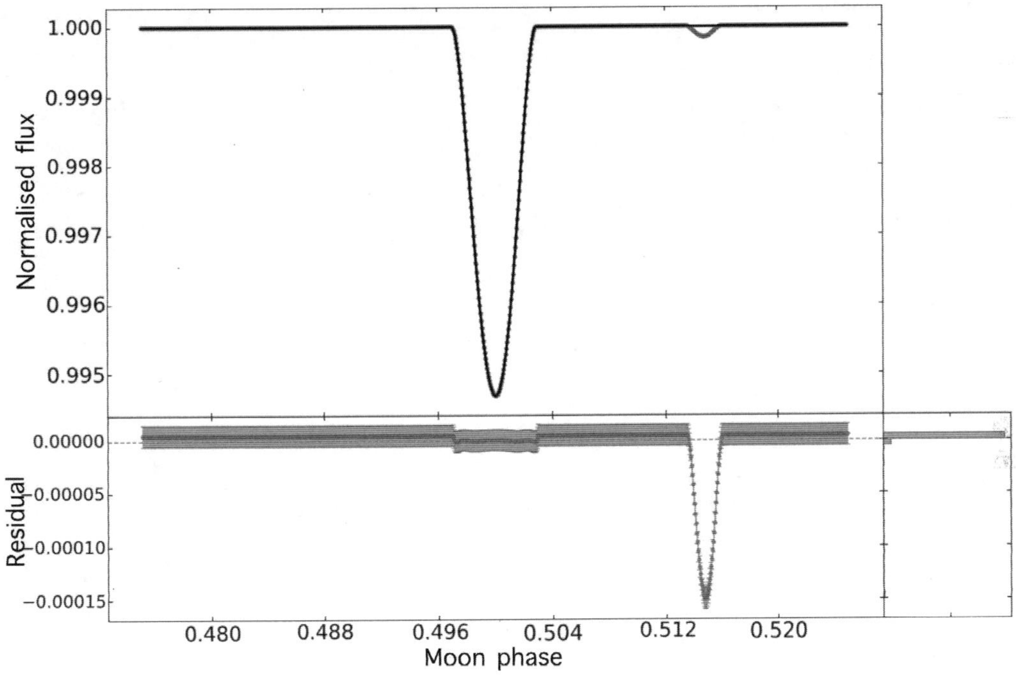

Figure 2. The star-planet-moon transit light curve with best-fit transit model (top) and its residuals from `TransitFit`. The bigger dip shows a planet transit and the smaller dip show a moon transit.

exomoon transit ($\phi_m = 0.15 - 0.35$) and secondary exomoon transit ($\phi_m = 0.65 - 0.85$) as discussed in section 1. Therefore, the phases around these transits are cut. The cut signals are also fitted with sine function (figure 4 and 5).

The theoretical model of TDV signals are sinusoidal functions with the same period and phase shift of $\pi/2$ of TTV sinusoidal function [7]. Neglecting exomoon transit, the transit depths are constant. For non-cutting moon phase model, the number of system with the reduced χ^2 are less than critical values at 99% significance level are 1% and 6% for TDV and transit depth

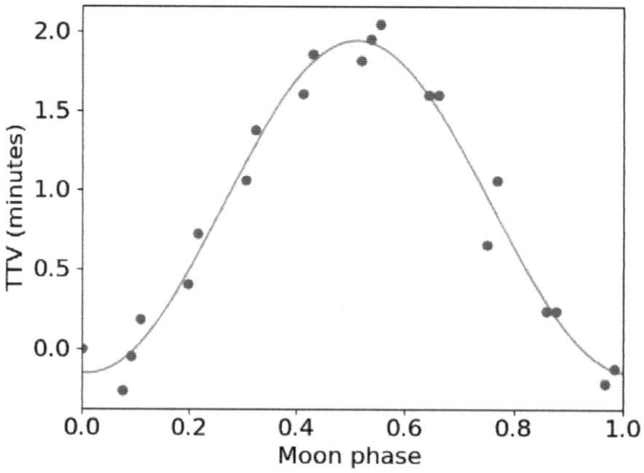

Figure 3. The phase-folded TTV signals from `TransitFit` (Blue dot) with its sinusoidal model (Orange line).

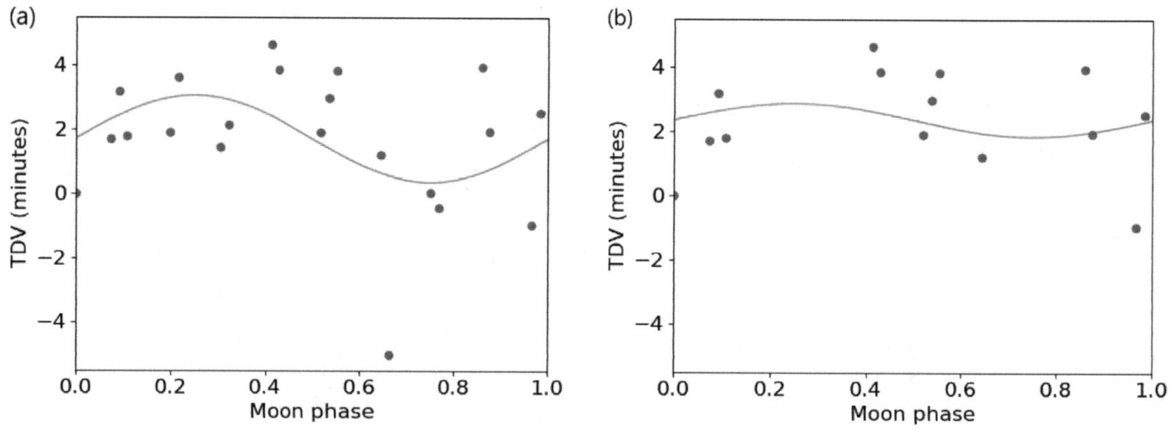

Figure 4. The phase-folded TDV signals from `TransitFit` (Blue dot) with its sinusoidal model (Orange line) of non-cuting moon phase model (a) and cutting moon phase model (b). The reduced χ^2 values of the models are 3.01 and 2.15, respectively.

data, respectively. However, for the cutting moon phase model, the number of system with the reduced χ^2 are less than critical values at 99% significance level increase to 11% and 85% for TDV and transit depth, respectively (figure 6). Therefore, cutting moon phase around exomoon transits can improve the exomoon detectability by a factor of ten.

As the number of systems which pass the TDV critical values is less than those with transit depth critical values. Only the systems which pass the TDV critical values are focused. The distribution between four parameters: moon orbital semi-major axis, planet inclination, stellar radius and stellar mass, with their reduced chi-square values are shown in figure 7. The systems with large moon orbital radius (> 0.01 AU) and high inclination ($> 87°$) around low mass stars can be detected in this work.

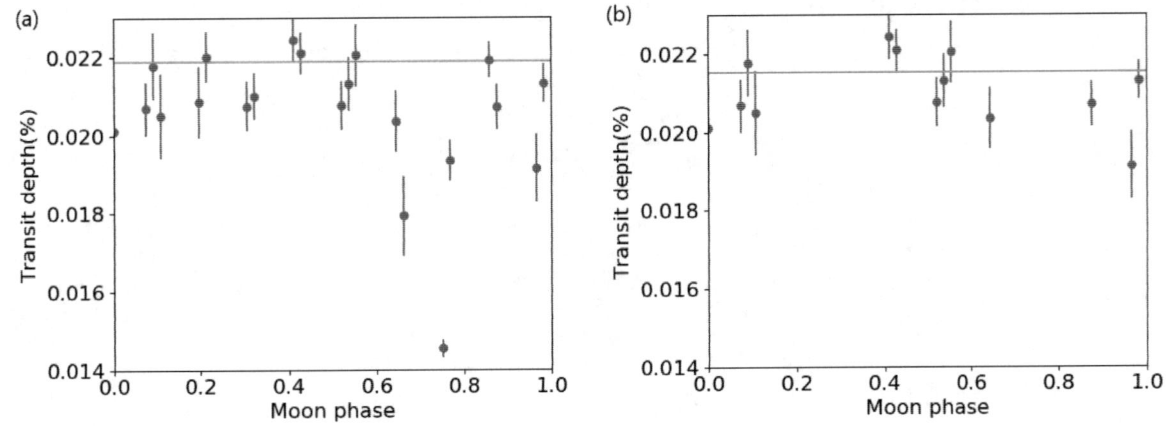

Figure 5. The phase-folded transit depth signals from `TransitFit` (Blue dot) with a constant transit depth value (Orange line) of non-cuting moon phase model (a) and cutting moon phase model (b). The reduced χ^2 values of the models are 68.8 and 2.15, respectively.

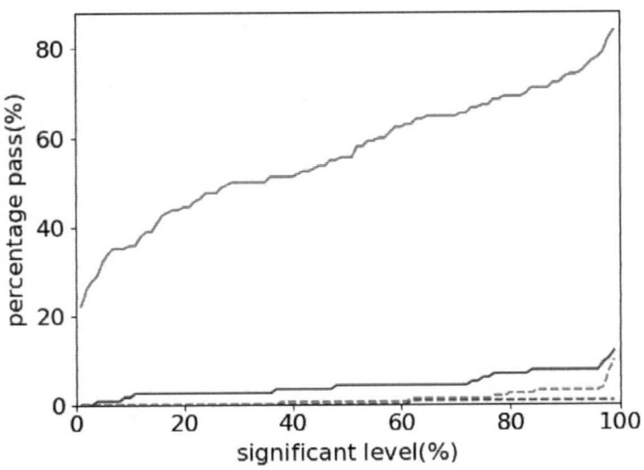

Figure 6. The significant level and the number of system which below their critical value of TDV (Blue) and transit depth (Red) data with non-cutting moon phase (Dashed line) and cutting moon phase (Solid line) models.

5. Conclusions

Neptune-mass exomoon orbiting a Jupiter-mass exomoon around main sequence stars light curves are simulated. Searching exomoon with TTV and TDV techniques using the photocentric fitting model can improve the the exomoon detectabilities by cutting the TDV and transit depth data around primary and secondary moon transits. The TDV still be the weakest signals to detect with photocentric fitting model. Only the systems with large moon orbital radius (> 0.01 AU) with nearly edge-on orbit ($> 87°$) around low mass stars can be detected.

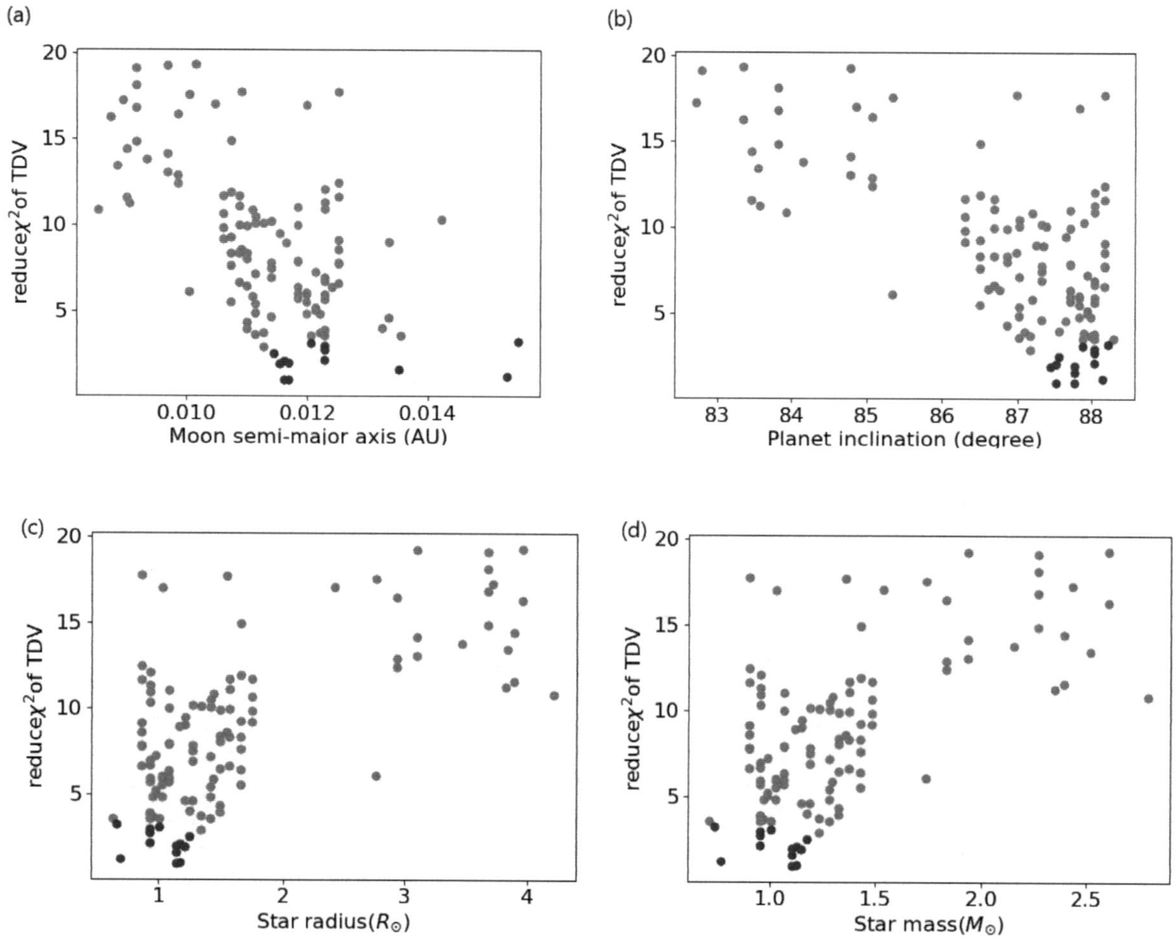

Figure 7. The relations between four physical parameters: moon orbital semi major axis (a), planet inclination (b), star radius (c) and star mass (d), and reduced χ^2 of TDV data below (Blue) and above (Red) the critical values at 99% significance level.

Acknowledgment
We acknowledge the financial support from National Astronomical Research Institute of Thailand (NARIT). SA acknowledge the financial support from a research funding project for young scientist researchers, The Coordinating Center for Thai Government Science and Technology Scholarship Students, National Science and Technology Development Agency.

References
[1] Bennett D P *et al* 2014 MOA-2011-BLG-262Lb: A sub-earth-mass moon orbiting a gas giant primary or a high velocity planetary system in the galactic bulge *Astrophys. J.* **785** 155
[2] Teachey A and Kipping D M 2018 Evidence for a large exomoon orbiting Kepler-1625b *Sci. Adv.* **4** eaav1784
[3] Simon A, Szatmary K and Szabo G M 2007 Determination of the size, mass, and density of "exomoons" from photometric transit timing variations *Astron. Astrophys.* **470** 727–31

[4] Han C and Han W 2002 On the feasibility of detecting satellites of extrasolar planets via microlensing *Astrophys. J.* **580** 490–93
[5] Sartoretti P and Schneider J 1999 On the detection of satellites of extrasolar planets with the method of transits *Astron. Astrophys. Suppl. Ser.* **134** 553–60
[6] Kipping D M 2008 Transit timing effects due to an exomoon *Mon. Notices Royal Astron. Soc.* **392** 181–9
[7] Kipping D M 2009 Transit timing effects due to an exomoon - II *Mon. Notices Royal Astron. Soc.* **396** 1797–804
[8] Kipping D M 2011 LUNA: an algorithm for generating dynamic planet-moon transits *Mon. Notices Royal Astron. Soc.* **416** 689–709
[9] Hayes J J C *et al* 2021 TransitFit: an exoplanet transit fitting package for multi-telescope data sets and its application to WASP-127, WASP-91, and WASP-126 *ArXiv* 2103.12139
[10] Chen J and Kipping D M 2016 Probabilistic forecasting of the masses and radii of other world *Astrophys. J.* **834** 17
[11] Eker Z *et al* 2018 Interrelated main-sequence mass–luminosity, mass–radius, and mass–effective temperature relations *Mon. Notices Royal Astron. Soc.* **479** 5491–511
[12] Claret A 2017 Limb and gravity-darkening coefficients for the TESS satellite at several metallicities, surface gravities, and microturbulent velocities *Astron. Astrophys.* **600** 6
[13] VanderPlas J T 2018 Understanding the Lomb-Scargle Periodogram *Astrophys. J. Suppl. Ser.* **236** 16

Retrieving exoplanet atmospheric parameters using random forest regression

Patcharawee Munsaket[1*], Supachai Awiphan[2], Poemwai Chainakun[1,3] and Eamonn Kerins[4]

[1]School of Physics, Institute of Science, Suranaree University of Technology, Nakhon Ratchasima 30000, Thailand

[2]National Astronomical Research Institute of Thailand (Public Organization), 260 Moo 4, Donkaew, Mae Rim, Chiang Mai, 50180, Thailand

[3]Centre of Excellence in High Energy Physics and Astrophysics, Suranaree University of Technology, Nakhon Ratchasima 30000, Thailand

[4]Jodrell Bank Centre for Astrophysics, University of Manchester, Oxford Road, Manchester, M13 9PL, United Kingdom

*E-mail: m6200343@g.sut.ac.th

Abstract. Understanding of exoplanet atmospheres can be extracted from the transmission spectra using an important tool based on a retrieval technique. However, the traditional retrieval method (e.g. MCMC and nested sampling) consumes a lot of computational time. Therefore, this work aims to apply the random forest regression, one of the supervised machine learning technique, to retrieve exoplanet atmospheric parameters from the transmission spectra observed in the optical wavelength. We discovered that the random forest regressor had the best accuracy in predicting planetary radius (R^2_{Fit} = 0.999) as well as acceptable accuracy in predicting planetary mass, temperature, and metallicity of planetary atmosphere. Our results suggested that the random forest regression consumes significantly less computing time while gives the predicted results equivalent to those of the nested sampling PLATON retrieval.

1. Introduction

Characterization of planetary atmospheres is a rapidly developing area in the exoplanet field. Transmission spectroscopy is the method commonly used to study exoplanetary atmospheres by measuring the variation of transit depths with wavelengths that depends on properties of elements in the planetary atmosphere [1]. The characteristics of the exoplanet's atmosphere can be obtained from the atmospheric retrieval process. However, the traditional retrieval processes (e.g. MCMC [2] and nested sampling [3]) is a time-consuming procedure. In order to reduce the computational time, machine learning techniques have been implemented for retrieving exoplanet atmosphere parameters. For example, random forest regression, a supervised machine learning technique, has previously been used to approximate the variables of an exoplanet's atmosphere using Hubble Space Telescope Wild Field Camera 3 data (WFC3) [4].

In this paper, random forest regression is employed to retrieve atmospheric properties from simulated photometric observational data of hot Jupiters. The PLanetary Atmospheric Transmission for Observer Noobs (`PLATON`) is used to generate exoplanetary transmission spectra [5]. The spectra then are weighted with the filter transmissions. The atmospheric parameters are estimated using random forest regression model in python `scikit-learn` package [6]. The results are compared with those obtained from nested sampling retrieval in `PLATON`. The workflow is illustrated in figure 1.

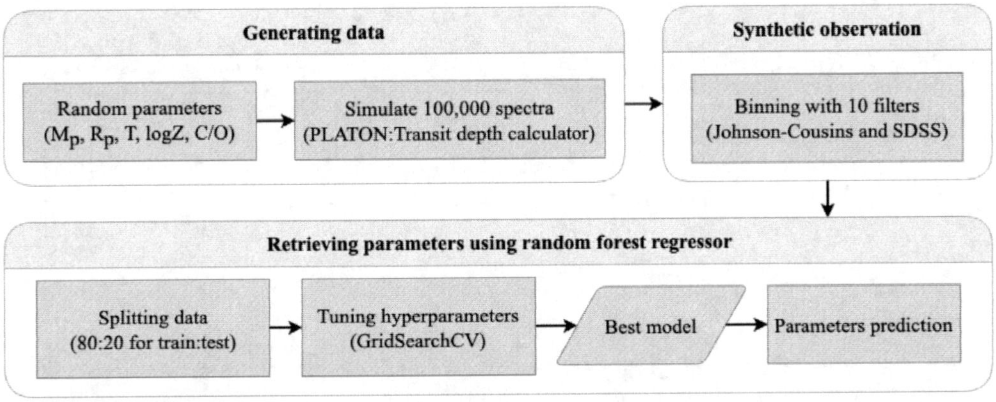

Figure 1. The flowchart for retrieving exoplanet atmospheric parameters from the synthetic transmission spectra in the optical wavelength using a random forest regressor.

2. Synthesize transmission spectra

In this work, only hot Jupiters, gas giant exoplanets with very short orbital periods ($P < 10$ days), are focused [7]. We investigated five atmospheric parameters including planet mass (M_p), planet radius (R_p), planet atmosphere temperature (T), metallicity of planetary atmosphere ($\log Z$), and atmospheric carbon to oxygen ratio (C/O). `PLATON` forward model is used to generate 100,000 hot Jupiter transmission spectra, by uniformly, randomly selecting the values of these parameters within the range shown in table 1. The adopted planetary mass and radius values are within 68 and 95 confident intervals of the mass-radius relationship of the Jovian exoplanets, respectively [8]. The planets are assumed to orbit a solar-analog star.

Table 1. The input parameters for generating the spectra using the transit depth calculator module in `PLATON`. Note that M_p is in Jupiter mass (M_J) and R_p is in earth radius ($R_\oplus$).

Parameters		Value	References
Planetary mass	(M_p)	$0.3M_J - 3M_J$	[9, 8]
Planetary radius	(R_p)	$8R_\oplus - 20R_\oplus$	[10, 8]
Planetary atmospheric temperature	(T)	1300 – 3000 K	[11]
Metallicity of planetary atmosphere	($\log Z$)	-1 – 3	[5]
Atmospheric Carbon-Oxygen ratio	(C/O)	0.35 – 1	[5]

To date, a number of transit observations are performed from the ground using broad-band optical filters. In this work, The five Johnson-Cousins (U, B, V, R, I) and five Sloan (u', g', r', i', z') optical filters are selected. To synthesize transmission spectra of optical observations, the transmission curve of the filters are used to weight the transmission spectra simulated from `PLATON`, as shown in figure 2. The Johnson-Cousins transmission curves are obtained from photometric transmission curves class in PyAstronomy package [12]. The Sloan filter profile is obtained from Sloan Digital Sky Survey Data Release 7 (SDSS Data Release 7: http://classic.sdss.org/dr7/instruments/imager/#filters). The weighted transmission curves are used as the transit depths that could be obtained from the corresponding filters.

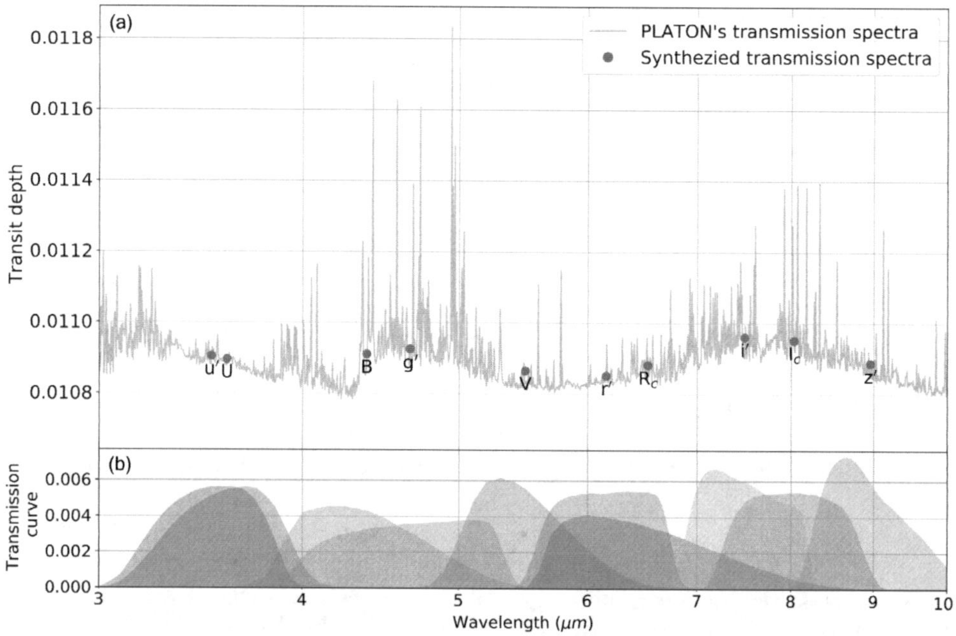

Figure 2. A synthesized transmission spectrum (red) obtained by simulated transmission spectra from `PLATON` (turquoise) weighted by the transmission curve of the filters (b).

3. Random forest regressor

In this work, the random forest regression is used to retrieve transmission spectra. From 100,000 synthetic transmission spectra of 10 Johnson-Cousins and Sloan filters, the spectra were split into two groups: 80,000 for training and 20,000 for testing. Ten binned transit depths are the model features while five planetary parameters (M_p, R_p, T, $\log Z$ and C/O) are the labels. The hyper-parameters tuning is performed using the `GridSearchCV` package in `scikit-learn` to optimize the appropriate values of the number of trees (number of estimators) and the maximum depth of each tree [6]. The best cross validation score is $R^2 = 0.5949$ for the maximum depth of 320 and the number of estimators of 2,560 (figure 3), hence these values are used in our regressor model to predict the planetary parameters from 20,000 spectra in our test data set.

The comparisons between real and guess values of five atmospheric variables are shown in light blue points in figure 4. Their coefficients of determination (R^2_{Fit}) are reported in table 2. The model has the best accuracy in predicting R_p with R^2_{Fit} of 0.999 and moderate accuracy in predicting M_p, T, and $\log Z$ with R^2_{Fit} of 0.742, 0.571 and 0.638, respectively. However, the model cannot predict the C/O ratio as the transmission features induced by the C/O variation are not significant enough in the optical bands.

4. Comparing results with the `PLATON` retrieval model

The results from our random forest regressor model is compared to those obtained from the nested sampling fitting model in the `PLATON`. A sample of 100 spectra is simulated and retrieved with the PLATON. The number of live points in nested sampling model are selected to be 50 and 100. In order to compare the computaional time, all models perform on a machine with a Core i9-10900 CPU and 32GB of RAM that runs Ubuntu 18.04.2 LTS. The results are shown in figure 4 and table 2. The random forest regressor performs 160,000 and 270,000 times faster than the PLATON nested sampling model with 50 and 100 live points, respectively. Judging by R^2_{Fit}, the accuracy obtained from the regressor model are comparable to that of the PLATON retrievals.

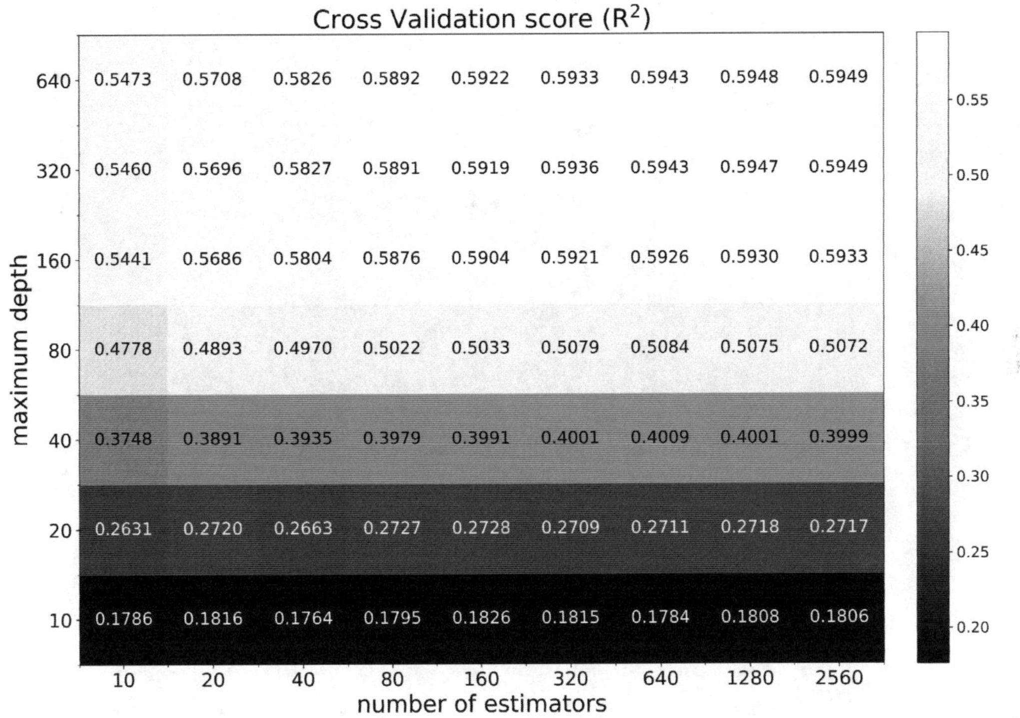

Figure 3. Cross validation score (R^2) obtained from hyper-parameter tuning using GridSearchCV package.

Table 2. Performance of our random forest regressor model compared to that of the PLATON retrievals with 50 and 100 live points. Note that the models are run on a machine with a Core i9-10900 CPU and 32GB of RAM that runs Ubuntu 18.04.2 LTS.

Approach	**Data sets**	**Time spent** (Hr:Min:Sec)	**Coefficient of determination** (R^2_{Fit}) M_p	R_p	T	$\log Z$	C/O
Random forest regressor	20,000	0:0:14	0.742	0.999	0.571	0.638	0.129
PLATON (nlive=50)	100	3:12:8	0.780	0.999	0.546	0.659	-0.842
PLATON (nlive=100)	100	5:16:45	0.791	0.999	0.616	0.591	-0.604

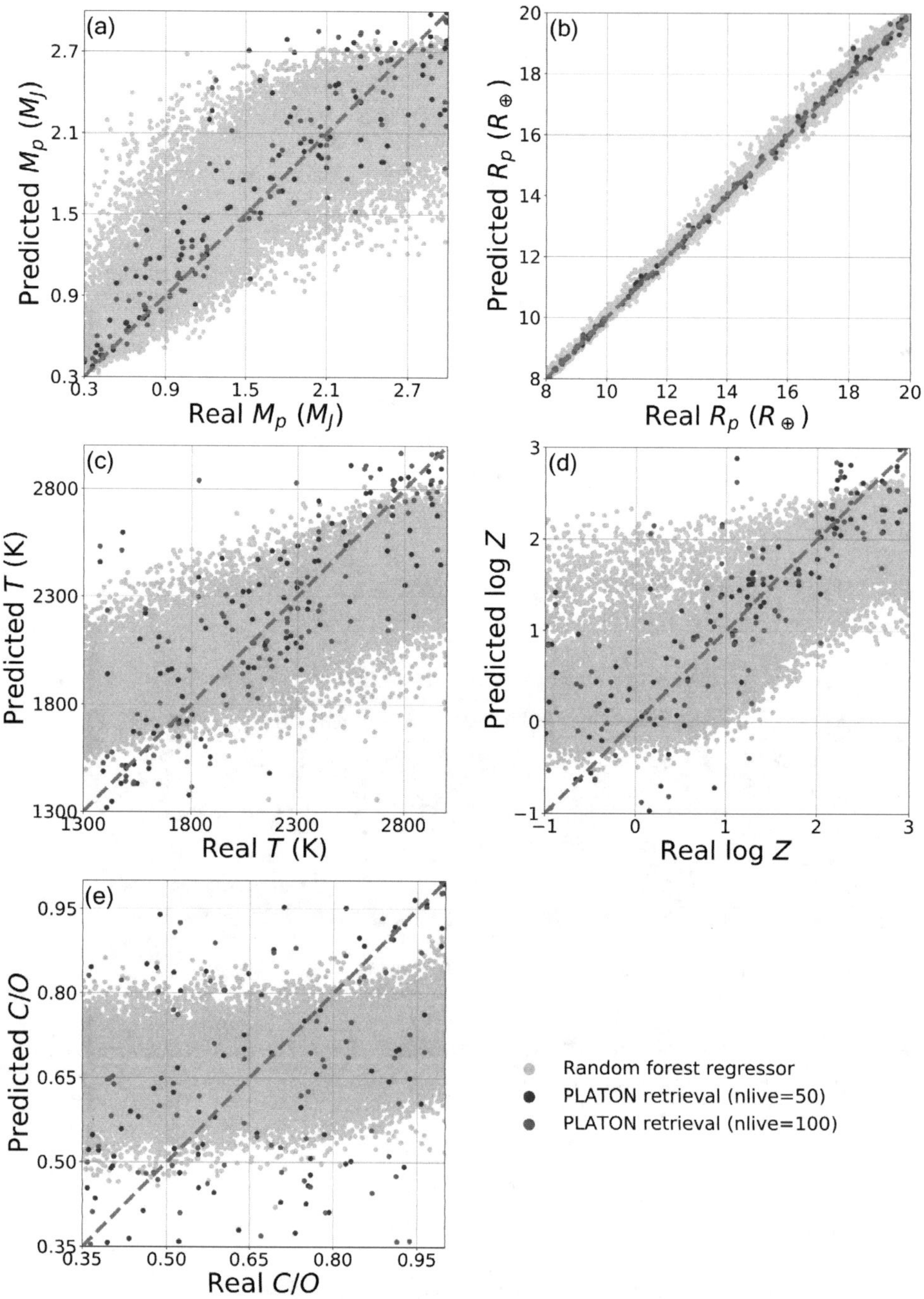

Figure 4. Real and predicted values of five planetary parameters: (a) planetary mass, (b) planetary radius, (c) planetary atmospheric temperature, (d) metalicity of planetary atmosphere and (e) Carbon to Oxygen ratio in the atmosphere, of 20,000 test set using random forest regressor (lightblue), the PLATON nested sampling retrival with number of live points of 50 (blue) and 100 (green). The red dashed lines show prefect prediction lines.

5. Conclusion

Transmission spectra of hot-Jupiter atmosphere are simulated using `PLATON` package. The spectra are weighted with 10 Johnson-Cousin and Sloan filter profiles. The random forest regressor are used to estimate five planetary parameters: M_p, R_p, T, $\log Z$ and C/O. The model provides the best cross validation score when the number of estimators is 2560 and the maximum depth is 320. The regressor can precisely predict the planetary radius with $R^2_{Fit} = 0.999$ and has intermediate accuracy in predicting M_p, T, and $\log Z$ values because of the planetary radius can be retrieved using the baseline of transit depths, while the other characteristics can be retrieved from the transit depth features in the specific filters. However, the model cannot predict the C/O correctly because different values of this parameter produce very small variation in optical spectra. We conclude that the random forest regressor can be used to estimate planetary parameters from optical transmission spectra with the prediction accuracy approximately as same as that of the traditional retrieval method, but with more than 100,000 times faster.

Acknowledgements

This work was supported by the Development and Promotion of Science and Technology Talents Project (DPST) - Royal Thai Government Scholarship, Suranaree University of Technology (SUT), and National Astronomical Research Institute of Thailand (NARIT). This research was supported by the Program Management Unit for Human Resources & Institutional Development, Research and Innovation, NXPO [grant number B16F630069].

References

[1] Seager S and Sasselov D 2000 *Astrophys. J.* **537** 916
[2] Foreman-Mackey D, Hogg D W, Lang D and Goodman J 2013 *Publ. Astron. Soc. Pac.* **125** 306
[3] Shaw J, Bridges M and Hobson M 2007 *Mon. Not. R. Astron. Soc.* **378** 1365–70
[4] Márquez-Neila P, Fisher C, Sznitman R and Heng K 2018 *Nat. Astron.* **2** 719–24
[5] Zhang M, Chachan Y, Kempton E M R, Knutson H A and Chang W H 2020 *Astrophys. J.* **899** 27
[6] Pedregosa F *et al.* 2011 *J. Mach. Learn. Res.* **12** 2825–30
[7] Gaudi B S, Seager S and Mallen-Ornelas G 2005 *Astrophys. J.* **623** 472
[8] Chen J and Kipping D 2016 *Astrophys. J.* **834** 17
[9] Stevens D J and Gaudi B S 2013 *Publ. Astron. Soc. Pac.* **125** 933
[10] Borucki W J *et al.* 2011 *Astrophys. J.* **736** 19
[11] Madhusudhan N, Knutson H, Fortney J and Barman T 2014 *Exoplanetary atmospheres* (Arizona: University of Arizona Press) p 739
[12] Czesla S, Schröter S, Schneider C P, Huber K F, Pfeifer F, Andreasen D T and Zechmeister M 2019 Pya: Python astronomy-related packages (*Preprint* `1906.010`)

Astrometric observations of water maser sources toward the Galactic Center with VLBI

D Sakai[1*], T Oyama[2], T Nagayama[2], H Kobayashi[2] and M Honma[2]

[1]National Astronomical Research Institute of Thailand (Public Organization), 260 Moo 4, T. Donkaew, A. Maerim, Chiangmai, 50180, Thailand
[2]Mizusawa VLBI Observatory, National Astronomical Observatory of Japan, 2-12 Hoshi-ga-oka, Mizusawa-ku, Oshu-shi, Iwate 023-0861, Japan

*E-mail: daisuke@narit.or.th

Abstract. The Central Molecular Zone (CMZ) in the Galactic Center region shows outstanding non-circular motion unlike the Galactic disk. Although several models describing this non-circular motion are proposed, an uniform kinematic model of the CMZ orbit is not appeared. Three dimensional velocity information including proper motions will be critical to constrain the orbital models of the CMZ because most of models proposed are devised to reproduce the line-of-sight velocity profiles of the molecular clouds in this region. To reveal the dynamics of the Galactic center region, we conducted VLBI astrometric observations of 22 GHz water maser sources toward the Galactic center with VERA. By measuring parallaxes and proper motions, we can identify whether each source is actually located in the CMZ or not, and identify the three dimensional positions and velocities in the non-circular orbit if the source is located in the CMZ. We show the results of astrometric study for several maser sources associated with molecular clouds toward the Galactic center including Sgr B2 complex and Sgr D HII region. The parallax measurement toward Sgr B2 obtained the parallax of 0.133 ± 0.038 mas, and its proper motions indicated that Sgr B2 complex is moving toward the positive Galactic longitude with $V = 100\,\mathrm{km\,s^{-1}}$ relative to Sgr A*.

1. Introduction

The Central Molecular Zone (CMZ), which ranges from -2 to $+2$ degree of the Galactic longtude, has very peculiar properties. The molecular gas density is very high ($n = 10^3 - 10^5\,\mathrm{cm^{-3}}$), the line width of the molecular emission is larger ($\Delta V = 20 - 50\,\mathrm{km\,s^{-1}}$), and the star formation rate (SFR) is quite low despite such high density. Recently, this inefficient SFR is considered to be come from the dynamical feature of the CMZ. Then, two representative models are proposed to explain these properties in the CMZ. First model is closed orbit model proposed by [1]. In this mode, the orbital velocity is constant, and the star formation is triggered by cloud-cloud collision in the intersection between different orbits. Another model is open orbit model proposed by [2]. In open orbit model, the tidal compression in the orbit triggers the star formation. Understanding the dynamical model of the CMZ is important to study not only whole property of the CMZ but also star-formation of individual molecular cloud in the CMZ.

22 GHz water maser emission is known as an indicater of active star forming regions and regions of extremely dense gas. This is a useful tool to study the kinematics and environment around young, massive stars. In the Galactic center, 22 GHz water maser is the most appropriate

maser line because there are few sources which have sufficient flux densities that we can detect with VLBI in other maser lines such as 6.7 GHz methanol and 1.6 GHz OH masers.

The giant molecular cloud complex Sagittarius B2 (hereafter Sgr B2) is one of the most active star forming region in the Galaxy, which has a projected distance of about 100 pc from the Galactic center. This source consists of three major star formation sites, Sgr B2(N), Sgr B2(M), and Sgr B2(S) [3,4], which are aligned from north to south with separation of 45 arcsec with each other. A number of observations have been conducted to reveal the nature of this source in radio continuum [5-7], molecular lines [8-11], and recombination lines [12,13].

The parallax of Sgr B2(M) and B2(N) have already been measured with VLBA to be 0.130 ± 0.012 mas and 0.128 ± 0.015 mas, respectively [14]. They also measured the orbital velocity of Sgr B2 complex in the Central Molecular Zone (CMZ) using the measured proper motion and LSR velocity. However, the measured proper motion still contains a large uncertainty of $\sim 1\,\mathrm{mas\,yr^{-1}}$ because they did not take into account of the internal motions of H_2O maser spots. To determine more accurate position and orbital velocity of Sgr B2 in the CMZ, we should measure and correct the internal motions of maser spots.

Water maser source G0.21 is located near G0.253+0.016, which is a candidate of a progenitor of massive star cluster. Because the line-of-sight velocity of the water maser is close to that of molecular emission ($V_{\rm lsr}$ = 30–40 $\mathrm{km\,s^{-1}}$) [15]. This source is supposed to be located at the front side of the CMZ. In open orbit model, the location is just after passing the pericenter of the orbit, and molecular cloud suffers strong compression due to tidal force. Eventually, the formation of high mass star cluster is triggered.

In this proceeding, we show two astrometric results of water masers associated with Sgr B2 and G0.21 by using VLBI observations. Note that the results for G0.21 are preliminary, and it could be changed by future detail analysis.

2. Observations

We conducted VLBI observations of H_2O maser sources associated with Sgr B2 complex at a rest frequency of 22.235080 GHz using VERA during the period from 2014 to 2017. The number of observation epochs was two in 2014, two in 2015, nine in 2016, and two in 2017. In each epoch, the on-source time for Sgr B2 complex was about 1 hour. The reason for such short on-source time is that a series of observations is aimed to observe other target maser sources, and that Sgr B2 was observed as a phase referencing source. The target maser source Sgr B2 and the position reference source J1745−2820 were simultaneously observed using the dual-beam system of VERA. The separation angle between two sources is 0.33°.

We also conducted wide-band VLBI observation of H_2O masers associated with G0.21 from 2018 to 2020. We observed one epoch in 2018, three epochs in 2019, and one epoch in 2020.The position reference source J1745−2820 were observed with the wide-band mode of VERA, whose bandwidth is 2040 MHz, and it is eight times larger width than usual observing mode.

3. Results and duscussion

3.1. Sgr B2

From phase-reference observations, we measured the parallax and absolute proper motions relative to the extra-galactic source J1745-2820. A maser spot at $v_{\rm LSR}$ = 59.58 $\mathrm{km\,s^{-1}}$ was detected over ten epochs from 2014 to 2016. The positions of the maser spots in the phase-referenced map are affected by linear motions and the annual parallax. Then, we could derive the proper motions in right ascension $\mu_\alpha \cos\delta$ and in declination μ_δ, the initial positions in right ascension $\alpha_0 \cos\delta$ and in declination δ_0, and the parallax π for the maser spot, using least-squares fitting. Figure 1(a) shows the projected motion of the maser spot. The direction of absolute proper motion goes from northeast to southwest. This is almost parallel to the Galactic plane. In figure 1(b), the motion in the directions of RA and Dec are shown as a function of time.

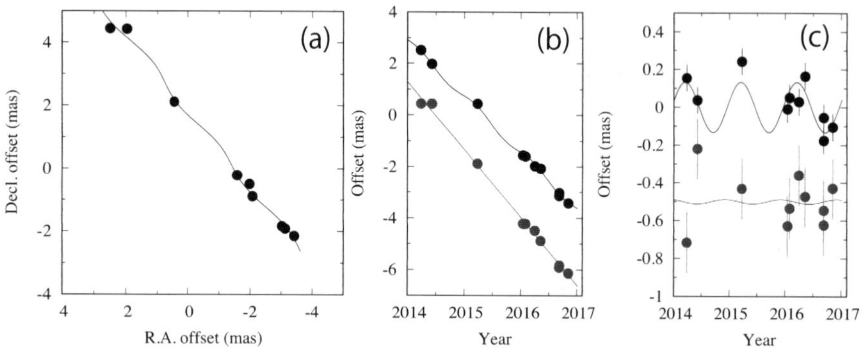

Figure 1. (a) Absolute proper motions of the maser spot at $v_{\rm LSR} = 59.58\,{\rm km\,s^{-1}}$ in Sgr B2(M) region. Filled circles show the observed points from phase referencing. (b) Motions towards R.A. and Dec. as a function of time. Black circles show the motion in the R.A. direction, and gray circles show the motion in the Dec. direction. (c) Result of parallax fitting. Error bars are evaluated so that a χ^2 value in the model fitting becomes unity.

By subtracting the expected linear motions from figure 1(b), we obtained the parallax shown in figure 1(c). The obtained parallax is 0.133 ± 0.038 mas, and proper motions in the Galactic coordinate is $(\mu_l, \mu_b) = (-4.17{\pm}0.19, -0.34{\pm}0.13)\,{\rm mas\,yr^{-1}}$. These parallax and proper motions are calcuated and fitted by using equations described in [16].

When we compare these velocities with the existing models, we found that Sgr B2 complex is located at the upstream of vertex of the orbit. This is consistent with the position of Sgr B2 complex suggested by the open orbit model in [2], in which the star-forming activity in the Sgr B2 complex is triggered by the tidal compression during their pericenter passage. Sgr B2 is located at the upstream of vertex of the orbit in the open orbit model, while in the closed orbit model, this is located at the vertex at which two different orbital families intersect. Also, the proper motions predicted in [2], $(\mu_l, \mu_b) = (-4.21, -0.95)\,{\rm mas\,yr^{-1}}$, is almost consistent with our results. It also supports the open orbit model in the CMZ.

3.2. G0.21

We succeeded to detect fringes from the phase reference source J1745−2820 thanks to the wide-band mode of VERA. Eventually, we could conduct astrometric observations for relatively weak water masers (2–3 Jy) associated with G0.21 region.

In figure 2, we show a preliminary results of astrometric observations for G0.21. For maser spots at $v_{\rm LSR} = 47.48\,{\rm km\,s^{-1}}$ and $v_{\rm LSR} = 47.06\,{\rm km\,s^{-1}}$, we measured their proper motions as $(\mu_l, \mu_b) = (-0.179, -0.975)\,{\rm mas\,yr^{-1}}$ and $(\mu_l, \mu_b) = (-0.568, -1.249)\,{\rm mas\,yr^{-1}}$, respectively. When we compare these value with the prediction from two representative models, $(\mu_l, \mu_b) = (-1.63, -0.56)\,{\rm mas\,yr^{-1}}$ for open orbit model, and $\mu_l = -3\,{\rm mas\,yr^{-1}}$ for closed orbit model, our result also prefers open orbit model to closed orbit model in G0.21 region.

4. Summary

The dynamics of the Central molecular zone is important to understand (i) the gas flow into the central region of the Galaxy, and (ii) the formation of stars in the Galactic center. Between two representative models, the open orbit model can reproduce the properties of star-formation in the CMZ especially fro G0.253 to Sgr B2. To increase the number of sources for reconstructing the orbital models in the CMZ, the developing of astrometry mode of EAVN, wide-band receiver, and accurate geodesy is important.

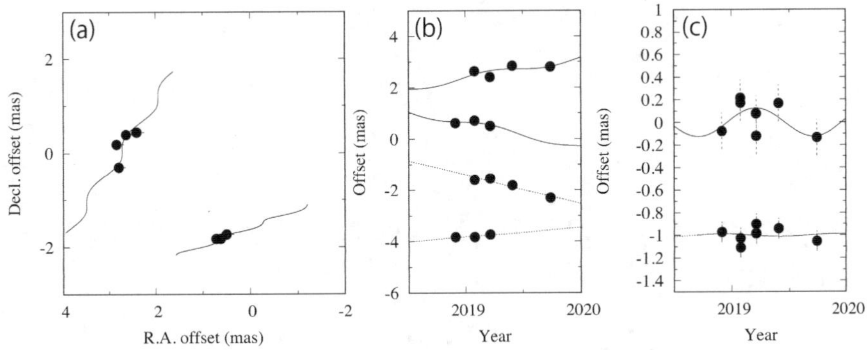

Figure 2. (a) Absolute proper motions of the maser spot at $v_{\rm LSR} = 47.48\,{\rm km\,s^{-1}}$ and $v_{\rm LSR} = 47.06\,{\rm km\,s^{-1}}$ in G0.21. Filled circles show the observed points from phase referencing. (b) Motions towards R.A. and Dec. as a function of time. Black circles show the motion in the R.A. direction, and gray circles show the motion in the Dec. direction. (c) Result of parallax fitting. Error bars are evaluated so that a χ^2 value in the model fitting becomes unity.

References

[1] Molinari S *et al.* 2011 *Astrophys. J. Lett.* **735** L33
[2] Kruijssen J M D, Dale J E and Longmore S N 2015 *Mon. Notices Royal Astron. Soc.* **447** 1059
[3] Huttemeister S, Wilson T L, Henkel C and Mauersberger R 1993 *Astron. Astrophys.* **276** 445
[4] Schmiedeke A *et al.* 2016 *Astron. Astrophys.* **588** A143
[5] De Pree C G *et al.* 2014 *Astrophys. J. Lett.* **781** L36
[6] Sánchez-Monge Á *et al.* 2017 *Astron. Astrophys.* **604** A6
[7] Ginsburg A *et al.* 2018 *Astrophys. J.* **853** 171
[8] Lis D C *et al.* 1993 *Astrophys. J.* **402** 238
[9] Qin S L *et al.* 2008 *Astrophys. J.* **677** 353
[10] Rolffs R *et al.* 2011 *Astron. Astrophys.* **529** A76
[11] Pols S *et al.* 2018 *Astron. Astrophys.* **614** A123
[12] De Pree C G *et al.* 1996 *Astrophys. J.* **464** 788
[13] Zhao J H and Wright M C H 2011 *Astrophys. J.* **742** 50
[14] Reid M J *et al.* 2009 *Astrophys. J.* **705** 1548
[15] Mills E A C *et al.* 2015 *Astrophys. J.* **805** 72
[16] Hachisuka K 2006 *Astrophys. J.* **645** 337

The study of the strongest solar event on a minimum of the 24th solar cycle

D Peldon[1], K Tshering[1], B Gurung[1], T Khumlumlert[1, 2]*and N Aiemsa-ad[3]

[1] Department of Physics, Faculty of Science, Naresuan University, Phitsanulok 65000, Thailand
[2] Research Center for Academic Excellence in Applied Physics, Faculty of Science, Naresuan University, Phitsanulok 65000, Thailand
[3] Faculty of Science and Technology, Rajabhat Rajanagarindra University, Chachoengsao, 24000, Thailand

*E-mail: thiraneek@nu.ac.th

Abstract. The strongest solar flares of the 24th solar cycle erupted on September 6, 2017, and it was the 8th strongest solar flare recorded since 1996. This extreme solar flare occurred at the minimum of the 24th solar cycle. The active region is located in the Western Hemisphere and produced the violent explosion of class X9.3 and X2.2 on September 6, X1.3 on September 7, and X8.2 on September 10, 2017. The injection duration of the solar energetic particles of the solar event was 17 minutes. All data for this solar event was collected from the Advanced Composition Explorer and simulated for particles' motion using the transport equation and solved by the numerical technique. We obtained the injection time of the solar energetic particle propagation by comparing fitting between the simulation results and the spacecraft data. Injection time taken by high-energy particles to travel from the Sun to the Earth was found to be in the range of 39 to 743 minutes. At the peak of this solar flare, the coronal mass ejection was detected, which increased the injection time. The Kp-index of this solar flare was 4; thus, there was no effect on the Earth. The Kp-index value increased to 8 on September 7-8, 2017, due to another solar event from the same sunspot region, indicating the effect of solar flare and CME, which resulted in the appearance of aurora.

1. Introduction

Space weather is directly related to the activity of the Sun, which affects the Earth's environment, are the solar energetic particles (SEPs). They have important effects on the Earth, such as interference radio communication, cause of electrical power failures, radiation warning on airplane flights, and aurora phenomena [1]. Most SEPs regard the study of eruption on the Sun, such as solar flares and coronal mass ejection (CME), which increase during the solar cycle. A solar flare is a violent explosion on the surface of the Sun, which releases the SEPs moving in interplanetary medium or space. Most flares occur in active regions around sunspots. The Sun goes through a cycle of activities called the solar cycle, during which the sunspot numbers increase during solar maximum and decrease during solar minimum. Obtaining various solar cycle phases (rise, maximum, and declining) is also important [2]. Numbering the solar cycle started in 1749 [3]. The present 25th solar cycle began around December

2019. We pass the 24th solar cycle (2008-2019) [4], in which there were no large solar flare events for the first two years. There were no solar flares of X-class intensity in 2008-2010. There were 8 solar flares of X-class in 2011, of which the highest solar flare was X6.9 on August 9. We can also note that the violent solar flares of X-class occurred around the time of solar maximum (2011-2014). This solar cycle had an interesting solar event at the end of the cycle in 2017. The violent solar flares on the minimum of the 24th solar cycle occurred on September 6-10, 2017, at same region 2673, which includes X9.3 flare on September 6, 2017, at 11.53 UT, X8.2 flare on September 10, 2017, at 15.35 UT, X2.2 flare on September 6, 2017, at 8.57 UT, and X1.3 on September 7, 2017, at 14.20 UT. Preliminary data of X-ray flux intensity illustrated the violence of the interested solar flare. The brightness of X-ray flux in watts/m^2 shows the intensity of erupted solar flares. It consists of 5 classes: A, B, C, M, and X class. The Geostationary Operational Environmental Satellite (GOES) has provided a platform for the solar x-ray imager and space environment monitoring instruments. The solar flare of X-class is a violent eruption and may affect the Earth. The solar flare of M-class is medium violent. A and B classes are of low intensity, which have little effect on the Earth. This research selected the violent solar flares of X-class during September 6-10, 2017.

2. Solar energetic particle propagation from the Sun to the Earth

Solar energetic particles or SEPs are high-energy particles, such as ions, protons, or electrons, released from the Sun. SEPs are one of the phenomena related to solar flares. Solar flares are a large explosion of electromagnetic radiation with intense variation in brightness on the Sun lasting from minutes to hours. A solar flare occurs when magnetic energy that has built up in the solar atmosphere is suddenly released. The intensity of a solar flare depends on the solar cycle. The solar cycle mainly refers to the cyclic variation of the number of sunspots visible on the photosphere of the Sun. The number of sunspots increases and regularly decreases over time, approximately 11- year cycle. More sunspots may be directly associated with increased solar activity, such as solar flares and CMEs. The highest number of sunspots in any given cycle is recorded and designated as the solar maximum, while the lowest is termed solar minimum. The solar cycle affects activities on the surface of the Sun, such as sunspots caused by the Sun's magnetic fields. Giant eruptions on the Sun, such as solar flares and coronal mass ejections, increase during the solar maximum. These eruptions send powerful bursts of energy and materials into space.

The information of particle acceleration on the surface of the Sun is uncertain. We selected the focused transport equation with the influence of the space environment to explain the SEPs propagation. Plasma from the outermost portion of the Sun continuously flows away and drags the irregular magnetic field out of the Sun. Magnetic field lines are bent spirally as the Sun rotates on its axis [2, 5, 6]. Most SEPs are charged particles; hence they are forced to move in circles around the magnetic field line with a constant speed. The intensity of the magnetic field line corresponds to the distance of the particle's trajectory from the Sun to the Earth. The magnetic field from the Sun to the Earth is irregular, so we focused on the longest distance that SEPs can move along the magnetic field line before getting scattered by the irregularity of the magnetic field, which is termed as the mean free path. The diffusion equation of particles in space and their effects explains the propagation of SEP. From the fundamental equation of Fokker-Plank and the transport equation of Ruffolo [1], we can explain the propagation of solar energetic particles from the Sun to the Earth as shown in equation (1). We simulated the particle propagation with a transport equation and solved it by the numerical method of Finite-differences.

$$\frac{\partial \vec{F}(t,\mu,z,p)}{\partial t} = \Bigg\{ -\frac{\partial}{\partial z}\mu v - \frac{\partial}{\partial z}(1-\mu^2\frac{v^2}{c^2})v_{sw}\sec\Psi + \frac{\partial}{\partial \mu}v_{sw}(\cos\Psi\frac{d}{dr}\sec\Psi)\mu(1-\mu^2)$$
$$-\frac{\partial}{\partial \mu}\frac{v}{2L(z)}\left[1+\mu\frac{v_{sw}}{v}\sec\Psi - \mu\frac{v_{sw}v}{c^2}\sec\Psi\right](1-\mu^2)$$

$$+\frac{\partial}{\partial\mu}\frac{\varphi(\mu)}{2}\frac{\partial}{\partial\mu}\left[1-\frac{\mu v_{sw} v\sec\Psi}{2L(z)}+\cos\Psi\frac{\mathrm{d}}{\mathrm{d}r}(\sec\Psi)\mu^2\right]$$

$$+\frac{\partial}{\partial p}pv_{sw}\left[\frac{\sec\Psi}{2L(z)}(1-\mu^2)+\cos\Psi\frac{\mathrm{d}}{\mathrm{d}r}(\sec\Psi)\mu^2\right]\Bigg\}\vec{F}(t,\mu,z,p), \quad (1)$$

where $\vec{F}$ is the distribution function; $\vec{F}(t,\mu,z,p)=\frac{\mathrm{d}^3N}{\mathrm{d}z\mathrm{d}\mu\mathrm{d}p}$, θ is the pitch angle, μ is the cosine of pitch angle, v_{sw} is the solar wind speed, Ψ is the angle between $\hat{r}$ and $\hat{z}$, c is the light speed, v is the particle speed (km/s), $\varphi(\mu)$ is the scattering coefficient, and p is the momentum of the particle. We focus the mean free path of particle from the relation of $\lambda\equiv\frac{3D}{v}$ where D is the diffusion coefficient; $D=\frac{v^2}{4}\int_{-1}^{1}\frac{(1-\mu^2)^2}{\varphi(\mu)}\mathrm{d}\mu$ in scattering term of $\varphi(\mu)$.

Preliminary data of X-ray flux intensity illustrate the violence of the interested solar flare. The brightness of X-ray flux in watts/m^2 shows the intensity of erupted solar flares. Solar flares of X-class are violent eruptions and may affect Earth. This research selected the violent solar flares of X-class during September 6-10, 2017. We used the particle data for the interested solar event from Solar Isotope Spectrometer (SIS) on the Advanced Composition Explorer (ACE spacecraft) as input data for simulation. The simulation result is the distribution of particles in time for various energy, which corresponds to the mean free path in the diffusion coefficient. The transport equation is a partial differential equation (PDE) consisting of many independent variables [3]. Thus, making us difficult to determine its solution directly. However, as the equation is in the form of PDE, we can apply boundary conditions to the Finite-differences method and determine its solution.

3. Methodology

This research used a quantitative methodology that involves the empirical investigation of observable and measurable variables used for theory testing. We collected data from spacecraft for the solar event from September 6-10, 2017. The data from spacecraft is the distribution function of time of Helium, Carbon, Nitrogen, and Oxygen of the solar event on September 6, 2017. Physical characteristics of the selected solar event at 2673 sunspot number are shown in table 1. The intensity of particles during that time was download from ACE. The space environment data correlated to the solar event of interest came from GOES. The main focus is given to how the data was collected, prepared for its initial values and how simulation was done to calculate the injection time and mean free path (λ) from the transport equation of Ruffolo solved by using the technique of numerical method.

Table 1. Physical characteristics of interested solar events.

Solar event	X-ray Class	Start time of flare	End time of flare	Shock wave time
6/9/2017	X2.2	08.57 UT	09.17 UT	
	X9.3	11.53 UT	12.10 UT	12.02 UT
7/9/2017	X1.3	14.20 UT	14.55 UT	
10/9/2017	X8.2	15.35 UT	16.31 UT	16.38 UT

We collected the data measured by SIS instrument on ACE spacecraft and calculated the initial values necessary for the program to simulate particle motion. The selected solar event has X-class solar flare, X9.3 and X8.2 erupted from S09W45 on the Sun with solar wind speeds of 494 and 560 km/s, respectively. The X9.3 was the largest flare in the 24th solar cycle. The increased solar activity periods during solar minimum is a rare occurrence. This unusual event occurred at the solar minimum in

September 2017, at the end of the 24th solar cycle [5]. On September 6, 2017, the most significant X-class flare in a decade erupted from active region 2673. Another X-class flare X8.2 was produced from the same position when X9.3 was just crossing the west limb. The sunspot region 2673 was one of the most active regions during the entire cycle, creating 4 X-class flares, including two of the largest flares in the cycle.

The data obtained from the ACE spacecraft was used in the simulation motion of particles by using the transport equation. The equation was solved using numerical methods to find the flux of particles in time at the best mean free path. We used the technique of the piecewise linear least-squares method for optimization of the injection duration. After the results are obtained from particle motion simulation, we analyzed the movement of high-energy particles and compared the data with actual data from the spacecraft. We used the linear least-squares fitting technique to find the mean free path and analyzed the injection time using the FWHM (full width at half maximum) method [6]. When the appropriate average distance is obtained, the result of fitting data will give the particle released in time, which can be graphed to determine the time of particles released from the Sun to the Earth.

4. Result

The Sun is continuously erupting charged particles in the form of solar flares, coronal mass ejections, or solar storms, which affect the Earth. So, we conducted a study of space conditions during the solar eruption in September 2017. There were 28 eruptions of M-class and 4 of X-class from September 6-10, 2017. We have found a rapid expansion of sunspots on 2673 on September 6, 2017, which was the most violent, X9.3, and on September 10, 2017, there was X8.2, the second most violent in the 24th cycle. The detail of this solar eruption from spacecraft is shown in table 2. The observed CME spread forward to the Earth and reached Earth's magnetosphere in the morning of September 7, 2017. There was a second CME from the same active region targeting the Earth on September 6, 2017, at 12:24 UT (This is the last CME in this series, most likely contributing to the complex ejecta later detected in situ) [7]. The second CME's shock wave reached Earth's magnetosphere in the afternoon of September 8, 2017. The simulation result in table 3 shows the mean free path of various elements and their injection time for the selected solar event on September 6, 2017, and the sample of fitting result for Nitrogen at 23.096 MeV/n as shown in figure 1.

Table 2. Data for the solar eruption in September.

Date	Start (UT)	Maximum (UT)	End (UT)	CME	CME time (UT)
6/9/2017	8:57	9.10	9:17	Not found	-
	11:53	12.02	12:10	Type II, IV	12.01, 12.24
7/9/2017	10:11	10.15	10:18	Type IV	10.35
	14:20	14.36	14:55	Type II	18.31
10/9/2017	15:35	16.06	16:31	Type II, IV	15.54

Table 3. The simulation results of the solar event occurred on September 6, 2017.

Element	Energy (MeV/n)	Mean free path (AU)	Injection time (min)	Element	Energy (MeV/n)	Mean free path (AU)	Injection time (min)
Nitrogen	8.009	0.805± 0.033	614	Carbon	7.443	1.350± 0.044	415
	13.317	0.508± 0.077	141		9.839	1.369± 0.053	379
	23.096	0.904± 0.022	339		15.505	1.795± 0.166	361
Oxygen	8.538	1.069± 0.079	281		21.229	0.827± 0.106	115
	11.427	1.717± 0.053	161		28.883	1.454± 0.208	39
	24.839	0.863± 0.153	74	Helium	5.390	0.914± 0.076	198
	49.832	0.960± 0.371	232		6.685	1.531± 0.097	171

The time for releasing particles from the Sun was between 39-743 minutes. The result shows that the duration of particles from the Sun decreases as the energy level of the particles increases. The trend of the mean free path is roughly constant for each element. The solar energetic particle with the high energy roughly has a high mean free path, and the diffusion of the particle will be less.

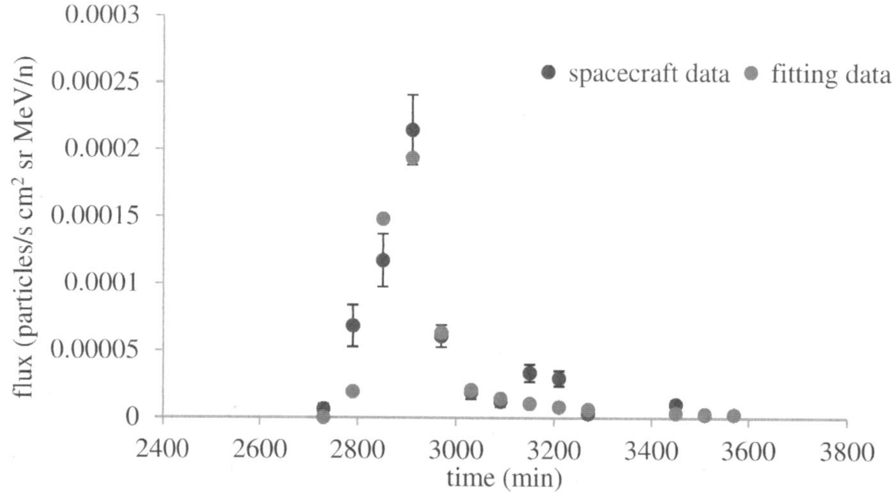

Figure 1. Shows the fitting result of Nitrogen at 23.096 MeV/n for the solar event on 6th September, 2017.

5. Conclusion

The study mainly focused on the solar event on September 6, 2017, where the most violent solar flare was observed during the solar minimum of the 24th solar cycle. The X9.3 was the strongest solar flare that started at 11:53 UT, peaked at 12:02 UT, and ended at 12:10 UT, which lasted for 17 minutes. It erupted at a position of S10W30 with a solar wind speed of 575 km/s. This flare was followed by CME, causing the acceleration of particles in space. CME in the profile of radio type IV and type II was detected close to the peak of this flare at 12.01 UT, which increased the injection time. The propagation of SEPs was simulated by the Ruffolo equation and solved by the Finite-differences method. The injection duration was estimated by the curve-fitting of piecewise linear least-squares fitting between the spacecraft data and simulation results in the profile of the distribution of particles in time. We found that the mean free path of this event was roughly constant for each particle which is different for the solar event at the maximum of the solar cycle. The Kp-index of this solar flare was 4; thus, there was no effect on the Earth. The Kp-index increased to 8 on September 7-8, 2017, due to another solar event from the same sunspot region indicating the effect of solar flare and CME, which resulted in the appearance of aurora in the United States, France, Kazakhstan, and Australia. The Kp-index is an excellent indicator for disturbances in the Earth's magnetic field. Kp-index is an integer of 0–9, with 1 being calm and 5 or more indicating a geomagnetic storm.

Acknowledgments

The authors wish to express sincere thanks to Prof. Dr.David Ruffolo, Mahidol University, Bangkok, Thailand, for the transport equation and simulation program. We are very grateful to Naresuan University in association with National Science, Research and Innovation Fund, Thailand (Project number FRB640025 contract number R2564B006) for the research grant. The Faculty of Science, Naresuan University, also supported this work.

References

[1] Roelof E C 1969 Propagation of solar cosmic rays in the interplanetary magnetic field *Lectures in High-Energy Astrophysics*, (NASA SP-199) (Washington D C: NASA Goddard Space Flight Center) p 111

[2] Ruffolo D, Khumlumlert T and Youngdee W 1998 Deconvolution of interplanetary transport of solar energetic particles *J. Geophys. Res.* **103** 20591

[3] Gopalswamy N 2012 Energetic particle and other space weather events of solar cycle 24 *AIP Conf. Proc.* **1500**(1) 14–9

[4] Kane R P 2014 Evolution of solar indices during the maximum solar cycle 24 *Indian J. Radio Space Phys.* **43** 151–5

[5] Tshering K and Khumlumlert T 2020 An analysis of the solar energetic particle propagation of the maximum solar flare on 24th solar cycle *NGRC2020 Conf. Proc.* pp 4324–30

[6] Baisri P, Khumlumlert T and Aiemsa-ad N 2017 Solar energetic particle propagation from solar flare *J. Phys.: Conf. Ser.* **901**(1) 012015

[7] Werner A L E, Yordanova E, Dimmock A P and Temmer M 2019 Modeling the multiple CME interaction event on 6-9 September 2017 with WSA-ENLIL+Cone *Space Weather* **17** 357–69

The spectral energy distribution of an X-ray pair halo from a gamma-ray source with a power-law continuum

A Eungwanichayapant[1,3]* and W Luangtip[2,3]

[1] School of Science, Mae Fah Luang University, Chiang Rai 57100, Thailand
[2] Department of Physics, Faculty of Science, Srinakharinwirot University, Sukhumvit 23, Bangkok 10110, Thailand
[3] National Astronomical Research Institute of Thailand, Chiang Mai 50180, Thailand

*E-mail: anant@mfu.ac.th

Abstract. Interactions between Very High Energy (VHE) gamma-rays from Active Galactic Nuclei (AGNs) and infrared photons from the Extragalactic Background Light (EBL) can start electromagnetic cascades. If the extragalactic magnetic field near a host galaxy is strong enough ($\sim$1 μG), the cascades would develop isotropically around the AGN. As a result, the electron/positron pairs created along the development of the cascades would create an X-ray halo via synchrotron radiation process. It is believed that the VHE gamma-ray spectra from the AGNs could be approximated by a power-law model which is truncated at high energy end (i.e. maximum energy). In this work we studied the X-ray Spectral Energy Distribution (SED) of the halo generated from the AGN spectra with different power indices and maximum energy levels. The results showed that the SEDs were slightly higher and broader, as they were obtaining higher flux if the power indices were lower. On the other hand, the SEDs were sensitive to the maximum energy levels between 100-300 TeV. More flux could be obtained from the higher maximum energy. However, we found that the SED becomes insensitive to the varied parameters when the maximum energy and the power index are $>$ 500 TeV and $<$ 1.5, respectively.

1. Introduction

Very High Energy (VHE) gamma-rays from extra-galactic sources such as Active Galactic Nuclei (AGNs) have been considered as an origin of an electron-positron pair halo [1–3]. The detailed explanation of the model were published many years ago [4, 5]. The pair halo model is shown in figure 1. The VHE gamma-rays that propagate in the cosmological scale can be absorbed by the Extra Galactic Background (EBL) photons via the $\gamma\gamma$ pair-production interaction. The interaction, in turn, creates electron/positron pairs. These electron/positron pairs can interact with the Cosmic Microwave Background (CMB) photons and up-scatter them to be gamma-ray photons via inverse Compton scattering. These cyclic interactions are called electromagnetic cascades. Moreover, these electromagnetic cascades can develop isotropically around the an AGN if the extragalactic magnetic field around the galaxy is strong enough (1 nG - 1 μG). In fact, the electron/positron pairs from the cascades can also emit X-ray photons via synchrotron radiation process [6].

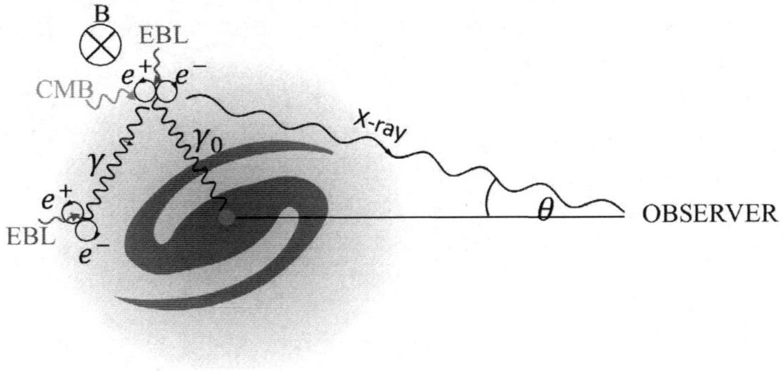

Figure 1. The diagram shows X-ray emission from a pair halo.

Observational data of the VHE gamma-ray data from AGNs can be simply described in the form of the power-law spectra. Some of them showed the absorption features [7] due to the EBL. These indicate the possibility of the existing of the electromagnetic cascades as well as the pair halos around the AGNs. The searches for the pair halo emission in the gamma-ray regime have been attempted during the last decades [8,9]. However, there has been no strong evidence of the halo emission confirmed yet. There have also been the alternative searches for the halo emission in the X-ray waveband. From these attempts, the results have begun to emerge [10, 11]. In our previous works [5,6], we studied the X-ray emission of the halo assuming that the intrinsic VHE gamma-rays from an AGN were in the form of mono-energetic gamma-ray spectra. These were for the sake of systematic studies. therefore, in this work, we used more realistic approaches. The X-ray pair halos from more realistic spectra were simulated and studied. the effects of the changes of power-index and maximum energy of the spectra on the SEDs of the X-ray pair halos were observed.

2. Methodology

The electromagnetic cascades were generated by intrinsic VHE gamma-rays from AGN, while the electron/positron pairs radiate synchrotron radiation during their gyrations. The synchrotron emission time distributions were computed during the simulations and used for calculating the spectral energy distribution (SED) of the halo.

In this work, it was assumed that an AGN had the gamma-ray luminosity of 10^{45} erg/s and located at $z \approx 0.1$ with an ambient magnetic field of 1 μG for all simulations. The intrinsic gamma-ray from the AGN was generated with power law spectra with power index, p, defined by the well known relation:

$$\frac{dN}{dE} \propto E^{-p}. \tag{1}$$

In the simulation, the gamma-ray photons have the energy between E_{min} and E_{max}. The E_{min} was fixed at 1 TeV, while the E_{max} and p were varied parameters. The energy of the gamma-ray with this spectrum was generated by transforming a random number, ξ, between 0 and 1 using the relation that we formulated,

$$E = \left(\xi(E_{max}^{-p+1} - E_{min}^{-p+1}) + E_{min}^{-p+1}\right)^{\frac{1}{-p+1}}. \tag{2}$$

The gamma-ray with this energy level was used as an input in the simulation as described in our previous works [5,6] to compute the SEDs of the X-ray photons from electron/positron pair halos. The effects of the power law spectra of the gamma-ray from the AGN on the X-ray SED from the pair halo were observed by varying the power index, p, and the maximum energy, E_{max}.

3. Results and discussion

In this work, we aimed to study responses of the SED and the flux of the X-ray pair halo to the power-law parameters – i.e the index p and maximum photons energy E_{max} – of the gamma-rays from the AGN. The parameters of the power-law spectra were varied as summarized in table 1. Thus, the changes of the X-ray SEDs were observed.

Table 1. The total synchrotron energy flux of X-ray pair halos with different power indices and maximum energy levels.

E_{max}	Energy flux (eV/cm^2 s)		
	$p = 1.5$	$p = 2.0$	$p = 2.5$
30 TeV	1.01	0.96	0.92
100 TeV	3.30	2.56	2.03
300 TeV	7.69	5.66	3.60
500 TeV	8.31	6.17	3.80

The results showed that the parameters did not affect the SEDs at the energy levels below 3 eV as shown in figure 2. However, at the energy above that level, the X-ray SEDs showed the responses to both parameters. The figure 2(a) shows how the SEDs changed in responses to three different values of the parameter E_{max} of the power-law. There are more fluxes in the higher energy bands when the E_{max} increased. However, it was found that the SEDs would reach the maximum level if the E_{max} reached ~300 TeV; from ~300 - 500 TeV, the SED as well as the total flux seems to be constant (or have only a small variability). This implies that, beyond the threshold level of E_{max} of 300 TeV, the X-ray pair halos would not depend on the parameter E_{max} of the power-law spectra.

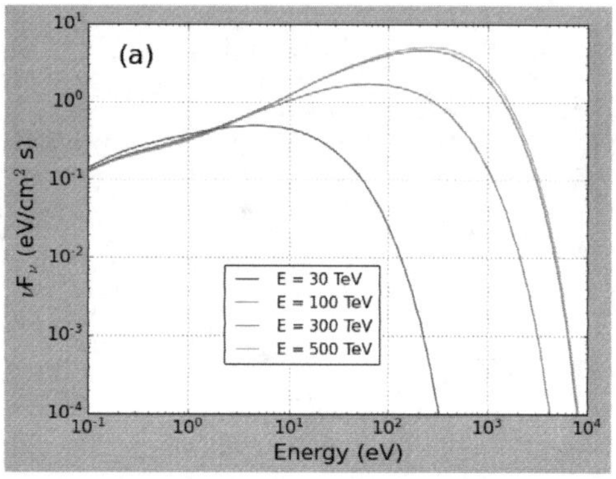

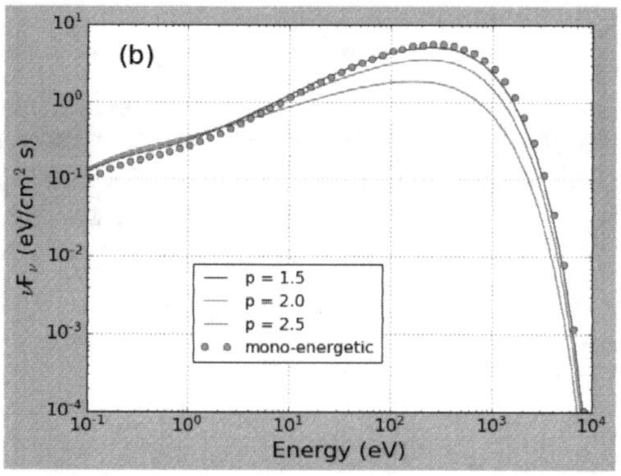

Figure 2. The SEDs of X-ray pair halo. (a) The comparison of the SEDs from different E_{max} levels at p =1.5. (b) The comparison of the X-ray SEDs from different power indices and a mono-energetic spectrum at E_{max} = 500 TeV.

On the other hand, the results also showed the effects of the power indices on the pair-halo X-ray fluxes (figure 2(b) and table 1) in which the lower power indices could yield the higher SEDs as well as the flux. We might explain this in the scenario that for the smaller power indices, there were more numbers of gamma-rays in higher energy bands that were responsible for triggering the electromagnetic cascades and, in turn, creating the electron/positron pair halos.

Similar to the case of varying the maximum energy, as the parameter p reach the value of 1.5 – i.e. the threshold value – it seems that the SEDs and the flux do not change significantly. Moreover, for comparison, in figure 2(b), we also over-plotted the SED obtained from mono-energetic spectrum of photon energy 500 TeV (see [6] for the detail calculation); we quote this SED as the halo SED that provide the maximum flux since all VHE photons have maximum energy (500 TeV). It could be seen that the SEDs from the values of $p = 1.5$ and $E_{max} = 500$ TeV, and that of the mono-energetic spectrum are well constitent. This implies that the X-ray pair halo SEDs would not depend on the power indices and maximum energy if their values exceed the their threshold values of 1.5 and 500 TeV, respectively, i.e. the similar SEDs and fluxes would be obtained.

Although the calculation of the observational feasibility of the halo is out of scope of this work, it could be worth to shortly mention this. The plots in figure 2 obviously show that, as the varied parameters change, the flux could vary mainly in the energy band of 10 – 5000 eV. Indeed, this covers in the observational band of current X-ray observatories such as *XMM-Newton* and *Chandra*. Nevertheless, since the halo flux is, actually, substantially low, comparing to the central AGN flux, in practise, it is difficult to distinguish the halo flux from the AGN flux (see e.g. [11]); one must be careful during the analysis to avoid the misleading result. However, given the high effective area and the good spatial resolution of the current and near future X-ray telescopes, the detection of the halo has been challenged.

4. Conclusion

In this work, we studied the SEDs in the X-ray waveband of the pair halo generated from the AGN which emits the VHE gamma-ray in the form of power-law continuum. Two varied parameters – power index and maximum photon energy – has been investigated how they affect the flux and shape of pair halo SEDs. It was found that these parameters could affect the SEDs of the X-ray pair halo, in which the halo flux could increase if the power index decreases and/or the maximum photon energy increases. However, if the values of the power index and the photon maximum energy exceeded their limits – i.e. $p < 1.5$ and $E_{\rm max} > 500$ TeV – the SEDs of the X-ray pair halos would seem to be source independent.

Acknowledgments

The authors would like to thank the National Science and Technology Development Agency for the research funding support that they offered.

References

[1] Aharonian F A, Coppi P S and Voelk H J 1994 *Astrophys. J.* **423** L5
[2] Eungwanichayapant A, Aharonian F A and Voelk H J 2005 *Proc. 9th Asian-Pacific Regional IAU Meeting* (Bandung: Institut Teknologi Bandung Press) pp 145–6
[3] Eungwanichayapant A and Aharonian F 2009 *Int. J. Mod. Phys.* D **18**(06) 911
[4] Eungwanichayapant A 2003 Ph.D. thesis Ruprecht-Karls Universitaet Heidelberg
[5] Eungwanichayapant A, Maithong W and Ruffolo D 2011 *Proc. Int. Astron. Union* vol 7 Symposium S284 (Cambridge: Cambridge University Press) pp 417–9
[6] Eungwanichayapant A, Luangtip W, Maithong W and Ruffolo D 2019 *Astrophys. J.* **880**(2) 124
[7] Costamante L 2013 *Int. J. Mod. Phys.* D **22**(13) 1330025
[8] Aharonian F A *et al.* 2001 *Astron. Astrophys.* **366** 746–51
[9] Neronov A, Semikoz D V, Tinyakov P G and Tkachev I I 2011 *Astron. Astrophys.* **526** A90
[10] Kueathan A, Luangtip W, Maithong W and Eungwanichayapant A 2018 *J. Phys.: Conf. Ser.* **1144** 012131
[11] Kueathan A, Luangtip W and Eungwanichayapant A 2019 *J. Phys.: Conf. Ser.* **1380**(1) 012117

Orbital stability of S-type circumbinary planets

N Nantanoi[1*], N Nantanoi[1†], S Awiphan[2‡], S Komonjinda[3] and T Bunfong[1]

[1] Chiang Mai University Demonstration School, 239, Huaykaew, Suthep, Chiang Mai, 50200, Thailand
[2] National Astronomical Research Institute of Thailand (Public Organization), 260 Moo 4, Donkaew, Mae Rim, Chiang Mai, 50180, Thailand
[3] Department of Physics and Materials Science, Faculty of Science, Chiang Mai University, Chiang Mai, 50200, Thailand

E-mail: *noppawith.n@satitcmu.ac.th, †noppawin.n@satitcmu.ac.th, ‡supachai@narit.or.th

Abstract. Nowadays, more than 4,000 exoplanets have been discovered, including a hundred of circumbinary planets. In the following work, the orbital variations of 67 S-type circumbinary planets have been studied. Their orbital evolutions for a thousand years are simulated using the REBOUND package. The published physical and orbital parameters of the systems are used to computed the systems' orbital instability limits: Roche limit and Hill's sphere. From 67 systems, there are two unstable circumbinary systems: Kepler-420 and GJ 86. Kepler-420 Ab orbit passes into the system's Roche limit due to its high orbital eccentricity. For GJ 86 Ab, the planet orbits outside its Hill's sphere. The instability of GJ 86 Ab might be caused by an inaccurate measurement of GJ 86 A physical parameters. Using the GJ 86 A mass obtained from Farihi et al., the planet orbits in the stable orbit zone.

1. Introduction

To date, a hundred of circumbinary planets, which are planets orbiting in the binary star systems, have been discovered. Circumbinary planet orbit is classified as a three-body problem, which is an exceptional case that the orbit cannot be solved precisely and remains a significant problem in astrophysics research. Although, the orbit cannot be precisely predicted, the systems can have long-term orbital stability. However, in the systems which the planets have high eccentricity orbits, the instability of planetary orbits might be detected [1].

From the Catalogue of Exoplanets in Binary Star Systems, the data of 88 circumbinary systems can be obtained. The systems can be classified into two categories: S-type (planet orbiting a single star) and P-type (planet orbiting two stars), totaling 67 and 21 systems, respectively. In this work, only S-type circumbinary system is focused. The orbital stability of S-type circumbinary planet is constrained by two limits: Roche limit and Hill's sphere. The inner orbital limit is Roche limit, the minimum distance where the planet is not ripped to pieces by its host's tidal forces. The Roche limit can be written as,

$$d = R_p \sqrt[3]{\frac{2M_h}{M_p}} , \quad (1)$$

where d is the distance of Roche limit, R_p is the planetary radius, M_h and M_p are masses of the host star and the planet, respectively. The maximum orbital radius is limited by the Hill sphere, r_H, which is given by

$$r_H = a(1-e)\sqrt[3]{\frac{M_h}{3M_c}}\,, \tag{2}$$

where a is the orbital separation between the binary, e is an eccentricity of the binary system and M_c is the binary companion mass.

In this work, 1,000 years orbit of 67 known S-type circumbinary planetary systems are simulated with `REBOUND` package [2].

2. Orbital stability of known S-type circumbinary systems

The physical and orbital parameters of 67 S-type circumbinary systems are obtained from Catalogue of Exoplanets in Binary Star Systems (`https://www.univie.ac.at/adg/schwarz/bincat_binary.html`). The systems are simulated using `REBOUND` package [2]. In this work, the short term orbital stability is interested. In each system, a thousand years orbit with a step size of 1 year is simulated. The planetary Roche limit and Hill's sphere are assigned to be orbital stability limit.

From the 67 systems, only two systems: Kepler-420 Ab and GJ 86 Ab, show unstable orbits. For a comprehensive study, the orbits of these two systems are re-simulated with one day step size.

Kepler-420 Ab (also known as KOI-1257 b) is a transiting exoplanet with very high eccentricity orbit ($e = 0.772$). The planet has a mass of 1.45±0.35 M_{Jupiter}, and a radius of 0.94±0.12 R_{Jupiter} with 86.647661±0.000034 days orbital period. The planet has been discovered by *Kepler* and radial velocity follow-up with the SOPHIE spectrograph [3]. The simulation shows that the planet orbits around its planet-star centre of mass (Kepler-420 Ab and Kepler-420 A) with a distance between 1-3 AU. The planet wobbles around its orbit with period of ~3 days (figure 1). Due to the wobbling, the planet passes inside the circumbinary Roche limit after 142 days (figure 1, 2 and 3). The orbital instability might be caused by the very high eccentricity planetary orbit.

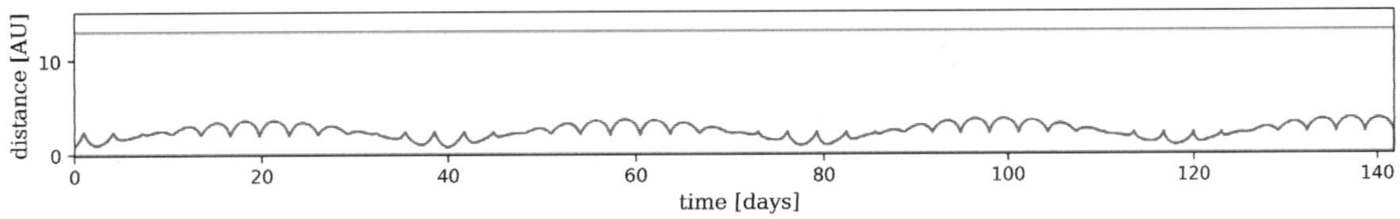

Figure 1. Distance between planet and planet-star center of mass of Kepler-420 Ab (Blue) with its Roche limit (Green) and Hill sphere (Orange). The simulation shows that Kepler-420 passes inside the Roche limit after 142 days.

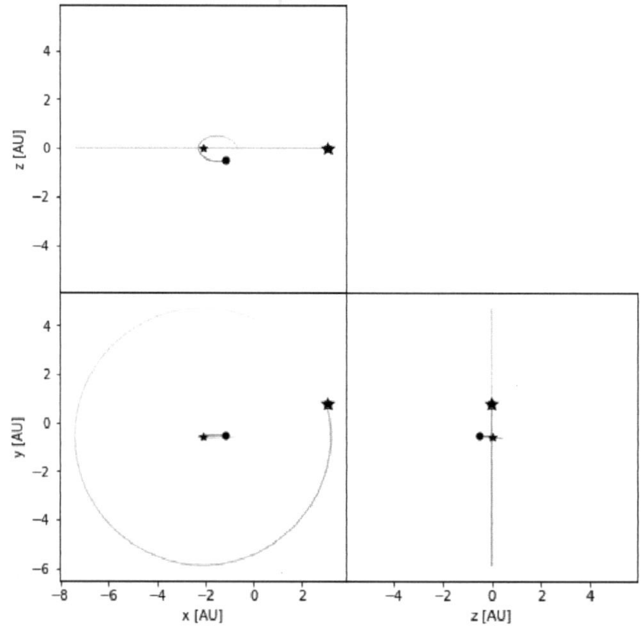

Figure 2. Top and side views of Kepler-420 system at 142nd day.

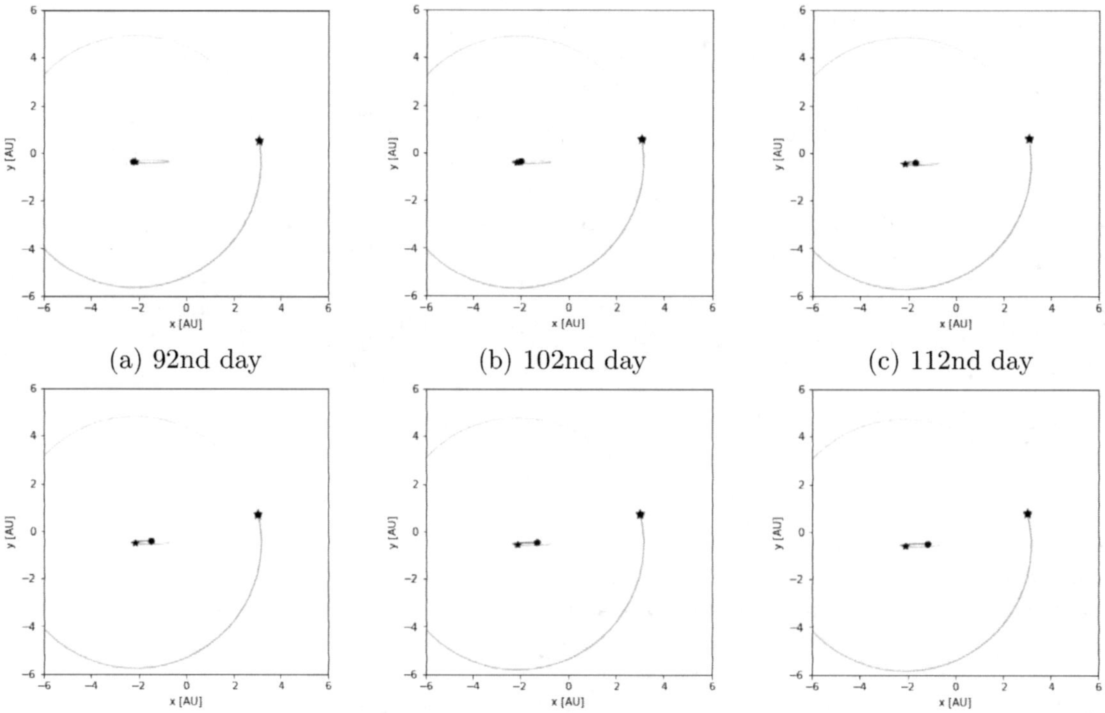

(a) 92nd day (b) 102nd day (c) 112nd day

(d) 122nd day (e) 132nd day (f) 142nd day

Figure 3. Top views of Kepler-420 system between 92nd and 142nd day.

GJ 86 Ab, a 4 M_{Jupiter} planet orbiting a nearby K main sequence star with orbital period of 15.8 days, was discovered by the CORALIE echelle spectrograph [4]. GJ 86 B, a 50 M_{Jupiter} companion orbiting GJ 86 at a distance of 18.75 AU, is discovered by Els. *et al.*, [5]. In 2005, Mugrauer and Neuhäuser proposed a model of faint T-dwarf companion with mass down to 35 M_{Jupiter} and separation between 1 and 23 AU [6]. From the Hubble Space Telescope observation, the mass of the binary is constrained to be 1.9 $M_\odot$ and 0.59 $M_\odot$ with orbital separation of >14 AU [7]. In this work, published mass 0.8 $M_\odot$ and 0.49 $M_\odot$ and separation of 21 AU from Catalogue of Exoplanets in Binary Star Systems are used.

GJ 86 Ab shows a stable orbit over a thousand years (figure 4). However, the planet orbits outside the Hill's sphere where the instability might be occurred. The instability of GJ 86 Ab cannot be confirmed due to the uncertainty of binary companion's mass and separation. The catalogue provides the binary with host star mass of 0.8 $M_\odot$ proposed by Queloz *et al.*, [4]. Using the update host star mass of 1.9 $M_\odot$ from Farihi *et al.* [7], the planet will orbit inside the Hill's sphere (figure 4).

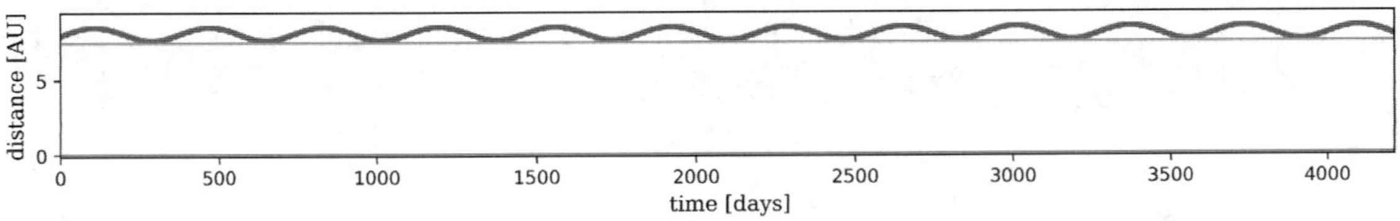

(a)

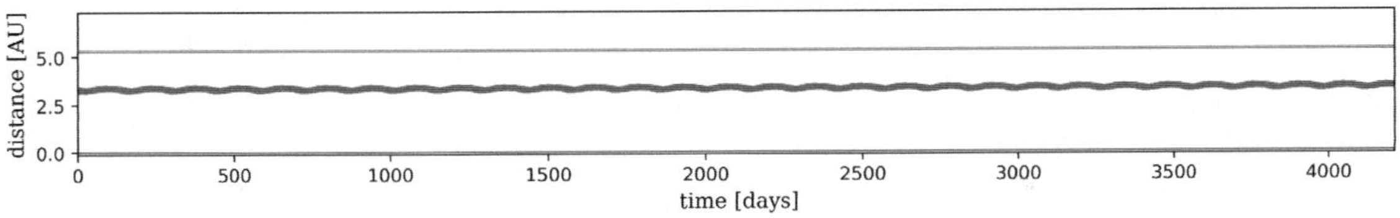

(b)

Figure 4. Distance between planet and planet-star center of mass of of GJ 86 Ab (Blue) with its Roche limit (Green) and Hill sphere (Orange) with parameters from (a) the Catalogue of Exoplanets in Binary Star Systems and (b) the Hubble space telescope observations of Farihi *et al.* [7].

3. Conclusion

The orbit of 67 S-type circumbinary systems are simulated using REBOUND package. Most of them have stable orbits over a thousand years. There are only two systems that show unstable orbits, namely, Kepler-420 Ab and GJ 86 Ab. In our simulation, Kepler-420 Ab has an orbit which passes into the Roche limit within 142 days. The planet has very high eccentricity orbit which might cause the instability of the system. GJ 86 Ab has a stable orbital characteristic but orbits outside the Hill's sphere. The instability of GJ 86 Ab might be caused by the uncertainty of host star mass. GJ 86 A mass varies between 0.8 and 1.9 $M_\odot$ while GJ 86 B mass varies between 0.05 and 0.59 $M_\odot$. The orbit is inside the Hill's sphere with the parameters from the Hubble space telescope observations Farihi *et al.* [7]. In the future work, further studies of these two systems are required.

Acknowledgements
This work was supported by the Science Classrooms in University-Affiliated school and the National Astronomical Research Institute of Thailand (NARIT).

References

[1] Popova E A and Shevchenko I I 2016 On the stability of circumbinary planetary systems *Astron. Lett.* **42(7)** 474
[2] Rein H and Liu S F 2012 REBOUND: an open-source multi-purpose N-body code for collisional dynamics *Astron. Astrophys.* **537** A128
[3] Santerne H *et al* 2014 SOPHIE velocimetry of Kepler transit candidates. XII. KOI-1257 b: a highly eccentric three-month period transiting exoplanet *Astron. Astrophys.* **571** A37
[4] Queloz D, Mayor M, Weber L, Blécha A, Burnet M, Confino B, Naef D, Pepe F, Santos N and Udry S 2000 The CORALIE survey for southern extra-solar planets. I. A planet orbiting the star Gliese 86 *Astron. Astrophys.* **354** 99
[5] Els S G, Sterzik M F, Marchis F, Pantin E, Endl M and Kürster M 2001 A second substellar companion in the Gliese 86 system. A brown dwarf in an extrasolar planetary system *Astron. Astrophys.* **370** L1
[6] Mugrauer M and Neuhäuser R 2005 Gl86B: a white dwarf orbits an exoplanet host star *Mon. Notices Royal Astron. Soc.* **361(1)** L15
[7] Farihi J, Bond H E, Dufour P, Haghighipour N, Schaefer G H, Holberg J B, Barstow M A and Burleigh M R 2013 Orbital and evolutionary constraints on the planet hosting binary GJ 86 from the Hubble Space Telescope *Mon. Notices Royal Astron. Soc.* **430(1)** 652

Astronomical seeing forecast for the Thai National Observatory

Ronald Macatangay* **and Somsawat Rattanasoon**

National Astronomical Research Institute of Thailand, Mae Rim, Chiang Mai, 50180, Thailand

*Corresponding author's Email: ronald@narit.or.th

Abstract. Forecasting the astronomical seeing above an observatory can assist astronomers plan their observations. In this study, the astronomical seeing above the Thai National Observatory (TNO) in Doi Inthanon, Chiang Mai, Thailand was simulated using the Weather Research and Forecasting (WRF) model. The model outputs were then compared to Polaris seeing observations and using the Differential Image Motion Monitor (DIMM). Results showed that the forecasts capture the variation of the astronomical seeing fairly well. However, bias correction is needed on the simulations due to lack of data from meteorological balloons to constrain the model.

1. Introduction

Ground-based astronomical observations are limited by two factors – diffraction and turbulence. Diffraction-limited point images have a point-spread function than can have a full-width at half maximum of $1.22\lambda/D$ at its best (when there is no effect from the atmosphere of the earth), where λ is the wavelength and D is the aperture size of the observing system, Considering the effects of the atmosphere of the earth, particularly turbulence in the air, the full-width at half maximum of an object's point spread function, or its astronomical seeing, can increase depending on the intensity of atmospheric turbulence. Turbulence can cause variations in the index of refraction of air causing plane wavefronts from distant light sources to become distorted. These distortions produce amplitude variations and positional shifts causing image degradation [1].

Aside from deducing astronomical seeing optically, it can also be calculated using meteorological parameters, particularly temperature, wind and atmospheric pressure vertical profiles. These variables can be forecasted, hence astronomical seeing can also be predicted [2].

In this study, the astronomical seeing is simulated over the Thai National Observatory (TNO) in Chiang Mai, Thailand using the Weather Research and Forecasting (WRF) model version 3.8.1 from February 10 – April 25, 2021 based on [3]. The model outputs are then compared to observations using the Polaris seeing monitor and the Differential Image Motion Monitor (DIMM) installed at TNO.

2. Methodology

According to [3], there are four variables that need to be determined prior to calculating the astronomical seeing from weather models. These are: (1) the mean potential temperature gradient; (2) the wind shear vertical profile; (3) the temperature structure vertical profile; and (4) the refractive index vertical profile.

2.1. Mean potential temperature gradient

The potential temperature, θ, is a theoretical temperature used in weather forecasting for comparing different air masses at different altitudes and different locations in the atmosphere without having the height dependence of air temperature. It is used for determining if there are warm or cold air masses being transported or advected over an observation site. The potential temperature is determined by bringing down an air mass to a standard pressure (1000 hPa) without losing energy to its surroundings. It generally has a value of around 300 K near the surface and increases with altitude to around 500 K at 20 km above the ground. The mean potential temperature gradient is then calculated by determining the change of the mean potential temperature with respect to height, z.

2.2. Wind shear vertical profile

Using the east-west and north-south horizontal wind components, $u(z)$ and $v(z)$, respectively, the square of wind shear vertical profile is calculated as the sum of squares of the change of $u(z)$ and $v(z)$ with respect to z. Physically, the wind shear vertical profile depicts changes in both the wind speed and wind direction from the ground to higher altitude levels.

2.3. Temperature structure vertical profile

Multiplying the mean potential temperature gradient with the square-root of the wind shear vertical profile and a parameter deduced from weather balloons, the temperature structure vertical profile can be calculated. In this study, we don't have the parameter deduced from weather balloons, therefore, a bias correction is necessary in the end.

2.4. Refractive index vertical profile

Using the temperature structure vertical profile, the atmospheric temperature vertical profile and the pressure profile, the refractive index vertical profile can be determined. The refractive index vertical profile shows how much the refractive index of air is perturbed by turbulence. It is usually high near the surface, at around 10 km altitude and at around 18 km from the ground [4]. High altitude turbulence is important since the astronomical seeing is an integrated profile from the ground to the top of the atmosphere.

2.5. Astronomical seeing

The astronomical seeing is calculated by integrating the refractive index vertical profile ideally from the ground to the top of the atmosphere divided by the fifth root of the wavelength (the visible wavelength of 532 nm was used in this study) [4].

3. Discussion

The results of the simulations are shown in figure 1. It can be seen that the astronomical seeing derived from the Polaris seeing monitor is generally higher than both the DIMM and the model simulations. The reason is that the Polaris seeing monitor points near the horizon (around 19 degrees from the horizon) all night where most of the turbulence occurs. In this case, the mean value of the DIMM observations are used for correcting the bias (11.72 arc seconds) in the model since DIMM tracks stars near the zenith and the model calculates the seeing along the zenith. The reason for the large bias of 11.72 arc seconds needs to be investigated further.

4. Conclusion

The astronomical seeing for TNO has been simulated for forecasting purposes. An offset of 11.72 arc seconds has been subtracted from the model output to match the DIMM observations. The reason for this bias will be investigated in future studies.

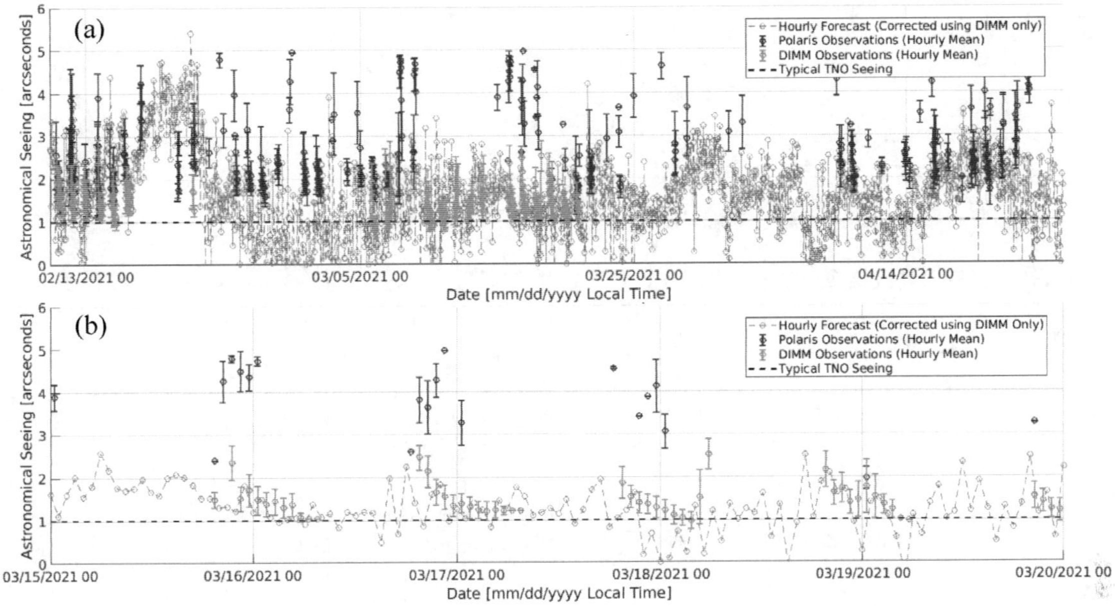

Figure 1. Astronomical seeing forecasts (red) as compared to the Polaris (blue) and DIMM (green) seeing observations for (a) the entire study period and (b) from March 15 – 20, 2021. Also shown is the typical TNO seeing of 1 arc second [5].

Acknowledgements
The authors would like to thank the Thailand Science Research and Innovation office for partially supporting this work.

References

[1] Bally J, Theil D, Billawala Y, Potter D, Loewenstein R F, Mrozek F and Lloyd J 1996 A Hartmann differential image motion monitor (H-DIMM) for atmospheric turbulence characterisation *Publ. Astron. Soc. Aust.* **13**(1) 22–7

[2] Ye Q Z 2011 Forecasting cloud cover and atmospheric seeing for astronomical observing: Application and evaluation of the global forecast system *Publ. Astron. Soc. Pac.* **123**(899) 113–24

[3] Giordano C, Vernin J, Vázquez Ramió H, Muñoz-Tuñón C, Varela A M and Trinquet H 2013 Atmospheric and seeing forecast: WRF model validation with in situ measurements at ORM *Mon. Not. R Astron. Soc.* **430**(4) 3102–11

[4] Trinquet H and Vernin J 2007 A statistical model to forecast the profile of the index structure constant CN^2 *Environ. Fluid Mech.* **7**(5) 397–407

[5] Dhillon V S et al. 2014 ULTRASPEC: a high-speed imaging photometer on the 2.4-m Thai National Telescope Mon. Not. R Astron. Soc. 444(4) 4009–21

Cumulants of the chiral order parameter in a nonequilibrium Bjorken expansion with a QCD critical point

Poramin Saikhum[1], Christoph Herold[1,*] and Mohamad Lukman Aidid Mohd Yusoff[2]

[1]School of Physics and Center of Excellence in High Energy Physics & Astrophysics, Suranaree University of Technology, Nakhon Ratchasima 30000, Thailand
[2]Theoretical Physics Laboratory, Department of Physics, University of Malaya, 50603 Kuala Lumpur, Malaysia

*E-mail: herold@g.sut.ac.th

Abstract. To understand experimentally obtained net-proton number cumulants in the search for the QCD critical point, we study a dynamical model based on an effective quark-meson Lagrangian with chiral symmetry. We investigate the evolution of the expanding medium created in a heavy-ion collision using a spatially homogeneous fluid and a time-dependent order parameter, the sigma field evolved by a Langevin equation. We extract cumulants of the sigma field along a parametrized freeze-out curve and match the obtained freeze-out points to corresponding beam energies. These cumulants can be related to cumulants of the net-proton number through the sigma-proton coupling to provide a qualitative comparison to experimental data from STAR's beam energy scan program. We demonstrate that the presence of the spinodal or mixed phase region around the first-order chiral phase transition allows for a wide interval of cumulants at the lowest beam energies.

1. Introduction

Quark–gluon plasma (QGP) is a new state of matter that was present at extremely high temperatures and densities in the early universe. Contrary to a hadronic medium (protons, neutrons, pions, etc.), it is characterized by deconfinement and chiral symmetry restoration.

The study of heavy ion collisions aims to explore the phase structure of nuclear matter as a function of baryochemical potential μ_B and temperature T [1,2]. At $\mu_B = 0$, a smooth crossover has been found while for large μ_B, a discontinuous first-order phase transition is widely expected which ends at a second-order critical point (CP) [3,4].

Event-by-event fluctuations of conserved quantities such as net-baryon, net-charge, and net-strangeness are sensitive to the correlation length and connected proxy to thermodynamic susceptibilities calculated in lattice QCD [5,6] or effective model calculations. The fluctuations of conserved quantities are defined in the form of cumulants. To compare with the experimental data, ratios of baryon number susceptibilities are used to eliminate the dependence on volume and temperature of the system which are notoriously difficult to access. Higher order cumulants of conserved quantities depend directly on higher powers of the correlation length [7], and the correlation length of the nonequilibrium system depends on expansion time and is limited by the system size. It has previously been shown that the correlation length increases to about 2 − 3 fm near the CP [8]. Experimental programs such as the beam-energy scan program of the Relativistic Heavy Ion Collider (RHIC) try to measure how e.g. net-proton skewness and kurtosis deviate from baseline calculations with e.g. UrQMD or a hadron resonance gas model to locate the QCD CP [9].

2. The NχFD Bjorken model

The spontaneous chiral symmetry breaking in vacuum as well as the restoration or chiral symmetry at large T or μ_B is described by the widely used quark-meson model whose Lagrangian reads

$$\mathcal{L} = \overline{q}\left(i\gamma^\mu\partial_\mu - g\sigma\right)q + \frac{1}{2}\partial_\mu\sigma\partial^\mu\sigma - U(\sigma) \ . \tag{1}$$

In QCD, the quark mass term explicitly breaks chiral symmetry which is introduced as a term $H\sigma$ in the chiral potential as follows,

$$U(\sigma,\vec{\pi}) = \frac{\lambda^2}{4}\left(\sigma^2 - {f_\pi}^2\right)^2 - H\sigma \ . \tag{2}$$

Here, q represents the light quark doublet $q = (u, d)$. The grand potential given in the mean-field approximation becomes

$$\Omega(T,\mu) = U(\sigma) + \Omega_{q\bar{q}} \ . \tag{3}$$

In this equation, the quark and antiquark contribution is given by

$$\Omega_{\bar{q}q} = -2N_cN_fT\int\frac{d^3\mathbf{p}}{(2\pi)^3}\left\{\ln\left[1+\exp\left(\frac{E-\mu}{T}\right)\right] + \ln\left[1+\exp\left(\frac{E+\mu}{T}\right)\right]\right\} \ . \tag{4}$$

This describes the negative pressure of a fermionic gas of quarks and antiquarks with quasiparticle energies $E = \sqrt{p^2+M^2}$, and effective mass of (constituent) quarks $M(\sigma) = g\sigma$ with the coupling g depending on the contributed quark mass in a vacuum. The parameters of this model are chosen as $f_\pi = 93$ MeV, $m_\pi = 138$ MeV and $H = f_\pi m_\pi^2$.

Note here that one can think of further extensions of this Lagrangian, most notably by including the Polyakov loop to account for a deconfinement transition as in [10]. Here, we focus on the criticality of the chiral phase transition which is expected to determine the singular behavior near the QCD critical point and which is captured with the quark-meson model. Although the model is admittedly crude, the addition of the Polyakov loop would shift the pseudocritical temperature at $\mu_{\rm B} = 0$ away from the value obtained in lattice QCD calculations [11]. For a realistic model and equation of state including both quark and hadronic degrees of freedom, a large number of fields is needed which requires a fine-tuning of many parameters [12]. This would render any nonequilibrium simulation based on such a model unnecessary complicated and highly sensitive to the numerics.

From the Lagrangian equation (1), the nonequilibrium chiral fluid dynamics model (NχFD) is constructed by propagating the chiral order parameter σ with the Langevin equation of motion [13–15],

$$\ddot{\sigma} + \left(\frac{D}{\tau} + \eta\right)\dot{\sigma} + \frac{\delta\Omega}{\delta\sigma} = \xi \ . \tag{5}$$

Hereby, the dots refer to derivatives with respect to proper time τ, $D = 1$ in the Hubble term for an expansion in one direction of the beam axis as assumed by the Bjorken model.

The damping coefficient η has been calculated as

$$\eta = \frac{12g^2}{\pi}\left[1 - 2n_{\rm F}\left(\frac{m_\sigma}{2}\right)\right]\frac{1}{m_\sigma^2}\left(\frac{m_\sigma^2}{4} - M^2\right)^{3/2} \ . \tag{6}$$

The white and Gaussian noise $\xi(t,x)$ in the above equation satisfies the fluctuation-dissipation theorem

$$\langle\xi(t)\xi(t')\rangle_\xi = \delta(t-t')\frac{m_\sigma\eta}{V}\coth\left(\frac{m_\sigma}{2T}\right) . \tag{7}$$

The mass of the field σ is equivalent to the curvature of the thermodynamic potential in equilibrium depending on temperature and chemical potential,

$$m_\sigma^2 = \left.\frac{\partial^2\Omega}{\partial\sigma^2}\right|_{\sigma=\langle\sigma\rangle} . \tag{8}$$

Under the assumption of the Bjorken model that the rapidity distribution of the charged particles is boost invariant, we obtain the hydrodynamic equations for energy density e and baryon number density n as [16]

$$\frac{\partial e}{\partial\tau} = -\frac{e+p}{\tau} + \left[\frac{\delta\Omega_{\bar{q}q}}{\delta\sigma} + \left(\frac{D}{\tau}+\eta\right)\frac{\partial\sigma}{\partial\tau}\right]\frac{\partial\sigma}{\partial\tau}, \tag{9}$$

$$\frac{dn}{d\tau} = -n/\tau. \tag{10}$$

We characterize fluctuations of the sigma field by cumulants C_n of the probability distribution defined as:

$$C_1 = \langle\sigma\rangle, \tag{11}$$
$$C_2 = \langle(\delta\sigma)^2\rangle, \tag{12}$$
$$C_3 = \langle(\delta\sigma)^3\rangle, \tag{13}$$
$$C_4 = \langle(\delta\sigma)^4\rangle - 3\langle(\delta\sigma)^2\rangle^2. \tag{14}$$

These are related to mean (M), variance (σ^2), skewness (S) and kurtosis (κ) via

$$M = C_1\ ,\quad \sigma^2 = C_2\ ,\quad S\sigma = \frac{C_3}{C_2}\ ,\quad \kappa\sigma^2 = \frac{C_4}{C_2}\ . \tag{15}$$

We investigate the higher order cumulant ratios of the sigma field along a freeze-out curve which has been obtained from thermal model fits to experimental data [17] over a wide range of beam energies. A polynomial fit yields a parametrization,

$$T(\mu_B) = a - b\mu_B^2 - c\mu_B^4\ , \tag{16}$$

where $a = 0.166$ GeV, $b = 0.139$ GeV^{-1} and $c = 0.053$ GeV^{-3}. We furthermore match the point where our Bjorken evolution hits this freeze-out curve with the corresponding beam energy via:

$$\mu_B(\sqrt{s}) = \frac{d}{1+e\sqrt{s}}\ , \tag{17}$$

where $d = 1.308$ GeV and $e = 0.273$ GeV^{-1} [18].

3. Fluctuations of the order parameter

We initialize the field and fluid in equilibrium at several starting points in the chirally restored phase, see the beginning of the trajectories in the phase diagram of the quark-meson model, figure 1. These points are conveniently characterized by a choice of initial values (T_0, μ_0). The

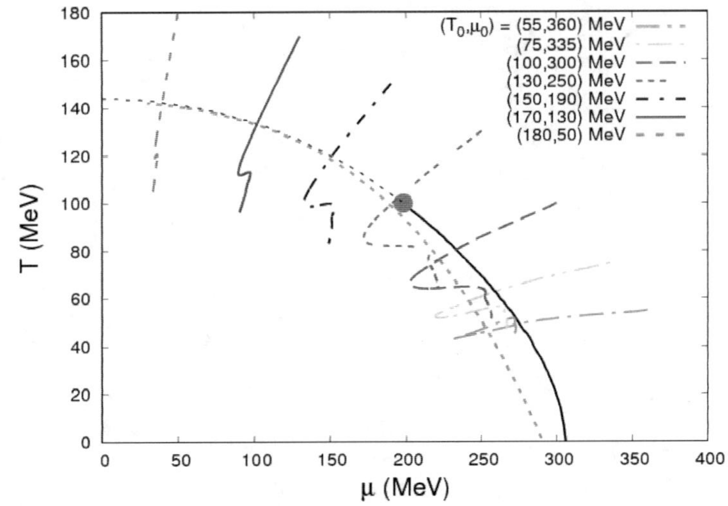

Figure 1. The phase structure of chiral phase transition in different initial conditions (T_0, μ_0) alongside trajectory of fireball.

coupled system of field and fluid is then evolved according to equations (5) and (10) which gives an evolution of T and μ that can be seen in the same figure.

The point where the trajectories hit the dashed freeze-out line is used to determine the event-by-event fluctuations. We hereby use a number of $N = 10^5$ events which are randomized through the stochastic noise term ξ that yields a non-deterministic evolution. The bending of the trajectories has been discussed in previous publications [16]. It is particularly worth noting here that, as a consequence of this behavior, the trajectories passing through the first-order phase transition cross the freeze-out curve twice. For these cases, we calculate the cumulants at both crossing points to give a range of possible values.

Figure 2 shows the so obtained σ^2, $S\sigma$ and $\kappa\sigma^2$ versus beam energy $\sqrt{s}$. We note that for all these cases, the consideration of the spinodal region and double-crossing of the freeze-out curve leads to an increased range of possible cumulant values for the lowest beam energies. The variance, which is more or less constant, decreases by a factor of 2 after passing through the second crossing, see figure 2(a). A similar effect is observed for the skewness, though somewhat less dramatic, see figure 2(b). The most dramatic impact is seen in the kurtosis. Although it increases monotonically within error bars, we see a clear sign change at beam energy $\sqrt{s} \sim 4$ GeV, see figure 2(c). As argued in [7] strong negative values of the kurtosis of the order parameter are understood as a direct consequence of a first-order phase transition.

As introduced previously, we ultimately aim to study fluctuations in the net-proton number N_p rather than the order parameter which is not experimentally accessible. However, assuming a σ-proton coupling, we can assume a relation between the fluctuations in N_p and the sigma field as

$$\delta N = \delta N^0 + V\, g\, \delta\boldsymbol{\sigma}\, d \int_{\boldsymbol{p}} \frac{\partial n_p}{\partial m}\,, \tag{18}$$

where g denotes the coupling constant, d the number of degrees of freedom, $f_{\boldsymbol{p}}$ the proton-number distribution function and m the dynamically generated mass [7].

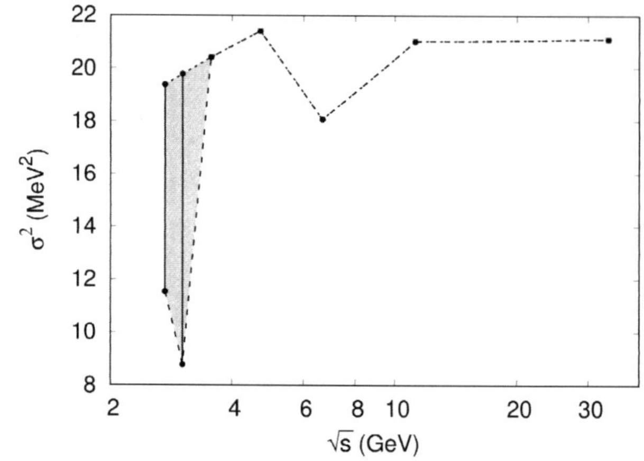

(a).

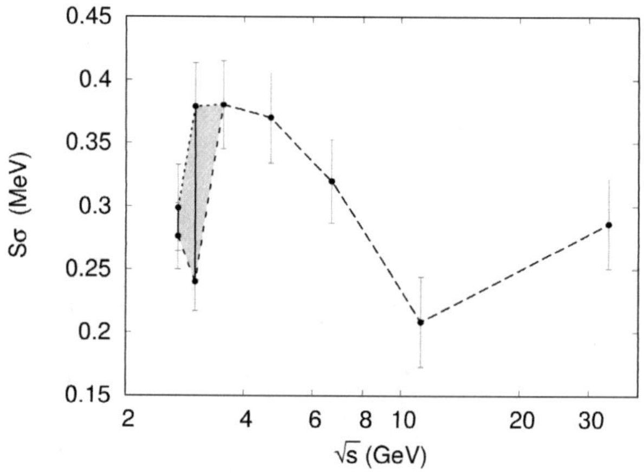

(b).

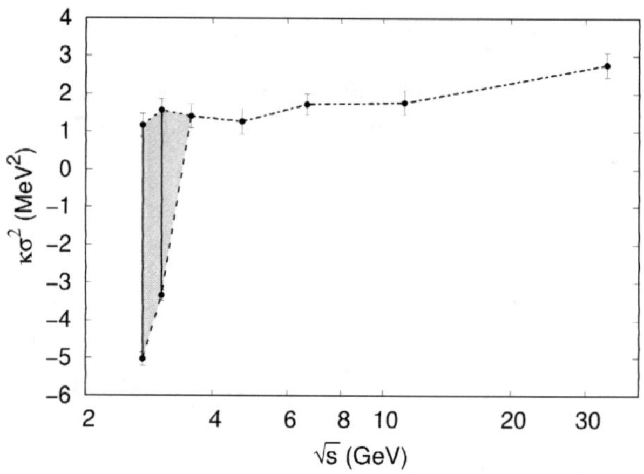

(c).

Figure 2. σ^2 (a) , Sσ (b) and $\kappa\sigma^2$ (c) of the sigma field as a function of beam energy $\sqrt{s}$.

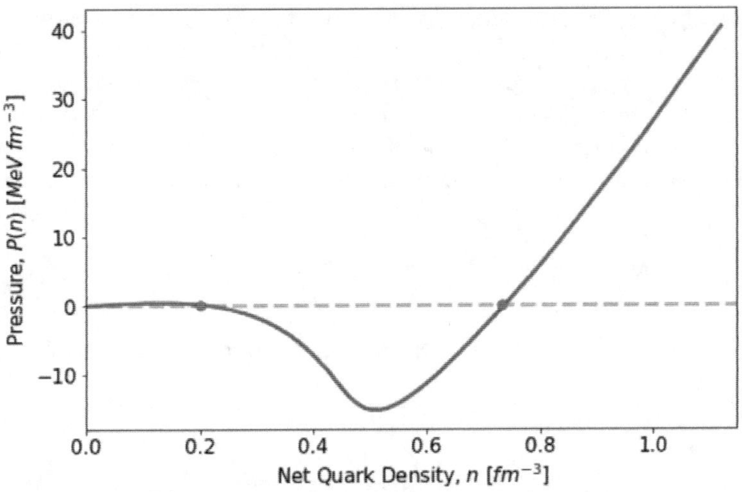

Figure 3. The equation of state at T = 0. The dashed orange line signifies the zero pressure threshold, which makes it easy to see the negative pressure portion. The red dots mark the negative pressure region in the x-axis.

For correct treatment of these particle number fluctuations, the effect of volume fluctuations needs to be considered which plays a major role especially for trajectories crossing the first-order phase transition. Work in this direction is currently in progress.

4. Equation of state at low T

The thermodynamics of strong interactions at low temperature and high density is important for heavy-ion collisions at low beam energies but also for astrophysics, specifically in understanding the inner structure of compact stars [19, 20]. Of particular importance for any hydrodynamic simulation is the equation of state (EoS), especially at ultra-low temperature ($T \approx 0$) [21] and in the presence of spinodal instabilities. As shown in [22], chiral models often suffer from the problem of negative pressures at low T which allows for the unphysical phenomenon of stable quark droplets in a vacuum and ultimately leads to a severe overestimation of fluctuation observables like e.g. net-baryon number moments.

The EoS is obtained by first determining the dynamical values of $\langle\sigma\rangle$ using the gap equation $\partial\Omega/\partial\sigma = 0$. After that, we determine the net-quark density by a derivative of the pressure P as $n = \frac{\partial P}{\partial \mu}$ and parametrize both $P = -\Omega$ and n through $\langle\sigma\rangle$.

An early result of this research work reveals several features of the obtained equation of state. Figure 3 shows that the pressure is only slightly above zero at $0.0 < n < 0.2$ fm^{-3}. Then, we have a negative pressure for $0.20 < n < 0.74$ fm^{-3}, before the pressure increases in the positive region for subsequently increasing density.

The negative pressure region poses the aforementioned problems to the nonequilibrium model. Therefore, more work needs to be done to find a reliable workaround. Currently, we are investigating comparisons with other models that include further hadronic degrees of freedom like the sigma-omega model.

5. Summary and conclusions

We have studied cumulant ratios of the sigma field for different beam energies within the nonequilibrium chiral Bjorken model based on the quark-meson Lagrangian. It must be stressed that although our theoretical framework is just a simple model it is nevertheless able to describe the complex nature of dynamical phenomena near a chiral CP. We found a large spread of possible values for variance, skewness, and kurtosis at the first-order phase transition. Especially the kurtosis with strongly negative values at low $\sqrt{s}$ can hint at the presence of a first-order phase transition. For comparison with experimental data, we need to evaluate the fluctuations in the net-proton number after a correct treatment of volume fluctuations and the possible need for an extension of the Lagrangian to cure the problem of negative pressure at lowest temperatures.

Acknowledgements
This work was supported by the Development and Promotion of Science and Technology Talents Project (DPST) - Royal Thai Government Scholarship, Suranaree University of Technology (SUT). MLAMY acknowledges support from the University of Malaya's Research Grant GPF044B-2018.

References

[1] Stephanov M, Rajagopal K and Shuryak E 1998 *Phys. Rev. Lett.* **81** 4816–9
[2] Alford M, Rajagopal K and Wilczek F 1998 *Phys. Lett.* B **422** 247–56
[3] Aoki Y, Endrődi G, Fodor Z, Katz S D and Szabó K K 2006 *Nature* **443** 675–8
[4] Steinbrecher P 2019 *Nucl. Phys.* A **982** 847–50
[5] Cheng M *et al.* 2009 *Phys. Rev.* D **79** 074505
[6] Gavai R V and Gupta S 2008 *Phys. Rev.* D **78** 114503
[7] Stephanov M A 2011 *Phys. Rev. Lett.* **107** 052301
[8] Stephanov M A 2009 *Phys. Rev. Lett.* **102** 032301
[9] Aggarwal M M *et al.* 2010 *Phys. Rev. Lett.* **105** 022302
[10] Herold C, Nahrgang M, Mishustin I and Bleicher M 2013 *Phys. Rev.* C **87** 014907
[11] Schaefer B J, Pawlowski J M and Wambach J 2007 *Phys. Rev.* D **76** 074023
[12] Steinheimer J, Dexheimer V, Bleicher M, Petersen H, Schramm S and Stöcker H 2010 *Phys. Rev.* C **81** 044913
[13] Nahrgang M, Leupold S, Herold C and Bleicher M 2011 *Phys. Rev.* C **84** 024912
[14] Herold C, Bleicher M, Nahrgang M, Steinheimer J, Limphirat A, Kobdaj C and Yan Y 2018 *Eur. Phys. J.* A **54** 19
[15] Nahrgang M, Herold C and Bleicher M 2013 *Nucl. Phys.* A **904-905** 899c–902c
[16] Herold C, Kittiratpattana A, Kobdaj C, Limphirat A, Yan Y, Nahrgang M, Steinheimer J and Bleicher M 2019 *Phys. Lett.* B **790** 557–62
[17] Adams J *et al.* 2005 *Nucl. Phys.* A **757** 102–83
[18] Cleymans J, Oeschler H, Redlich K and Wheaton S 2006 *Phys. Rev.* C **73** 034905
[19] Glendenning N 2012 *Compact Stars: Nuclear Physics, Particle Physics and General Relativity* (New York: Springer)
[20] Baym G, Hatsuda T, Kojo T, Powell P D, Song Y and Takatsuka T 2018 *Rep. Prog. Phys.* **81** 056902
[21] Oertel M, Hempel M, Klähn T and Typel S 2017 *Rev. Mod. Phys.* **89** 015007
[22] Steinheimer J, Randrup J and Koch V 2014 *Phys. Rev.* C **89** 034901

20-inch photomultiplier tube timing study for JUNO

Narongkiat Rodphai[1], Zhimin Wang[2], Narumon Suwonjandee[1,†] and Burin Asavapibhop[1,‡]

[1]Department of Physics, Faculty of Science, Chulalongkorn University, 10330, Bangkok, Thailand
[2]Institute of High Energy Physics, Chinese Academy of Sciences, Beijing 100049, China

E-mail: narongkiat.r@gmail.com, wangzm@ihep.ac.cn, †narumon.s@chula.ac.th, ‡burin.a@chula.ac.th

Abstract. Jiangmen Underground Neutrino Observatory (JUNO) is a large liquid scintillator neutrino detector now under construction at Jiangmen, Guangdong, China for determination of neutrino mass ordering with 3% energy resolution at 1 MeV, a precise measurement of neutrino oscillation parameters, and other neutrino physics. The central detector is made up of a 35.4-meter diameter acrylic sphere which contains 20 kton of liquid scintillator and is surrounded by about 18k 20-inch photomultiplier tubes (PMTs). The PMTs performance is one of the JUNO's key successes to reach the high resolution goal. In this study, the PMT characteristic and its timing related responses were determined via the PMT generated signals, extracted from the PMT in a scanning station system. About 2,400 of micro-channel plate PMTs (MCP-PMTs) and dynode PMTs were analyzed for their responses with LED source such as rise time, fall time, transit time spread (TTS), gain, etc., which relate to photon incident on different positions of PMT's glass surface. Furthermore, we also observed the fluctuation of PMT performance under magnetic field which can decrease the PMT photon detection efficiency (PDE).

1. Introduction

JUNO aims to study the neutrino mass hierarchy with 3% energy resolution at 1 MeV, a precise determination of neutrino oscillation parameters, and other neutrino physics. The JUNO detector will detect the interactions between the incoming neutrinos and 20 ktons liquid scintillator (LS), contained inside a 35.4-meter diameter acrylic sphere as shown in figure 1.

An incoming neutrino will interact with the LS via the inverse beta decay (IBD): $\overline{\nu}_e + p \rightarrow e^+ + n$ which produces a positron and a neutron as the result. The positron will annihilate with an electron, generating a pair of photons in a few ns and these photons will be defined as a prompt signal. About 200 μs later on average, a delayed signal will be produced, represented by another photon from neutron captured with hydrogen atom of LS. There are over 20,000 of 20" photomultiplier tubes (PMTs) (figure 2) and 25,000 of 3" PMTs on the surface of acrylic sphere with over 75% coverage to detect light signals of IBD interaction. JUNO PMT testing system was designed to determine the PMTs characteristic before the installation [1].

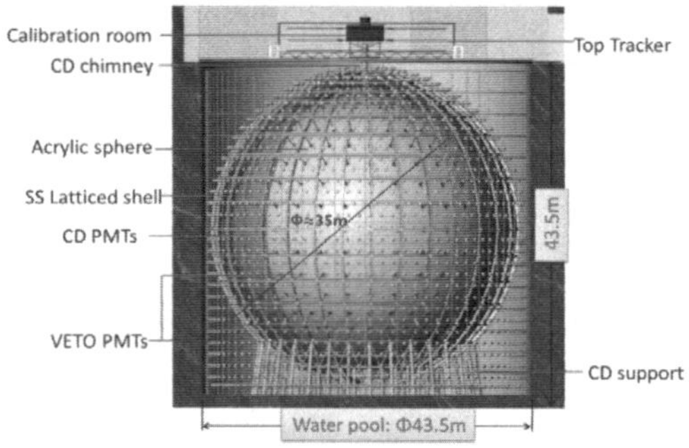

Figure 1. JUNO configuration [1].

Figure 2. 20" photomultiplier tubes.

2. PMT testing system

An incoming photon impinging on the PMT's glass surface interacts with the photocathode material and produces a photoelectron. This photoelectron will be accelerated under supplied voltage into the electron multiplier for amplifying the number of electron and then convert into an electrical pulse. The 20" PMTs used in this experiment were designed with different inner structure of electron multiplications: micro-channel plates PMT (MCP PMT) and dynode PMT. The structure of MCP PMT is formed as thin disk of tiny channels bundled in parallel which electron will pass through and increase a number of electron by striking with an inner surface of these tiny channels. However, the electron multiplication of dynode PMT is caused by an electron collision on several stages of dynode to gain a number of electrons [2].

To verify the PMT performance before the JUNO operation, PMTs have been tested and analyzed their performances since 2017. The PMT testing procedure is divided into two testing steps: PMT container station and PMT scanning station. In this study, we focus on the scanning station which was designed to study the PMT uniformity on its entire surface. In the scanning station, PMT will be placed among the rotating machine with 7 LED light sources located on the rotating frame which will be rotated 15° each step for 24 steps to cover its entire surface as shown in figure 3. Therefore, the PMT surface will be divided into 168 testing grids (figure 4), follows these LEDs and rotating steps [3].

Figure 3. The rotating machine with 7 LED sources located on the rotating frame in the scanning station [3].

Figure 4. The top view of PMT, divided into 168 testing grid over its surface.

Besides, the trajectory of photoelectron from photocathode might be bent by the external magnetic field before reaching the electron multiplier and this might decrease the PMT efficiency. The PMT scanning station is also designed to adjust the magnetic field of the experimental area which can help us to understand the PMT's characteristic in the presence of magnetic field [4].

3. Result and discussion

The PMT's pulses generate a signal waveform as shown in figure 5 which has been used to extract several timing parameters such as:

- Rise time : Time spent when signal rises from 10% to 90% of the amplitude
- Fall time : Time spent when signal drops from 90% to 10% of the amplitude
- Full width at half maximum (FWHM) : Time spent when signal rises from 50% of the amplitude to its maximum then drops to 50% of the amplitude
- Signal's amplitude and hit time
- Transit time spread (TTS) : This parameter represents the time fluctuation of PMT starting when the light is detected till the signal is generated. This parameter can be calculated by averaging the fluctuation of hit time distribution which contains the hit time of hundreds of waveform (not shown in figure 5)

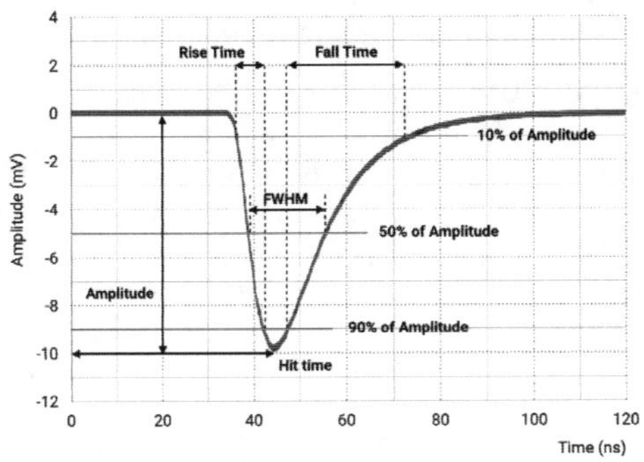

Figure 5. A signal waveform, labeled with parameter definitions.

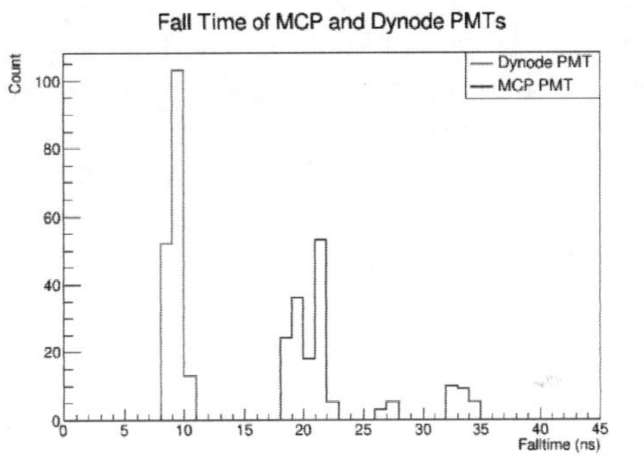

Figure 6. Fall time distribution of both PMT types.

We analyzed about 2,400 PMTs (2,131 MCP and 268 dynode PMTs), tested from the PMT scanning station to study the timing parameters and other information. We found that the signal characteristic of MCP PMT has a long-tailed waveform, compared to the signal measured from dynode PMT. This results in the fall time and FWHM from MCP PMT having higher values than dynode PMT. Figure 6 is an example of fall time distribution, averaged over 168 points on the surface of both PMT types. The result shows that the signal of MCP PMT can reach the fall time up to 35 ns at some position on its surface.

We studied the PMT's uniformity according to the LED light sources and the step of rotating frame around its surface, which will be determined as zenith angle and azimuthal angle, respectively. These timing parameters of both PMT types increase as PMT's zenith angle increases. The results suggests that the PMT responds better for forward incident photons. While the comparison on azimuthal angle shows the uniformity on these timing parameters over its surface as shown in figure 7.

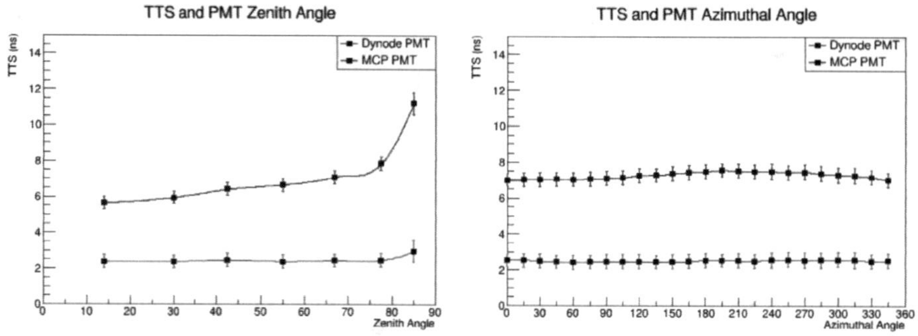

Figure 7. TTS of both PMT types, determining in PMT zenith and azimuthal angles.

The TTS is involved in the time likelihood function which is used to optimize the event vertex reconstruction inside the LS sphere. The time likelihood function is based on the time residual propagation of scintillating photon through the LS and the first hit time of each PMT which TTS is included [5]. Thus, PMT timing response can effect the performances of vertex reconstruction.

Furthermore, we also studied the PMT performance under the magnetic field by canceling the EMF down to 0-20 μT during the process of scanning station. The result shows that the PMT efficiency decreases since the residual magnetic field interferes the trajectory of photoelectron to reach the electron multiplier. Those timing parameters are significantly increased, resulting from the fluctuation of generated waveform by PMT which is spreading out.

To minimize the effect from magnetic field to PMT, a 16-pair-coil system will be installed to cancel the EMF in the detection region and help all PMTs to reach their excellent efficiency. Thus, the PMT performance and characteristic could be now used for the calibration and improvement for event energy and vertex reconstruction for JUNO experiment.

Acknowledgment
This project is funded by CUniverse research promotion project of Chulalongkorn University (grant CUAASC), Rachadapisek Sompot Fund, Chulalongkorn University. We would like to thank the JUNO collaboration and especially the JUNO PMT Instrumentation Group who dedicates to the PMT testing system and provides us the testing information and statistics.

References
[1] An F *et al.* 2016 *J. Phys. G: Nucl. Part. Phys.* **43** 030401
[2] Zhang H *et al.* 2020 *J. Phys.: Conf. Ser.* **1468** 012197
[3] Anfimov N 2017 *J. Instrum.* **12** C06017
[4] Payupol T, Suwonjandee N and Asavapibhop B 2018 *J. Phys.: Conf. Ser.* **1144** 012092
[5] Li Z *et al.* 2021 *Nucl. Sci. Tech.* **32** 1–12

Study of electroweak penguin B decays at Belle II experiment

J Ineead[1,*], S Nishida[2,3], B Asavapibhop[1,†] and N Suwonjandee[1]

[1]Department of Physics, Chulalongkorn University, 254 Phayathai Road, Pathumwan, Bangkok 10330, Thailand

[2]Institute of Particle and Nuclear Studies, High Energy Accelerator Research Organization, 1-1 Oho, Tsukuba, Ibaraki 305-0801, Japan

[3]Department of Particle and Nuclear Physics, The Graduate University for Advanced Studies, 1-1 Oho, Tsukuba, Ibaraki 305-0801, Japan

E-mail: burin.a@chula.ac.th[†], jittapan.i@student.chula.ac.th[*]

Abstract. The electroweak $b \to sll$ ($l = e, \mu$) transition is a flavor-changing neutral current process that mediates through a one-loop penguin diagram. The decay is considered to be a good probe for the New Physics as particles predicted in the beyond Standard Model theories can enter into the loop. The exclusive decay $B \to K^{(*)}l^+l^-$ was first observed by the Belle experiment and it provides many observables such as the branching fraction, CP asymmetry, forward-backward asymmetry, and other angular observables. Recently, the LHCb experiment has reported some clue of a lepton flavor universality violation from the branching fraction ratio of the $B \to K\mu^+\mu^-$ and $B \to Ke^+e^-$ decays. In this presentation, we report the status of the $B \to Kl^+l^-$ decay analysis at the Belle II experiment which started the data taking in 2019. We also, present an activity at the Belle II Chulalongkorn University group where we study the $B \to KJ/\psi$ decay which has the same topology as the $B \to Kl^+l^-$.

1. Introduction

The electroweak penguin (EWP) of B decay is a flavor-changing neutral current (FCNC) process in the Standard Model (SM). In this process, a b quark transits into an s quark and is followed by a pair of leptons, i.e. $b \to sll$, where l can be e or μ. The EWP process is expressed via a one-loop Feynman diagram which the mediators involving in this process can be the virtual SM particles as a t quark, a W^-, and a Z^0 or a γ, see figure 1, or even the virtual non-SM particles which are predicted by some beyond SM (BSM) theories such as charged Higgs bosons predicted by the two-Higgs-doublet model [1, 2], or leptoquark particles predicted by some extended models of the effective field theory [3], etc. The typical exclusive process, $B \to K^{(*)}l^+l^-$ with its branching fraction, CP asymmetry, forward-backward asymmetry, or other angular observables, is considered as some good probes to test the SM or to search for new physics (NP).

2. Previous $B \to K^{(*)}l^+l^-$ analysis

The first observation of the decay $B \to K^{(*)}l^+l^-$ was done by Belle [4] and later by BaBar [5]. The measured branching fraction from both experiments had been refined several times when increasing datasets. At present, the branching fractions of these B decay modes obtain

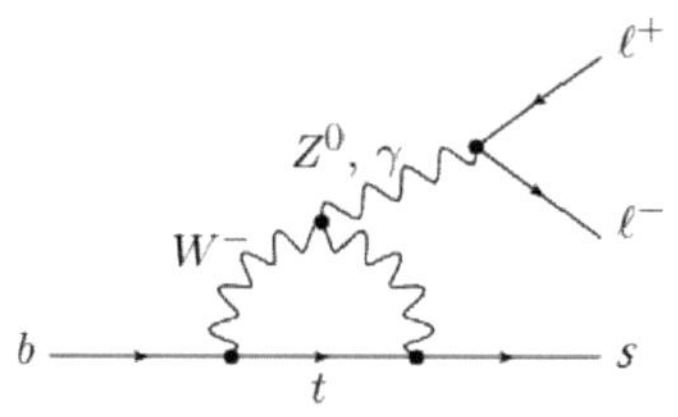

Figure 1. The $b \to sll$ loop diagram.

by combining results from BaBar, Belle, and LHCb measurements are $(4.51 \pm 0.23) \times 10^{-7}$ for $B^+ \to K^+l^+l^-$, $(1.01 \pm 0.11) \times 10^{-6}$ for $B^+ \to K^*(892)^+l^+l^-$, $(3.1^{+0.8}_{-0.7}) \times 10^{-7}$ for $B^0 \to K^0l^+l^-$, and $(9.9^{+1.2}_{-1.1}) \times 10^{-7}$ for $B^0 \to K^*(892)^0l^+l^-$ [6]. Remarkably, these results are consistent with the SM prediction. Later, some observables such as CP asymmetry, forward-backward asymmetry, and other angular observables are also measured. Most of these measurements are compatible with the SM prediction, except some observables such as $P_5^{'}$, measured by Belle [7] and LHCb [8] for which departures from SM predictions have been observed. Recently, LHCb has reported a new result of lepton flavor universality (LFU) violation from this EWP B process by measuring the branching fraction ratio between the decays $B \to K\mu^+\mu^-$ and $B \to Ke^+e^-$ [9].

3. Belle II experiment

The Belle II experiment is a high-energy physics experiment that is mainly designed to study the properties of B meson and is set up at the High Energy Accelerator Research Organization (KEK) in Tsukuba, Japan. The experimental facility consists of the SuperKEKB accelerator and the Belle II detector. The SuperKEKB accelerator is a 3-km circumference circular accelerator with two rings for 7 GeV of electron and 4 GeV of positron beams. The collisions take place at a center of mass energy of 10.58 GeV, as a result, $\Upsilon(4S)$ resonances along with $q\bar{q}$ and $\tau^+\tau^-$ are produced. And the $\Upsilon(4S)$ continues decaying to a B meson pair, both charged and neutral. The Belle II detector is a cylindrical shape detector with a 7-m diameter and a 7.5-m length. It is installed covering a crossing beam pipe of the electron and the positron for detecting particles from the electron-positron interaction. The Belle II detector is composed of several subdetectors, i.e. a vertex detector (VXD), the innermost subdetector which consists of a silicon pixel detector (PXD) and a silicon vertex detector (SVD), a central drift chamber (CDC), a time-of-propagation counter (TOP), an aerogel ring imaging Cherenkov counter (ARICH), an electromagnetic calorimeter (ECL), and a kaon (K_L^0) and muon (μ) detector (KLM), the outermost subdetector which is installed outside a 1.5 Tesla solenoid coil. The Belle II experiment started taking data in early 2019. It is designed to operate at a maximum luminosity of 6×10^{35} cm^{-2}s^{-1} with planning to collect data of 50 ab^{-1} during an 8-years operational period [10, 11].

4. Analysis status of the decay $B \to Kl^+l^-$

With the initial data of 62.8 fb^{-1} collected by the Belle II experiment in 2019, the decay of $B \to Kl^+l^-$ was studied through the $B^+ \to K^+e^+e^-$ and $B^+ \to K^+\mu^+\mu^-$ final states. The particles as K^+, $e^\pm$, and $\mu^\pm$ are selected by applying the criteria of (i) the distance from e^+e^- interaction point (d_0) in xy-plan is less than 2.0 cm, (ii) the distance along the z-axis from the interaction point ($|z_0|$) is less than 4.0 cm, these will confirm that the selected charged particles come from the interaction region, (iii) the momentum ($|\overrightarrow{p}|$) of $e^\pm$ and $\mu^\pm$ is more than 0.5 and 0.8 GeV/c, respectively, and (iv) the particle identification for e, μ, and K. In addition, the $e^{\pm}$ are selected by considering the radiated Bremsstrahlung in the ECL. The B candidate is reconstructed from these charged particles and is selected by examining two variables [12], the energy difference, $\Delta E = E^*_{\text{beam}} - E^*_B$, and the beam-energy constrained mass, $M_{\text{bc}} = \sqrt{E^{*2}_{\text{beam}} - \vec{p}^{\,*2}_B}$, where $E^*_{\text{beam}} = \sqrt{s}/2 \cong 5.29$ GeV is the beam energy in the center-of-mass (CM) frame or $\Upsilon(4S)$ rest frame and E^*_B, $\vec{p}^{\,*}_B$ are the measured energy and momentum of the B candidate in the CM frame, respectively, and the criteria for selecting B candidate are $|\Delta E| < 0.3$ GeV and $5.20 < M_{\text{bc}} < 5.29$ GeV/c^2. The background events from the decays $B \to J/\psi(\to l^+l^-)K$ and $B \to \psi(2S)(\to l^+l^-)K$, the Dalitz decay as $\pi^0 \to e^+e^-\gamma$, and the photon conversion, $\gamma^* \to e^+e^-$ for electron mode, the misidentified particle events as $B^+ \to \overline{D}^0(\to K^+\pi^-)\pi^+$ and $B^+ \to K^+\pi^+\pi^-$, and the K^* mode as $B^+ \to K^*l^+l^-$, including the $q\overline{q}$ continuum backgrounds are all suppressed. Finally, the B signal yield is extracted by performing a 2D un-binned extended maximum likelihood fit on the ΔE and M_{bc} distributions, the results are shown in figure 2 [13], with obtaining $8.6^{+4.3}_{-3.9} \pm 0.4$ events, the first and second errors are statistical and systematic, respectively, and most of the systematic errors are from the $B^+ \to K^+\pi^+\pi^-$ background and the shape of probability density function during fitting.

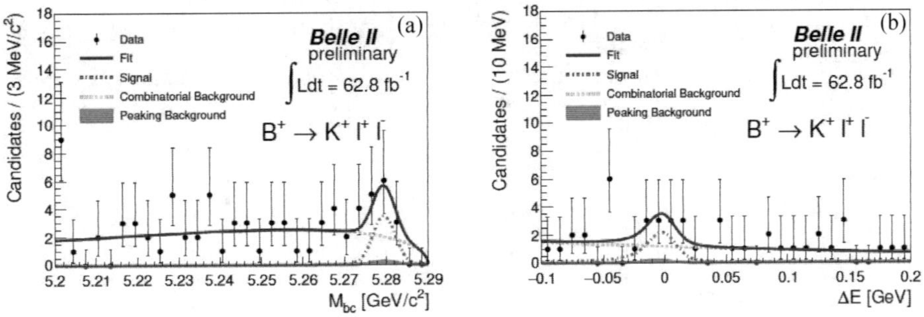

Figure 2. The fitted M_{bc} (a) and ΔE (b) distributions for the decay $B^+ \to K^+l^+l^-$, where combinatorial backgrounds are from $q\overline{q}$ and the other B decay events, while the peaking background is from $B^+ \to K^+\pi^+\pi^-$ event only.

5. Monte Carlo study for the decay $B \to KJ/\psi$

The decay $B \to KJ/\psi$ is considered as a control sample or a physics background of the $B \to Kl^+l^-$ process since it has the same final states as $B \to Kl^+l^-$ when J/ψ decays to a lepton pair. The Belle II Chulalongkorn University group study this B decay mode by using the Belle II analysis software framework (`basf2`) [14] and analyze the decays $B^{\pm} \to K^{\pm}J/\psi(\to e^+e^-)$ and $B^{\pm} \to K^{\pm}J/\psi(\to \mu^+\mu^-)$ by using Monte Carlo (MC) samples. The signal MC sample used to estimate the reconstruction efficiency consists of a pure of $B^{\pm} \to K^{\pm}J/\psi(\to e^+e^-)$ and $B^{\pm} \to K^{\pm}J/\psi(\to \mu^+\mu^-)$ datasets with 1,000,000 events each. The MC sample simulated for the whole $e^+e^- \to \Upsilon(4S) \to B\overline{B}$ and continuum process as $e^+e^- \to q\overline{q}$ ($q = u, d, s, c$) corresponds to 10 fb^{-1} for each dataset used in this analysis as experimental data. We begin the analysis by selecting the charged final state particles, $K^{\pm}$, $e^{\pm}$, and $\mu^{\pm}$ and apply the criteria from the $B \to Kl^+l^-$ analysis, detailed in section 4. The J/ψ is reconstructed from a lepton pair with selecting its invariant mass in the range between 2.91 and 3.19 GeV/c^2 for electron mode and 2.96 and 3.19 GeV/c^2 for muon mode, these ranges cover all J/ψ mass region. The B candidate is reconstructed from the $K^{\pm}$ and J/ψ by considering two essential variables described in section 4, the ΔE and M_{bc}. The selection criteria for ΔE and M_{bc} are $|\Delta E| < 0.1$ GeV/c and $5.20 < M_{\text{bc}} < 5.29$ GeV/c^2, respectively. This ΔE range can suppress some peaking backgrounds from $B^0\overline{B}^0$ events. Finally, we extract the B signal yield by fitting the $M_{\rm bc}$ distribution with the Crystal Ball function for a signal peak and the ARGUS function for continuum backgrounds. Using this MC sample, we measure 255 ± 16(stat) signal events for $B^\pm \to K^\pm J/\psi(\to e^+e^-)$ decay mode and 283 ± 17(stat) signal events for $B^\pm \to K^\pm J/\psi(\to \mu^+\mu^-)$ decay mode.

To calculate the branching fraction ($\mathcal{B}$) of B^+, we use the formula as,

$$\mathcal{B}(B^+ \to K^+J/\psi) = \frac{\text{no. of observed } B^\pm \text{ signal events}}{(2)(\text{no. of produced } \Upsilon(4S) \to B^+B^- \text{ events})(\mathcal{B}(J/\psi \to l^+l^-))(\varepsilon)}, \quad (1)$$

where the number of produced B^+B^- events from the $B\overline{B}$ MC dataset equals 5,205,442 events, the branching fraction of the decays $J/\psi \to e^+e^-$ and $J/\psi \to \mu^+\mu^-$ equal $(5.971 \pm 0.032)\%$ and $(5.961 \pm 0.033)\%$, respectively [6], and the reconstruction efficiency (ε) extracted from the signal MC equals $(38.473 \pm 0.062(\text{stat}))\%$ and $(44.315 \pm 0.067(\text{stat}))\%$ for the decays $B^\pm \to K^\pm J/\psi(\to e^+e^-)$ and $B^\pm \to K^\pm J/\psi(\to \mu^+\mu^-)$, respectively. We obtain the branching fraction as,

$$\begin{aligned}\mathcal{B}(B^+ \to K^+J/\psi) &= (1.066 \pm 0.067(\text{stat})) \times 10^{-3} \text{ (for } e \text{ mode)},\\ \mathcal{B}(B^+ \to K^+J/\psi) &= (1.029 \pm 0.062(\text{stat})) \times 10^{-3} \text{ (for } \mu \text{ mode)}.\end{aligned} \quad (2)$$

These measured values are consistent with the branching fraction specified for generating the $B\overline{B}$ MC event which is equal to 1.014×10^{-3} for $B^+ \to K^+J/\psi$ decay mode [15].

6. Conclusion

The decay $B \to K^{(*)}l^+l^-$ is one of the other B decay modes that is the main target for investigating at the Belle II experiment. The observable, such as the branching fraction, will be measured at the first stage of the experiment, followed by other observables with improved precision respecting previous measurements. The Belle II group at Chulalongkorn University contributes to this study by analyzing the $B \to KJ/\psi$ decay mode which is considered as one of the critical control samples for $B \to Kl^+l^-$ studies.

Acknowledgments

We would like to thank the Belle II Collaboration for providing the MC and data samples. We thank Saurabh Sandilya and Elisa Manoni for the valuable comments and suggestions. This work is supported by CUniverse research promotion project of Chulalongkorn University (grant CUAASC), Ratchadapisek Sompot Fund, Chulalongkorn University.

References

[1] Burdman G 1995 *Phys. Rev.* D **52** 6400
[2] Hewett L and Wells J D 1997 *Phys. Rev.* D **55** 5549
[3] Cornella C, Faroughy D A, Fuentes-Martin J, Isidori G and Neubert M 2021 *Preprint* hep-ph/2103.16558
[4] Abe K *et al.* 2002 *Phys. Rev. Lett.* **88** 021801
[5] Aubert B *et al.* 2003 *Phys. Rev. Lett.* **91** 221802
[6] Zyla P A *et al.* 2020 *Prog. Theor. Exp. Phys.* **2020** 083C01
[7] Wehle S *et al.* 2017 *Phys. Rev. Lett.* **118** 111801
[8] Aaij R *et al.* 2021 *Phys. Rev. Lett.* **126** 161802
[9] Aaij R *et al.* 2021 *Preprint* hep-ex/2103.11769v1
[10] Abe T *et al.* 2010 *Belle II Technical Design Report* (Tsukuba: Belle II Experiment)
[11] Kou E *et al.* 2019 *Prog. Theor. Exp. Phys.* **2019** 123C01
[12] Browder T E and Honschied K 1995 *Prog. Part. Nucl. Phys.* **35** 81
[13] Halder S, Tiwary R, Sandilya S and Mohanty G B 2021 *Approved Plots: Study of* $B^+ \to K^+l^+l^-$ *at Belle II* https://docs.belle2.org/record/2310
[14] Kuhr T, Pulvermacher C, Ritter M, Hauth T and Braun N 2019 *Comp. Softw. Big Sci.* **3** 1
[15] Nakamura K *et al.* 2010 *J. Phys.* G **37** 075021

Assessment of contamination of natural and anthropogenic radionuclides in rice samples collected from Songkhla province, Thailand

P Kessaratikoon[1*], D Riyapunt[2], R Boonkrongcheep[3] and N Changkit[4]

[1]Department of Basic Science and Mathematics, Faculty of Science, Thaksin University, Songkhla, 90000, Thailand
[2]Undergraduate Student, B.Ed. (Science-Physics) Program, Faculty of Education, Thaksin University, Songkhla, 90000, Thailand
[3]Research Assistant, Nuclear and Material Physics Research Unit, Department of Basic Science and Mathematics, Faculty of Science, Thaksin University, Songkhla, 90000, Thailand
[4]Thailand Institute of Nuclear Technology (Public Organization), Nakhon Nayok, 26120, Thailand

[*]Corresponding author's e-mail address: prasong@tsu.ac.th

Abstract. Study on assessment of contamination of natural and artificial radionuclides in agricultural products samples are very important to all human being as a consumer. In the present study, we have measured and evaluated the specific activities of natural (^{40}K, ^{226}Ra and ^{232}Th) and anthropogenic (^{137}Cs) radionuclide in rice samples. The 30 rice samples were collected from general and department stores at Songkhla province in the south of Thailand. The high-purity germanium (HPGe) detector and gamma-ray spectrometry analysis system which was set-up in advanced laboratory in Thailand Institute of Nuclear Technology (public Organization) or TINT were employed to perform all of measurements and analysis for this study. The frequency distribution of specific activities of ^{40}K, ^{226}Ra, ^{232}Th and ^{137}Cs for this study were also studied and found to be asymmetrical distribution with the skewness of 1.29, 1.43, 2.32 and 0.82, respectively. For this reason, the median values of specific activities of ^{40}K, ^{226}Ra and ^{232}Th which were 620.04 ± 44.30, 3.73 ± 0.54 and 2.44 ± 0.54 Bq/kg respectively, should be selected and also used to calculate some related radiological hazard indices in this study. Furthermore, the excess lifetime cancer risk (ELCR) would be also evaluated and presented. Moreover, the results of present study were taken to compare with some data and studies in Thailand and global measurement and calculations. It was found that the outcomes satisfied the standards of UNSCEAR and IAEA.

1. Introduction

A few decades ago, there are some concerns with radioactivity buildup in food, vegetables and fruits around the world [1-3]. Radioactive materials from soil and fertilizers utilized to the soil for enrichment would be removed to vegetables and plants. Owing to the increase of up to one order of magnitude or more in the concentration of radionuclide levels in food, it is evaluated that the use of phosphate fertilizer has doubled the extended exposure of humans to radioactive materials from ingestion of food [4]. Hence, many food items contain tiny amounts of uranium (U), thorium (Th), potassium (^{40}K), and their progeny products. Anthropogenic radionuclides such as ^{137}Cs can be presented in soil in consequence of nuclear accidents [5-7]. For this reason, there is a high probability for natural (^{226}Ra, ^{232}Th and ^{40}K) and anthropogenic (^{137}Cs) radionuclides entering plants and later into foods consumed by humans. Radionuclides mostly enter the human body by way of ingestion and inhalation. The ingested radionuclides could be accumulated in some parts of the body [8]. Usually, Thai people like to eat rice as a staple food for three meals a day. One of the most well known main dishes among Thais and foreigners is "fried rice". The fired rice or "Kao Pad" is generally consist of rice with pork, beef or chicken meat, vegetables, egg and then stir fire together with some typical spice source. In addition, some other common dishes such as noodles, use rice flour as the main ingredient. In Thailand, not many researches has been carried out so far to assess the amount of the concentration of some concerned radionuclides present in the normal Thai main staple food, which has caused difficulty to decide whether there is a risk for Thai population due to the ingestion of radioactive materials. Hence, it is very importance to create background levels and specific activities of natural and anthropogenic radionuclides in local foods in order to secure the community from future ingestion. However, there are only few researches that have been conducted to assess the concentrations of interested radionuclides of Thai rice. Therefore, the main objectives of this study were to study and evaluate the specific activity of gamma emitting radioactive elements and some related radiological hazard indices and the excess lifetime cancer risk (ELCR) present in rice samples which collected randomly from some local and department stores in Songkhla province, Thailand. Moreover, the study results should be used for discovering the effects caused by the consumption on the Thai community.

2. Materials and methods

2.1. Sample collection and processing

All of 30 rice samples which are Jasmine rice (16 samples), Hom Prathum rice (8 samples) and white rice (6 samples) were collected randomly from some local and department stores in Maueng district in Songkhla province, Thailand. Basically, we do not know which provinces and regions of Thailand have been cultivated for all rice samples. Each sample was collected, treated and prepared by following the International Atomic Energy Agency (IAEA) sampling and preparation of vegetables samples [9]. For example, the rice samples would be grinded with a blender machine and then spread on plastic or aluminium trays and let to dry at room temperature for a few days. Then, all rice samples would be sieved through a 2 mm mesh-sized sieve to remove some coarse portion. A slow-airflow, temperature (60°C) for 5 hours in drying closet will accelerate the drying process without loss of radionuclides from the rice samples. Each sample (about 154 cm^3 which approximately 139-158 gms) was placed in a PVC cylindrical container of diameter 7.0 cm and height 7.0 cm. The containers were sealed tightly with a thick cellophane tape around their necks to avoid any gas escape from them, and kept for a minimum period of 4 weeks to ensure equilibrium between ^{222}Rn and its progeny products.

2.2. Gamma-ray spectrometry with high-purity germanium (HPGe) detector

All rice samples were taken to measure and analyze by using a gamma spectrometry analysis system with a measuring time 10,800 s. The HPGe detector (EG&G ORTEC Model GEM 20 P4) and gamma spectrometry analysis system at advanced laboratory, Thailand Institute of Nuclear Technology (Public Organization) (TINT) were employed with sufficient lead shielding which reduced the background by a factor about 95 %. The specific activities of various radionuclides of interest in 30 rice samples were determined in Bq/kg using the counted gamma-ray spectra. The gamma-ray photopeaks corresponding to 351.9 keV (^{214}Pb), 583.2 keV (^{208}Tl), 1460.8 keV (^{40}K) and 661.7 keV (^{137}Cs) were evaluated in achieving of the specific activities of ^{226}Ra, ^{232}Th, ^{40}K, and ^{137}Cs in the rice samples, respectively. The IAEA-473 Milk Powder reference materials (International Atomic Energy Agency (IAEA), Vienna, Austria) was used to determine the geometric efficiency for rice sample matrices in the container.

2.3. Frequency distribution of specific activities

The frequency distribution of specific activities of natural (^{40}K, ^{226}Ra and ^{232}Th) and anthropogenic (^{137}Cs) radionuclides in 30 rice samples for the present study were studied and analyzed by using the SPSS computer program. The results would be presented and shown in the subsection 3.1.

2.4. Calculation of related radiological hazard indices and excess lifetime cancer risk

Three related radiological hazard indices which are annual radionuclide intake (D_a) in Bq/yr, annual effective dose (D_{eff}) in Sv/yr, total annual effective dose (D^{total}) in Sv/yr and excess lifetime cancer risk (ELCR) [10] were evaluated and presented for the studied area by using the appropriate medium values of specific activities of ^{40}K, ^{226}Ra and ^{232}Th which were analyzed, determined and chosen from section 3.1. In addition, the annual consumption of rice for Thai population in 2020 (83 kg/yr) [11], the dose conversion factor for ^{40}K, ^{226}Ra and ^{232}Th as shown in ICRP 1996 [12], the lifetime span of Thai population in year 2019 (75 years) [13] and the mortality risk coefficient for ^{40}K, ^{226}Ra and ^{232}Th [14] were also used to evaluate all results in this study. Moreover, the comparison between the results with some research data in Thailand and global recommended values were shown in subsection 3.2

3. Results and discussions

3.1. Frequency distribution of specific activities and statistic values

The frequency distribution of specific activities of ^{40}K, ^{226}Ra, ^{232}Th and ^{137}Cs in 30 rice samples collected from Songkhla province were studied, analyzed and presented in the following figures 1– 4 by using the SPSS computer program. In addition, all statistic values which were calculated from this study, were also presented in table 1.

3.2. Radiological hazard indices values and comparison

From section 3.1, it was found that the frequency distribution of specific activities of ^{40}K, ^{226}Ra, ^{232}Th and ^{137}Cs in 30 rice samples collected from the studied area, were asymmetrical distribution with the skewness of 1.29, 1.43, 2.32 and 0.82, respectively. For this reason, the median values of ^{40}K, ^{226}Ra, ^{232}Th and ^{137}Cs which were 620.04 ± 44.30, 3.73 ± 0.54, 2.44 ± 0.54, and 1.94 ± 0.37 Bq/kg for this present study, were the appropriate medium value and should be chosen for calculation four radiological hazard indices and the ELCR value in this study. Moreover, some researches data in Thailand, foreign countries, Office of Atoms for Peace (OAP) annual report data and the recommended values informed by United Nations Scientific Committee on the Effects of Atomic Radiation (UNSCEAR) and IAEA were chosen to compare with the studied results and their average values as presented in table 2.

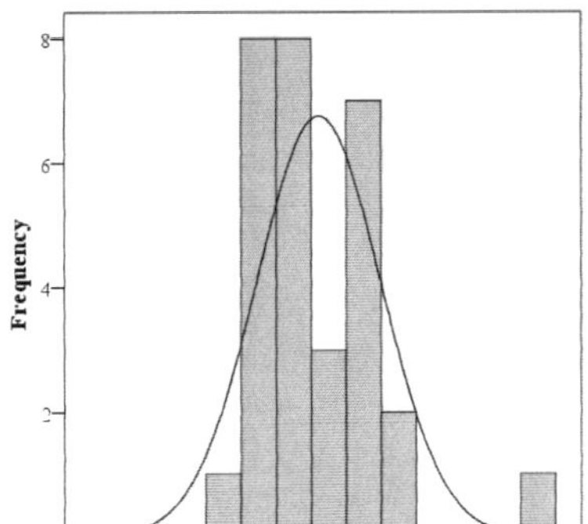

Figure 1. Frequency distribution of Specific activities of ^{40}K in 30 rice samples of Songkhla province

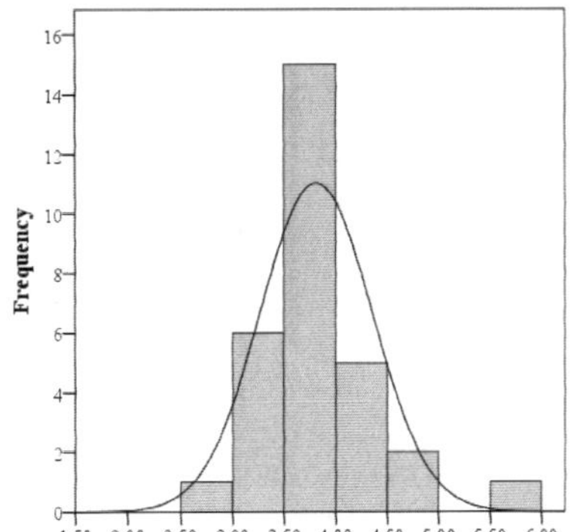

Figure 2. Frequency distribution of Specific activities of ^{226}Ra in 30 rice samples of Songkhla province

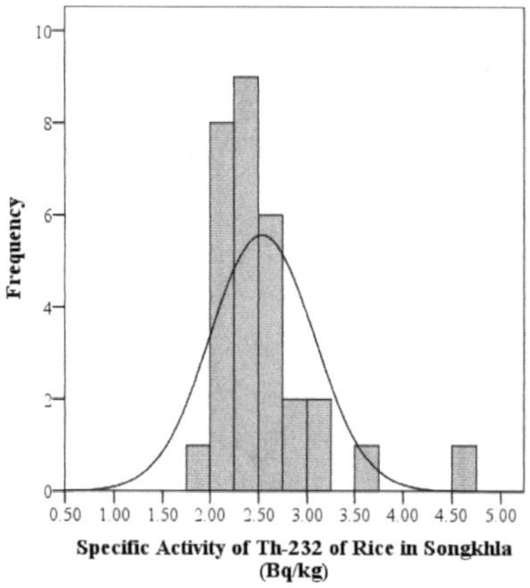

Figure 3. Frequency distribution of Specific activities of ^{232}Th in 30 rice samples of Songkhla province

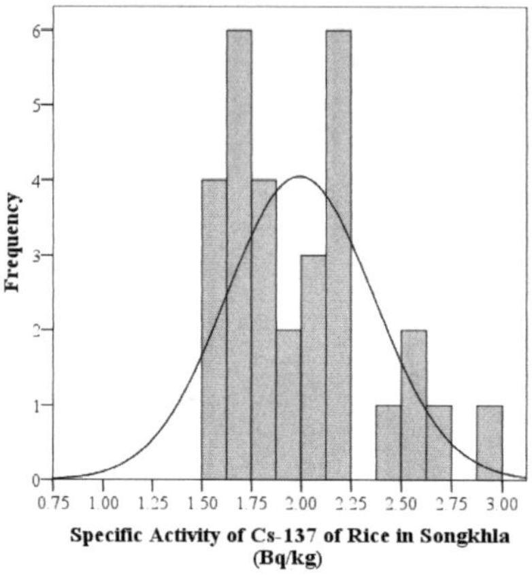

Figure 4. Frequency distribution of Specific activities of ^{137}Cs in 30 rice samples of Songkhla province

Table 1. Statistic values of the frequency distribution of specific activities of ^{40}K, ^{226}Ra, ^{232}Th and ^{137}Cs in 30 rice samples collected from Songkhla province, Thailand.

Statistic values	Analyzed values			
	^{40}K	^{226}Ra	^{232}Th	^{137}Cs
Mean (Bq/kg)	630.34	3.81	2.54	1.99
Median (Bq/kg)	**620.04**	**3.73**	**2.44**	**1.94**
Mode (Bq/kg)	568.95	3.36	2.09	1.55
Std deviation	44.30	0.54	0.54	0.37
Skewness	1.29	1.43	2.32	0.82
Kurtosis	2.75	3.83	7.01	-0.02
Minimum value (Bq/kg)	568.95	2.90	1.97	1.55
Maximum value (Bq/kg)	777.62	5.69	4.60	2.91

Table 2. Related radiological hazard indices and ELCR values for the present investigation and compare to some research studies in Thailand, foreign counties, UNSCEAR and IAEA.

Literatures	D_a (kBq/yr)	D_{eff} (Sv/yr) × 10^{-5}			D^{total} (Sv/yr) × 10^{-5}	ELCR × 10^{-3}
		D_{eff} (^{40}K)	D_{eff}(^{226}Ra)	D_{eff}(^{232}Th)		
Rice samples in Nakhon Nayok [15]	355.70 ± 3.44	218.40 ± 2.13	66.37	24.72	309.49 ± 2.13	17.46 ± 0.15
Sungyod rice in Phatthalung [16]	97.52 ± 14.19	59.81 ± 8.59	18.41 ± 5.60	8.95 ± 3.17	87.17 ± 17.36	4.80 ± 0.78
Jasmin rice in Songkhla [17]	37.82 ± 12.87	20.38 ± 6.60	13.11 ± 4.51	102.99 ± 47.53	136.48 ± 58.64	2.61 ± 0.97
Hom Pratum rice in Songkhla [17]	22.68 ± 13.17	9.48 ± 6.00	13.27 ± 6.04	158.89 ± 75.23	181.64 ± 87.28	2.28 ± 1.18
Organic rice in Phatthalung [18]	50.53 ± 2.69	30.81 ± 1.61	14.87 ± 2.63	7.03 ± 0.00	52.71 ± 4.23	2.63 ± 0.18
Rice in Penang Malasia [19]	6.19 ± 1.73	3.71 ± 1.03	3.30 ± 1.28	1.97 ± 0.59	8.98 ± 2.90	0.36 ± 0.11
Canned rice in Iraq [20]	7.18 ± 3.28	4.00 ± 1.98	6.48 ± 0.95	11.56 ± 1.11	22.04 ± 4.04	0.54 ± 0.17
Amber rice samples in Iraq [21]	2.11	0.87	13.80	5.06	19.73	0.46
Rice samples in Egypt [10]	3.10	1.85	1.39	1.47	4.72	0.18
Rice samples in Saudi Arabia [22]	11.50	7.10	0.93	0.38	8.41	0.53
Rice samples in Italy [23]	10.30	6.14	6.74	5.35	18.22	0.65
Rice samples in India [24]	13.13	6.22	7.13	65.48	78.83	1.15
Rice samples in Ghana [25]	9.41	5.37	10.97	8.27	24.61	0.73
Rice in Songkhla*	**51.98 ± 3.77**	**31.91 ± 2.28**	**8.67 ± 1.25**	**4.66 ± 1.03**	**45.23 ± 4.57**	**2.53 ± 0.20**
UNSCEAR [26-28]	32.29	15.95	155.71	22.91	194.57	5.31
IAEA [29]	-	-	-	-	100	-

*Present study

From table 2, the D_a values of the present study was 51.98 ± 3.77 kBq/yr which was lower than the research data in rice samples from Nahkon Nayok and Sungyod rice from Phatthalung provinces but higher than in Jasmine and Hom Pratum rice samples from Songkhla and Organic rice from Phatthalung provinces, all of international research data and the recommended value (32.29 kBq/yr) as reported by UNSCEAR. The D^{total} value $(45.23 \pm 4.57) \times 10^{-5}$ Sv/yr) from this study which is comprised of D_{eff} (^{40}K), D_{eff}(^{226}Ra) and D_{eff}(^{232}Th), was lower than 194.57×10^{-5} Sv/yr which is reported by UNSCEAR, all of research data in Thailand, but higher than all research data of foreign countries except in India. Furthermore, the D^{total} value was approximately 0.55 times lower than the IAEA dose constraints for public exposure in planned exposure situations and reference level for public exposure in specific existing exposure situations, e.g. exposure due to radionuclides in commodities such as food, drinking water or construction materials (< 1 mSv/yr or 100×10^{-5} Sv/yr). In addition, the ELCR value was also calculated and equal to $(2.53 \pm 0.20) \times 10^{-3}$ which lower than all research data in the Thailand, India and UNSCEAR recommended values but higher than all of international research data. Moreover, we can see that the specific activity of ^{40}K in this investigation which was equal to 620.04 ± 44.30 Bq/kg, should be the main factor of the high value of evaluated related radiological hazard indices and ELCR value for this present study. These high value of the concentration of ^{40}K might be caused from the regular using fertilizer of some group farmers in Thailand with is corresponded to the research in Sri lanka [30]. The results from this study should be repeated and increase more rice samples in different region of Thailand kingdom. Consequently, the concentration level of natural and anthropogenic radionuclides in rice samples for every kind of rice which are produced in Thailand should be randomly measured and monitored in every year of production. Moreover, the results of this study could be considered and used to be the baseline reference of background radiations for the daily food of Thai people and for green food and environment in the future.

4. Conclusions

The evaluated median values of specific activity of natural (^{40}K, ^{226}Ra and ^{232}Th) and anthropogenic (^{137}Cs) radionuclides in 30 rice samples collected randomly from Songkhla province were 620.04 ± 44.30, 3.73 ± 0.54, 2.44 ± 0.54 and 1.94 ± 0.37 Bq/kg, respectively. The median values of specific activity of ^{40}K, ^{226}Ra and ^{232}Th were chosen to assess three related radiological hazard indices (D_a, D_{eff}, and D^{total}) and the ELCR value. The results obtained in this study fall within the range of values reported in similar studies conducted nationwide and worldwide. The D_a value and the D_{eff} (^{40}K) was higher than the recommended values 32.29 kBq/yr and 15.95×10^{-5} Sv/yr as reported by UNSCEAR respectively. The D_{eff} (^{226}Ra) and D_{eff} (^{232}Th) values was lower than the recommended values 155.71 and 22.91×10^{-5} Sv/yr as presented by UNSCEAR. Furthermore, the D^{total} value which evaluated from all three of D_{eff} values, was lower than recommended values 194.57×10^{-5} Sv/yr as announced by UNSCEAR and lower than the IAEA dose constraints and reference level for public exposure (< 1 mSv/yr). In addition, the ELCR values was found below the recommended values 5.31×10^{-3} as announced by UNSCEAR. It can be seen that Thai people and consumers of Thai rice will receive the effective annual dose due to some radioactive substances present in Thai rice within the safe range. However, the results of this study show directly the effect of the specific activity of ^{40}K to the ELCR value. The specific activities of ^{40}K might be accumulated from the original paddy soil, water and fertilizers used by groups of rice farmers then transfer to the stem, leave and seed of rice plant. According to the present study results, we should concern to monitor, measure and assess the concentration of natural radionuclides especially ^{40}K in all kinds of rice which usually are the daily food for Thai people in every harvesting year. Hence, this data may contribute a general background level for the rice plant and may also support as a guideline for future measurement and assessment of possible radiological risks to Thai people health. Therefore, we should to study, measure and assess the specific activities of natural and anthropogenic radionuclides in all kinds of rice samples collected from cultivated area around the Thailand Kingdom and also study the transfer factor of those radionuclides from their paddy soil to a rice plant in the next studies and evaluation.

Acknowledgements
This study was carried out under partially sponsored by Department of Basic Science and Mathematics, Faculty of Science, Thaksin University, Songkhla Campus. The authors would like to express our deep gratitude to Thailand Institute of Nuclear Technology (Public Organization) (TINT) for providing the various facilities needed to use and accomplish this work. A special thanks goes to our energetic undergraduate and graduate students for their keen help, patience and constant encouragement in samples collection, preparations, measurements and evaluation of some parts of these data.

References

[1] Chibowski S 2000 *Pol. J. Environ. Stud.* **9**(4) 249–53
[2] Shanthi G, Maniyan C G, Allan Gnana Raj G and Thampi Thanka Kumaran J 2009 *Curr. Sci.* **97**(9) 1331–5
[3] Abojassim A A, Hady H N and Mohammed Z B 2016 *J. Bioen. Food Sci.* **3**(3) 113–23
[4] ICRP 1999 *Protection of the Public in Situations of Prolonged Radiation Exposure* (ICRP Publication 82) (New York: USEPA)
[5] Mikami S *et al.* 2015 *J. Environ. Radioact.* **139** 320–43
[6] Akata N, Hasegawa H, Kawabata H, Kakiuchi H, Chikuchi Y, Shima N, Suzuki T and Hisamatsu S 2015 *J. Radioanal. Nucl. Chem.* **303** 1217–22
[7] Niimura N, Kikuchi K, Tuyen N D, Komatsuzaki M and Motohashi Y 2015 *J. Environ. Radioact.* **139** 234–9
[8] Jayasinghe C, Molligoda V, Attanayaka T and Waduge V 2019 *Enviro. Geochem. Health* **41** 1123–9
[9] IAEA 1989 *Measurement of Radionuclides in Food and the Environment: A Guidebook* (IAEA Technical Reports Series No. 295) (Vienna: IAEA)
[10] Alrefae T, Nageswaran T N, Alshemaly T and Demir N 2018 *Kuwait J. Sci.* **45**(2) 42–52
[11] Department of Internal Tread, Ministry of Commerce 2020 *Department of Internal Trade Revealing that Thai People Consume Rice Down to 83 kg per Year and Move forward to Campaign for Thais to Consume Rice* (Marketing News) (Bangkok: Ministry of Commerce)
[12] ICRP 1996 *Age-dependent Doses to Members of the Public from Intake of Radionuclides: Parts 5 Complilation of Ingestion and Inhalation Dose Coefficients* (ICRP Publication 72) (New York: ICRP)
[13] Thailand Development Research Institute (TDRI) 2019 What will change when Thailand became a longevity society? *TDRI Rep.* **152** 1–8 (In Thai)
[14] USEPA 1999 *Cancer Risk Coeffcients for Environmental Exposure to Radionuclides* (Federal Guidance Report No. 13) (Wasington, DC: USEPA)
[15] Office of Atoms for Peace (OAP) 1991–2003 *OAP 1991–2003 Annual Reports.* (Bangkok: Ministry of Science and Technology)
[16] Sungthong U 2009 *Measurement of Specific Activities of Natural (^{40}K, ^{226}Ra and ^{232}Th) Radionuclides in Sungyod Rice Samples Collected from Phatthalung Province, Thailand* (Undergraduate Science Project) (Phatthalung: Thaksin University)
[17] Boonkrongcheep R, Maeha A and Kessaratikoon P 2015 *Thaksin Univ. J.* **18**(3) 241–7
[18] Chauychoo R 2018 *Measurement and Analysis of Specific Activities of Natural (^{40}K, ^{226}Ra and ^{232}Th) and Anthropogenic Radionuclides in Organic Rice Collected from Don Pradu Sub-district in Pak Phrayun District in Phatthalung Province, Thailand* (Undergraduate Science Project) (Songkhla: Thaksin University)
[19] Alsaffar M S, Jaafar M S, Kabir N A and Ahmad N 2015 *J. Radiat. Res. Appl. Sci.* **8**(3) 300–10
[20] Abass A H and Muttaleb M K 2018 *Plant Arch.* **18**(2) 1847–50

[21] Younis A S and Tawfiq N F 2019 *E3S Web Conf.* **122** 05004
[22] Al-Hassan A A, Abdel-Salam A M and El-Taher A 2014 *Life Sci. J.* **11**(11) 829–36
[23] Desideri D, Meli M A, Roselli C, Forini N, Rongoni A and Feduzi L 2014 *J. Radioanal. Nucl. Chem.* **299**(3) 1461–67
[24] Al-Zahrani J H 2016 *Int. J. Curr. Res.* **8**(06) 32892–97
[25] Awudu A, Faanu A, Darko E, Emi-Reynolds G, Adukpo O, Kpeglo D and Otoo F, Lawluvi H, Kpodzro R and Ali I 2012 *J. Radioanal. Nucl. Chem.* **291**(3) 635–41
[26] UNSCEAR 1988 *Source, Effects and Risks of Ionizing Radiation* (UNSCEAR 1988 Report to the General Assembly, with Scientific Annexes) (New York: UNSCEAR)
[27] UNSCEAR 1993 *Sources and Effects of Ionizing Radiation* (UNSCEAR 1993 Report to the General Assembly, with Scientific Annexes) (New York: UNSCEAR)
[28] UNSCEAR 2000 *Sources and Effects of Ionizing Radiation* (UNSCEAR 2000 Report to the General Assembly, with Scientific Annexes) (New York : UNSCEAR)
[29] IAEA 2018 *Radiation Protection of the Public and the Environment General Safety Guide Jointly Sponsored by The International Atomic Energy Agency and United Nations Environment Programme* (IAEA Safety Stadard Series No GSG-8) (Vienna: IAEA)
[30] Chandrajith R, Seneviratna S, Wickramaarachchi K, Attanayake T, Aturaliya T N C and Dissanayake C B 2010 *Environ. Earth Sci.* **60** 193–201

Effect of electric potential across electrodes for trapping joss smoke particles in DBD plasma atmospheric pressure

P Tangjitsomboon[1,4*], D Ngamrungroj[2,4] and R Mongkolnavin[3,4]

[1] Department of General Education, Faculty of Science and Health Technology, Navamindradhiraj University, Bangkok, 10300, Thailand
[2] Department of Social and Applied science, College of Industrial Technology, King Mongkut's of University Technology North, Bangkok, 10820, Thailand
[3] Department of Physics, Faculty of Science, Chulalongkorn University, Bangkok 10330, Thailand
[4] Plasma Technology and Nuclear Fusion Research Unit, Chulalongkorn University, Bangkok 10330, Thailand

*E-mail: prajya.t@nmu.ac.th

Abstract. Particulate matters (PM) in air pollution have been known to be the cause of respiratory diseases. Many researchers have investigated methods of trapping the particulate matter. In this work, the trapping of smoke particles generated from a joss stick by using a dielectric barrier discharge (DBD) system operated under the atmospheric pressure condition was investigated. DBD system consists of an inner electrode which is made of aluminum wire filaments that are placed inside the acrylic cylindrical tube, and the outer electrode is made of metallic wrap around the tube. The electrodes were connected to a 50 Hz high voltage AC source which was adjusted to 0 V, 5kV, 7kV, and 10kV. A ventilating fan was used for draining the smoke particle from the joss stick through the inner electrode with an airflow velocity of 2.68 m/s. The effect of electric field and plasma trapping the smoke particles was investigated. Results from the experiment were further compared with a study by simulation. It was found that the smoke particle density measured by applying an electric potential difference of 0 V and 5 kV was similar; both conditions showed the highest smoke density values. On the other hand, when the electric potential difference was adjusted to 7 kV and 10kV, it was found that the smoke particles density decreased by 90%. The experiment also illustrated when the electric potential difference was increased high enough such that plasma was produced at 7 kV and 10 kV, the smoke particle density released from the tube was similar. Nevertheless, when comparing the mass of particles collected from the inner electrode with the plasma condition, it was found that the mass collected increased more than the operating condition with an electric potential difference of 0 kV and 5 kV without plasma.

1. Introduction

One of many causes of respiratory diseases has come from particulate matter (PM) in air pollution [1], and it is a major concern in Bangkok and many other cities [1]. Combustion of fuel from vehicles can generate PM are the cause of health risk, therefore; investigation on means to reduce these particles has been gaining interest.

There are many ways to reduce the particles in the air one of which is trapping. The particles can be trapped by using nano-fiber filters [2-6] or using an electrical discharge system [7-9]. In this research, a type of electrical discharge system based on dielectric barrier discharge (DBD) is investigated. For example, removal of PM in a hen house [7] using a dry filter and electrostatic precipitator (ESP) trap were compared. It was found that the efficiency of the dry filter in trapping PM 10 was 40.1%, but its efficiency was not significant for trapping PM 2.5. In comparison, the trapping efficiency of ESP was 57% for PM 10 and it was 45.3% for PM 2.5. Huang *et al.* [8] also studied the trapping of PM 2.5 using the DBD system. They showed that 53% of PM 2.5 was trapped by the DBD system and both their experimental results and their simulation results were compared and are in good agreement. Jaworek *et al.* [9] studied the efficiency of a two-stage electrostatic plate to trap PM 2.5. The first plate was used to discharge the gas that is flowing through the plate. The second plate electric field was applied to collect the charged particles. The efficiency of PM 2.5 trapping using DC and AC power supply was compared. The results showed that the number of the particle collected by using AC was more than 90% and from DC was more than 95%.

From the reviews, an electrics discharge system can efficiently trap the particulate matter. In this investigation, we investigate also the effect of plasma from a DBD system that may be an additional cause of PM removal.

2. Simulation and experiment

The dielectric barrier discharge (DBD) system for the experiment consists of a transparent acrylic cylindrical tube shown in figure 1. The outer of the tube is covered by an outer metallic cylindrical electrode. The aluminum filament is used as an inner electrode and is placed inside the acrylic tube. The length of the acrylic cylindrical tube is 10 cm, the diameter of the cylindrical tube is 3 cm and both the outer and inner electrode is 6 cm long. A 50 Hz ac power supply is used in this work and is connected to the electrodes. The operation is at atmospheric pressure condition.

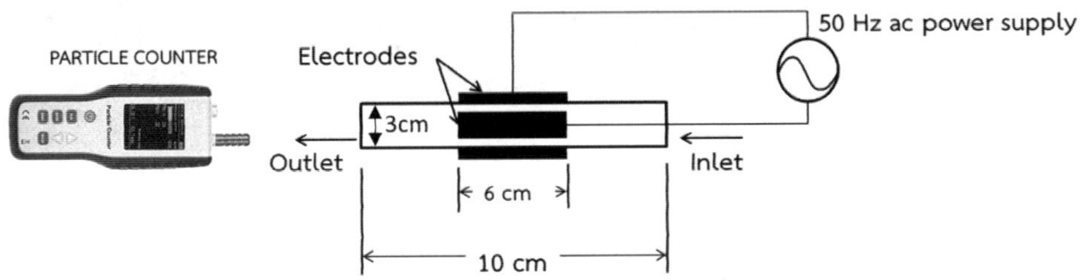

Figure 1. Diagram showing experiment setup of DBD system.

2.1. Simulation of plasma and smoke particle's motion

In this research, plasma characteristics and motion of smoke particles are simulated by using a simulation program (COMSOL) [10]. The dynamics of plasma can be explained by using both transport theory and electrostatics theory. A laminar flow model was conducted to replicate the airflow through the inner electrode placed inside the acrylic tube. Particulate matters were included in the airflow, and Newton's law of motion was used to estimate the movement of the particles.

The electric field between the electrodes was calculated by using an electrostatic theory that this electric field affects the particles in plasma. The equation is shown as follows;

$$\vec{\nabla}\cdot\ \varepsilon\vec{E}\ = \rho_f \tag{1}$$

where $\vec{E}$ is an electric field in medium (N/C), ρ_f is the density of free charge in the space (C/m^3), ε is the permittivity of the medium (C^2/N·m^2).

The characteristics of plasma such as the number density of electrons and the temperature of plasma were estimated by using transport theory. The transport equations of the electron can be shown as;

$$\frac{\partial n_e}{\partial t} + \vec{\nabla}.\vec{\Gamma}_e = R_e - \vec{u}\cdot\vec{\nabla}n_e \tag{2}$$

$$\frac{\partial n_\varepsilon}{\partial t} + \vec{\nabla}.\vec{\Gamma}_{en} + \vec{E}.\vec{\Gamma}_e = S_{en} - \vec{u}\cdot\vec{\nabla}n_\varepsilon + Q \tag{3}$$

where n_e is the number density of electrons (m^{-3}), n_ε is electron energy density (J/m^3), R_e is an electron creation rate (m^{-3}·s^{-1}), S_{en} is an energy loss/gain from inelastic collision (W/m^3), $\vec{u}$ is a vector field velocity (m/s), Q is a total heat source (W/m^3), $\vec{\Gamma}_e$ is an electron flux (m^{-2}·s^{-1}), and $\vec{\Gamma}_{en}$ is an electron energy flux (W/m^2).

A laminar flow model was used to calculate dynamics parameters of the particles such as the velocity of particles, and pressure. The equation that was used to represent the model is shown as follow;

$$\rho(\vec{u}\cdot\vec{\nabla})\vec{u} = \vec{\nabla}\cdot(p\boldsymbol{I}+\boldsymbol{K}) + \vec{F} + \rho\vec{g} \tag{4}$$

where $\boldsymbol{K} = \eta(\vec{\nabla}\vec{u} + \vec{\nabla}\vec{u}^{\,T})$, η is dynamics viscosity of air (Pa·s) , ρ is the density of air (kg/m^3) , $\vec{u}$ is a vector field velocity (m/s), p is a pressure (Pa), $\vec{F}$ is an external force density (N/m^3), $\boldsymbol{I}$ is an identity matrix, and $\vec{g}$ is gravitation field (m/s^2).

For the simulation, the surface of the acrylic tube was set to be a dielectric surface. Both applied voltage and ground were set on the surface of the electrodes. The recombination process of ions and the lost electron on the wall due to the collision was included as the boundary condition of plasma. All boundary surfaces were set to be sticky walls, so the smoke particles were trapped on the wall. The 2D simulation was considered and the phenomenon was analyzed only the cross-section of the tube which is shown in figure 2. The simulation results are shown in figures 3, 4, and 5 respectively.

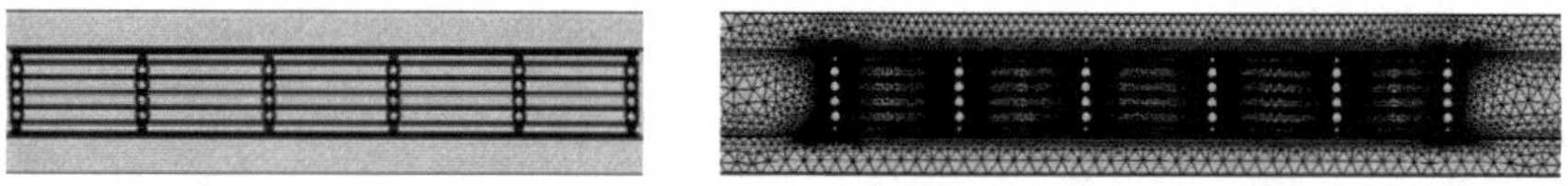

(a) (b)

Figure 2. (a) Design of the simulation (b) Mesh of system used for finite element method.

2.2. Experiment method

For the experiment, a particle counter, as shown in figure 1, was used to detect the particulate matters which were emitted from a joss stick. The meter could count the particles whose sizes are 0.3 μm, 0.5 μm, 1.0 μm, 2.5 μm, 5.0 μm, and 10.0 μm. The smoke particles flowed through the electrodes which were connected to the high voltage AC power supply. The operating voltage was varied to 5 kV, 7 kV, and 10 kV. The numbers of the smoke particles were investigated for each different operating condition.

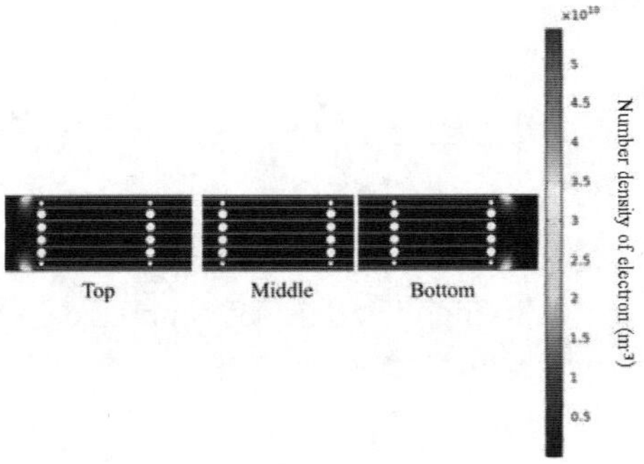

(a)

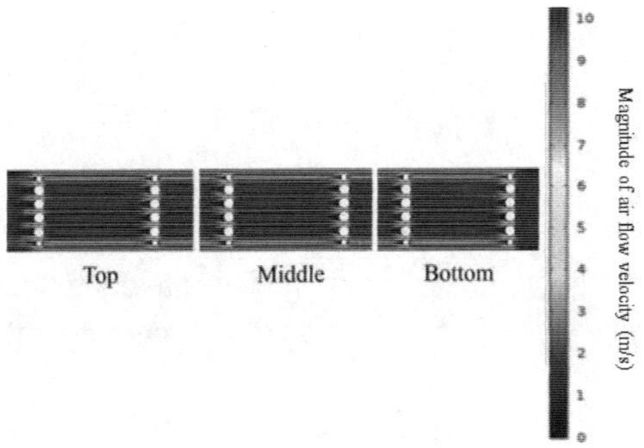

(b)

Figure 3. (a) Distribution of electron density of plasma from the simulation (b) Air flow velocity distribution obtained from the simulation.

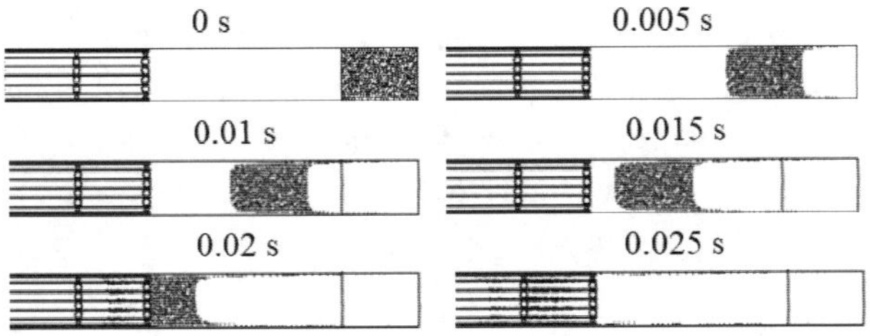

Figure 4. Picture showing simulated particle trajectories from the simulation.

3. Results and discussion

From the simulation, figure 5 shows a line plot of the relationship between the number density of electrons across the horizontal and vertical position of the tube. The number density varies with positions along the axis of the tube which has the maximum value of the number density at surfaces of the electrodes. The number density of electrons is homogeneous between the electrodes. In the vertical axis, the number density of electrons is the maximum at the center of the electrode. From this result, both electrons and ions suggested that they cause the change to the charge property of the particulate matters which are affected by the electric field between the electrodes. The particulate matter is trapped between the electrodes.

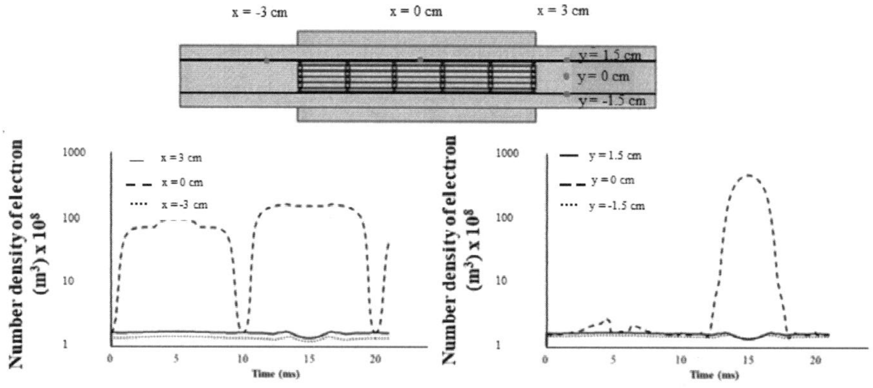

(a) (b)

Figure 5. (a) Number density of electron along horizontal axis varying with time (b) Number density of electron along vertical axis varying with time.

The relationship between the maximum number density of electrons and the operating voltage is plotted and shown in figure 6. The maximum number density also increases due to the increase of the operating voltage. The number of particulate matters detected under each operating voltage condition is compared to the number of the particles under non-operating voltage conditions. The percentages of the particles from various operating voltages are shown in table 1. From the experimental results, the percentage of detected particles decreases when the operating voltage is increased. Both particles with 1.0 μm and 2.5 μm in size were trapped the most in the system. However, most particles escaped when the operating voltage is the lowest at 5 kV. It is interesting to note that the plasma was not generated at an operating voltage of 5 kV. Since the particulate matter are not charged particles therefore it is not surprising that the electric field alone did not have any effect on these particles. In figure 6, it can be seen that there is increasing in the number of electrons when the operating is increased.

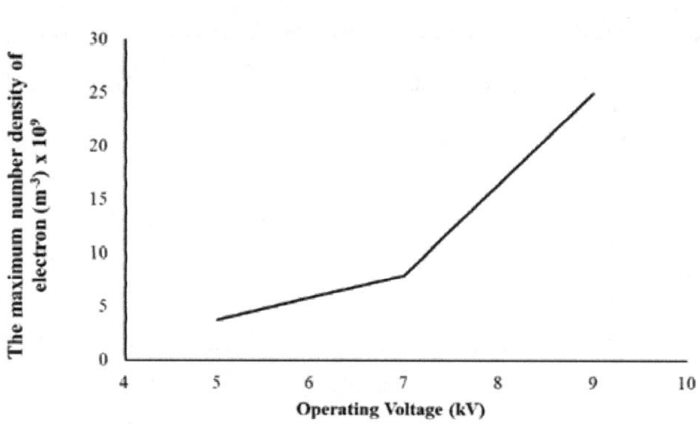

Figure 6. The relationship between the operating voltage and the maximum number density of electrons.

Table 1. The percentage of detected PM obtained by experiment under operating voltage.

Particle sizes (μm)	Percentage of detected particulate matters			Number of the counted particle under non-operating voltage
	Operating Voltage			
	5 kV	7 kV	10 kV	
0.3	96.05	19.76	13.22	774,194 ± 130,867
0.5	96.63	10.41	7.98	373,980 ± 29,523
1.0	100.00	6.99	6.27	67,814 ± 18,295
2.5	89.21	6.07	6.16	6,625 ± 4,071
5.0	93.38	10.55	9.03	443 ± 353
10.0	97.44	18.11	11.08	117 ± 96

From the simulation results and the experimental results, it can be deduced that particulate matter is affected by charged particles from the plasma. Electrons generated in plasma affect the charge property of the particulate matter and therefore they are moved by the electric force toward the electrodes. By increasing the operating voltage, the strength of the electric field is increased as well as raising the number of electrons. With both of these effects, the numbers of microparticles detected by the particle counter device are expected to reduce as shown by the result. The size of particles generated most from combustion is 0.3 μm. The smaller particles respond to the attractive force better than the bigger particle, so the particles that were trapped by the system are small in sizes which are 1.0 μm, and 2.5 μm.

In this work, we put our emphasis on the effect of plasma in trapping PM, in practice, the heating effect associate with plasma production must be considered. The effect of heat is not seen here as the operation is limited to a small time scale of 20 seconds. For longer operating time, materials with better heat tolerance that has similar dielectric property must be considered such as glass or quartz. Another important issue that must be considered for practical use is the cleaning process of the system which was not investigated here, but it is possible to design a cleaning cycle such that the PM is flushed out by clean air when DBD plasma is idle to a trapping water reservoir.

4. Conclusions

Dielectric barrier discharge plasma can trap smoke particles from a joss stick. The experimental results were confirmed by the result from the simulation. The dynamics of the plasma and the trajectory of particles were simulated by using the finite element method. The number of particles was measured by a particle counter. It was founded that;

(1) Increasing the number density of electrons and the strength of the electric field by increasing the operating voltage can cause the number of particulate matters to reduce. Both ionic charge and electron in the plasma could change the charge properties of the smoke particles. By rising the electric field, the electric force is increased which can attract charged particulate matters to the inner electrode.

(2) Sizes of particles that were the most affected were 1.0 μm, and 2.5 μm due to lower inertia.

Further work on system management, appropriate materials, and design are required for the practical implementation of this concept.

Acknowledgment
The authors would like to extend their great appreciation to the Plasma Technology and Nuclear Fusion Research Unit, Chulalongkorn University and College of industry technology, King Mongkut's University of Technology North Bangkok for initiating the plasma research for this work, and Navamindradhiraj University for their financial support.

Reference

[1] Pongpiachan S, Liu S, Huang R, Zhao Z, Palakun J, Kositanont C and Cao J 2017 *Arch. Environ. Contam. Toxicol.* **72** 364
[2] Yun K M, Suryamas A B, Iskandar F, Bao L, Niinuma H and Okuyama K 2010 *Sep. Purif. Technol.* **75** 340
[3] Zhang L, Luo J, Menkhaus T J, Varadaraju H, Sun Y and Fong H 2011 *J. Membr. Sci.* **369** 499
[4] Sambaer W, Zatloukal M and Kimmer D 2012 *Chem. Eng. Sci.* **82** 299
[5] Wang N, Raza A, Si Y, Yu J, Sun G and Ding B 2013 *J. Colloid Interface Sci.* **398** 240
[6] Xu B, Wu Y and Cui P 2014 *Particuology* **13** 60
[7] Winkel A, Mosquera J, Aarnink A J A, Koerkamp P W G and Ogink N W M 2015 *Biosyst. Eng.* **129** 212
[8] Huang X, Guo F, Chen J, Wang S, Hu Z, Wang L, Chen Z and Liu M 2016 *IEEE T. Plasma Sci.* **44** 1341
[9] Jaworek A, Marchewicz A, Sobczyk A, Krupa A and Czech T 2017 *J. Clean. Prod.* **164** 1645
[10] COMSOL 1998–2016 *COMSOL Multiphysics Reference Manual (Version COMSOL 5.2a)* (Sweden, Stockholm: COMSOL Inc.)

Journal of Physics: Conference Series **2145** (2022) 012021 doi:10.1088/1742-6596/2145/1/012021

Simulation of 3D vortex jets in plasma torch application

W Kongpiboolkid[1,3*] and R Mongkolnavin[2,3]

[1] Department of General Education, Faculty of Science and Health Technology, Navamindradhiraj University, Bangkok, 10300, Thailand
[2] Department of Physics, Faculty of Science, Chulalongkorn University, Bangkok 10330, Thailand
[3] Plasma Technology and Nuclear Fusion Research Unit, Chulalongkorn University, Bangkok 10330, Thailand

*E-mail: watcharapon@nmu.ac.th

Abstract. The gas tunnel type plasma jet is an effective heat source for thermal processing applications such as plasma spraying. The key concept of gas tunnel plasma is its torch configuration, especially the role of the vortex gas flow. This is very important for the stability and energy density of the plasma jet produced. This work studied the flow of gas vortex in 3 dimensions using a finite element simulation. The simulation is based on solving partial differential equations where the incompressible Navier-Stokes equation is used as a governing equation that describes the laminar flow. The geometry of the plasma torch investigated is based on the design by A. Kobayashi. Key parameters investigated were gas pressure, velocity and profile of the vortex. It can be shown that the simulation produced results that are better matched to the experimental result than the calculation done in previous work. The simulation can also show detailed pictures of the vortex and its properties within the plasma chamber. This study could be useful in the design optimization of the plasma torch in the future.

1. Introduction

The gas tunnel type plasma jet generator has been proposed as a heat source suitable for any kinds of applications such as coating, synthesizing of functional material, welding and gasification of waste disposal. The generator produces a plasma jet that has high speed and high energy density. Currently, it has also gained interest in the field of biomedicine and environment where a more precise and lower energy plasma is needed. To be able to achieve this specification, the use of vortex flow is introduced to a typical gas tunnel type plasma generator. Figure 1 shows a conceptual design of a part of a gas-tunnel generator proposed by Arata and Kobayashi [1] that the gas is flowing through small cylindrical tubes tangentially placed around the chamber to produce a vortex flow in the chamber. For plasma production purposes, a gas tunnel provides several advantages. Ionization of the gas is very easy, due to the low-pressure region in the gas tunnel. Moreover, a "gas wall" with a sharp pressure gradient in the radial direction effectively stabilizes the plasma beam.

The key concept of the gas tunnel plasma jet generator is its torch configuration and the role of the vortex gas flow. These factors affect the stability and energy density of the plasma jet produced [1-3]. In previous studies of vortex flow, it was found that the pressure in the middle of the vortex is lower than the outer edge of the vortex [4-7]. The research also investigated a model of the vortex flow based on a model of water draining down the sink. This is commonly known as the free vortex flow model as shown by figure 2. The model assumes that the gas is drawn from the circumference of the cylindrical chamber into a sink at the centre axis and a radial velocity is produced. The equation of motion is expressed as follow;

$$\frac{d\rho V_r}{dt} = \frac{dp}{dr} - \frac{V_\theta^2}{r} \tag{1}$$

where p is the pressure, r is the radius, ρ is the density of the gas, V_θ is the velocity in the tangential direction, V_r is the velocity in the radial direction and t is time. By further applying the continuity equation of and the law of conservation of angular momentum to the first equation, the velocity and the pressure distribution can be solved analytically from equation (2) as follow;

$$\frac{(p/p_w)^2 - V_{w*}^2}{1 - V_{w*}^2} = \exp\{-\frac{V_{\theta w}^2}{RT}[(\frac{r_w}{r})^2 - 1]\} \tag{2}$$

where p_w is the pressure at the chamber wall and r_w radius of the chamber wall, V_{w*}($V_{w*} = V_{rw} / V_{\theta w}$)is the ratio of the two velocities at the wall, with $V_{\theta w}$ in the tangential direction and V_{rw} in the radial direction, R is the gas constant and T is the temperature.

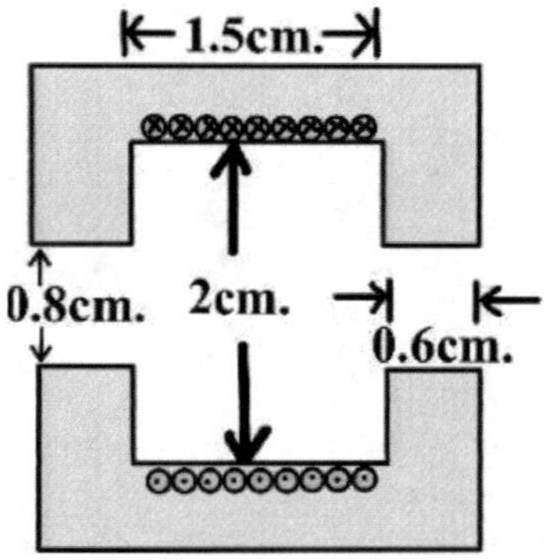

Figure 1. Conceptual diagram of a gas tunnel generator.

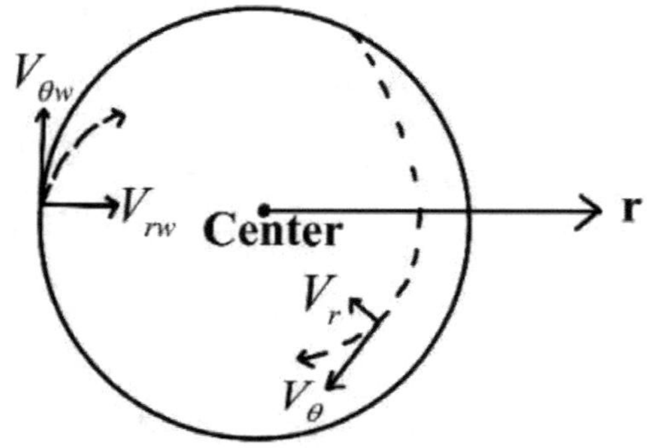

Figure 2. Vortex flow model.

However, a simple mechanical fluid flow theory cannot represent the whole picture of the vortex produced by a more complex system. The calculated tangential velocity and the calculated radial velocity do not match the experimental measurement. This study is, therefore, looking at simulating the vortex in 3 dimensions using a finite element method to look at a pressure profile and velocities that are a better match to the experimental measurement.

2. Simulation

The finite element method (FEM) is used extensively for numerically solving differential equations arising in any mathematical or engineering model. It is commonly used for analysing physical properties such as structural, heat transfer, fluid flow, mass transport, and electromagnetic potential [8-9]. In this research, the vortex flow was simulated using FEM software (COMSOL) [10]. The schematic of the vortex cavity investigated is shown in figure 3. The inlet gas nozzles are in the tangential direction on the cavity wall and evenly distributed every 90° apart. The cavity has an opening at both ends that allow the air to flow out freely to space with an ambient pressure of 1 atm. The software solves appropriate partial differential equations of discrete elements defined by a mesh of a 3D geometry as shown in figure 4.

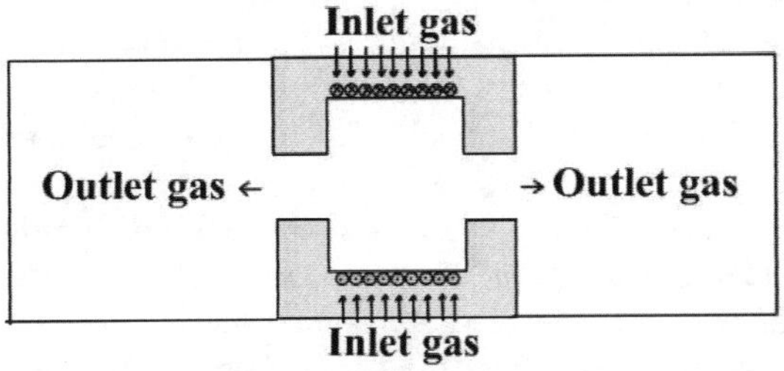

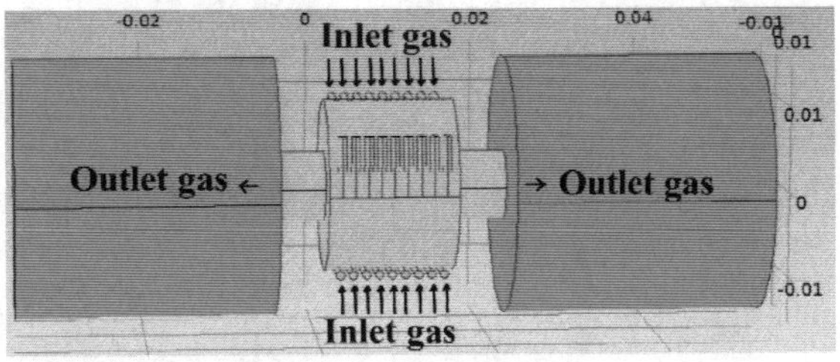

Figure 3. Schematic showing vortex nozzles of a part of the gas tunnel generator.

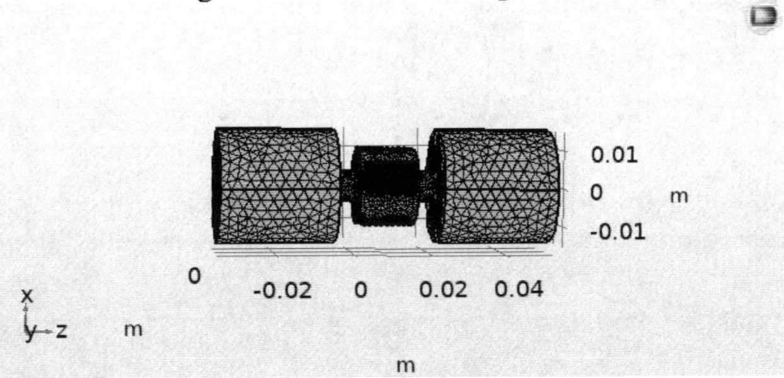

Figure 4. Schematic of a mesh used for FEM simulation.

The gas flow simulation was based on a laminar flow model where the flow begins at the inlet gas nozzles. Various gas flow rates were simulated. The compressible Navier-Stokes equation was used as the governing equation that describes the flow of air in the vortex cavity, as shown in equation (3) as well as the continuity equation as shown in equation (4).

$$\rho\frac{\partial u}{\partial t}+\rho(u\cdot\nabla)u=\nabla\cdot[-p\mathrm{I}+\mu(\nabla u)+(\nabla u)^{\mathrm{T}}]+F \quad (3)$$

$$\frac{\partial\rho}{\partial t}+\rho\nabla\cdot(u)=0 \quad (4)$$

where ρ is the density (kg/m^3), u is the velocity (m/s), p is pressure (Pa), I is an identity matrix, T is the transpose of a matrix, μ is the viscosity (kg/m.s) and F is the volume force vector (N/m^3).

3. Result and discussion

From the simulation, the relationship of relative tangential velocity and radial direction distance is shown in figure 5 (a). Even though the simulation result resembles the result obtained by the experiment done by Arata and Kobayashi [1], it does not totally agree with the analytical model proposed by them. Figure 5 (b) shows the relative radial velocity with radial direction distance. Again, a similar discrepancy between the simulation and calculation in previous research in the central region can be seen. However, both experimental results done by Arata and Kobayashi [1] and the simulation show good agreement where the radial velocity reduces towards the centre of the vortex. By simulating the vortex in 3D, the velocity in the axial direction could be deduced. Figure 5 (c) shows a variation of relative axial velocity of gas flow with the radial position. The result shows that the axial velocity has its direction out of the tunnel along the centre region. To complete the result, figure 5 (d) shows the comparison plot of relative pressure at different radial positions. The results show a similar trend between the simulation and the experimental results obtained by Arata and Kobayashi [1]. The lowest pressure can be found in the middle and on the axis of symmetry. However, the pressure found in the simulation is much higher than that of the experiment which can be explained possibly by the difference in the actual experimental setup used by Arata and Kobayashi [1]. Their experimental setup has gas diverter valves at the opening which allow the excess gas to exit freely.

The effect seen at the outer central region can be explained by a free vortex flow which is the model that assumes the fluid is drawn to a sink. This is the model that was assumed by the previous research done by Arata and Kobayashi. However, in reality, the actual vortex is produced by injecting air tangentially which is more consistent with the forced vortex flow model. Therefore, to accurately simulate the vortex that was produced in the experimental set-up, it is essential to consider both effects of the sink and the high-speed airflow. This is why the result from the FEM simulation has a better agreement with the experimental result than the theoretical model assumed before. The result also agrees with Rankine (1858) model explained by Azarpira and Zarrati [7] where he assumed that the vortex is composed of two parts of a forced vortex and a free vortex zone where the turning point defines the boundary of the two zones. In this case, the turning point is at 0.4 cm which corresponds to the radius of the opening of 0.4 cm that was in the simulation and the experiment.

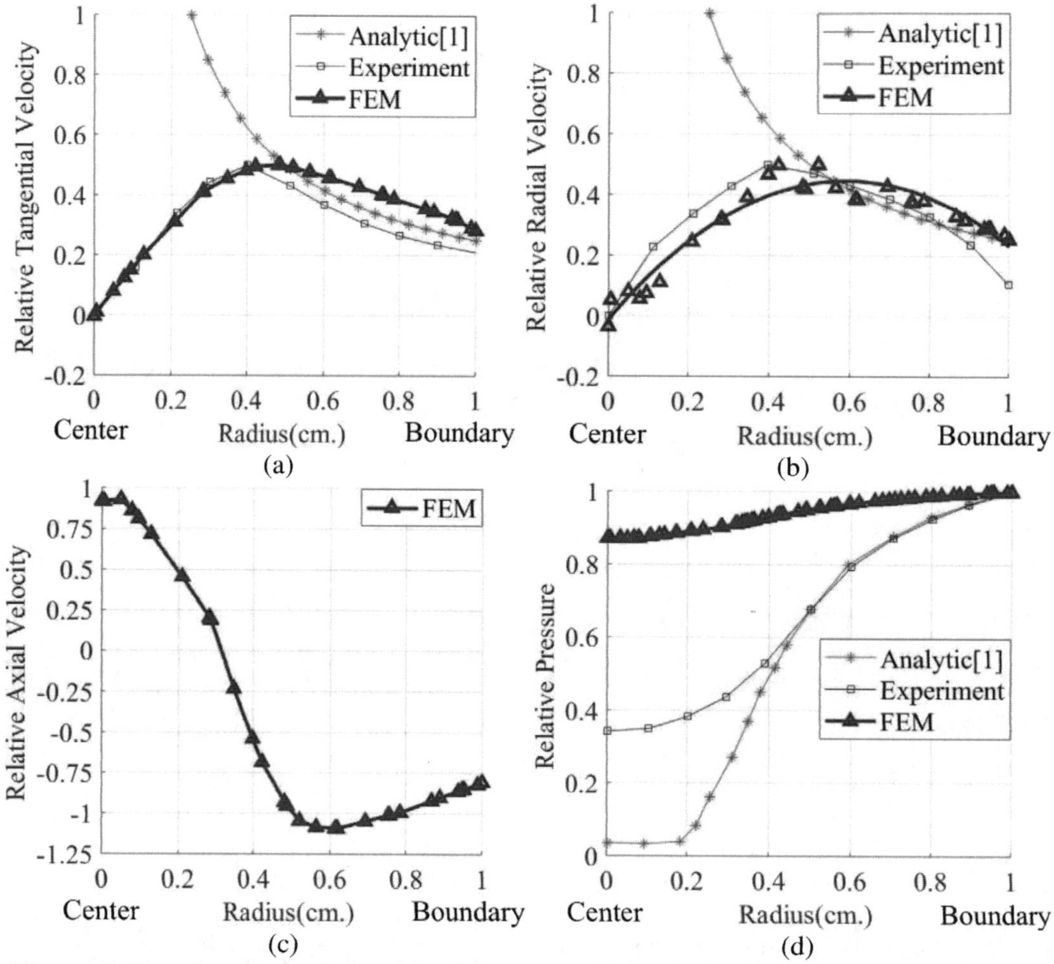

Figure 5. Plot showing variation (a) relative tangential velocity (b) relative radial velocity (c) relative axial velocity and (d) relative pressure with the radial position.

Additionally, the simulation results can also show a heatmap of the pressure profile of the vortex flow in the gas tunnel generator. Figure 6 shows the 3D profile of the velocity. It shows the gas flows through the gas tunnel generator and the direction in which the gas travels with vortex motion. The velocity inside the gas tunnel generator is higher in the cavity than that on the centre axis region.

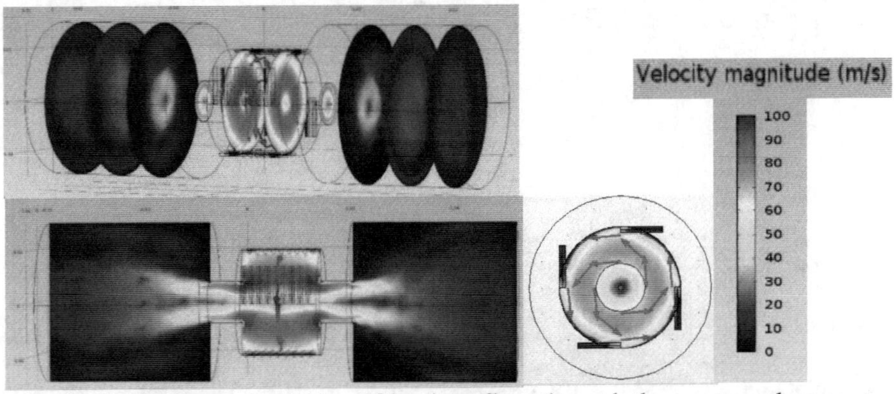

Figure 6. Heat map of velocity from simulation of laminar flow through the gas tunnel generator.

Furthering the investigation by varying the gas inlet flow rate (Q in litre/minute), the tangential velocity profile, as well as the pressure profile, can be shown in figure 7 and figure 8 respectively. It can be seen that the vortex tangential velocity increased as the gas inlet flow rate was increased. This is expected as the inlet flow rate Q relates to the initial pressure in equation (3) according to Poiseuille's Law. The pressure envelope due to the pressure gradient between the centre and the outer region inside the cavity became more distinct when the gas flow increased. This is due to the increasing difference between the pressure at the edge and the middle. As a result, it is possible to design and optimize the gas inlet flow rate and the dimensions of the plasma generator with the information obtained from the simulation. A suitable geometry and gas flow rate can be determined such that it produces the required low-pressure cavity for lower power input and a more stable plasma production.

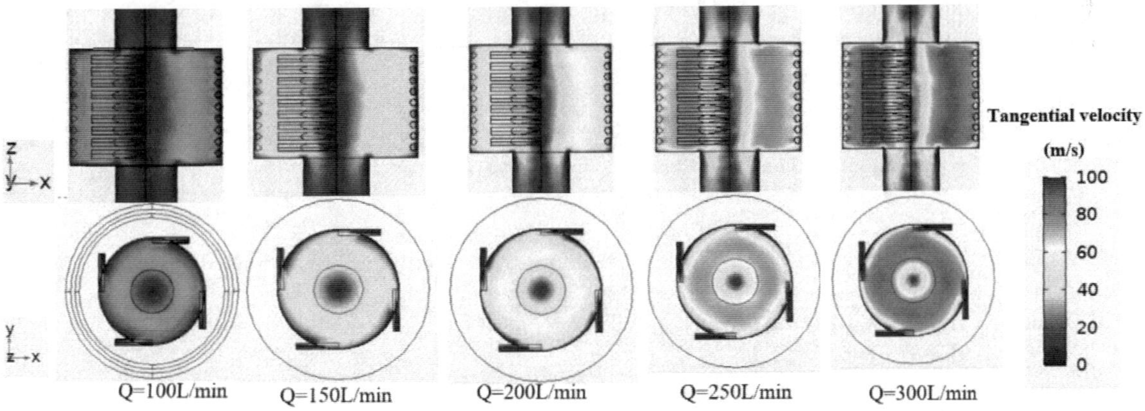

Figure 7. Tangential velocity simulation of laminar flow through a gas-tunnel generator.

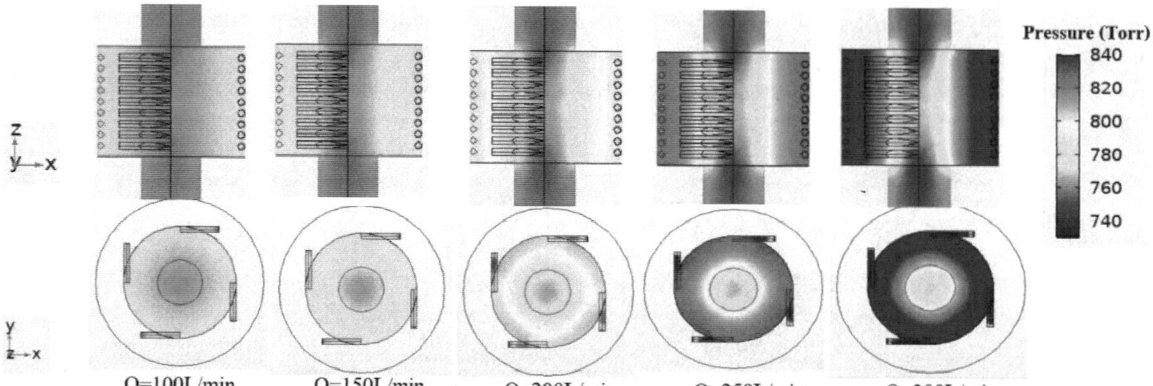

Figure 8. Pressure simulation of laminar flow through a gas-tunnel generator.

4. Conclusion

A vortex flow based on a gas tunnel type plasma jet generator designed by Arata and Kobayashi [1] has been simulated using FEM software. The compressible Navier-Stokes equation was used as the governing equation for FEM to describe the airflow in the vortex cavity. Various gas flow rates ranging from 100 L/m to 300 L/m were simulated. It was found that the velocity inside the gas tunnel generator was higher within the cavity than in the centre region. The result from the FEM simulation has a better agreement with the experimental result than the theoretical model assumed before. The discrepancy can be explained by the combined effect of the sink and the high-speed airflow that should be taken into account. With the good agreement of the result with the experiment, the simulation was extended to show a detailed picture of the vortex and its properties within the chamber. It was also found that the pressure envelope of the vortex inside the cavity became more distinct when the gas flow was increased. This study could be useful for designing and optimizing the gas tunnel type plasma torch in the future.

Acknowledgements

The authors would like to extend their great appreciation to Prof. Akira Kobayashi for his guidance and Navamindradhiraj University and Plasma Technology and Nuclear Fusion Research Unit, Chulalongkorn University for their full support of the work done in this research.

References

[1] Arata Y and Kobayashi A 1986 *J. Appl. Phys.* **59** 3038–44
[2] Arata Y, Kobayashi A, Habara Y and Jing S 1984 *Trans. JWRI* **15** 227–31
[3] Szałatkiewicz J 2014 *Pol. J. Environ. Stud.* **23** 277–81
[4] Burgers J M 1948 *A Mathematical Model Illustrating the Theory of Turbulence Advances in Applied Mechanics* vol **1** eds R Von Mises and T Von Karman (New York: Academic Press) pp 171–99
[5] Vatistas G H 1998 *J. Propuls. Power* **4** 462–9
[6] Vatistas G H 2006 *J. Aircr.* **5** 1577–9
[7] Azarpira M and Zarrati AR 2019 *Water Sci. Eng.* **12** 244–52
[8] Akter F and Deb U K 2017 *Am. J. Comput. Math.* **7** 58–69
[9] Bottani E, Rizzo R and Vignali G 2007 *Chem. Prod. Process Model.* **2** 58–69
[10] COMSOL 1998–2016 *COMSOL Multiphysics Reference Manual (Version COMSOL 5.2a)* (Sweden, Stockholm: COMSOL Inc.)

Estimation of impurity release from planar liquid surface in plasma

N Somboonkittichai

Department of Physics, Faculty of Science, Kasetsart University, Chatuchak, Bangkok 10900, Thailand

E-mail: fscinrso@ku.ac.th

Abstract. A liquid metallic surface exposed to a plasma interacts with ions and electrons numerously. The surface becomes charged. Subsequently, the electric field is formed via the balance of incoming ion and electron currents. The kinetic energies and the momentum of the incoming ions and electrons are transferred to the surface. This is enhanced by the self-generated electric field and gives rise to notable pressures and energy fluxes on the surface. By adopting several plasma and liquid parameters to the physical models, the time evolution of the surface temperature can be characterized with the net energy flux, the net pressure, and the net impurity outgoing flux. The study suggests that electron temperature and ion mass significantly govern, but the ion to electron temperature ratio does not, the trend of the surface temperature. Besides, it is found that at long elapsed time at which evaporation becomes strong, the energy and impurity fluxes reduce in conjunction with the rate of change of the surface temperature lessens.

1. Introduction

The implementation of liquid metal as a part of plasma facing surfaces in tokamaks has been so attractive because this improves the overall fusion operation in several aspects, for examples reducing wall recycling and plasma instability suppression [1]. It permits in-vessel impurities to be sank on, as a result, the liquid metal surface introduces the new possibility to flush those impurities away. However, it is unpreventable for plasma-surface interactions. The plasma heats the liquid surface up, and this leads to the metallic impurity release. Such impurities are counted in the inventory of the undesired impurities. Subsequently, fusion operations are ceased. For example, it has been reported that placing a liquid lithium (Li) surface too deep from the inner tokamak wall caused disruption [2].

The physical models describing the determination of the surface floating potential and the trends of temperature, pressures, and energy fluxes on and impurity fluxes from the liquid surface are reviewed in section 2. Section 3 discusses the trends of an electrostatic potential of the surface in various plasmas against their electron temperatures and ion to electron temperature ratios. The trends of the surface temperature, the net pressure, and the energy flux on and the net impurity flux from the surface versus time and plasma parameters are also illustrated with discussion.

2. Physical models

The surface charging by plasma particles, the determinations of temperature, pressures and energy fluxes on a liquid surface and the calculation of net impurity outgoing flux from the surface are reviewed in this section. Several assumptions were implemented for simplicity as follows. The liquid is in static equilibrium with uniform temperature distribution. No external electric and magnetic fields are considered. Its planar surface and its vapor are in thermal equilibrium. Soon after a vapor particle releases, it then moves away from the surface without collision due to the dilution of a plasma. Plasma particles are perfectly absorbed by the liquid. The bulk plasma is a heat reservoir, so it is negligibly affected by thermal radiation. The adsorption mechanism is not included.

2.1. *Surface charging*

At equilibrium, plasma particles thermally move in a bulk plasma and harmoniously form quasi-neutrality. Each plasma species can be described by Maxwell-Boltzmann distribution [3-4], $f^{q}_{M\text{-}B}(v_x,v_y,v_z) = n_q\left(\frac{m_q}{2\pi k_B T_q}\right)^{3/2}\exp\left(-\frac{m_q(v_x^2+v_y^2+v_z^2)}{2k_B T_q}\right)$, where (v_x, v_y, v_z) is a set of the velocity components, $v = \left(v_x^2+v_y^2+v_z^2\right)^{1/2}$ represents a resultant speed, and n_q, m_q and T_q is the number density, the single particle mass and the temperature of a plasma species (q), which represents both positive ions (i) and electrons (e). A surface exposed in the bulk plasma is very fast to be negatively biased by electrons. During the charging by electrons, more ions are attracted, and more electrons are repelled. Eventually, a new equilibrium of the plasma with the exposed surface is formed with the condition of zero net current at the surface, that is $\sum_q I_q = 0$, where $I_q = q_q\Gamma_q$, q_q represents an elementary charge with a sign related to the type of charge: $+q$ for a positive charge; and $-q$ for a negative charge and Γ_q is the particle flux of a plasma species (q) without reflection at a target surface. The electron flux on the surface is determined by the 1-D Maxwellian flux [4]. The ion flux on the surface is estimated by the multiplication of ion number density and ion sound speed [4]. Using the condition, the electrostatic potential, namely *floating potential* (ϕ_f), of a planar surface [4] can be determined as follows,

$$\phi_f = \frac{k_B T_e}{2q}\ln\left(2\pi\frac{m_e}{m_i}(1+\gamma\beta)\right), \tag{1}$$

where $\gamma = \frac{5}{3}$, $\beta = \frac{T_i}{T_e}$, the potential in the bulk plasma (ϕ) is assumed to be zero. The normalized form of the floating potential is $\Phi_f = \frac{q|\phi_f|}{k_B T_e}$.

2.2. *Surface temperature*

It was assumed that net incoming particles are perfectly absorbed in the liquid. In addition, their kinetic energies were completely lost on and nearby its surface. Meanwhile, the surface was quenched mainly by thermal radiation, evaporation, and physical sputtering. The temperature of the surroundings (T_r), apart from the plasma and the target surface, was assumed to be 300 K as a constant.

With regards to the heating flux by ions (Ξ_i) and electrons (Ξ_e), their expression are obtained by the integration of their kinetic energy with respect to their flux on the surface [4], and this results in

$$\Xi_i = \left(2\beta k_B T_e + q|\phi_f|\right)\Gamma_i = \left(\frac{1}{4}n_i\left(\frac{8\beta k_B T_e}{\pi m_i}\right)^{1/2}\right)\left(2\beta k_B T_e + q|\phi_f|\right), \tag{2}$$

$$\Xi_e = 2k_B T_e \Gamma_e = \frac{1}{2}n_e k_B T_e\left(\frac{8k_B T_e}{\pi m_e}\right)^{1/2}\exp(-\Phi_f), \tag{3}$$

In terms of the cooling mechanism, thermal radiation is described by Kirchhoff's and Stefan-Boltzmann laws [5] as follows,

$$\Xi_{rad} = \varepsilon\sigma\left(T^4 - T_r^4\right), \tag{4}$$

where ε and σ represent the emissivity of the surface and the Stefan-Boltzmann constant, respectively. The evaporation and the physical sputtering contribute the energy quenching flux (Ξ_{evap}+Ξ_{sp}), that is [4]

$$\Xi_{evap}+\Xi_{sp} = (2k_B T+L)(\Gamma_{evap}+\Gamma_{sp}) , \tag{5}$$

where L is the surface binding energy of a liquid particle (or the latent heat of evaporation per particle) ([6] for lithium (Li), [7] for aluminum (Al) and [8] for tin (Sn)), Γ_{evap} is the particle flux by evaporation and Γ_{sp} is the particle flux by physical sputtering. The liquid surface temperature (T) is determined by solving $\Xi_i+\Xi_e-\Xi_{rad}-\Xi_{evap}-\Xi_{sp} = \rho_{liq} H c_p \frac{\Delta T}{\Delta t}$, where ρ_{liq}, H and c_p is the liquid density, the liquid thickness and the specific heat capacity of the liquid at a constant pressure [7], and $\frac{\Delta}{\Delta t}$ is the average time rate of change.

2.3. Pressure on surface

A pressure is caused by momentum ($\boldsymbol{p}_q$) transferring to a surface. For each plasma species (q), the associated pressure (P_q) can be calculated by exploiting $P_q = \boldsymbol{p}_q \Gamma_q$, the general form of which [4-5] is $P_q = \int_{\text{all possible } v} \boldsymbol{p}_q \, v f^q_{M\text{-}B}(v) \, \mathrm{d}^3\boldsymbol{v}$. The integration corresponds to the attraction of ions and the repulsion of electrons influenced by the electric field of the planar surface. All ions but only electrons, which occupy kinetic energies more than $q|\phi_f|$, bombard on the surface. Therefore, the pressures of ions (P_i) and electrons (P_e) are written as

$$P_i = \frac{1}{2} n_i \beta k_B T_e \left(\exp\left(\frac{\mathcal{q}_f}{\beta}\right) \operatorname{erfc}\left(\left(\frac{\mathcal{q}_f}{\beta}\right)^{1/2}\right) + 2\left(\frac{\mathcal{q}_f}{\pi\beta}\right)^{1/2} \right), \tag{6}$$

$$P_e = \frac{1}{2} n_e k_B T_e \exp(-\mathcal{q}_f) , \tag{7}$$

where $n_i = n_e$ due to quasi-neutrality.

Once each surface particle gains an energy, equal to a binding energy, it releases from the surface as a vapor particle. Evaporation and physical sputtering are involved and macroscopically contribute to pressures (P_{evap} and P_{sp}, respectively) on the surface due to reverse momentum transfer [5]. From this,

$$P_{evap}+P_{sp} = m_{vap} \left(\frac{8k_B T}{\pi m_{vap}}\right)^{1/2} (\Gamma_{evap}+\Gamma_{sp}) , \tag{8}$$

where m_{vap} is the mass of a vapor particle.

On a charged surface, an electrostatic pressure is determined by $P_{es} = \frac{1}{2}\epsilon_0 |\boldsymbol{E}|^2$ [9], where $\boldsymbol{E}$ is an electric field originated from the surface charging by plasma. It is assumed that $|\boldsymbol{E}|$ is approximately decreased by the characteristic length, which is an electron Debye length $\left(\lambda_D = \left(\frac{\epsilon_0 k_B T_e}{n_e q^2}\right)^{1/2}\right)$ in a plasma [10], near the surface. Furthermore, figure 2.6 in [4] suggests that the electrostatic potential spatially and drastically decay from the surface (ϕ_f) to the bulk plasma (ϕ), normally $\phi = 0$ as a reference, by approximately $10\lambda_D$. From this,

$$P_{es} = \frac{1}{2}\epsilon_0 |\boldsymbol{E}|^2 = \frac{\epsilon_0 \phi_f^2}{2(10\lambda_D)^2} . \tag{9}$$

At equilibrium, the equation describing the net pressure, as a vapor pressure (P_{vap}), at a surface is written as

$$P_{vap} = P_i + P_e + P_{evap} + P_{sp} - P_{es} \,. \tag{10}$$

2.4. Net impurity outgoing flux

The impurity release is considered from the net outgoing flux contributed from evaporation and physical sputtering. The expression of the net impurity outgoing flux is $\Gamma_{net} = \Gamma_{evap} + \Gamma_{sp}$, where Γ_{evap} is determined by the Hertz-Knudsen-Langmuir equation, that is

$$\Gamma_{evap} = P_{vap}\left(2\pi m_{vap} k_B T\right)^{-1/2} , \tag{11}$$

[5], and the expression of Γ_{sp} is

$$\Gamma_{sp} = \Gamma_i Y(E_i) , \tag{12}$$

where $Y(E_i)$ is the physical sputtering yield, provided in [11] as follows $Y(E_i) = QS_n(E_r)\frac{\left(\frac{E_i}{E_{th}}-1\right)^{\mu}}{\lambda+\left(\frac{E_i}{E_{th}}-1\right)^{\mu}}$, where Q, λ and μ are the numerical factors of the curve fitting (tables 1, 2 and 7 in [11] used in this study), E_i is an incoming ion energy, assumed as a mean ion energy, that is $E_i = 2\beta k_B T_e + q|\phi_f|$, E_{th} is a threshold energy of physical sputtering at a target surface, $S_n(E_r)$ is a nuclear stopping power and E_r is a reduce energy.

3. Result and discussion

By quasi-neutrality, ϕ_f depends only on β, T_e and m_i of a plasma, as the main parameters, to characterize the plasma interactions on a surface. As a result, P_i, P_e, P_{es}, Ξ_i and Ξ_e are the parameters solely corresponding to the plasma. Figure 1 shows the trends of ϕ_f of each plasma species, Hydrogen (H), Deuterium (D), Helium (He) and Argon (Ar), versus T_e and β. It is clearly seen in figure 1 that the surface is more negatively biased with respect to the reference potentials of the bulk plasmas ($\phi = 0$), when their T_e and m_i are larger. Larger T_e causes more electrons reaching the surface, while larger ion mass prohibits more ions to reach the surface. Both results in the surface being relatively less positively biased. In contrast, larger β leads to being greater positively biased. Also, it represents larger ion speed. This means that larger ion flux reaches the surface. For $T_e < 1$ eV, $|\phi_f|$ is very small relative to the reference potential. By increasing T_e, ϕ_f can be negatively increased to several hundreds volts for $T_e \sim 10$ - 10^2 eV and to a few thousands volts for $T_e > 10^2$ eV approximately. This implies that within the characteristic length of the order of λ_D near the charged surface, the electric field is great in magnitude. The main trends of ϕ_f versus T_e and β are similar among different plasmas. Therefore, the D plasma is selected to be considered for its P_i, P_e, P_{es}, Ξ_i and Ξ_e as shown in figures 2 and 3, respectively. It is clearly seen in the figures that P_i, P_e, P_{es}, Ξ_i and Ξ_e are risen by an increase in T_e, as well as ϕ_f. To put it simply, ions and electrons with higher T_e transfer their momentum and kinetic energies greater to the target surface. This leads to larger P_i, P_e, Ξ_i and Ξ_e due to the larger kinetic energy and the stronger electric field of the surface, and larger P_{es} due to larger ϕ_f. However, considering figure 2 with figure 1, it turns up that larger P_{es} requires lower β. This is undeniable because larger β implies larger ion flux onto the surface, so at equilibrium, this relieves the negative bias, by fast electrons. Furthermore, P_i, P_e and P_{es} of various plasmas, assumed to be in steady state with a fixed number density ($n = 10^{19}$ m^{-3}), on the surface are comparable to the pressures of the neutral gases with the same order of number density if $T_e \sim 1$ - 10 eV, and are even greater if $T_e > 10^2$ eV approximately. Consequently, P_i, P_e and P_{es} play an important role on the plasma-surface interactions at higher T_e.

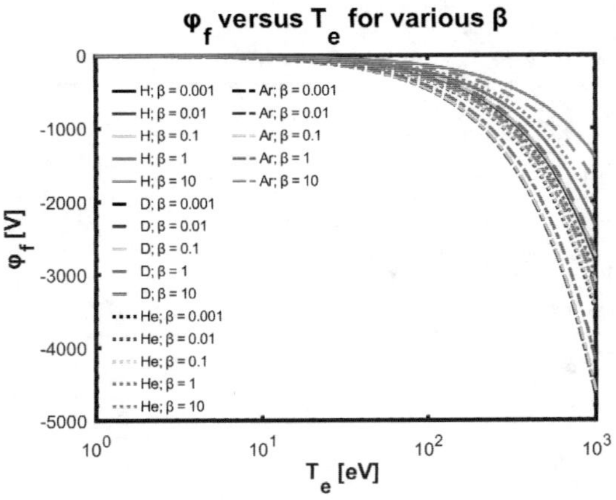

Figure 1. ϕ_f of H, D, He and Ar plasmas, varied with T_e and β.

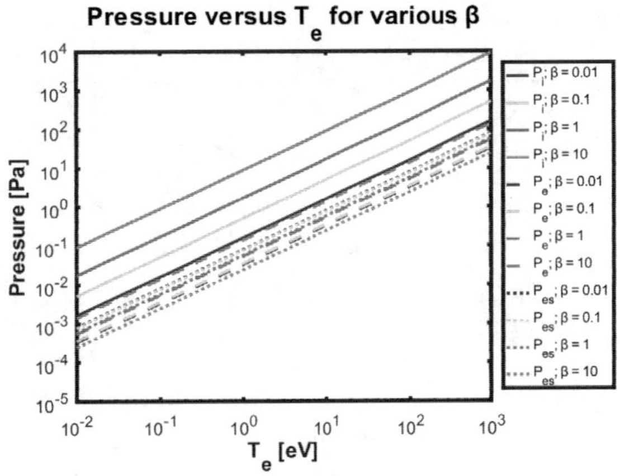

Figure 2. P_i, P_e and P_{es} of D plasma, varied by T_e and β.

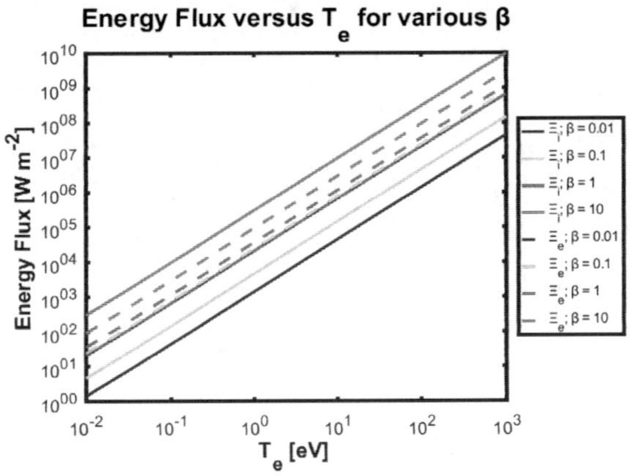

Figure 3. Ξ_i and Ξ_e of D plasma, varied by T_e and β.

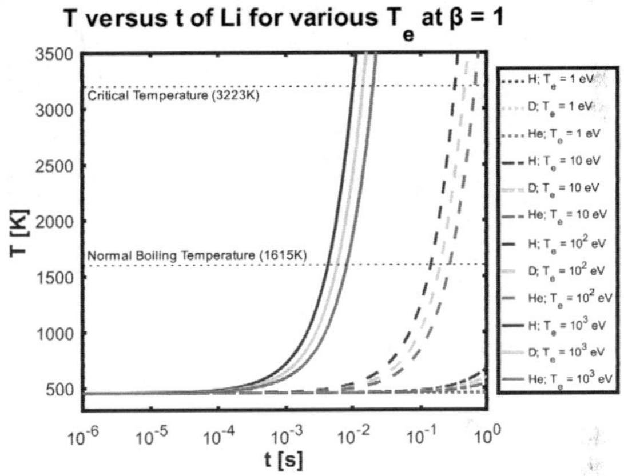

Figure 4. T vs t of Li surface bombarding by H, D, He and Ar plasmas at $\beta = 1$.

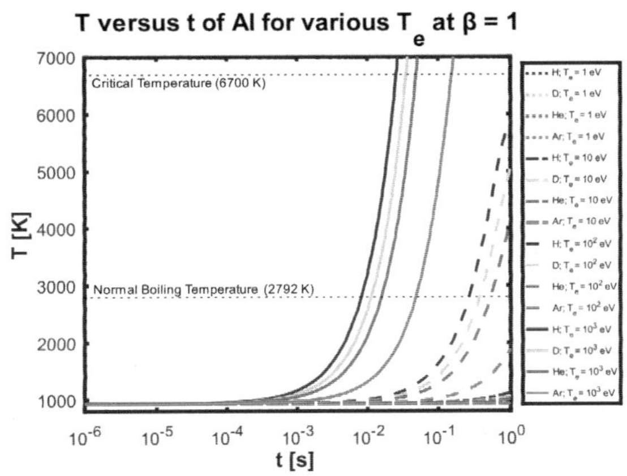

Figure 5. T vs t of Al surface bombarding by H, D, He and Ar plasmas at $\beta = 1$.

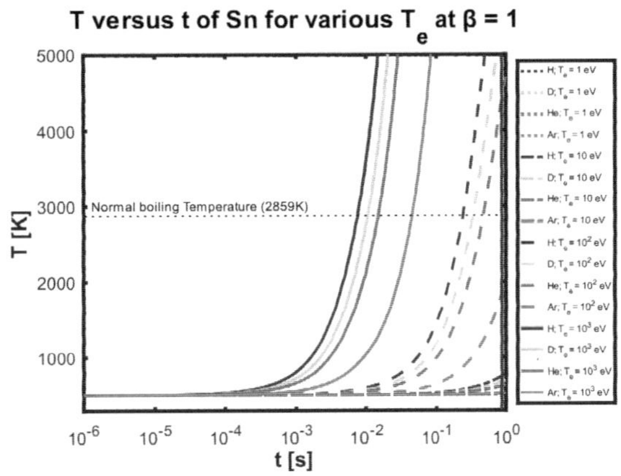

Figure 6. T vs t of Sn surface bombarding by H, D, He and Ar plasmas at $\beta = 1$.

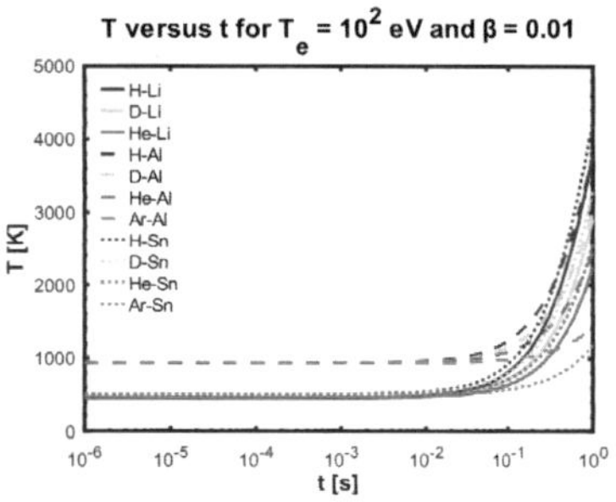

Figure 7. T vs t of various plasmas and surfaces for $T_e = 10^2$ eV and $\beta = 0.01$.

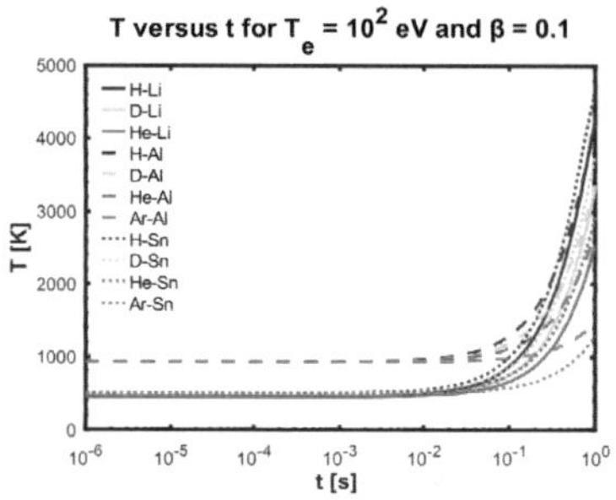

Figure 8. T vs t of various plasmas and surfaces for $T_e = 10^2$ eV and $\beta = 0.1$.

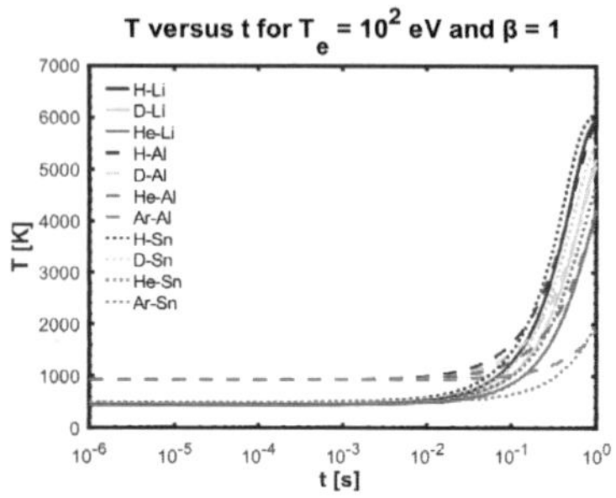

Figure 9. T vs t of various plasmas and surfaces for $T_e = 10^2$ eV and $\beta = 1$.

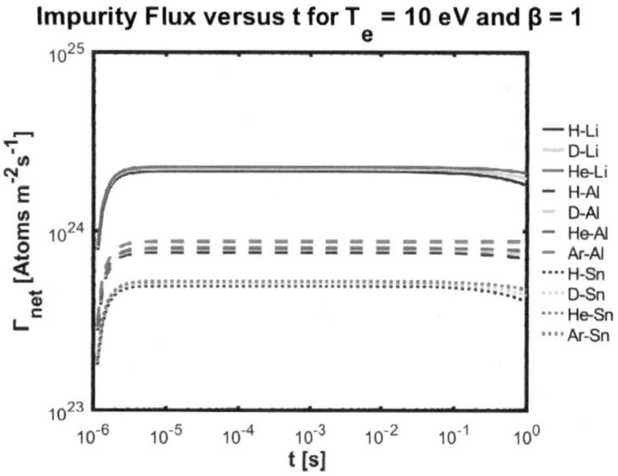

Figure 10. Γ_{net} vs t of various plasmas and surfaces for $T_e = 10$ eV and $\beta = 1$.

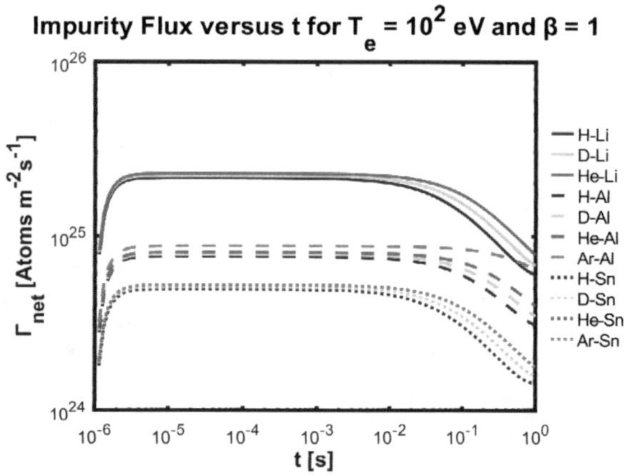

Figure 11. Γ_{net} vs t of various plasmas and surfaces for $T_e = 10^2$ eV and $\beta = 1$.

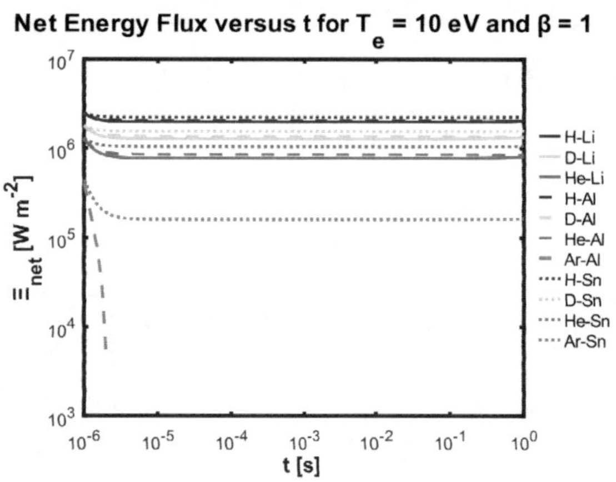

Figure 12. Ξ_{net} vs t of various plasmas and surfaces for $T_e = 10$ eV and $\beta = 1$.

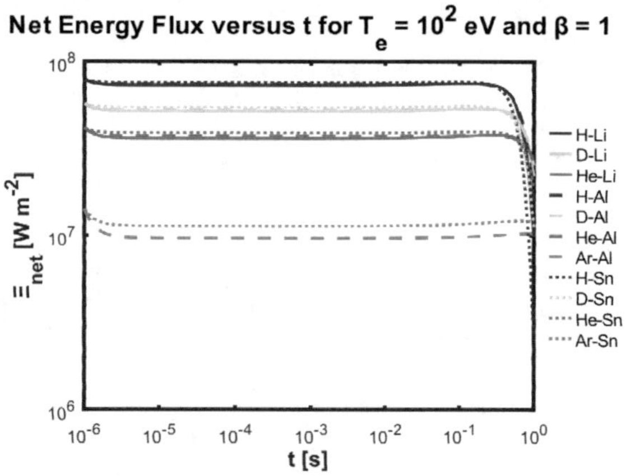

Figure 13. Ξ_{net} vs t of various plasmas and surfaces for $T_e = 10^2$ eV and $\beta = 1$.

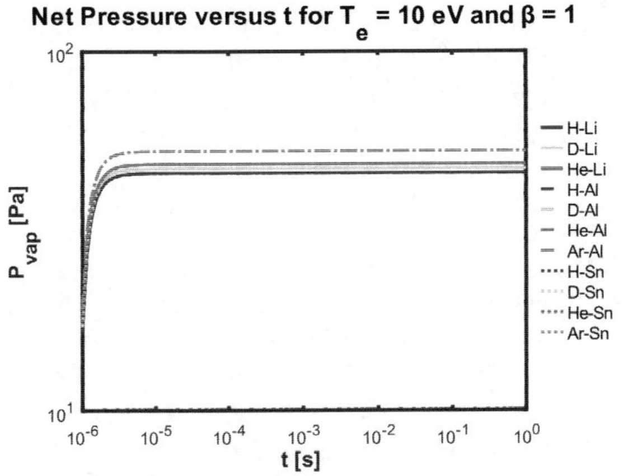

Figure 14. P_{vap} vs t of various plasmas and surfaces for T_e = 10 eV and β = 1.

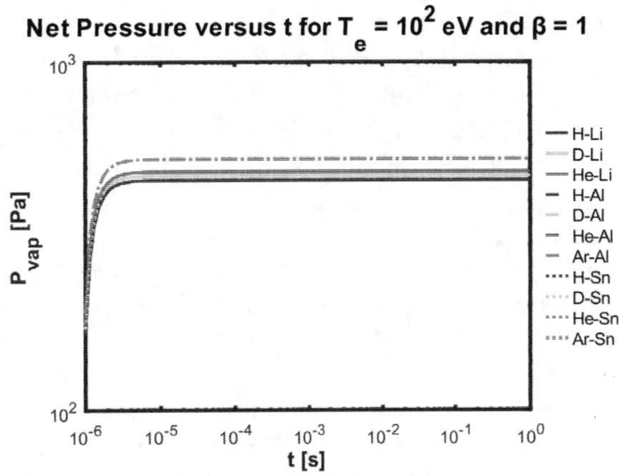

Figure 15. P_{vap} vs t of various plasmas and surfaces for T_e = 10^2 eV and β = 1.

Initially, no impurity vapor releases by any mechanism, that is Γ_{evap} = 0 and Γ_{sp} = 0. The impurities have no contribution to heating and pressure on the surface. The liquid surfaces are made of molten lithium (Li), aluminum (Al) and tin (Sn), whose initial temperatures are their melting temperatures, T = 454, 933 and 505 K [7], respectively, and thickness (H) is 0.005 m. In further instant, T is updated, so $\Gamma_{evap} > 0$. Γ_{evap} is proportional to P_{vap} but inversely depends on $T^{1/2}$. Therefore, P_{evap} and Ξ_{evap} are not time-independence. Moreover, $\Xi_{rad} \propto T^4$. Thus, solving time-dependence heat equation is required.

Figures 4-6 illustrate the trends of surface temperatures (T) varied by elapsed time (t) of H, D, He and Ar plasmas interacted with the liquid Li, Al and Sn surfaces, whose ε were assumed to be 1. Figures 7-9 show the trends of surface temperatures (T) of the mentioned plasma-surface interactions versus elapsed time (t) at which β is varied. Figures 10-11 and 12-13 demonstrate the trends of the net outgoing flux and the net energy flux versus elapsed time (t) for T_e = 10 and 10^2 eV with β = 1, respectively. Overall, considering all pairs of plasma-surface interactions, large T_e strongly increases T from the initial temperature in a relatively short period. The temperature rising timescale can be approximated as follows: 0.1 - 10 ms for $T_e \sim 10^3$ eV; 10 ms - 1 s for $T_e \sim 10^2$ eV; and a few seconds for $T <$ a few tens eV approximately. Reducing β, as illustrated in figures 7-9, does not distort the timescales and the trends of an increase in T, except the degree of T for each β at any instant may not be the same. This only lessens the rate of an increase in T due to less incoming ion flux. It is noticeable that at long elapsed time, the rate of a rise in T is slightly reduced as clearly seen in figure 9. This satisfies the drop of the net energy flux, Ξ_{net}, from ~ 10^7 - 10^8 W m^{-2} to lower, as seen in figure 13 for all pairs of plasma surface interactions, during the same period. This simultaneously corresponds to the drop of the net impurity outgoing flux, Γ_{net}, from a steady quantity of ~ 10^{25} particles m^{-2} s^{-1} by half an order, as shown in figure 11, for T_e = 10^2 eV. The trend of the decreases of Γ_{net} and Ξ_{net} for lower T_e, as exemplified in figures 10 and 12, is similar to that of figures 11 and 13 of higher T_e, except that the drop-triggering timescale of lower T_e is much longer than that of higher T_e. The timescale is varied from ~ 1 ms of T_e = 10^2 eV to ~ 0.1 s of T_e = 10 eV. The drop of Γ_{net} and Ξ_{net} is mainly caused by the evaporation, whose effect appears at long elapsed time. This also implies that it is a negative feedback preventing T rising too much. Moreover, the cooling by physical sputtering is enhanced for all pairs of plasma surface interactions at which $\beta T_e >$ a few tens eV approximately.

Between their boiling to critical temperatures, illustrated by the black dotted lines in figures 4-6, the current model cannot anticipate the trends of T and t clearly. This is because several thermally related situations, for examples, superheating and some kinds of thermal instabilities, may enhance net impurity outgoing flux. Furthermore, the reduction of the boiling temperature by lowering vapor pressure (P_{vap}) may allow such thermally related situations to occur sooner. Figures 14-15 show that P_{vap} of T_e = 10 and 10^2 eV ramp up at the beginning of the period due to the buildup of net impurity outgoing flux, subsequently level off over the rest of the period. The steady P_{vap} is roughly in the order of 10^2 Pa. This leads to the reduction of the actual boiling temperature with respect to the normal one approximately by one third [12]. This implies that the trends of T may stray from their looks in figures 4-6 earlier but still at the order of the temperature rising timescales.

4. Conclusion

Floating potential is a succinct parameter describing plasma surface interactions. This is associated with plasma particle flux on a surface, which is subsequently in direct and indirect connection with various pressures and energy fluxes on the surface. This supports the determination of net impurity outgoing flux. With the simple physical models, the magnitude of floating potential inversely depends on the ion to electron temperature ratio, but directly on electron temperature (T_e) and ion mass (m_i) of a plasma. The pressure and the energy flux provided by the incoming ions are the largest and dominated from the starting point of the consideration. It appears that after the surface temperature is appropriately risen, the pressure and energy flux due to evaporation is comparable to those of ions so that they are in competition for net heating and pressure. At long elapsed time, the evaporation introduces the decay of the net energy flux and the net impurity outgoing flux. Large electron temperature ($T_e > 10$ eV) should increase the surface temperature to reach a boiling point in a short period. Overall, the ion to electron temperature ratio (β) is not as effective as T_e and m_i to change the surface temperature. The liquid Li, Al and Sn surfaces produce the most, the second most and the third most quantities in the vapor particle inventory. The release flux of each vapor species is around 10^{23} - 10^{25} particles m^{-2} s^{-1} throughout most of the period, except at long elapsed time where the cooling by the evaporation becomes comparable to the ion heating.

Acknowledgements

N. Somboonkittichai deeply appreciates S. Aiemkong and my family for encouragement. N. Somboonkittichai thanks to GNU Octave (5.2.0) and QtiPlot (0.9.8.9) for being available to be used in research without charge. This work is aimed to promote the research activities of plasma and fusion physics in Thailand.

References

[1] Nygren R E and Tabarés F L 2016 *Nucl. Mater. Energy* **9** 6–21
[2] Zuo G Z, Ren J, Hu J S, Sun Z, Yang Q X, Li J G, Zakharov L E and Ruzic D N 2014 *Fusion Eng. Des.* **89** 2845–52
[3] Kauzmann W 1966 *Kinetic Theory of Gases* (New York: W.A. Benjamin Inc.)
[4] Stangeby P C 2000 *The Plasma Boundary of Magetic Fusion Devices (Plasma Physics Series)* (London: Institute of Physics Publishing)
[5] Chambers A 2004 *Modern Vacuum Physics (Master Series in Physics and Astronomy)* (Florida: CRC Press)
[6] Velasco S, Román F L, White J A and Muleco A 2006 *Fluid Ph. Equilibria* **244** 11–5
[7] Haynes W M, Lide D R and Bruno T J 2014 *CRC Handbook of Chemistry and Physics* (Florida: CRC Press)
[8] Zhang Y, Evans J R G and Yang S 2011 *J. Chem. Eng. Data* **56** 328–37
[9] Jackson J D 1998 *Classical Electrodynamics* (New York: John Wiley & Sons, Inc.)
[10] Freidberg J 2007 *Plasma Physics and Fusion Energy* (New York: Cambridge University Press)
[11] Behrisch R and Eckstein W 2007 *Sputtering by Particle Bombardment* (Berlin Heidelberg: Springer-Verlag)
[12] National Institute of Standards and Technology (NIST). *NIST Chemistry WebBook, SRD 69.* Available from: https://doi.org/10.18434/T4D303 [Accessed 21st July 2021]

The new evidence from the fingerprint region in FT-IR spectra to indicate the heat treatment of blue sapphire sample

A Phlayrahan* and H Homkhajorn

Faculty of Gems, Burapha University Chanthaburi Campus, Chanthaburi, 22170, Thailand

*E-mail: aumaparn.ph@go.buu.ac.th

Abstract. FT-IR spectroscopy is a significant method to detect the heat treatment of gemstones, especially ruby and sapphire. There are a set of certain peaks, i.e., 3309, 3232, and 3185 cm^{-1} in the FT-IR spectra used as an indicator to determine whether the samples have been undergone heat treatment. In this study, however, new evidence has emerged. The O-Al-O bending vibration peak at the fingerprint region around 600-700 cm^{-1} has been suggested as new clues to define the heated stones. The blue sapphire samples were prepared, then the heat treatment was performed separately at 800, 1000, 1200, 1400, and 1650 °C under an oxidizing atmosphere with a soaking time of 1 hour. The energy dispersive X-ray fluorescence (EDXRF) shows that there is no significant difference in the chemical composition of Al_2O_3, Fe_2O_3, TiO_2, and Ga_2O_3 between unheated and heated samples. The alteration of blue color and the UV-Vis absorption spectra are also difficult to classify. When comparing the samples at each heating temperature, the alteration of blue color was not related to one another because the color of the samples was decreased depending on those of the unheated ones. The FT-IR spectra revealed that the broad absorption peak of O-Al-O bending at approximately 650 cm^{-1} was slightly shifted to a lower wavenumber (630 – 635 cm^{-1}) after the samples were heated at higher temperatures. The broad absorption peak also turned into a sharper one when the sample has undergone heating at ≥1000 °C. It is suggested that the peak is related to the rearrangement of the Al_2O_3 structure of the blue sapphire sample after heat treatment.

1. Introduction

Heat treatment is one of the most efficient methods to improve the quality (color and clarity) of gemstones. The process is carried out by heating the sample in a controlled furnace at different ranges of temperatures, depending on the condition of the raw sample [1]. Normally, after heat treatment, the color of the stone is still caused by its internal chemical composition, thus this technique is accepted in the gem market. However, the undergone heat-treated stone is normally valued less than those of the natural ones [1]. Therefore, the heating process must be revealed at all costs.

Blue Sapphire is an alpha-alumina (α-Al_2O_3) mineral and one of the most popular stones in the gem market. The blue sapphire is shown in various shades of blue. The expensive one usually contains the deep and vibrant blue color, which nowadays, is rarely found in nature. Thus, heat treatment plays an important role in improving the blue color of the stones. Normally, the heat treatment can be detected using the internal features as well as the alteration of the absorption spectra [2]. However, those indicators would be accurate when compared with the unheated sample. Recently, the FT-IR spectra were also applied to indicate the undergone heat treatment of the sample, especially, the specific peaks at a -OH stretching region (3000-4000 cm^{-1}) [3]. Even though the peaks have been useful, they have not seemed to appear in every sample [2].

Therefore, in this study, new evidence for using another set of peaks in FT-IR spectra has been proposed. The peaks in the fingerprint region around 600-650 cm^{-1} related to the Al_2O_3 structure have been verified to apply the FT-IR spectra for indicating heat treatment of blue sapphire samples.

2. Experiments

Rough blue sapphire samples were collected and prepared. The samples were cut in half, one for referencing (un-heating), another for heating. The cutting direction of the samples is perpendicular to their c-axis and the thickness is polished to less than 2 mm (figure 1). Then the samples were heated separately under the oxidizing atmosphere for an hour at the temperature of 800, 1000, 1200, 1400, and 1650 °C representing the low to high temperature.

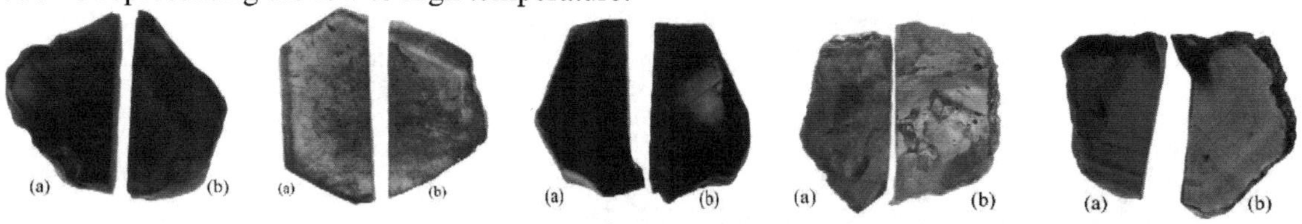

Figure 1. Blue sapphire samples at different heating temperatures from 800 to 1650 °C (a) with their unheated counterpart (b).

The alteration of the blue color before and after heat treatment at each temperature was determined using the Perkin Elmer Lamda 850+ UV-Vis spectrophotometer at the Faculty of Gems, Burapha University Chanthaburi Campus. The absorption spectra were detected in the range of 350 – 800 nm. Then the CIE L* a* b*color measurements were calculated using the color master software. The chemical composition of Al_2O_3 and the trace elements in the blue sapphire samples before and after heating were analyzed by an Energy Dispersive X-ray Fluorescence (EDXRF) at the Burapha Gemological Laboratory, Chanthaburi.

To investigate the behavior of the peaks in FT-IR spectra throughout the heating process, the spectra were acquired by the Thermo Scientific Nicolet iS50 FT-IR spectrometer in the absorption mode with the diffuse reflectance accessory at the Faculty of Gems, Burapha University, Chanthaburi Campus. The spectra were recorded from 400-4000 cm^{-1} with a spectral resolution of 4 cm^{-1} and 32 number of scans. To avoid the polarization effect, the FT-IR spectra were measured in a parallel direction to the c-axis of the samples.

3. Results and discussion

The chemical composition of blue sapphire samples before and after heating at different temperatures obtained by EDXRF is shown in table 1. The result indicates that Al_2O_3 is a main component of the blue sapphire. The trace elements responsible for the blue color of these samples are Fe_2O_3 and TiO_2. The Ga_2O_3 is also presented in every sample. It is noteworthy that the chemical composition of the samples before and after heat treatment is slightly different. Therefore, only the chemical composition cannot be used as evidence for identifying the undergone heating.

Table 1. Chemical composition obtained by the EDXRF of main and trace element of blue sapphire samples.

Chemical Composition	Percentage by weigth (%wt)									
	un	800	un	1000	un	1200	un	1400	un	1650
Al_2O_3	99.24	99.32	98.93	98.94	99.36	99.42	98.18	98.09	97.54	97.74
TiO_2	0.15	0.16	0.02	0.02	0.10	0.09	0.09	0.10	0.07	0.05
Fe_2O_3	0.58	0.49	1.03	1.02	0.52	0.47	1.71	1.79	2.37	2.19
Ga_2O_3	0.03	0.03	0.02	0.02	0.02	0.02	0.02	0.02	0.02	0.02

The alteration of the color and clarity of the samples after heat treatment can be observed by the naked eyes as shown in figure 1. The unheated samples possessed a greenish-blue color. However, after experiencing heat treatment, the green color vanished, and the blue color was increased in the samples. The clarity of the samples was also improved. It became clearer than those of the unheated ones. This result is correspondingly well with the L* a* and b* factors of color measurement as shown in table 2. The lightness (L*) possesses a higher value after heat treatment except at 800 °C, corresponding when observed with naked eyes. The blue color representing as -b* also increased when the samples were heated. However, at a high temperature (≥ 1200 °C), the -b* value does not show a significant change.

Table 2. CIE L* a* b* color measurement of blue sapphire samples comparing between unheated and heated samples.

Heating Condition (°C)	CIE color measurement					
	unheated			heated		
	L*	a*	b*	L*	a*	b*
800	18.845	-8.256	-6.391	10.809	-0.095	-23.062
1000	18.143	-8.681	-8.489	30.145	-5.025	-25.084
1200	3.751	-1.408	-0.587	20.735	-4.954	-1.198
1400	30.069	-1.022	-5.305	61.235	0.113	-5.685
1650	38.829	-3.871	-12.325	52.892	-2.823	-13.704

The UV-Vis absorption spectra of unheated and heated samples at different heating temperatures were measured to investigate the mechanisms of trace elements related to the blue color in sapphire during the heating processes. The UV-Vis absorption spectra of the representative blue sapphire samples are shown in figure 2. The spectra of unheated and heated samples at 800 and 1650 °C show the absorption peaks at 387 nm and 450 nm defining the single Fe^{3+} [4]. The broad absorption band at 560 nm also appears in the spectra. This band has been assigned as Fe^{3+}/Ti^{4+} pair [5]. The intensity of the band depends on the amount of Fe_2O_3 and TiO_2 relating to the cause of the blue color of the sapphire [6]. However, the UV-Vis absorption spectra were hardly changed throughout the heating process. It could be more reliable when compared only with the unheated counterpart. Thus, the UV-Vis absorption spectra are only useful as a supporting clue to verify the heat treatment.

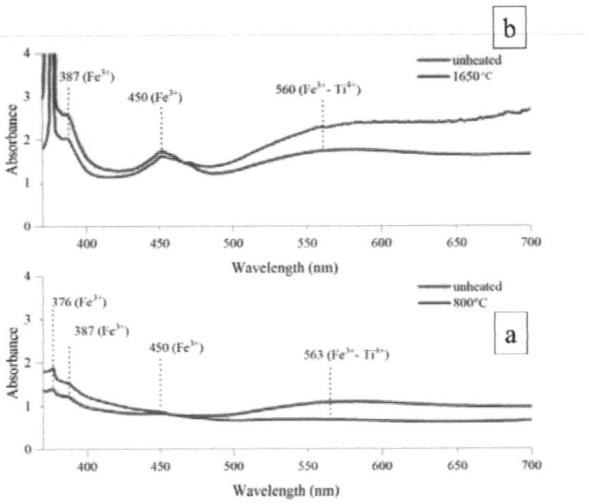

Figure 2. UV-Vis spectra of the representative blue sapphire samples comparing between unheated and heated sample at 800 (a) and 1650 °C (b).

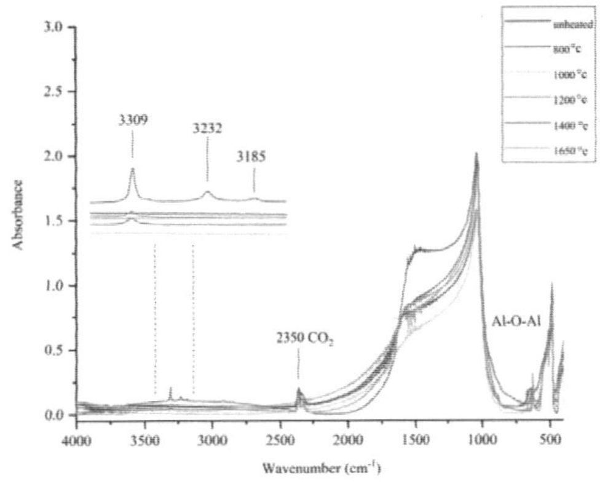

Figure 3. FT-IR spectra measured at 400-4000 cm^{-1} of blue sapphire samples at different heating temperatures.

The FTIR spectra of blue sapphire samples are shown in figure 3. The spectra were recorded at 400-4000 cm^{-1} the common range for analyzing a blue sapphire. The spectra in the -OH stretching region (3000-3500 cm^{-1}) present the peaks at 3309, 3232, and 3185 cm^{-1} assigned to -Ti-OH stretching vibration [2-3]. The peaks are normally used as one of the important indicators to classify the undergone heating of the samples. It should be noted that the peaks are not present in every sample but seem to appear only in the sample heating at low temperatures ($\leq$ 800 °C) and disappear when the samples are heated at a higher one. Thus, it is not suitable for identifying the heat treatment at high temperatures, which is a usual method to improve the quality of the blue sapphire.

The absorption peaks in the fingerprint region around 600 - 700 cm^{-1} (figure 4) in the FT-IR spectra are related strongly to the Al-O structure of the blue sapphire. The spectra show the peak at 668 cm^{-1} of Al-O stretching in every heating temperature corresponding to the vibration of Al in the octahedral structure and of the blue sapphire. Simultaneously, in this study, the alteration of the peaks around 625-650 cm^{-1} has been proposed to be applied as a new indicator to classify whether the blue sapphire has been undergone heating. The peaks have been defined as O-Al-O bending vibration relating to the arrangement of the Al_2O_3 in the blue sapphire structure [7]. The unheated sample possesses a broad absorption band centered at 648 cm^{-1}. After the samples were heated at a higher temperature, the peak had been slightly shifted to a higher wavenumber. The feature of the peak also slightly changed from broad absorption to a shaper one. The alteration of the peak responsible for the main structure of blue sapphire in the fingerprint region depends on the temperature of heating. After heating, the disordered atoms, particularly, the Al and O in the blue sapphire structure were rearranged and caused the shift of the peak [6]. Since the Al-O is always present in the blue sapphire as the main component, therefore, the peaks relating to them are more reliable than those in the -OH stretching region, which is associates with the trace elements.

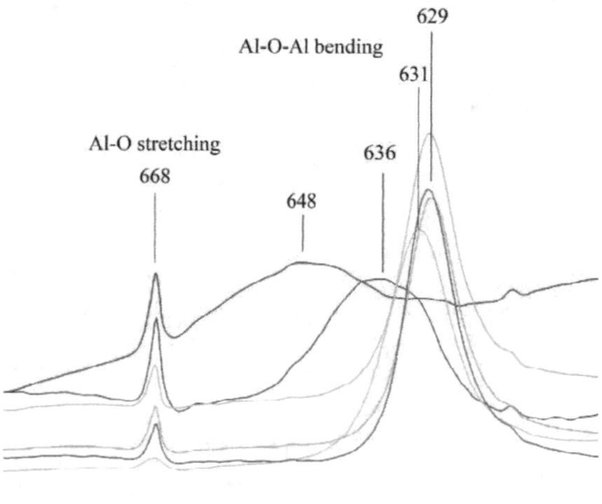

Figure 4. The expanded FT-IR spectra of blue sapphire samples showing the shifted position of Al-O-Al stretching at different heating temperatures; red line = unheated, blue line = 800 °C, yellow line = 1000 °C, green line = 1200 °C, violet line = 1400 °C, and gray line = 1650 °C

4. Conclusion

The alteration of the O-Al-O bending vibration peak around 625-650 cm^{-1} in FT-IR spectra of blue sapphire has been suggested as one of the new evidence for identifying whether the samples have been undergone heat treatment. The peak is related to the rearrangement of the disordered Al and O atoms caused by heat treatment. This result can be applied in the blue sapphire heating at every range of temperatures.

Acknowledgments

The authors would like to thank, the Faculty of Gems, Burapha University Chanthaburi Campus for providing a research fund and for the use of their instrument facilities. The Gem and Mineral Science SRU, Department of Earth Sciences, Faculty of Science, Kasetsart University for providing the blue sapphire samples.

Journal of Physics: Conference Series **2145** (2022) 012023 doi:10.1088/1742-6596/2145/1/012023

References

[1] Hughes R W 1997 *Ruby and Sapphire* (Colorado:RWH Publishing)
[2] Phlayrahan A, Monarumit N, Satitkune S and Wathanakul P 2018 *J. Appl. Spectros.* **85** 385–90
[3] Phlayrahan A, Monarumit N, Boonmee C, Satitkune S and Wathanakul P 2018 *J. Phys.: Conf. Ser.* **1144** 012057
[4] Lehmann G and Harder H 1970 *Am. Mineral.* **55** 98–105
[5] Wongrawang P, Monarumit N, Thammajak N, Wathanakul P and Wongkokua W 2016 *Mater. Res. Express* **3** 026201
[6] Phlayrahan A, Monarumit N, Lhuaamporn T, Satitkune S and Wathanakul P 2019 *J. Appl. Spectros.* **86** 810–6
[7] Adamczyk A and Dludon E 2012 *Spectrochim. Acta A Mol. Biomol. Spectrosc.* **89** 11–7

Heuristic compactness maximization algorithm for two-dimensional single-atom traps rearrangement

T Mamee[1,2,3], W Anukool[1,2,4] , N Thaicharoen[1], N Chattrapiban[1,2,4] and P Sompet[1,*]

[1] Research Center for Quantum Technology, Faculty of Science, Chiang Mai University, Chiang Mai 50200, Thailand

[2] Department of Physics and Materials Science, Faculty of Science, Chiang Mai University, Chiang Mai 50200, Thailand

[3] Graduate School, Chiang Mai University, Chiang Mai 50200, Thailand

[4] Thailand Center of Excellence in Physics, Commission on Higher Education, Bangkok 10400, Thailand

[*] E-mail: pimonpan.sompet@cmu.ac.th

Abstract. We establish an algorithm and computational results based on heuristic rearrangement of randomly filled array toward a defect-free and compact array. In this approach, the vacancies are filled from the inner layer that is related to the distance from the center of each loading site. By rearranging the position of atoms that maximize the compactness of the system layer by layer, the algorithm is set to iterate until the compactness reaches its local maximum. The results show that by applying the algorithm, the compactness of the system converges up to ~97% of the theoretical maximum.

1. Introduction

The construction of defect-free arrays of single atoms is a basis towards synthesizing a fully controllable and scalable quantum system in terms of interaction and a number of interacting systems. The system is promising for applications such as quantum information processing [1], quantum simulation [2], quantum error correction [3] and quantum communication [4]. However, there is a limitation of constructing a perfectly filled deterministic array of single-atoms due to the collisional pair-loss induced by light-assisted collision [5] during a single-atom loading process. In this case, the loading efficiency is bounded to the success probability of ~50% in the case of a large red-detuning of the assisted-collision light [5]. This probabilistic loading undoubtedly creates some vacancies in an initially loaded array. Thus, in order to obtain a defect-free array, ones can increase the loading efficiency via a more costly and complicated techniques e.g. those that employ blue-detuned laser light to induce light assisted collisions [6]. Another method is to rearrange filled traps to remove defected sites from the target area (pre-identified by fluorescence imaging) [7]. The rearrangement could be accomplished by transporting each atom one-by-one with a deep optical envelope via a 2d acousto-optic deflector (AOD) [7–9] or by changing the position of each individual trap inside an array with a spatial light modulator (SLM) via an optical-phase control [10]. The main focus of this work is in the later approach, the single-atom trap rearrangement.

Regarding to the capability of creating a fully-filled array, there are three key limitations constraining the rearrangement process, which are 1) a finite vacuum-limited lifetime of a trapped atom, 2) the heating mechanism during the trap translation, and 3) the trap size and separation. In particular, the vacuum-limited lifetime limits the upper bound of the total time an algorithm takes to calculate and rearrange traps. Concomitantly, the heating mechanism limits the number of all moves an atom can undergo before kinetic energy overcomes the trapping potentials. Lastly, the trap size and separation prohibit the translation along diagonals of the trapping array due to the number loss affected by a merged-trap geometry. Therefore, the atom-rearrangement algorithm is restricted to only translations along the edge of the array (a line connecting trap centers) and all the mentioned limitations must be taken into consideration.

In principle, an optimization algorithm [11] can be designed from a brute-force method, i.e. comparing total costs of all possible trajectories regardless of computation costs. However, due to the key limitations mentioned, the calculation of all possible trajectories is highly time consuming and therefore impractical [9]. Other alternatives that offer more practical calculation time and obtain nearly optimized solutions, concern a heuristic paradigm to solving problems of rearrangement from a specific set of instructions. Examples are the short-move first algorithm [7], the compression algorithm [9], the A* searching algorithm (ASA) and the heuristic cluster algorithm (HCA) [12].

In this work, we show a preliminary demonstration of a heuristic algorithm that we developed called the heuristic compactness maximization algorithm. The algorithm focuses on compressing a trap distribution to maximally packed with filled traps to minimize a designated spatial cost function and thus maximize the "compactness" of the system. As a results, the algorithm can be used to maximize compactness of the system in any desired region by minimizing the variable where the performance depends on the array's size.

2. Theory and background: Heuristic compactness maximization algorithm

2.1. Overall idea of the algorithm

The main task of the heuristic compactness maximization algorithm is to maximize the compactness of the system defined through a moment matrix of filled-site $\mathcal{P} = \mathcal{O} \odot \mathcal{D}$, the element-wise multiplication between the occupancy matrix $\mathcal{O}$ and the distance matrix $\mathcal{D}$. The element of $\mathcal{P}$ at each array site $\{x, y\}$ is described by $\mathcal{P}_{x,y} = \mathcal{O}_{x,y}\mathcal{D}_{x,y}$, where $\mathcal{O}_{x,y}$ is the occupancy of a trap with value 1(0) for the occupied(unoccupied) trap, and the element $\mathcal{D}_{x,y}$ of $\mathcal{D}$ is defined as $\mathcal{D}_{x,y} = \sqrt{(x - x_0)^2 + (y - y_0)^2}$, where $\{x_0, y_0\}$ is the array center. The summation of all elements of $\mathcal{P}$ is

$$\mathcal{P}_s = \sum_x \sum_y \mathcal{P}_{x,y}, \tag{1}$$

which represents the strength of the moment related to an average spreading of the occupied site from the array's center. Therefore, when the algorithm minimizes $\mathcal{P}_s$, the compactness is maximized. As a result, the atoms seek the sites that have lower $\mathcal{P}_{x,y}$, which allows one to define the final shape of the filled system via the $\mathcal{P}$ matrix accordingly.

2.2. Algorithm flowchart

The flowchart of the algorithm is shown in figure 1(a). In the beginning, an initial traps loading is constructed from the Monte-Carlo method with 50-percent success probability to obtain a single atom for each trap. As a result, the temporal distribution of each single trap follows to a sub-Poissonian distribution. In this case, the filling factor is 0.5 [13] and this filling factor is used for all of the subsequent calculations throughout this work. To illustrate how the algorithm works, figure 1(b)–(d) are selected as examples. In figure 1(b) the algorithm generates a random distribution of an initial occupancy ($\mathcal{O}$ matrix). Afterward, it defines the filling layer shown in figure 1(c). In this example, there are 6 layers labeled in numbers. Each layer has the same value of $\mathcal{P}_{x,y}$. The algorithm starts its iteration loop from the smallest unfilled layer. Then it seeks to fill the vacancies in the layer with the nearest-neighbor atoms from the nearest outer layer. At this step, the algorithm aims to minimize the $\mathcal{P}_s$ parameter. After the first layer is fully occupied, the next layer is considered. For an example as shown in figure 1(d), the vacancy at position $\{2, 4\}$ needs to be filled. From here, the atom located in the next layer, i.e. at $\{2, 5\}$, is selected because $\mathcal{P}_s$ is the most minimized comparing to the other nearest neighbor atoms in the same layer. Finally, because the allowed trajectories for the atom is restricted to a move along the edges between traps (the step size is one between adjacent traps), the algorithm may need several steps until $\mathcal{P}_s$ reaches the minimum.

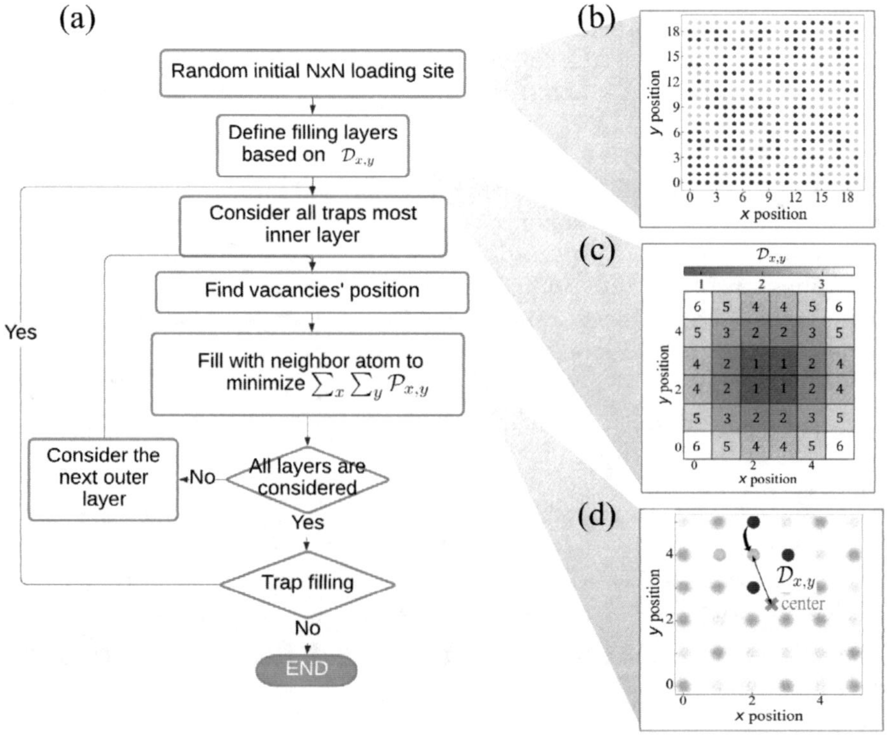

Figure 1. A schematic diagram of compactness maximization algorithm. From the flowchart (a), initial loaded array is randomized as shown in (b) and then the algorithm defines filling layers (shown in (c)) corresponding to each trap's specific spatial property called $\mathcal{D}_{x,y}$. To maximize the compactness, the partially filled innermost layer is assigned to the lowest $\mathcal{D}_{x,y}$. Once a vacancy is found, the only one nearest neighbor atom that can minimize $\mathcal{P}_s = \sum_x \sum_y \mathcal{P}_{x,y}$, as shown in (d), is moved. From here, the algorithm tends to fill the inner vacancies with atoms from outer layer where $\mathcal{D}_{x,y}$ is higher. (run over 500 loading samples)

3. Central occupancy evolution

In this section, the algorithm based on $\mathcal{P}_s$ as defined in equation (1), is applied to the array of 20×20 loading sites. The calculation repeats over $k = 500$ randomly loaded $\mathcal{O}$ matrices to inquire about the averaged occupancy matrix, $\langle \mathcal{O} \rangle = \sum_{i=1}^{k} O_i/k$. From figure 2, at an initial setup (0th iteration), the average occupancy $\langle \mathcal{O}^{(0)} \rangle$ of the system is nearly 0.5 and equally distributed over the loading sites. As the number of iterations increases to j, the average occupancy $\langle \mathcal{O}^{(j)} \rangle$ near the center of the array become higher and reach the saturated value of one, which is the maximum obtainable averaged occupancy because each site can only trap zero or one atom. After several iterations, the algorithm drive the system to form an x-shape pattern of saturated occupancy around the center. This is because the atoms are allowed to move only in either the x- or y- direction and the occupancy along the array's diagonals are populated from the moves in both directions. For a larger number of iterations, $\langle \mathcal{O} \rangle$ gradually increases toward uniform saturation, starting from the inner layer to the outer layer. The average occupancy is almost depleted near the boundary, forming a circular pattern respective to the defined $\mathcal{D}$ matrix. This compact round shape can be seen in the single-shot sample showing that all vacancies in the inner layer are eliminated while there is no atom left in the outer layer. This demonstrates the mass transportation driven by the algorithm from the wings toward the low $\mathcal{D}_{x,y}$ region around the center of the system.

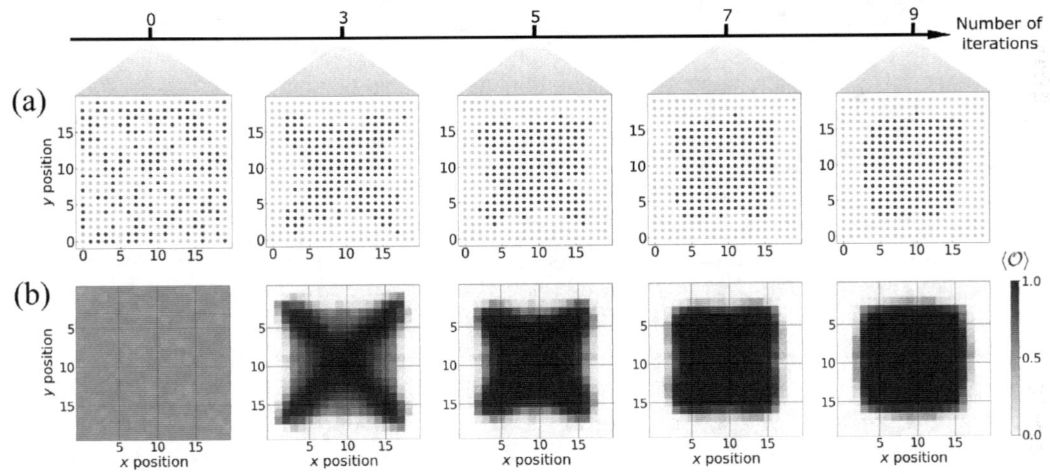

Figure 2. (a) Single-shot examples of $\mathcal{O}$ and (b) average occupancy matrices $\langle \mathcal{O} \rangle$ for each chosen number of iteration, showing compactness progression of a single-atoms array of 20×20 loading site, performed rearranging via $\mathcal{P}_\mathrm{s}$ based algorithm for radial $\mathcal{D}$.

4. Convergence of $\mathcal{P}_\mathrm{s}$ parameter

In this section, the behavior of the algorithm based on $\mathcal{P}_\mathrm{s}$ is established. The evolution of $\mathcal{P}_\mathrm{s}$ is shown in figure 3. Figure 3(a) shows that the average of $\mathcal{P}_\mathrm{s}$ exponentially decreases toward a saturated value as the number of iterations increases. This behavior agrees well with figure 2 where the system's compactness rises. For the 0th iteration, the initial $\mathcal{P}_\mathrm{s}$ parameter is calculated from the condition that the atom is randomly distributed. For the later iterations, the value of $\mathcal{P}_\mathrm{s}$ gradually decreases until it converges to a local minimum and the system reaches local maximum of compactness. The ratio between the local maximum compactness $\mathcal{C}_\mathrm{l}$ and the theoretical compactness $\mathcal{C}_\mathrm{th}$ is calculated as follows:

$$\frac{\mathcal{C}_\mathrm{l}}{\mathcal{C}_\mathrm{th}} = \frac{\max(\mathcal{P}_\mathrm{s}) - \min(\mathcal{P}_\mathrm{s})}{\max(\mathcal{P}_\mathrm{s}) - \min(\mathcal{P}'_\mathrm{s})}, \tag{2}$$

where $\min(\mathcal{P}'_s)$ is the theoretical minimum of $\mathcal{P}_s$. From the calculation of over 500 loading samples, $\langle \mathcal{C}_l/\mathcal{C}_{th} \rangle = 0.97$. The time constant τ of the decay is plotted in the insets inferring that the saturation time depends on the the system size. This could be further explained in figure 3(b) in which the inset shows the average of accumulated moves $\langle M_a \rangle$ for each iteration. The value of $\langle M_a \rangle$ moderately increases until it reaches an asymptotic value, indicating that the total number of the required moves for every atom in the system reaches the maximum compactness. Meanwhile, the plot between the maximum averaged value of the accumulated moves and the loading array's size in the main plot in figure 3(b) shows a nonlinear trend. For a larger loading array's size, more moves are needed to compact the system. This agrees well with a smaller decay rate for a larger array in figure 3(a).

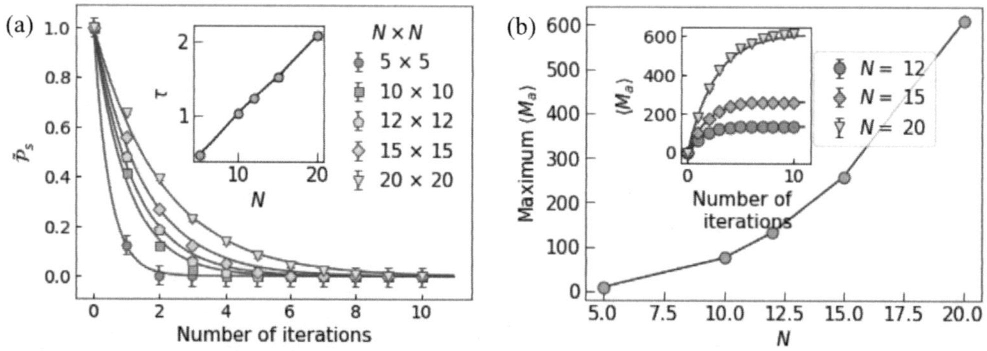

Figure 3. Evolution of $\mathcal{P}_s$ and the effect of the system size. (a) The scaled $\mathcal{P}_s$ (from 0 to 1), $\tilde{\mathcal{P}}_s$, is plotted as a function of number of iterations for different array sizes. In the plot, $\tilde{\mathcal{P}}_s$ is scaled with respect to the maximum value of $\mathcal{P}_s$, $\max(\mathcal{P}_s)$, and the local minimum, $\min(\mathcal{P}_s)$, for each array size condition. Thus $\tilde{\mathcal{P}}_s$ does not show the discrepancy between $\min(\mathcal{P}_s)$, and the expected theoretical minimum, $\min(\mathcal{P}'_s)$. The inset shows the decay constant τ from the exponential fit, plotted against the array lateral length N. (b) Maximum number of steps needed in the atom rearrangement algorithm for different lateral array lengths. The inset shows an accumulated number of moves, M_a, increasing for each iteration. The data are averaged over 500 samples and each error bar represents one standard deviation, which is especially small in (b).

5. Arbitrary occupancy shaping

To demonstrate applicability of our algorithm, we provide three spatial distributions. In particular, the arrangement of trap potentially implies different physics that experimenters convey. In this section, to examine if the algorithm could give different trap configurations, three different $\mathcal{D}$ matrices are used in the algorithm. Figure 4 shows the averaged occupancy matrices $\langle \mathcal{O} \rangle$, the distribution matrices $\mathcal{D}$ and their line cross-section from left to right respectively. In the same figure, from top to bottom row, the $\mathcal{D}$ matrices are radially linear distribution, radially quadratic distribution and horizontal triangular distribution respectively. For the case of linear and quadratic distribution, the averaged occupation matrices $\langle \mathcal{O} \rangle$ are similar and have shapes between a circular and a rectangular. This is because the algorithm restriction on the direction of moves compresses $\mathcal{O}$ toward a rectangular shape at an early stage and stall at the local maximum compactness as mentioned in section 4. Nonetheless, for the triangular $\mathcal{D}$ matrix, the averaged occupancy drastically follow the triangular shape.

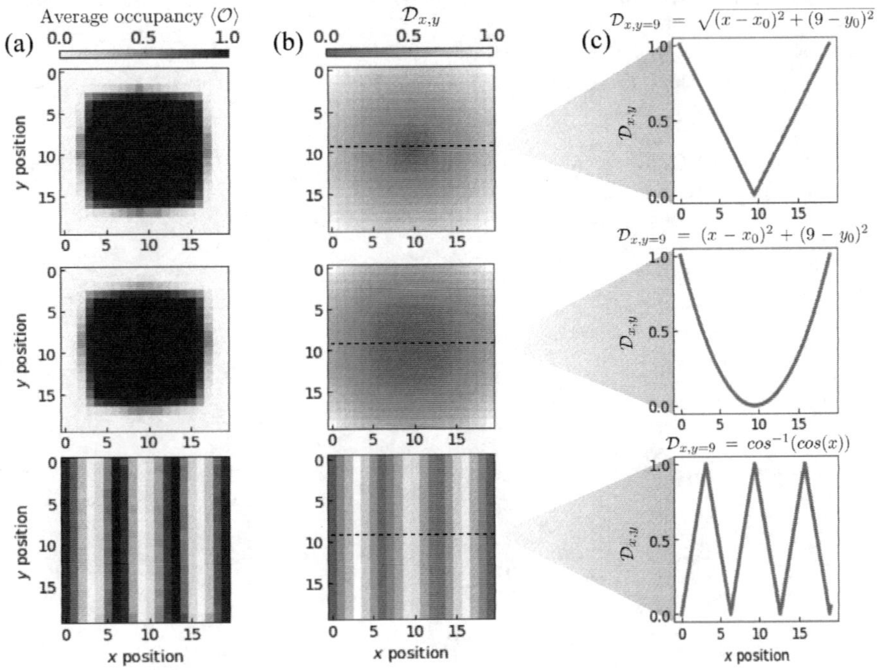

Figure 4. Averaged occupancy matrix $\langle\mathcal{O}\rangle$ (a) and $\mathcal{D}$ matrix (b) after performing the algorithm for three different $\mathcal{D}_{x,y}$ which alter the pattern of the layer structure. The cross section at $y = 9$ for each $\mathcal{D}$ matrix is shown in (c).

6. Conclusions

We propose the compactness maximization algorithm to rearrange atoms in a Cartesian array. The algorithm works by filling atom layer by layer from the innermost layer (based on $\mathcal{P}_s$) to the outermost one while the averaged occupancy at the target region increases. The results shows that the system compactness could be maximized up to ~97% of the theoretical maximum. For a larger array, more iteration is needed to reach the local maximum of compactness. Finally, the $\mathcal{D}$ matrix could be arbitrarily designed to give a desired atom distribution.

Several tasks are required to improve the performance of the algorithm. The main tasks relates to the optimization of moves and rearrangement time. In particular, coding efficiency and the local minimum escape can be improved by allowing complex trajectories via a reinforcement learning model.

References

[1] Saffman M, Walker T G and Mølmer K 2010 *Rev. Mod. Phys.* **82** 2313–63
[2] Feynman R P 1982 *Int. J. Theor. Phys.* **21** 467–88
[3] Fowler A G, Mariantoni M, Martinis J M and Cleland A N 2012 *Phys. Rev.* A **86**(3) 032324
[4] Duan L M, Lukin M D, Cirac J I and Zoller P 2001 *Nature* **414** 413–18
[5] Sompet P, Carpentier A V, Fung Y H, McGovern M and Andersen M F 2013 *Phys. Rev.* A **88**(5) 051401
[6] Carpentier A V, Fung Y H, Sompet P, Hilliard A J, Walker T G and Andersen M F 2013 *Laser Phys. Lett.* **10** 125501
[7] Barredo D, Léséleuc S D, Lienhard V, Lahaye T and Browaeys A 2016 *Science* **354** 1021–3
[8] Ebadi S *et al* 2021 *Nature* **595** 227–32

[9] Schymik K N, Lienhard V, Barredo D, Scholl P, Williams H, Browaeys A and Lahaye T 2020 *Phys. Rev.* A **102**(6) 063107

[10] Kim H, Lee W, Lee H g, Jo H, Song Y and Ahn J 2016 *Nat. Commun.* **7** 13317

[11] Lee W, Kim H and Ahn J 2017 *Phys. Rev.* A **95**(5) 053424

[12] Sheng C, Hou J, He X, Xu P, Wang K, Zhuang J, Li X, Liu M, Wang J and Zhan M 2021 *Phys. Rev. Res.* **3**(2) 023008

[13] Ohl de Mello D, Schäffner D, Werkmann J, Preuschoff T, Kohfahl L, Schlosser M and Birkl G 2019 *Phys. Rev. Lett.* **122**(20) 203601

Dynamics of spiral waves and bubble formation in a closed chemical system of thin photosensitive excitable media

Kritsana Khaothong, Vikanda Chanchang, Jarin Kanchanawarin, Malee Sutthiopad and Chaiya Luengviriya*

Department of Physics, Kasetsart University 50 Phaholyothin Road, Jatujak, Bangkok 10900, Thailand

*E-mail : fscicyl@ku.ac.th

Abstract. Spiral waves have been observed in a thin layer of excitable media. Especially, electrical spiral waves in cardiac tissues connect to cardiac tachycardia and life-threatening fibrillations. The Belousov-Zhabotinsky (BZ) reaction is the most widely used system to study the dynamics of spiral waves in experiments. When the light sensitive $Ru(bpy)_3^{2+}$ is used as the catalyst, the BZ reaction becomes photosensitive and the excitability of the reaction can be controlled by varying the illumination intensity. However, the typical photosensitive BZ reaction produces many CO_2 bubbles so the spiral waves are always studied in thin layer media with opened top surfaces to release the bubbles. In this work, we develop new chemical recipes of the photosensitive BZ reaction which produces less bubbles. To observe the production of bubbles, we investigate the dynamics of spiral waves in a closed thin layer system. The results show that both the speed of spiral waves and the number of bubbles increase with the concentration of sulfuric acid (H_2SO_4) and sodium bromate ($NaBrO_3$). For high initial concentrations of both reactants, the size of bubbles increases with time until the wave structures are destroyed. We expect that the chemical recipes reported here can be used to study complicated dynamics of three-dimensional spiral waves in thick BZ media where the bubbles cannot escape.

1. Introduction

Spiral waves have been found in a thin layer of different excitable media, including CO oxidation on a platinum surface [1], the slime mold Dictyostelium discoideum [2], cardiac tissue [3], and the Belousov-Zhabotinsky (BZ) reaction [4]. Such spiral patterns of an electrical signal (action potential) in heart tissues can occur that also have an important effect on human health. It is found some cardiac arrhythmia (e.g., atrial and ventricular tachycardia [5]). Questionability of the spiral waves pattern expanded into waves turbulence [6] which corresponded to ventricular fibrillation [5] and can lead to a sudden cardiac death. The waves turbulence can be investigated only in a three-dimensional (3D) system. It has observed in anisotropic myocardium tissues with thickness of 3-10 mm [6].

In very different systems, spiral waves are governed by a reaction-diffusion mechanism [7]. The BZ reaction is the most popular reaction to study the dynamics of spiral waves because it is easy to prepare and convenient to detect the wave structures. Moreover, the system can be easily controlled by the light intensity when the light sensitive catalyst ($Ru(bpy)_3^{2+}$) is filled in the BZ reaction. The control of spiral waves by light are reviewed in [8].

The mechanism of the BZ reaction is very complex and involve many different steps. The overall reaction: $3\ CH_2(COOH)_2 + 4\ BrO_3^- \rightarrow 4\ Br^- + 9\ CO_2 + 6\ H_2O$ shows CO_2 as one of the end products [9].

In the course of time, CO_2 increases and appear as bubbles in close unstirred BZ systems. The dynamics of spiral waves is disturbed when the bubbles are large and act as unexcitable defects [10]. The occurrence of CO_2 bubbles is an important problem in the BZ reaction, especially in a thick media. The effect of CO_2 bubbles not only perturbs the structure of spiral waves but also obstruct observations of 3D structures [11]. The typical photosensitive BZ reaction usually [12-15] use a high initial concentration which produce plenty bubbles. Therefore, photosensitive spiral waves are studied in thin layer media with opened top surfaces to release the bubbles.

In this research, we modify the chemical recipes of photosensitive BZ recipes to reduce the bubble production. We use agarose to reduce the hydrodynamic effect and observe them in a closed thin layer system to study the dynamics of spiral waves.

2. Methods

In all experiments, 10 ml of the photosensitive BZ solutions are prepared from $NaBrO_3$, H_2SO_4 and Malonic acid (MA) all purchased from Merck. Moreover, we use Tris(bipyridine)ruthenium (II) chloride ($Ru(bpy)_3Cl_2$) as a catalyst (Sigma). Stock solutions of $NaBrO_3$ (1M), $Ru(bpy)_3Cl_2$ (25mM) and MA (1M), are prepared by dissolving the chemicals in deionized water (conductivity ~ 0.056 $\mu S\ cm^{-1}$) while stock solution of H_2SO_4 (2.5 M) is purchased. To prevent the hydrodynamic perturbation, the reaction is embedded in a 1.0 % w/w agarose gel (Sigma). The BZ solution is prepared with a heat solution of agarose then the temperature is allowed to reduce until it reaches 36 °C and finally the reactants is filled. These solutions are prepared with different initial reagent concentrations: $[NaBrO_3]$ = 0.05-0.20 M, and $[H_2SO_4]$ = 0.4 – 1 M, whereas [MA] = 0.05 M and $[Ru(bpy)_3Cl_2]$ = 0.625 mM. In addition, a surfactant, sodium dodecyl sulfate (SDS, Fluka), is added to the solution to decrease CO_2 bubble production [9].

Dynamics of spiral waves in photosensitive BZ solution are investigated in a flat closed reactor constructed from transparent Plexiglas. After the BZ solution is filled into the reactor with 60 x 60 x 0.6 mm^3 of volume, propagating waves occur spontaneously, usually at the boundary of the reactor. We shine a strong spotlight (power =150 watts, diameter = 5 mm) to a segment of wave front so its propagation is entirely suppressed. After that those two free ends of fronts start curling and develop to spiral waves.

The reactor is placed between a projector and a color CCD camera (Super-HAD, Sony) to record the images every second with a resolution of 0.03 mm $pixel^{-1}$. All positions of the reactor are homogeneously illuminated with 350-lux blue light (a blue image with RGB value of [0 0 255]) using a video projector (BenQ XGA). Note that the projector and the light beam produce heat that causes the temperature of the reaction increases in the absence of cooling system. Therefore, we carefully control the temperature of the reactor by using an air conditioner and a fan blowing air direct to the reactor. After a transient interval at the beginning of experiments, the temperature of the reactor is well controlled to 22 ± 1 °C.

3. Results

Figure 1 shows examples of the spiral waves in the photosensitive BZ reaction with $[NaBrO_3]$ in the range of 0.05-0.20 M whereas $[H_2SO_4]$ = 0.4 M and the concentrations of the other reagents are described in the methods. Spontaneous waves occur and the spiral wave generation can be started shortly after the solution is poured into the reactor, except for the lowest $[NaBrO_3]$ in figure 1(a), they appear about 10 minutes slower than the others.

In cases of low concentrations, $[NaBrO_3] \leq 0.1$ M, the media look clear without CO_2 bubbles from the beginning [figures 1(a) and 1(b)] until the end of the experiments [figures 1(a′) and 1(b′)]. In contrast, tiny bubbles appear in those with higher concentrations, $[NaBrO_3] \geq 0.15$ M [figures 1(c) and 1(d)]. In the course of time, the size and amount of such bubbles increases and disrupt the wave structures [figures 1(c′) and 1(d′)]. It is worth noting that we also observe appearances of the bubbles and the disruption of the waves in the other series of experiments with $[H_2SO_4] \geq 0.70$ M where $[NaBrO_3]$ is kept constant at 0.05 M (images not shown).

For all spiral waves in this study, their shapes are unchanged after they evolve about 10 rotations, so we start to analyse their dynamics at this state. The spiral tips at the center of the wave structures rotate around small circles namely the spiral cores. Far from the spiral cores, the period *T*, wavelength λ, average speed *v* ($v = \lambda/T$) of the wave fronts are measured.

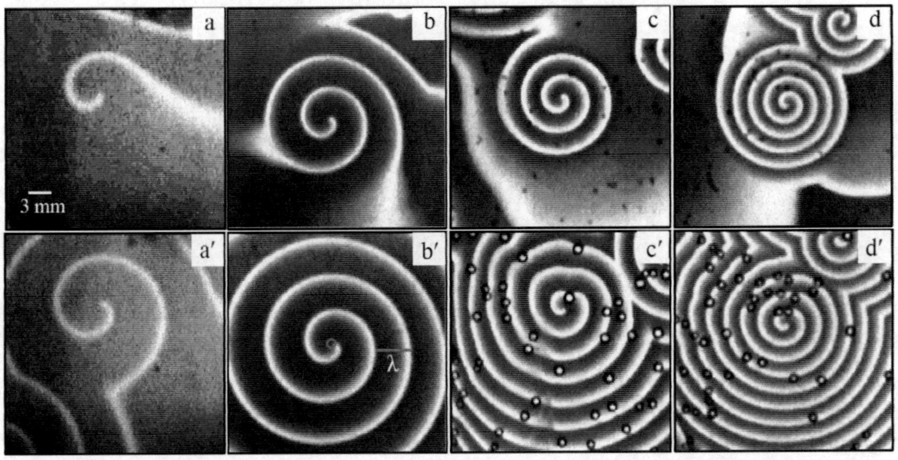

Figure 1. Spiral waves in the photosensitive BZ reaction with different [$NaBrO_3$]: (a, a′) 0.05, (b, b′) 0.1, (c, c′) 0.15, and (d, d′) 0.2 M. Top and bottom rows show the spiral waves after about 10 and 30 rotations, respectively. The other initial concentrations are [H_2SO_4] = 0.4 M, [MA] = 0.05 M and [$Ru(bpy)_3Cl_2$] = 0.625 mM. In (b′), The small circle represents the spiral core and the wavelength λ is depicted.

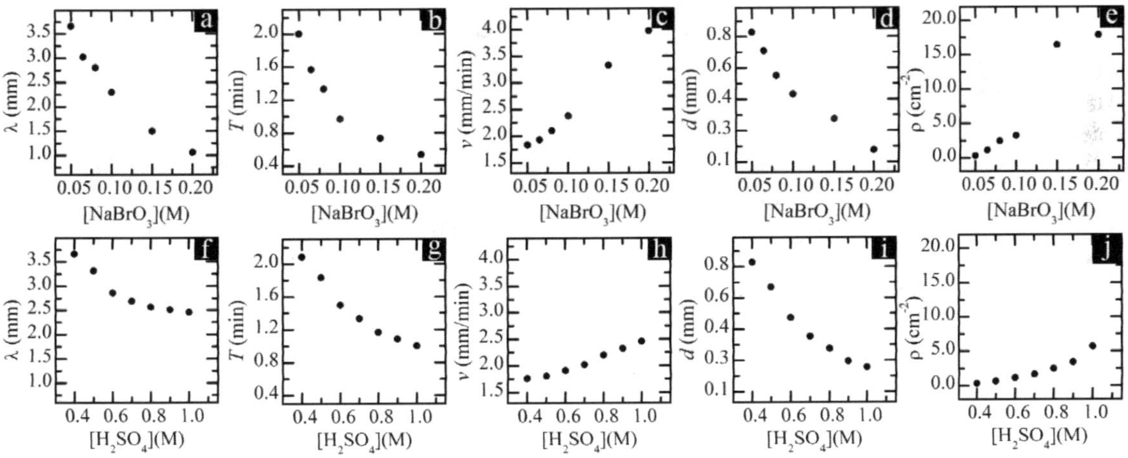

Figure 2. The dynamics of spiral waves in the photosensitive BZ reaction and the density of CO_2 bubbles with a variation of (a-e) [$NaBrO_3$] or (f-j) [H_2SO_4]. The standard concentrations are [$NaBrO_3$] = 0.05 M, [H_2SO_4] =0.4 M, [MA] = 0.05 M and [$Ru(bpy)_3Cl_2$] = 0.625 mM.

The analysis of the dynamics of spiral waves and the density of CO_2 bubbles are summarized in figure 2. They are divided into two series of experiments with a variation of [$NaBrO_3$] or [H_2SO_4]. Figures 2(a) – 2(e) show that when [$NaBrO_3$] is increased, the period *T*, wavelength λ and the spiral core diameter *d* decrease but the average speed *v* increases. As shown in figures 2(f) – 2(i), the effect of [H_2SO_4] on the dynamics of spiral waves is similar to that of [$NaBrO_3$], i.e., when [H_2SO_4] is increased, the period *T*, wavelength λ and the spiral core diameter *d* decrease but the average speed *v* increases.

To compare CO_2 bubbles in different experiments, we measure the density ρ of CO_2 bubbles in the images during 20-30 wave periods where the bubbles have a similar size of about 1 mm (e.g., as in figures c′ and d′). The number n of such bubbles in the observed area A of 6.15 cm^2 is counted and the density ρ is calculated as $\rho = n/A$. Figures 2(e) and 2(j) show that the density ρ of CO_2 bubbles increases with [$NaBrO_3$] and [H_2SO_4], especially it overshoots when [$NaBrO_3$] reaches 0.15 M.

4. Conclusion and discussion

We have presented an investigation of the dynamics of spiral waves in a closed thin layer of the photosensitive BZ reaction with new chemical recipes to reduce the production of CO_2 bubbles. For low concentrations of $NaBrO_3$ and H_2SO_4, the reaction has very few CO_2 bubbles. In contrast, for high concentrations of these reagents, the amount and size of bubbles increase with time. The solutions with chemical recipes in this study support rigidly rotating spiral waves whose tips rotate around circular spiral cores. When the concentration of $NaBrO_3$ or H_2SO_4 is increased, the period, wavelength and spiral core diameter decrease but the wave speed increases.

Traditionally, experiments have been done using the photosensitive BZ reaction with [$NaBrO_3$] = 0.2 - 0.55 M, [H_2SO_4] = 0.3 – 0.77 M, and [MA] = 0.17 – 0.3M [11-14]. The reactions with such high concentrations of reagents produce plenty of CO_2 bubbles so the investigations were limited to perform using thin layer media with opened top surface which can release the bubbles to the atmosphere.

In our study, the experiments are intentionally performed using the photosensitive BZ solutions with closed top surface to observe the bubble production in the course of time. The results suggest that the reaction with low initial concentrations of $NaBrO_3$ and H_2SO_4, produces low bubble density (see figures 2e and 2j) so we expect that some chemical recipes reported here are suitable for a study of 3D spiral waves in thick BZ media, where complicated and exotic wave dynamics can occur. We plan to perform 3D experiments in the future.

Acknowledgment

We thank the Research and Development Institute (KURDI), Kasetsart University, the Thailand Research Fund (Grant No. RSA6280088) and the National Research Council of Thailand (Grant No. NRCT5-RGJ63002-029) for financial support.

References

[1] Nettesheim S, Von Oertzen A, Rotermund H H and Ertl G 1993 *J. Chem. Phys.* **98** 9977
[2] Siegert F and Weijer C J 1989 *J. Cell Sci.* **93** 325
[3] Davidenko J M, Pertsov A M, Salomonz R, Baxter W and Jalife J 1992 *Nature* **335** 349
[4] Winfree A T 1972 *Science* **175** 634
[5] Cherry E M and Fenton F H 2008 *New J. Phys.* **10** 125016
[6] Winfree A T 1994 *Science* **266** 1003
[7] Cross M C and Hohenberg P C 1993 *Rev. Mod. Phys.* **65** 851
[8] Mikhailov A S and Showalter K 2006 *Phys. Rep.* **425** 79–194
[9] Lister T, O'Driscoll C and Reed N 1995 *Classic Chemistry Demonstrations* (London: The Royal Society of Chemistry) pp 1–69.
[10] Storb U, Neto C R, Bär M and Müller S C 2003 *Phys. Chem. Chem. Phys.* **5** 2344
[11] Porjai P, Sutthiopad M, Khaothong K, Phantu M, Kumchaiseemak N, Luengviriya J, Showalter K and Luengviriya C 2019 *Phys. Chem. Chem. Phys.* **21** 2419
[12] Kheowan O, Gáspár V, Zykov V S and Müller S C 2001 *Phys. Chem. Chem. Phys.* **3** 4747
[13] Nakata S, Matsushita M, Sato T, Suematsu J N, Kitahata H, Amemiya T and Mori Y 2011 *J. Phys. Chem.* A **115** 7406
[14] Steinbock O and Müller S C 1993 *Int. J. Bifurcat. Chaos* **2** 437
[15] Amemiya T, Kettunen P, Kádár S, Yamaguchi T and Showalter K 1998 *Chaos* **8** 872

Stability and electronic structure of magnesium hydride and magnesium deuteride under high pressure

Chayaphon Boonchot[1*], Prutthipong Tsuppayakorn-aek[2,3], Udomsilp Pinsook[1,3] and Thiti Bovornratanaraks[2,3]

[1]Department of Physics, Faculty of Science, Chulalongkorn University, 10330 Bangkok, Thailand
[2]Extreme Conditions Physics Research Laboratory (ECPRL) and Physics of Energy Materials Research Unit (PEMRU), Department of Physics, Faculty of Science, Chulalongkorn University, Bangkok, 10330, Thailand
[3]Thailand Center of Excellence in Physics, Ministry of Higher Education, Science, Research and Innovation, 328 Si Ayutthaya Road, Bangkok, 10400, Thailand

*E-mail: 6270129823@student.chula.ac.th

Abstract. Metal polyhydrides have attracted considerable attention because some of them become a metal under high pressure, and some undergo a phase transition into a superconductor. Some superconducting metal polyhydrides have recently been discovered with a high value of critical temperature (T_c) under pressure. In this research, we calculated the structures of MgH_2, MgH_3 and MgD_3 under pressure between 0-300 GPa in order to determine the formation enthalpy and electronic property of their structures under high pressure by using density functional theory (DFT) based on the Quantum Espresso code. We found that the band structures reveal the metallic character of the compounds under high pressure. The energy band structures of MgH_x and MgD_x are exactly the same. However, their phonon dispersions are different due to the so-called isotope effect. We determined the composition stability by using the convex hull of Mg, H and the compounds. We found that MgH_3 becomes thermodynamically more stable than MgH_2 at around 150 GPa. The results of phonons confirm that they are dynamically stable. This finding is served as a basis for future superconducting calculations.

1. Introduction

Metal polyhydrides have received much of attention in condensed matter research because some of them become a superconductor with a high critical temperature (T_c) under higher pressure [1]. The Mg/Ca substituted hexahydride at 200 GPa, exhibiting superconductivity with T_c = 288 K [2]. In this research, we are interested in magnesium hydrides (MgH_x). MgH_2 is widely used for the application of the hydrogen storage in electric vehicles with high gravimetric and volumetric capacities of H_2 (7.7 wt% and 109 gH_2) [3,4]. Under high pressure, the hydrogen content can be increased and the compound composition can be evolved to MgH_x, where x > 2. We reported the relative structural stability of MgH_2 and MgH_3 in the pressure range of 0-300 GPa. We also substituted the hydrogen atoms with deuterium (D) to determine the dynamical stability under high pressure. This finding will provide a basis for future superconducting critical temperature calculations.

2. Computational methods

In this work, all the calculations were performed by using the density functional theory (DFT) [5, 6] and the self-consistent field method, as implemented in the Quantum-ESPRESSO code [7]. We used the Perdew-Burke-Ernzerhof (PBE) functional for the generalized gradient approximation (GGA) [8,9] and the projector-augmented wave (PAW) method [10] for the pseudopotentials. The kinetic energy cut off of the wave functions in the plane wave basis set was chosen at 80 Ry, and the k-point grid was chosen to be 12×12×12 meshes for geometry optimization.

3. Results and discussion

The lattice parameters of MgH_2 and MgH_3 are presented in table 1. The structures of MgH_2 and MgH_3 are exhibited in figure 1. All structures were optimized under pressure between 0–300 GPa. The convex hulls in figure 2 showed that the Fm-3m structure of MgH_3 is thermodynamically stable between 100–300 GPa and 0–200 GPa for MgH_2. Furthermore, MgH_3 becomes relatively more energetically favourable than MgH_2 at 150 GPa onwards. We also substituted the D atoms instead of the H atoms. Nevertheless, the enthalpies per atom of MgH_3 and MgD_3 are the same because the pseudopotentials are the same, only the masses of H and D are different.

Table 1. Lattice parameters of MgH_2, MgH_3 at the atmospheric pressure (0 GPa).

	Space group	a (Å)	b (Å)	c (Å)	Density (g/cm^3)	Volume (Å^3)
MgH_2	Fm-3m	3.367	3.367	3.367	1.62	26.999
MgH_3	Fm-3m	3.327	3.327	3.327	1.74	26.042
MgD_3	Fm-3m	3.327	3.327	3.327	1.74	26.042

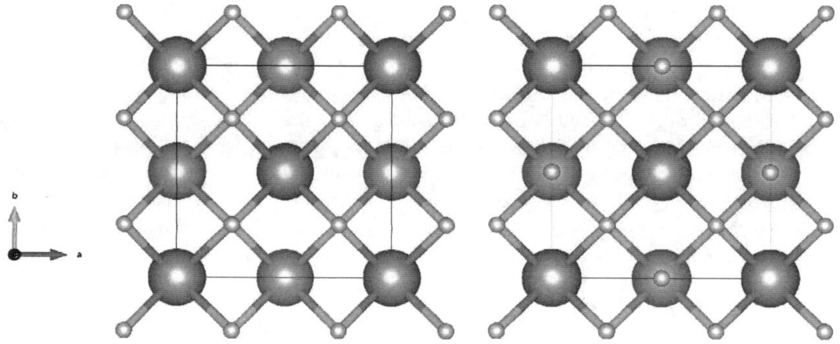

(a) (b)

Figure 1. The crystal structures of fcc MgH_2 (a) and fcc MgH_3 (b) (draw by VESTA [11]). The structure of MgD_3 is the same as of MgH_3.

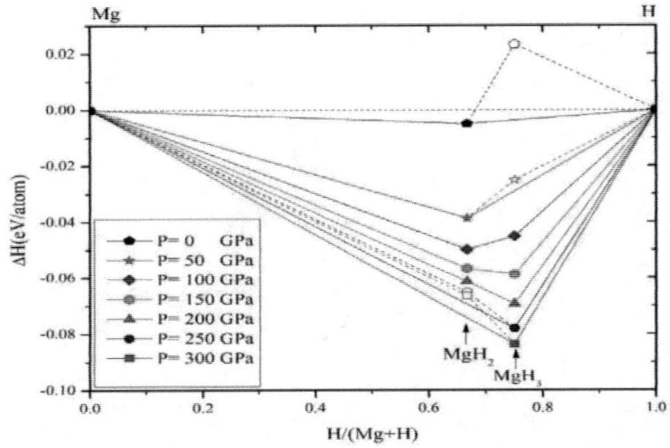

Figure 2. The enthalpy per atom of MgH_2 and MgH_3 are presented in the form of the convex hull. (solid markers refer to a stable structure, whereas transparent markers refer to a meta-stable structure.)

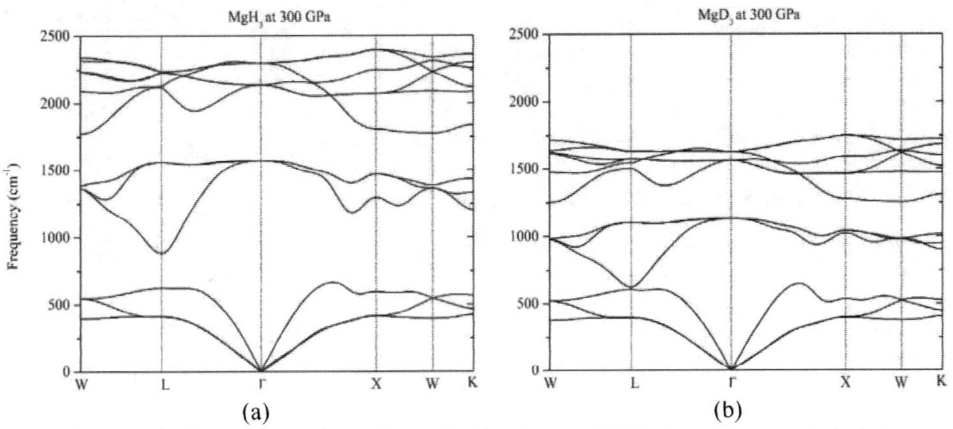

Figure 3. The phonon dispersions of MgH_3 (a) and MgD_3 (b) at pressure 300 GPa.

From the results of figure 2, we calculated the phonon dispersions of MgH_3 and MgD_3 in order to confirm the dynamical stability between P = 150–300 GPa. The examples of the phonon dispersion at 300 GPa are shown in figure 3. The phonon frequencies involving H of MgH_3 are higher than those involving D of MgD_3 at the same k point. This is the so-called isotope effect [12], which can be observed from the replacement of the hydrogen atoms (H) with the deuterium atoms (D). In fact, the mass of deuterium is twice heavier than the mass of hydrogen. Therefore, the phonon frequencies involving the vibrations of the hydrogen atoms are $\sqrt{2}$ times higher than those of deuterium atoms as $\omega \propto \frac{1}{\sqrt{m}}$. By closer inspection, we found that the phonon frequencies below 750 cm^{-1} in the left panel of figure 3 belong to the vibrations of the Mg atoms. These frequencies are mostly intact in the right panel of figure 3. However, the phonon frequencies above 750 cm^{-1} in the left panel of figure 3 belong to the vibrations of the hydrogen atoms. The corresponding modes in the right panel have $\frac{1}{\sqrt{2}}$ times lower frequencies than in the left panel of figure 3. These are due to the vibrations of the heavier deuterium atoms. These phonon dispersion results will be used as a basis for future superconducting critical temperature calculations.

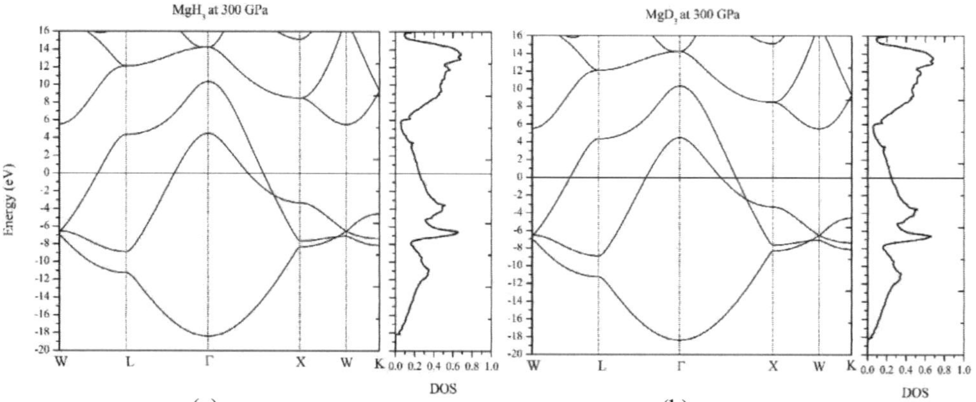

(a) (b)

Figure 4. The electronic band structures of MgH_3 (a) and MgD_3 (b) at pressure 300 GPa.

In addition, we calculated the electronic band structure of MgH_3 and MgD_3 between P = 150–300 GPa. The example of the electronic band structure of MgH_3 and MgD_3 at 300 GPa is shown in figure 4. The associating electronic density of states (DOS) is also shown. The electronic structures and their DOS show that MgH_3 and MgD_3 are metallic under pressure. However, at this level of DFT calculations, there is no distinction between the electronic structures of MgH_3 and MgD_3. In the future work, the van der Waals interaction will be included in order to distinguish between H and D, as suggested in the case of solid hydrogen under high pressure [13]. For simplicity of discussion, the BCS theory showed that $T_c \propto \omega_D e^{-\frac{1}{N_F V}}$ [1], where ω_D is the Debye frequency of the material (in units of temperature, or to be precise the Debye temperature), N_F is the density of electron at the Fermi level, and V is the electron-phonon interaction. Our calculations of the phonon dispersion and electronic structure provide the knowledge of ω_D and N_F. However, V is quite computationally expensive and will be pursued in our future work.

4. Conclusion

In this work, we have calculated the convex hull of MgH_2 and MgH_3 under pressure from 0–300 GPa by using DFT. We found that MgH_3 is thermodynamically stable between 100–300 GPa, and becomes relatively more energetically favourable than MgH_2 at 150 GPa onwards. The phonon calculations confirmed that MgH_3 is dynamically stable under high pressure, and the phonon frequencies of MgD_3 is $\frac{1}{\sqrt{2}}$ times lower than those of MgH_3 due to the isotope effect. We also reported the electronic band structures for MgH_3 and MgD_3, which appear to be exactly the same.

Acknowledgments

C. Boonchot would like to acknowledge the department of Physics, Faculty of Science, Chulalongkorn University, Bangkok, Thailand for the financial support. In addition, we would like to thank A. Sukserm, T. Pakornchote, W. Sukmas, and members of Extreme Conditions Physics Research Laboratory (ECPRL) for supports.

References

[1] Pinsook U 2020 *J. Met. Mater. Miner.* **30** 31–41
[2] Sukmas W, Tsuppayakorn-aek P, Pinsook U and Bovornratanaraks T 2020 *J. Alloys Compd.* **849** 156434
[3] Jain I P, Lal C and Jain A 2010 *Int. J. Hydrog. Energy.* **35** 5133–44

[4] Pluengphon P, Bovornratanaraks T, Tsuppayakorn-aek P, Pinsook U and Inceesungvorn B 2019 *Int. J. Hydrog. Energy.* **44** 21948–54
[5] Hohenberg P and Kohn W 1964 *Phys. Rev.* B **136** 864
[6] Kohn W and Sham L J 1965 *Phys. Rev.* A **140** 1133–8
[7] Giannozzi P *et al* 2009 *J. Phys. Condens. Matter* **21** 395502
[8] Perdew J P, Burke K and Ernzerhof M 1996 *Phys. Rev. Lett.* **77** 3865
[9] Perdew J P, Chevary J A, Vosko S H, Jackson K A, Pederson M R, Singh D J and Fiolhais C 1992 *Phys. Rev.* B **46** 6671
[10] Blöchl P E 1994 *Phys. Rev.* B **50** 17953
[11] Momma K and Izumi F 2008 *J. Appl. Crystallogr.* **41** 653–8
[12] Hinks D G and Jorgensen J D 2003 *Physica* C *Supercond.* **385** 98–104
[13] Azadi S and Ackland G J 2017 *Phys. Chem. Chem. Phys.* **19** 21829–39

Fabrication of SnO_2 by RF magnetron sputtering for electron transport layer of planar perovskite solar cells

R Thanimkan[1], B Namnuan[2] and S Chatraphorn[1,3 *]

[1]Physics of Energy Materials Research Unit, Department of Physics, Faculty of Science, Chulalongkorn University, Phyathai Rd., Bangkok 10330, Thailand
[2]Department of Physics, Faculty of Science, Silpakorn University, Nakhon Pathom 73000, Thailand
[3]Thailand Center of Excellence in Physics, CHE, 328 Si Ayutthaya Rd., Bangkok 10400, Thailand

[*] Corresponding author's e-mail: Sojiphong.C@chula.ac.th

Abstract. The requirements of electron transport layer (ETL) for high efficiency Perovskite solar cells (PSCs) are, for example, appropriate band energy alignment, high electron mobility, high optical transmittance, high stability, and easy processing. SnO_2 has attracted more attention as ETL for PSCs because it has diverse advantages, e.g., wide bandgap energy, excellent optical and chemical stability, high transparency, high electron mobility, and easy preparation. The SnO_2 ETL was fabricated by RF magnetron sputtering technique to ensure the chemical composition and uniform layer thickness when compared to the use of chemical solution via spin-coating method. The RF power was varied from 60 - 150 W. The Ar sputtering gas pressure was varied from 1×10^{-3} - 6×10^{-3} mbar while keeping O_2 partial pressure at 1×10^{-4} mbar. The thickness of SnO_2 layer decreases as the Ar gas pressure increases resulting in the increase of sheet resistance. The surface morphology and optical transmission of the SnO_2 ETL were investigated. It was found that the optimum thickness of SnO_2 layer was approximately 35 - 40 nm. The best device shows $J_{sc} = 27.4$ mA/cm^2, $V_{oc} = 1.03$ V, fill factor = 0.63, and efficiency = 17.7%.

1. Introduction

Perovskite is the next generation of photovoltaic devices because it has achieved high power conversion efficiency (PCE). The PCE has been evolved from 3.8 % to more than 23 % in less than 10 years. The fabrication process is also relatively simple. The normal structure of PSCs consists of five layers, i.e., transparent electrode (FTO or ITO), electron transport layer (ETL), perovskite absorber layer, hole transport layer (HTL), and metal electrode [1]. The requirements of ETL for high efficiency PSCs are (i) appropriate energy band alignment, (ii) high electron mobility, (iii) high optical transmittance, and (iv) high stability and easy processing [2]. The ETLs that have been proposed for PSCs are, for example, TiO_2, ZnO, SnO_2, etc. TiO_2 is usually used for ETL as a compact layer and a mesoporous layer. Both layers give relatively higher efficiency PSCs. However, TiO_2 layer has some limitations for PSCs such as it needs high temperature process and yields low electron mobility. The effect of TiO_2 layer negatively affects the device stability under ultraviolet (UV) illumination. In addition, ZnO is another choice of ETL. It can be deposited easily with low temperature process. However, ZnO is environmentally unstable due to the hydroxide (OH) residue on the ZnO surface causing the decomposition of perovskite layers. Recently, SnO_2 has attracted great attentions as an ETL for PSCs because it has diverse advantages, e.g., wide bandgap energy (3.6 - 4.0 eV), excellent optical and chemical stability, high transparency, high electron mobility (~ 240 cm^2/Vs), and easy preparation [3-4]. In this work, we demonstrated normal planar structure of PSCs with optimized room-temperature-processed SnO_2 as ETL prepared by spin-coating and RF magnetron sputtering process [5-7]. Two types of process were investigated and found that the devices based on SnO_2 film by RF magnetron sputtering showed superior performance. We considered the effects of RF sputtering parameters such as RF power, Ar and O_2 gas pressure in the performance of PSCs.

2. Experimental

2.1. Preparation of SnO_2 films

The 3 cm × 3 cm fluorine-doped tin oxide (FTO) substrates were etched with Zn powder and HCl for transparent electrode. The FTO substrates were ultrasonically cleaned with detergent (Micro 90), deionized (DI) water, acetone, and 2-propanol (IPA) for 30 min, respectively. The surface treatment of the FTO substrates were done by UV-ozone for 10 min prior to the deposition of ETL. In this work, SnO_2 films as ETL were prepared by spin-coating and RF magnetron sputtering techniques. For spin-coating, the SnO_2 film was deposited using $SnCl_4.5H_2O$ dissolved in IPA (0.15 M), as seen in figure 1 (a). The SnO_2 film was spin-coated at 3000 rpm for 40 s and then annealed on a hotplate at 200 °C for 60 min in air. For RF sputtering technique, Ar and O_2 gas were used during sputtering process as shown in figure 1 (b). RF sputtering power was varied from 60 - 150 W and the Ar sputtering gas pressure was varied from 1×10^{-3} - 6×10^{-3} mbar while keeping O_2 partial pressure at 1×10^{-4} mbar.

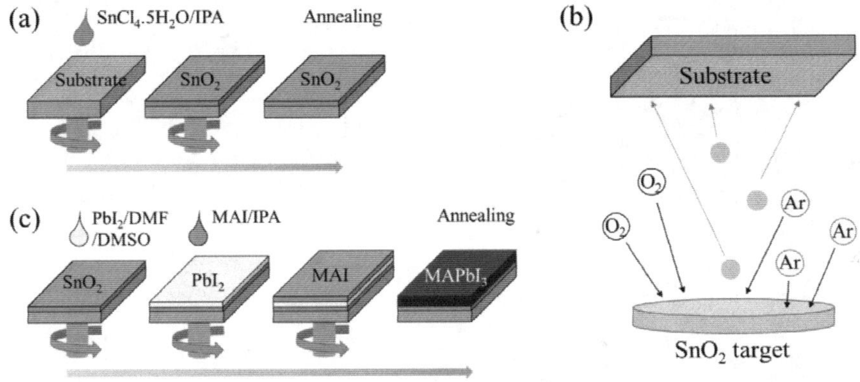

Figure 1. (a) spin-coating method of SnO_2 films, (b) RF magnetron sputtering deposition process, and (c) two-step spin-coating method to deposit $MAPbI_3$ perovskite films.

2.2. Device fabrication

The SnO_2 film was treated by UV-ozone for 10 min to improve its hydrophilicity. The methylammonium lead iodide ($MAPbI_3$ or $CH_3NH_3PbI_3$) perovskite thin film was deposited by the two-step spin-coating method in N_2-filled glovebox as shown in figure 1 (c). The PbI_2 solution was prepared from 0.461 g PbI_2 dissolved in 1 ml of DMF (*N,N*-Dimethylformamide) and DMSO (*N,N*-Dimethylsulfoxide) with 4:1 by volume and stirred at 70 °C for 12 hrs. Before depositing perovskite layer, the SnO_2 substrates were preheated on a hotplate at 70 °C for 10 min. The PbI_2 solution was deposited on the preheated SnO_2 by spin-coating at 3000 rpm for 30 s, and then annealed on a hotplate at 70 °C for 20 min. After that, the solution of methylammonium iodide (MAI) was spin-coated on the PbI_2 layer at 2000 rpm for 20 s, which consists of 0.010 g MAI dissolved in 1 ml of IPA and stirred for 12 hrs. In order to obtain a high quality perovskite absorber layer, the anti-solvent technique was employed while spin-coating the MAI layer. All the perovskite layers were annealed on a hotplate at 120 °C for 10 min. The HTL was deposited on top of the perovskite layer using Spiro-OMeTAD solution by a spin-coating at 3000 rpm for 30 s, which was prepared by dissolving 0.0723 g of Spiro-OMeTAD into 1 ml chlorobenzene (CB) and 28.8 μl of 4-tert-butylpyridine (t-BP) and 17.5 μl of lithium bistrifluoromethanesulfonimidate (Li-TFSI) (predissolved as 520 mg/ml solution in acetonitrile). It was worthy to note that the perovskite absorber layer and HTL process were carried out inside the N_2-filled glovebox. Finally, the Au metal electrode was deposited on the HTL by thermal evaporation.

2.3. Characterization
The surface morphology of the SnO_2 and perovskite films was investigated using a field-emission scanning electron microscopy (FESEM, JEOL Model JSM-7001F). The transmittance and absorbance spectra were investigated in the range of 330 - 1100 nm by a UV-Vis-NIR spectrophotometer (Shimadzu Model UVPC1600). The J-V curves of the PSCs were measured under AM 1.5 using a Xe-lamp solar simulator and a Keithley 238 source-meter unit. The PSC devices under investigation have an active area of 0.06 and 0.2 cm^2.

3. Results and discussion
3.1. Spin-coated SnO_2 films
The morphology of SnO_2 film deposited on soda-lime glass (SLG) substrates by spin-coating method is shown in figure 2 (a). The SnO_2 films can fully cover the SLG substrates. However, it can be seen that when SnO_2 precursor was deposited by spin-coating method, some particulates from recrystallization of $SnCl_4$ precursor was obtained on some area of SnO_2 surface. Figure 2 (b) shows the J-V curves of the best PSC by spin-coating of SnO_2 film measured under both forward and reverse voltage scan directions. The PCE was about 12.9 % and the open-circuit voltage (V_{oc}), the short-circuit current density (J_{sc}), and the fill factor (FF) of the best device are 1.04 V, 18.4 mA/cm^2, and 0.68, respectively. The J-V curves show some slight hysteresis between forward and reverse directions.

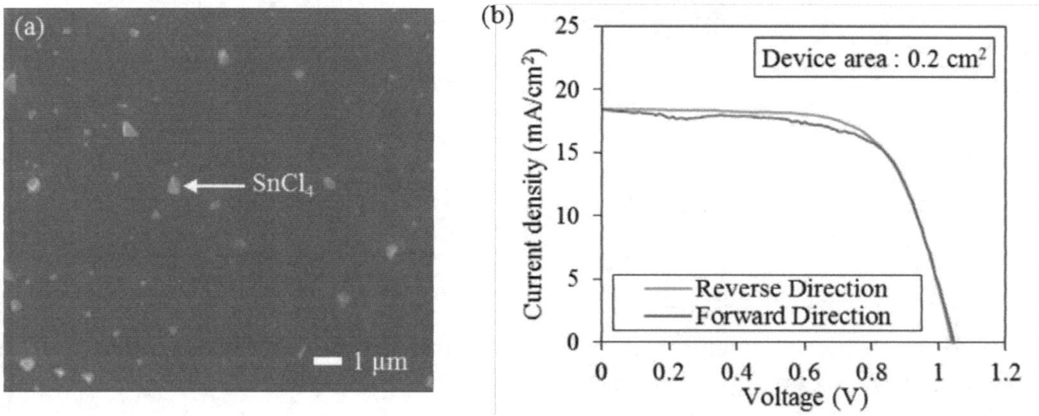

Figure 2. (a) SEM image shows surface morphology of spin-coated SnO_2 film on SLG substrates. (b) J-V curves of the best PSC based on spin-coated SnO_2 film measured under forward and reverse voltage scans.

3.2. Sputtered SnO_2 films
SEM image of surface morphology of sputtered SnO_2 film on SLG substrates shows very smooth surface as shown in figure 3 (a), The cross-section of a complete PSC is shown in figure 3 (b). The J-V curves of four conditions are compared for the sputtered SnO_2 films under various RF sputtering power of 60, 90, 120 and 150 W. It is shown in the J-V curves in figure 4 (a) that the performance of PSCs become worse with increasing RF sputtering power. Table 1 summarizes the photovoltaics parameters of PSCs with different sputtering power. The best device was obtained with 60 W RF sputtering power with champion PCE of 17.7 %, V_{oc} of 1.03 V, J_{sc} of 27.4 mA/cm^2, and fill factor of 0.63.

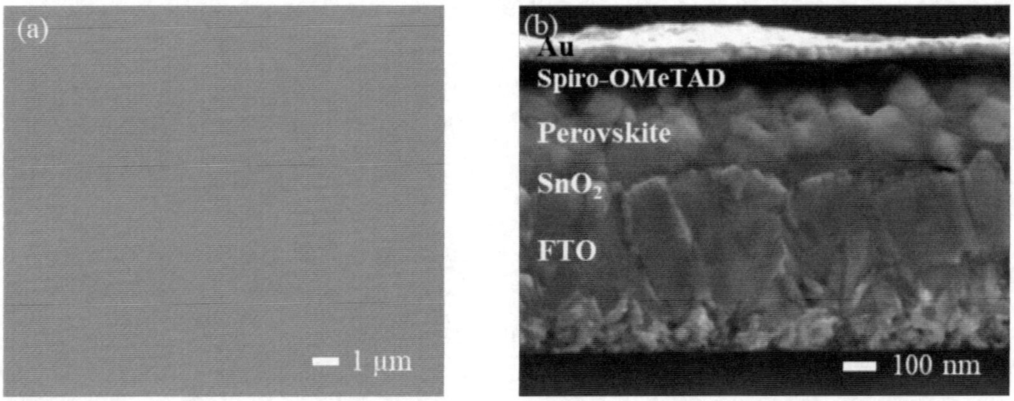

Figure 3. SEM image shows (a) smooth surface morphology of sputtered SnO_2 film on SLG substrate, and (b) cross-section of a complete PSC.

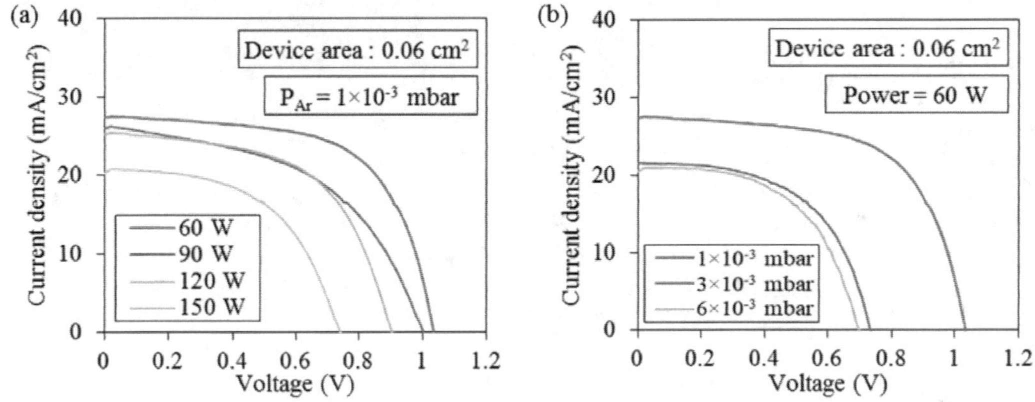

Figure 4. J-V curves of PSCs based on sputtered SnO_2 film; (a) with different RF sputtering power, and (b) RF sputtering power of 60 W at different Ar gas pressures while keeping O_2 gas pressure.

Table 1. The solar cell parameters of PSCs based on different RF sputtering power of sputtered SnO_2 film.

RF power (W)	V_{oc} (V)	J_{sc} (mA/cm^2)	Fill factor	PCE (%)
60	1.03	27.4	0.63	17.7
90	1.00	26.0	0.50	12.9
120	0.90	25.2	0.57	12.9
150	0.74	20.3	0.55	8.2

In order to investigate the effect of Ar sputtering gas pressure on the device performance. The complete devices were fabricated on the sputtered SnO_2 films under various Ar gas pressure (1×10^{-3}, 3×10^{-3}, and 6×10^{-3} mbar) while keeping O_2 gas pressure at 1×10^{-4} mbar. As shown in figure 4 (b), with decreasing the Ar gas pressure, the device performance was gradually improved. The optical transmission of the sputtered SnO_2 films were greater than 85% in the range of solar spectrum as shown in figure 5. The best device was obtained when the Ar gas pressure was 1×10^{-3} mbar, showing high V_{oc} = 1.03 V, J_{sc} = 27.4 mA/cm^2, fill factor = 0.63, and PCE of 17.7 %. The detailed photovoltaics parameters were listed in table 2. The increase in Ar gas pressure results in the decrease of mean free path, leading to the increase in the deposition time to achieve the optimum thickness about 35 – 40 nm. In addition, decreasing the thickness of SnO_2 layer corresponded to increasing of sheet resistance.

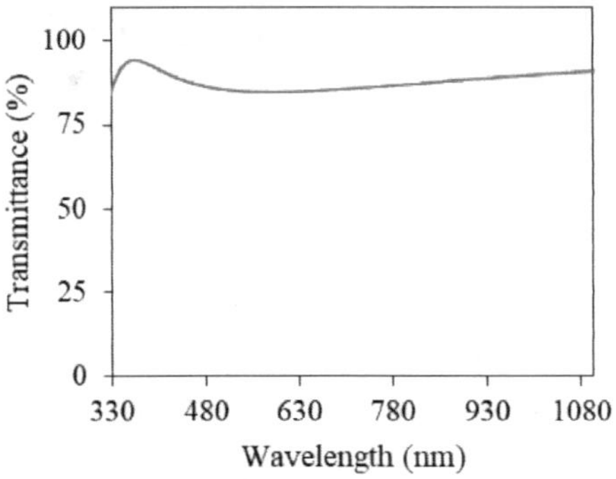

Figure 5. Optical transmission spectrum of the sputtered SnO_2 film.

Table 2. Photovoltaic parameters of PSCs with sputtered SnO_2 ETL with different Ar gas pressure.

Ar pressure(mbar)	V_{oc} (V)	J_{sc} (mA/cm^2)	Fill factor	PCE (%)
1×10^{-3}	1.03	27.4	0.63	17.7
3×10^{-3}	0.74	21.0	0.58	9.0
6×10^{-3}	0.70	20.5	0.56	8.0

4. Conclusion

SnO_2 films for use as an ETL layer in planar PSCs were fabricated and compared for the performance of PSCs employing SnO_2 films by spin-coating and RF magnetron sputtering techniques. It was found that sputtered SnO_2 based devices were demonstrated to have better device performance and stability than spin-coated SnO_2 based devices. Wet chemical processes were avoided to minimize the particulates from recrystallization of SnO_2 precursor. This also leads to uneven surface of the ETL layer. It can also be observed that improper sputtering parameters strongly affect the V_{oc}, J_{sc} and thus the PCE. In this work, the PSC based on sputtered SnO_2, with the RF sputtering power of 60 W and Ar gas pressure of 1×10^{-3} mbar with O_2 gas partial pressure of 1×10^{-4} mbar showed champion PCE of 17.7 %.

Acknowledgments

This work was financially supported by Thailand Center of Excellence in Physics, Thailand. The authors would like to thank P Panchawirat (ThEP Center, Thailand) for helping with FESEM measurements.

References

[1] Lee J and Park N 2015 *MRS Bull.* **40** 654–9
[2] Jiang Q, Zhang X and You J 2018 *Small* **14** 1801154
[3] Xiong L, Guo Y, Wen J, Liu H, Yang G, Qin P and Fang G 2018 *Adv. Funct. Mater.* **28** 1802757
[4] Yi H, Wang D, Mahmud M, Haque F, Upama M, Xu C, Duan L and Uddin A 2018 *ACS Appl. Energy Mater.* **1** 6027–39
[5] Kam M, Zhang Q, Zhang D and Fan Z 2019 *Sci. Rep.* **9** 6963
[6] Bai G, Wu Z, Li J, Bu T, Li W, Huang F, Zhang Q, Cheng Y and Zhong J 2019 *Sol. Energy* **183** 306–14
[7] Otoufi M, Renjbar M, Kermanpur A, Taghavinia N, Minbashi M, Forouzandeh M and Ebadi F 2020 *Sol. Energy* **208** 697–707

Journal of Physics: Conference Series **2145** (2022) 012028 doi:10.1088/1742-6596/2145/1/012028

Improving the triple-cation perovskite solar cells by two-step deposition methods with perovskite seeds

P Phiromruk[1,2] **and S Chatraphorn**[1,2*]

[1] Physics of Energy Materials Research Unit, Department of Physics, Faculty of Science, Chulalongkorn University, Phyathai Rd., Bangkok 10330, Thailand
[2] Thailand Center of Excellence in Physics, CHE, 328 Si Ayutthaya Rd., Bangkok 10400, Thailand

*Corresponding author's e-mail: sojiphong.c@chula.ac.th

Abstract. As of recent years, triple-cation perovskite solar cells have received immense attention due to its superior efficiency and better stability comparing to the classic single-cation perovskite solar cells such as $MAPbI_3$ or $FAPbI_3$. A triple-cation perovskite layer which has been used most recently is cesium-containing $FAPbI_3$-based perovskite. One of decent approaches to fabricate the layer is spin-coating technique by using two-step deposition process in which mixed lead-halide and CsI precursor is firstly spin-coated onto a substrate, then organic cation solution is deposited on the lead-halide layer. In this work, the results show that the performance of the devices from this process is lower than expected that could be due to difficulty of cesium ion incorporation as a stabilizer for $FAPbI_3$-based perovskite. Perovskite seeding growth is introduced to solve the problem where the process is slightly modified from conventional two-step deposition methods by adding small amount of perovskite seed precursor into PbI_2 solution. The concentration of the perovskite seed in PbI_2 solution was varied for 0, 7, 14 and 20% v/v. The highest average efficiency of 11.9% was obtained from 7% v/v seeding concentration. Furthermore, the device performance could be improved by using proper amount of chlorobenzene (CB) as an anti-solvent. The highest efficiency of 18.4% was achieved by using 30 µl of chlorobenzene.

1. Introduction

Over the past decade, the improvement of perovskite solar cells has been incredibly astounding in which the power conversion efficiency (PCE) rose from 3.8% in 2009 to >20% in the recent years because of high absorption over the visible spectrum, long charge carrier diffusion length, as well as engineering composition [1,2]. Perovskite materials for solar cells have ABX_3 structure where A is an organic monovalent cation such as $CH_3NH_3^+$ (methylammonium: MA^+), $CH_3(NH_2)_2^+$ (formamidinium: FA^+), and Cs^+, B is typically Pb^{2+} and X is a halide ion ,e.g., I^- and Br^-.

The conventional perovskite materials ($MAPbI_3$ and $FAPbI_3$) have been studied and shown that the disadvantage of $MAPbI_3$ is that it is sensitive to heat and humidity [3-5], whereas $FAPbI_3$ appears to be a better option in comparison with $MAPbI_3$ due to lower band gap and greater thermal stability [5]. However, $FAPbI_3$ lacks structural stability, namely, it has two different structures at room temperature. One is photo-inactive hexagonal structure (yellow phase) and the other one is photo-active cubic structure (black phase). In addition, the black phase of $FAPbI_3$ only stabilizes at high temperature [6]. These problems could be fixed by adding small amount of MA ions as a stabilizer for black phase of $FAPbI_3$ [1] and the efficiency of double-cation perovskite, such as $FA_{0.83}MA_{0.17}Pb(I_{0.83}Br_{0.17})_3$, reached

18% [7]. Nonetheless, there is the trace of yellow phase left in case of double-cation perovskite [1]. Triple-cation perovskite was introduced to solve the problem. The triple-cation perovskite $Cs_{0.05}(FA_{0.83}MA_{0.17})_{0.95}Pb(I_{0.83}Br_{0.17})_3$ was fabricated by one-step deposition method and the PCE from this process exceeded 20% and remained stable at 18% for at least 250 hours at maximum power point tracking [1]. Latterly, some reports demonstrated that the performance of $FAPbI_3$-based devices from one-step deposition process are inferior to that of two-step deposition method [8].

In this work, the two-step deposition method is a process in which mixed lead-halide and CsI precursor is firstly spin-coated onto a substrate, then organic cation solution is deposited on the lead-halide layer. Instead of mixing CsI with organic cation salts in IPA, CsI was dissolved in DMF because of the low solubility of CsI. As a result, the efficiency from this process was considerably lower than expected due to the difficulty of cesium ion incorporation as a stabilizer for $FAPbI_3$-based perovskite and CsI might assemble with PbI_2 and acquire $CsPbI_3$. Though $CsPbI_3$ has an acceptable band gap of 1.73 eV, the photoactive perovskite phase is only stable at very high temperature (more than 300 °C) [1]. One also suggested that adding CsI directly with lead-halide solvent could lead to discrepancy in performance of the devices among batches resulting from poor perovskite nucleation [8]. Two-step deposition method with perovskite seeds could be used to assist cesium ions to stabilize the black phase of $FAPbI_3$ [8]. This process resembles to the combination of one-step and two-step deposition methods. Perovskite seed precursor was separately prepared from PbI_2 solution, then the perovskite seed precursor is added in PbI_2 solution and the rest of the steps are the same as the conventional two-step deposition method.

2. Methodology

2.1. Materials preparation

For conventional two-step deposition method; 1 M of PbI_2 and 0.14 M of $PbBr_2$ were dissolved in mixed solvent (DMF: DMSO = 4:1 volume ratio) and stirred overnight at 70 °C. 1 M of CsI was dissolved in mixed solvent (DMF: DMSO = 4:1 volume ratio) then stirred overnight at room temperature and mixed in lead-halide solution. MAI and FAI organic cation salts were dissolved in 2-propanol with a concentration of 10 mg/ml and stirred overnight at room temperature. For two-step deposition method with perovskite seeds, 1.8725 g of PbI_2 was dissolved in 1 mL DMF and 160 μL DMSO and stirred overnight at 70 °C. Perovskite seed precursor was prepared by mixing PbI_2 (1.2 M), $PbBr_2$ (0.21 M), FAI (1.08 M), MABr (0.18 M), and CsI (0.14 M) in DMF: DMSO = 4:1 volume ratio then stirred at about 60 °C for 10 min and mixed in PbI_2 solution to form $Cs_{0.09}FA_{0.77}MA_{0.14}Pb(I_{0.85}Br_{0.15})_3$ perovskite seed. The seeding concentration in this work was 7, 14, and 20% v/v. The organic cation solution was obtained from FAI (8.3 mg), MABr (0.83 mg), and MACl (0.87 mg) salts dissolved in 1 ml of IPA and stirred overnight at room temperature. Also, 0.0723 g of Spiro-MeOTAD was dissolved in 1 ml of chlorobenzene and stirred overnight at room temperature. 28.8 μl of 4-tert-butylpyridine(t-BP) and 17.5 μl of Lithium bistrifluoromethanesul formamide (Li-TFSI) of which 0.52 g dissolved in 1 ml of acetonitrile were mixed in 1 ml of Spiro-MeOTAD and kept stirring for 3 hours before use.

2.2. Device fabrication

Fluorine-doped tin oxide (FTO)-coated glasses were cleaned by using detergent, deionized water, acetone, and 2-propanol, sequentially. Compact-TiO_2 film (~50 nm) was spin-coated on the FTO-glass at 3000 rpm for 30s, then the film was annealed for 30 min at 500 °C. Mesoporous-TiO_2 film (~250 nm), after that, was deposited on the cp-TiO_2 layer at 3000 rpm for 30s, the film was then heated at 500 °C for 30 min. Next, lead-halide layer was spin-coated on the substrate at the rate of 3000 rpm for 30s, then annealed at 70 °C for 20 min and cool down. To form the perovskite layer (~600 nm), organic cation solution was then spin-coated at 2000 rpm for 20s, and repeated again, then followed by 50 μl of chlorobenzene (CB) while the substrate was still spinning. Next, the perovskite layer was placed on a hot plate at 140 °C for 10 min. After cooling down, Spiro-MeOTAD layer (~150 nm) was spin-coated on perovskite layer at 3000 rpm for 30s and kept overnight for oxidation of the Spiro-MeOTAD. All the processes were carried out in N_2 filled glove box. Finally, 120 nm thick of Au contact was thermally evaporated on top of the spiro-MeOTAD layer.

2.3. Characterization

To study optical transmission and morphology of perovskite films, perovskite layer was deposited on FTO/Cp-TiO_2/mp-TiO_2. $Cs_{0.03}(FA_{0.90}MA_{0.10})_{0.97}Pb(I_{0.9}Br_{0.1})_3$ and $Cs_{0.03}(FA_{0.85}MA_{0.15})_{0.97}Pb(I_{0.9}Br_{0.1})_3$ films were obtained from conventional two-step deposition methods. For two-step deposition methods with perovskite seeds, the samples were prepared by PbI_2 solution with 0, 7, 14, 20% v/v seeding concentration. The optical transmission (%T) of perovskite films was examined by a spectrophotometer in the range of 350 – 1100 nm. Morphology of the perovskite films was characterized by a field-emission scanning electron microscope (FESEM; JEOL Model JSM-7001F). Solar cell parameters, e.g., PCE, fill factor (FF), short-circuit current density (J_{SC}) and open-circuit voltage (V_{OC}) were investigated under AM1.5 illumination by using Xe-lamp solar simulator and Keithley 238 source meter-unit and the active area of devices was 0.06 cm^2.

3. Results and discussion

3.1. Optical transmission

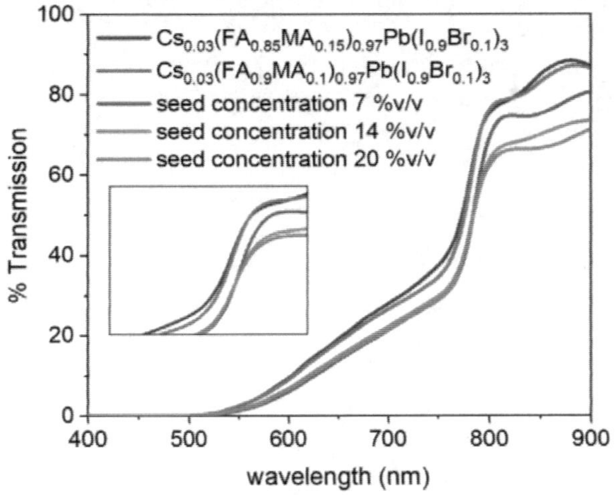

Figure 1. Optical transmission of perovskite films fabricated by two-step deposition method (black and red lines) and two-step deposition method with perovskite seeding concentration 7 (blue line), 14 (green line), and 20% v/v (purple line). [Inset shows the zone of absorption of perovskite films]

The optical transmission of perovskite films fabricated by traditional two-step deposition process (black and red lines) and two-step deposition process with perovskite seeds (blue, green, and purple lines) are plotted in figure 1. For the conventional process, the onset of optical absorption was observed at around 790 nm while it was slightly red shifted for the seeded-samples with the onset of optical absorption around 795 nm. The optical transmission of the samples from the new method are almost identical regardless of seeding concentration because very small amount of perovskite seeds was incorporated when compared to the content of the primary cations and halide ions. Thus, band gap energy among the seeded perovskite samples was unaltered.

3.2. Morphology

Figure 2 shows the surfaces of perovskite films fabricated by two-step deposition process (a) $Cs_{0.03}(FA_{0.85}MA_{0.15})_{0.97}Pb(I_{0.9}Br_{0.1})_3$ and (b) $Cs_{0.03}(FA_{0.90}MA_{0.10})_{0.97}Pb(I_{0.9}Br_{0.1})_3$ and two-step deposition process with seeding concentration (c) 7, (d) 14, and (e) 20% v/v. The grain shape and size from both methods were quite similar since the content of seeding concentration was very small as previously discussed. Even so, the seeded perovskite films looked denser and more compact causing smoother surface coverage for better contact between perovskite layer and hole transport layer and enhancing the

Figure 2. SEM images of perovskite films fabricated by two-step deposition method (a) $Cs_{0.03}(FA_{0.85}MA_{0.15})_{0.97}Pb(I_{0.9}Br_{0.1})_3$ and (b) $Cs_{0.03}(FA_{0.90}MA_{0.10})_{0.97}Pb(I_{0.9}Br_{0.1})_3$ and 2-step deposition method with seeding concentration (c) 7, (d) 14, and (e) 20% v/v, (f) cross-section image of triple-cation perovskite solar cell.

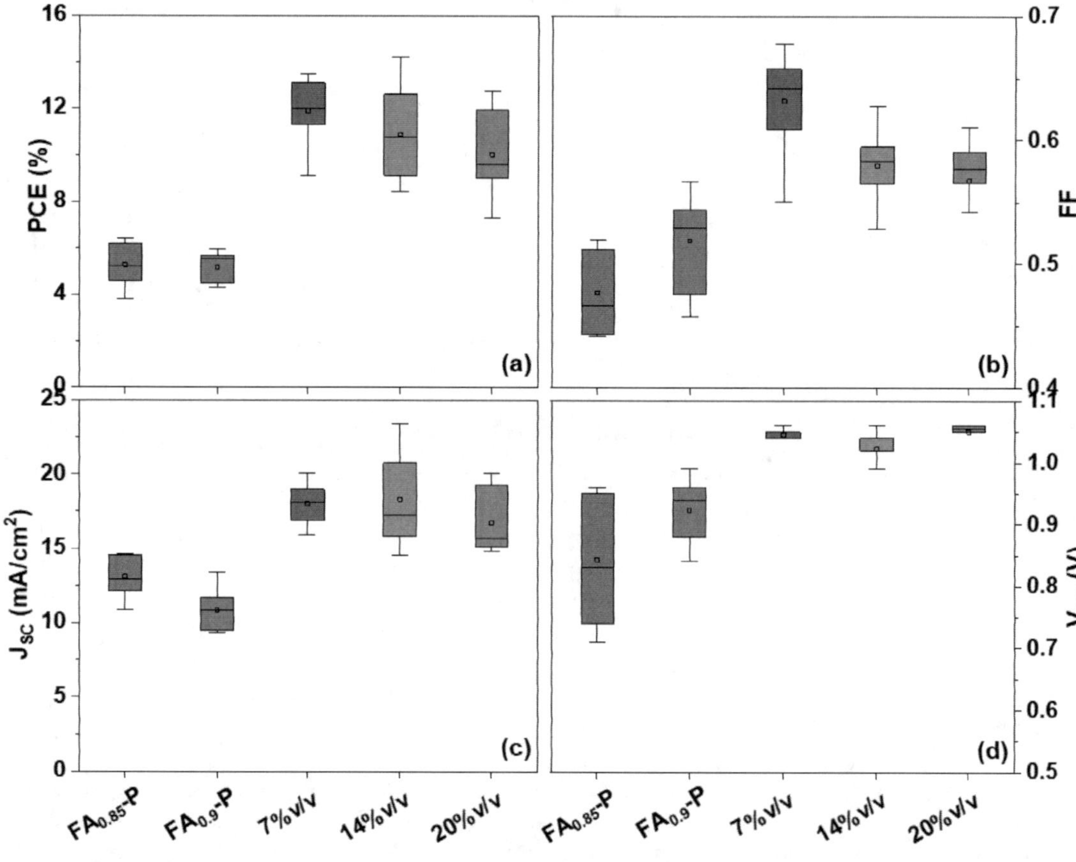

Figure 3. Box chart plots of solar cell parameter: (a) PCE, (b) FF, (c) J_{SC}, and (d) V_{OC} from two-step deposition method: $Cs_{0.03}(FA_{0.85}MA_{0.15})_{0.97}Pb(I_{0.9}Br_{0.1})_3$ (denoted as $FA_{0.85}$-P) and $Cs_{0.03}(FA_{0.90}MA_{0.10})_{0.97}Pb(I_{0.9}Br_{0.1})_3$ (denoted as $FA_{0.9}$-P) and two-step deposition method with different seeding concentration. 50 µl of CB is applied for both processes.

performance of the devices. The small amount of perovskite seed in PbI_2 could help the nucleation of perovskite crystals and thus improved the crystal formation of the perovskite layer.

3.3. Solar cells parameters

J-V characteristics were investigated to obtain the solar cell parameters and are illustrated in figure 3. The grey and red box charts represent $Cs_{0.03}(FA_{0.85}MA_{0.15})_{0.97}Pb(I_{0.9}Br_{0.1})_3$ (denoted as $FA_{0.85}$-P) and $Cs_{0.03}(FA_{0.90}MA_{0.10})_{0.97}Pb(I_{0.9}Br_{0.1})_3$ (denoted as $FA_{0.90}$-P) devices, respectively. The blue, green, and purple box charts represent devices from two-step method with different seeding concentration. The box charts indicate that the overall performance from conventional two-step deposition method was considerably poor comparing to that from two-step deposition method with perovskite seeds. For instance, the average PCE from the old method was only at 5.2%. On the other hand, the average PCE of 11.9% was from the two-step deposition method with seeding concentration 7% v/v. Also, short-circuit current density and open-circuit voltage, from perovskite seed process were clearly superior. The enhanced parameters of the seeded samples could be due to the improvement of the surface of perovskite films. Figure 4 shows the J-V curves of devices with best efficiency from each process. Furthermore, the amount of anti-solvent was found to be of importance as well. The use of chlorobenzene as anti-solvent of 30, 50, and 100 µl were studied in the two-step deposition process with perovskite seeds. As shown in figure 5, the devices with 30 µl of chlorobenzene yielded somewhat higher efficiency than devices with 50 and 100 µl of chlorobenzene. The highest efficiency of 18.4% was attained from two-step deposition method with 7% v/v seeding concentration and the amount of 30 µl of anti-solvent applied.

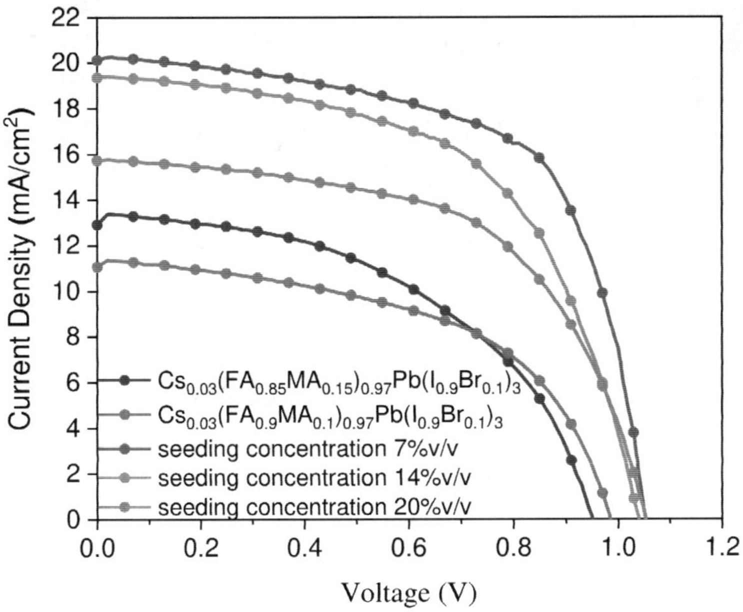

Figure 4. J-V curves of the best PCEs from two-step deposition method (black and red lines) and two-step deposition method with perovskite seed concentration 7 (blue line), 14 (green line), 20% v/v (purple line). 50 µl of CB is applied for both processes.

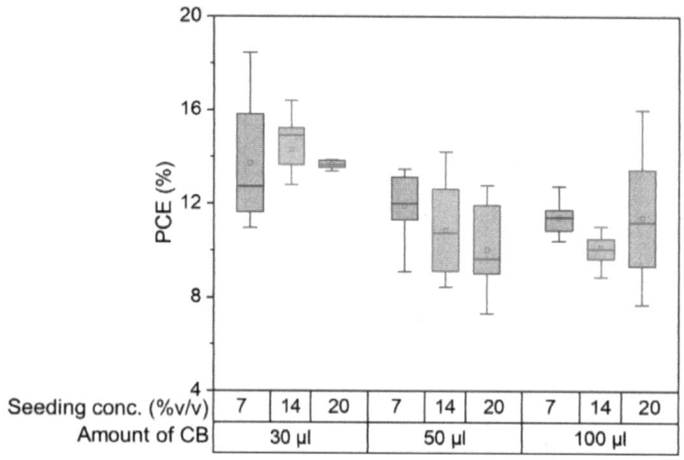

Figure 5. Box chart plots of PCE from two-step deposition method with seeding concentration 7 (grey), 14 (red), and 20% v/v (blue) and different amount of anti-solvent.

4. Conclusion

$FAPbI_3$-based triple-cation perovskite solar cells were of interest because of its performance and stability and two-step deposition method was an intriguing option to fabricate such devices. The triple-cation perovskite films on ETL layer from two-step deposition method and two-step deposition method with perovskite seeds were studied. The optical transmission and morphology of perovskite films were not evidently affected by such processes apart from the fact that perovskite films from two-step deposition method with perovskite seeds were denser resulting in improved surface coverage. The PCE of devices from the seeding method yielded higher efficiencies compared to the conventional one because the crystal formation from PbI_2 to perovskite was enhanced by using two-step deposition method with perovskite seeds and the improved contact between perovskite layer and HTL. In addition, Cs in perovskite seeds could better stabilize the black phase of $FAPbI_3$. It was also found that the suitable amount of anti-solvent could enhance the efficiency of devices from two-step deposition process with perovskite seeds. The best efficiency of 18.4% was obtained from two-step deposition method with 7% v/v seed concentration and the use 30 μl of chlorobenzene as anti-solvent.

Acknowledgement

This work was financially supported by Thailand Center of Excellence in Physics, Thailand. The author would like to also thank P Panchawirat (ThEP Center, Thailand) for SEM measurement.

References

[1] Saliba M *et al.* 2016 *Energy Environ.* Sci. **9** 1989–97
[2] Jena A K, Kulkarni A and Miyasaka T 2019 *Chem. Rev.* **119** 3036–103
[3] Conings B *et al.* 2015 *Adv. Energy Mater.* **5** 1500477
[4] Noh J H, Im S H, Heo J H, Mandal T N and Seok S I 2013 *Nano Lett.* **13** 1764–9
[5] Eperon G E, Stranks S D, Menelaou C, Johnston M B, Herz L M and Snaith H J 2014 *Energy Environ. Sci.* **7** 982–8
[6] Yi C, Luo J, Meloni S, Boziki A, Ashari-Astani N, Grätzel C, Zakeeruddin S M, Röthlisberger U and Grätzel M 2016 *Energy Environ. Sci.* **9** 656–62
[7] Baena J P C *et al.* 2015 *Energy Environ. Sci.* **8** 2928–34
[8] Zhao Y *et al.* 2018 *Nat. Commun.* **9** 1607

Facile synthesis of high purity silica xerogel from rice straw

Oranut Yosma and Chalad Yuenyao*

Department of Physics, Faculty of Science and Technology, Phetchabun Rajabhat University, 83 Moo 11 Saraburi-Lom Sak Road, Sadiang Sub-district, Muang, Phetchabun 67000, Thailand

*Corresponding author's e-mail address: chalad.yue@pcru.ac.th

Abstract. This research aims to synthesize a silica xerogel from rice straw that is a residue biomass generating from agriculture. Purity and morphological structure of synthesized silica xerogel are also studied. The first step of the synthesis is the preparation of sodium silicate from rice straw ash that is then used as silica source. To prepare a silica source, pretreated rice straw was burnt at 700 °C for 2 h to obtain a rice straw ash. After that resulted rice straw ash is washed and reacted with 1.0 M HCl and 2.0 M NaOH aqueous solution at 80 °C for 1 h, respectively. The reacted solution is then filtrated two times by a no.41 filter paper and ion exchange resin, respectively. The obtained sodium silicate is mixed with 1.0 M HCl under stirring for 6 h to produce the nano-silica. To increase the purity of nano-silica xerogels, as-synthesized silica was washed with deionized water for 3 times. Washed silica is dried in electric oven at 85 °C overnight and calcined at 500 °C for 5 h. Scanning electron microscopy/energy dispersive x-ray spectrometry (SEM/EDS) and % whiteness are employed to evaluate the morphology and purity of particles. Experimental results showed that nano-silica with purity up to 99.0 wt% was completely synthesized. Different morphological structure of silica synthesized under pH of 7, 8 and 9 were obtained.

1. Introduction

Nowadays, farmers in Thailand produce large amount of agricultural biomass residue such as sugarcane straw, maize straw and rice straw. Some of these residues are used as fuel in the power plant and as a raw materials in organic fertilizer production. However, much of the rest biomass residues are eliminated by burning in the fields. This burning of biomass residues is one of the most important cause of PM. 2.5 issue and others air pollution. One of the most important ways to overcome this problem is the reduction of inefficient biomass residues burning. Utilization of biomass residues to produce a high value added products likes activated carbon and silica nanoparticles, is the other ways to mitigate this problem. From literatures, silica is one of the composition of ash generating from biomass such as rice husk, rice straw and maize straw. In case of rice straw, it was found that a silica content is about 14 and 82 wt% for rice straw and rice straw ash (RSA) [1,2], respectively. In addition, content of silica increase up to 91.0 wt% by pretreatment of rice straw with acid solution [3]. Many research groups pay attention to synthesis the silica gels due to their show promising applications in a new range of new technologies such as functional nanomaterials, nanoelectronics, environmental treatment and energy storage [4]. Generally, the produced silica gel is classified into three types depending on the amount of water in the silica gel structure that is (1) Aqua-gel (rich of water gel) (2) Xerogel (drying gel) (3) Aerogel (super-critical drying gel) [5]. The preparation of silica xerogel from the ash of wheat requires the preparation of sodium silicate as the starting material of silica, and then the preparation of gel by reducing the state of acidity to the surface of the silica. Before being neutralized by salt titrate, the results of this process were found to be highly effective in improving the purity of gel. In this study, we focused on the easily, low energy and low cost method to synthesize a silica xerogel from rice straw.

In this work, experimental results from sintering of silica xerogel extracted from rice straw ash was presented. Influence of gelation pH on the morphological structure of the produced silica xerogel was studied through the micrograph from Scanning Electron Microscope (SEM). The study hope to increase the value of agricultural waste and expecting to extract silica from rice straw ash with purity up to 95%. This research is also interested to studying the physical structure and purity of silica xerogels by physical techniques.

2. Materials and methods

Hydrochloric acid (HCl, 37%) and sodium hydroxide in the pellet forms (NaOH, 98.0 % purity) were purchased from RCI LabScan and Sigma-Aldrich, respectively. Ashless filter papers no. 41 (Whatman™) were supplied from GE Healthcare Life Sciences.

In this research, rice straw was divided into two groups. The first group was just washed by tap water and RO-water, respectively while the second group of rice straw was washed by pure water for three times and treated with 0.1 M HCl at 80 °C for 1 h, respectively. After treated with HCl, 2nd group of rice straw was washed again for three times with pure water before drying at 90 °C for 15 h in electric oven. To obtain a rice straw ash (RSA), pretreated rice straw was then burnt at 700 °C for 2 h.

To prepare sodium silicate to be used as silica source from RSA, 20 g of RSA was treated with 100 ml of 1.0 M HCl and reacted with 60 ml of 2.0 M NaOH at 80 °C for 1 h. Resultant mixture was filtrated two times by ashless filter papers (No.41) and strong acid ion exchange resin, respectively. Afterwards, sodium silicate was tritrated with 1.0 M HCl to adjust a pH to 7, 8 and 9. After the adjusted pH mixtures were placed at room temperature for 18 h, phase of mixture change from liquid form to gel form. Gel was then exploded and washed with pure water for 3 times and dried at 85 °C for 12 h, respectively. These processes can increase the purity of produced-gel from biomass residue [3,6]. Steps of preparation of silica xerogel from RSA was illustrated in figure 1. Morphological structure and purity in term of % whiteness of synthesized silica gels were characterized by SEM and spectrophotometer, respectively.

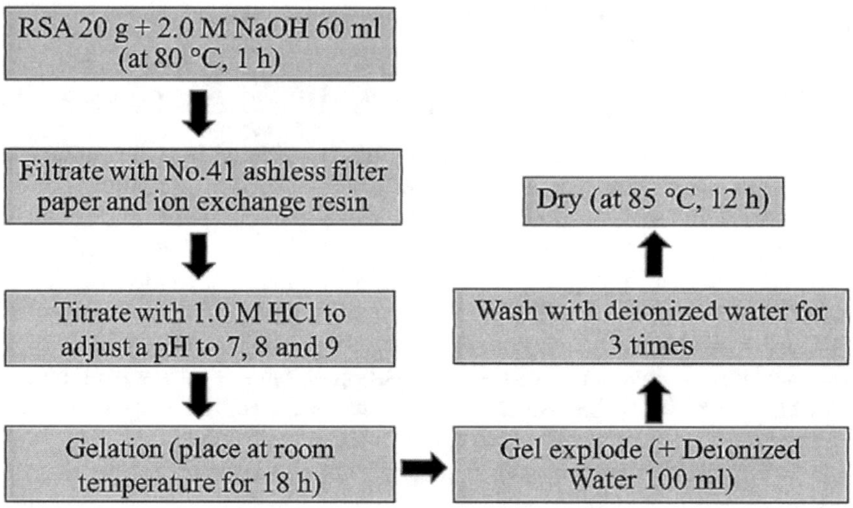

Figure 1. Synthesis process of silica xerogels from RSA.

3. Results and discussion

Figure 2 illustrates the difference of rice straw ash obtained from rice straw. It was showed that the color of ash influenced from the acid pre-treatment process of rice straw. The color of ash from rice straw washed with 0.1 M HCl is quite white. The study of SEM micrograph of silica xerogel that is synthesized from sodium silicate preparing from RSA by controlling the level of pH at 7, 8 and 9 as shown in figure 3 found that silica xerogels have the characteristics of a mixture of porous and dense gels or amorphous. A porous and dense gels were indicated by a red and yellow arrows, respectively. SEM micrographs at the magnification of 10,000x and 20,000x, it clearly show the difference of morphological structure of silica xerogels synthesized at pH 7, 8 and 9. Porous structure more appeared with increase of pH.

Figure 2. Color of rice straw ash from rice straw, (a) un-treated and (b) treated, by 0.1 M HCl.

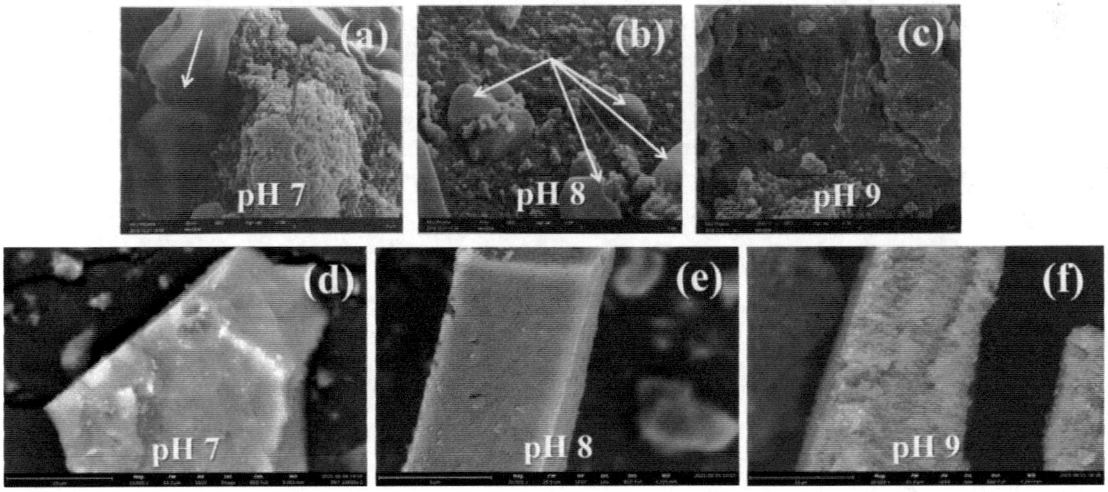

Figure 3. SEM micrograph of silica xerogels synthesized at (a,d) pH 7, (b,e) pH 8 and (c,f) pH 9.

As shown in figure 3, particle sizes of produced gel is varied from nanometers to micrometers. Silica xerogel which is prepared at pH 7 have more amorphous than silica prepared at pH of 8 and 9 due to the high solubility properties of amorphous silica in basic conditions. Silica prepared at pH 9 are almost all particles. It can be said that when pH is increased the resulting silica xerogel has a lower amorphousness. Additionally, SEM micrograph confirmed that particle size of silica xerogels smaller than 0.3 μm and the particle size distribution is fairly uniform. In this research, the synthesized silica was also subjected to color measuring to estimate the purity. Generally, if the color of the synthesized material has a high percentage of white this implies that the material is high purity [7]. From the color measuring as shown in table 1, it was found that the % whiteness of silica xerogel is quite a lot; 76.69, 81.25 and 79.69 for silica xerogel synthesized at pH of 7, 8 and 9, respectively. Comparing the silica xerogels prepared at pH 7, 8 and 9, found silica xerogel which prepared at pH 8 have higher % whiteness than the xerogel prepared at pH 7 and 9. It can be said that silica xerogels prepared in this research. relatively high purity. From literature, percent of silicon dioxide in the rice husk ash that produced from acid treated rice husk and have a % whiteness in the range of 70-75 is about 95.0-99.5 [7]. Additionally, results from EDS confirmed the purity of resulted xerogels.

Table 1. % whiteness of silica xerogels measured by ColorFlex EZ.

Name of substance	L* (% whiteness)
Silica xerogel (pH 7)	76.69
Silica xerogel (pH 8)	81.25
Silica xerogel (pH 9)	79.69

4. Conclusion

In summary, RSA prepared from acid treated rice straw is whiter than RSA prepared from untreated rice straw. Generally if the color of the ashes from the straw is very white, it means that straw ash contains a lot of silica. Amorphous silica was completely synthesized in this research. Amorphousness of produced silica xerogels decrease with increase of pH level. Percent whiteness of produced gels is quite high and it can conclude that silica xerogels synthesized in this research relatively high purity.

Acknowledgements

This work was financially supported by the revenue budget fund of Phetchabun Rajabhat University (PCRU) . We would like to thank to the Science Center and Department of Physics, Faculty of Science and Technology PCRU for chemicals, scientific instruments and infrastructure support. We also thank to Mr. Sorawich Asapha and Mr. Wirachai Meesuk for preparation and pretreatment of rice straw.

References

[1] Abdel-Mohdy F A, Abdel-Halim E S, Abu-Ayana Y M and El-Sawy S M 2009 Rice straw as a new resource for some beneficial uses *Carbohydr. Polym.* **75** 44–51

[2] El-Sayed M A and El-Samni T M 2006 Physical and chemical properties of rice straw ash and its effect on the cement paste produced from different cement types *J. King Saud Univ. Eng. Sci.* **19**(1) 21–30

[3] Kalapathy U, Proctor A and Shultz J 2000 A simple method for production of pure silica from rice hull ash *Bioresour. Technol.* **73** 257–62

[4] Falk G, Shinhe G P, Teixeira L B, Moraes E G and Novaes de Oliveira A P 2019 Synthesis of silica nanoparticles from sugarcane bagasse ash and nano-silicon via magnesiothermic reactions *Ceram. Int.* **45** 21618–24

[5] Anderson M L, Stroud R M, Morris C A, Merzbacher C I and Rolison D R 2000 Tailoring advanced nanoscale materials through synthesis of composite aerogel architectures *Adv. Eng. Mater.* **2**(8) 481–88

[6] Affandi S, Setyawan H, Winardi S, Purwanto A and Balgis R 2009 A facile method for production of high-purity silica xerogels from bagasse ash *Adv. Powder Technol.* **20** 468–72

[7] Kongkajun N, Thongruang R, Saisin N and Kaewngeun S 2013 Characterization of high purity silica from rice husk by acid treatment *J. Sci. Technol.* **2** 159–66

Effect of Zr content on the structure and morphology of CrZrN thin films prepared by reactive DC magnetron co-sputtering method

Adisorn Buranawong[1,2] and Nirun Witit-Anun[1,2*]

[1] Department of Physics, Faculty of Science, Burapha University, Chonburi 20131, Thailand
[2] Thailand Center of Excellence in Physics (ThEP), MHESI, Bangkok 10400, Thailand

*E-mail: nirun@buu.ac.th

Abstract. In this research, nanostructured chromium zirconium nitride (CrZrN) thin film has been deposited on Si(100) substrates by reactive DC magnetron co-sputtering method without in situ substrate heating and post-deposition annealing. The effects of Zr content on thin film structure and morphology were investigated. The Zr content in the films were varied by applied the sputtering current of Zr target (I_{zr}) in the range of 300 to 900 mA, whereas the current of Cr target was kept at 300 mA. The crystal structure, microstructure, morphology, thickness, and chemical composition were characterized by glancing angle X-ray diffraction (GA-XRD), field emission scanning electron microscopy (FE-SEM), and energy-dispersive X-ray spectroscopy (EDS) techniques, respectively. The results showed that the increase of I_{zr} not only increased the deposition rate, but also increased the Zr content of the as-deposited film ranging from 3.9 to 26.5 at%. The as-deposited thin films were formed as a (Cr,Zr)N solid solution, with fcc structure in (111) and (200) plane, where Cr atoms were replaced by Zr atoms in the CrN lattice. The 2θ diffraction peaks were shifted to the lower value as increase of Zr content which was obtained by increased I_{zr}. The nanocrystalline CrZrN structure with crystal sizes smaller than 10 nm structure were calculated for as-deposited thin films. The lattice parameters increased from 4.187 to 4.381 Å, whereas the crystal size decreased from 8.3 to 6.4 nm. The FE-SEM images of all the CrZrN films exhibited compact columnar with dense morphology as a function of Zr content. Moreover, the thickness of the CrZrN thin films was increased of 302 – 421 nm.

1. Introduction

For many years, surface engineering have been widely studied in research area because of the modern industrial requirement for novel and excellent surface properties for many applications such as machine parts, forming, and cutting tools. The transition metal nitrides for example TiN, VN, and CrN are good candidated of hard coating applications. These metal nitrides' film possesses a higher hardness than the traditional steel and cabides because of improve both mechanical and tribological properties [1].

Compared to the other nitride films, CrN has become the good choice as coating in many industrial applications due to their superior hardness, corrosion resistance, wear behavior, and easy to deposit. Therefore, CrN film was investigated in many industrial applications [2]. A drawback of using CrN thin films as high-speed machining processes was occured when operated at high-temperature (> 700 °C), the formation of porous oxides on film surface was found resulting the degradation of their mechanical properties [3]. To overcome these problems, it can be done by alloying the other elements, such as Ti, Al, and Zr, in the CrN structure to form the ternary nitride thin films, for example, CrTiN, CrAlN, and

CrZrN, which have excellent properties. Among these films, CrZrN showed unique characteristics as mechanical properties enhancement and a achieving low surface roughness through the Zr content in the film [4], so it could be one of the most promising solution for practical applications.

The CrZrN thin film can be synthesized by various techniques, especially sputtering. Many research groups have been dedicated to synthesized the CrZrN thin films, used substrate heating, and biased to improve the film structure These processes cause the substrate to heat up, increasing manufacturing costs and was unsuitable for thermally sensitive substrates. Therefore, it is important to synthesized these films without heating and biasing. In addition, most previous studies have focused on the tribological property, which is limited to hardness, friction, and wear. However, the study of the effect of film's composition on the structure and morphology of the CrZrN thin films is still limited.

In this research, as a part of the ongoing research projects that effort to develop the ternary nitride thin films of the Cr-based hard coating for hard coating applications, CrZrN films with various Zr content were synthesized by reactive DC unbalanced magnetron co-sputtering at room temperature with unheated and no substrate bias process, and characteristics such as the crystalline structure, surface morphology and microstructure of the as-deposited films were investigated at different Zr content.

2. Materials and methods

The CrZrN coatings were deposited on Si(100) substrates at room temperature without heating and biasing using separate metallic disc sputtering target of Cr (99.95%) and Zr (99.95%) with a 75° magnetron cathode arrangement by the reactive DC magnetron co-sputtering method. The high purity processing sputtering gases of Ar (99.999%) and reactive gases of N_2 (99.999%) were injected into a chamber. The deposition chamber pressure was reduced to a base pressure of 5×10^{-5} mbar before each deposition. The total pressure was 5×10^{-3} mbar with fixed of Ar and N flow rate were 10 sccm and 12 sccm, respectively. In the deposition process, the sputtering current of the Cr target (I_{Cr}) was kept at 300 mA, whereas the sputtering current of the Zr target (I_{Zr}) were 300 mA, 600 mA and 900 mA with constant deposition time of 60 minutes. The target to substrate separation was 150 mm

The thin films were characterized by various techniques. The element composition analysis was carried out by EDS (EDAX). The phase and crystal structures were examined by a glancing angle X-ray diffractometer (BRUKER D8) using Cu Kα radiation at grazing incidence angle of 3°. The crystallite size is calculated by Scherrer's equation. The microstructure, cross-sectional morphology, and thickness were studied by Field Emission Scanning Electron Microscope (FE-SEM, Hitachi s4700).

3. Results and discussions

The CrZrN films were successfully deposited by reactive DC magnetron co-sputtering technique using separate metallic Cr and Zr targets on Si substrates. The chemical composition of the as-deposited CrZrN thin films results from the various sputtering current of the Zr target was evaluated by the EDS technique is listed in table 1. It was showed that the enhancement of Zr content in films but the reduction of Cr content was found as a function of sputtering current on the Zr target. Although film No.1 used the same sputtering current of Zr and Cr target, the Cr content is obviously superior compared to Zr content. This can be attributed to the fact that the sputtering yield of Cr (~ 0.87) is higher than Zr (~ 0.41) [5]. It also shows that the atomic ratio of the Zr and N content is defined as x = Zr/(Cr+zr) and y = N/(Cr+zr). It was found that the atomic ratio of Zr content (x) increased from 0.15 to 0.78 at% as sputtering current of Zr target increased, while the N content (y) in all deposited films decreased from 2.76 to 1.93 at%. Moreover, the atomic ratio of nitrogen to metals (Cr+Zr) (y) was higher than 1 implied that the over stoichiometry of CrZrN thin films were investigated.

Table 1. Thickness, composition, crystal size, and lattice constant of the CrZrN thin films.

No.	I_{Zr}	Thickness	Composition (at.%)			Atomic ratio		Crystal size (nm)		Lattice constants (Å)	
	(mA)	(nm)	Cr	Zr	N	x=Zr/(Cr+Zr)	y=N/(Cr+Zr)	(111)	(200)	(111)	(200)
1	300	302	22.7	3.9	73.4	0.15	2.76	8.2	8.3	4.210	4.187
2	600	363	9.3	20.5	70.2	0.69	2.35	7.8	7.4	4.263	4.242
3	900	421	7.6	26.5	65.9	0.78	1.93	6.4	6.9	4.381	4.357

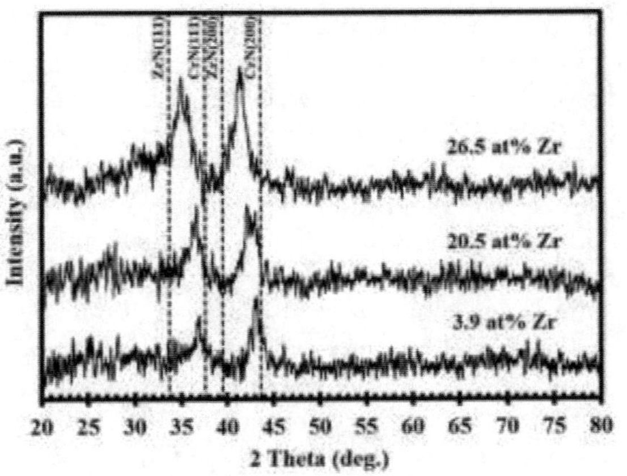

Figure 1. XRD diffraction patterns of the CrZrN thin films with different Zr content.

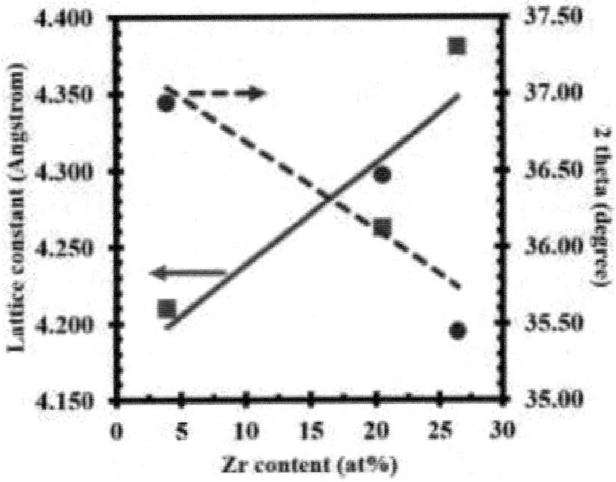

Figure 2. 2θ position and lattice constants of the CrZrN thin films with different Zr content.

XRD patterns of the as-deposited CrZrN films deposited on Si substrates through different Zr content are illustrated in figure 1. For comparison, the standard diffraction peaks for the (111) and (200) planes of ZrN (JCPDS No. 78-1420) and CrN (JCPDS No. 77-0047) were included. The results showed that the diffraction peaks are located between the standard diffraction peaks of ZrN and CrN. This result implied that the solid solution of (Cr,Zr)N with FCC structure formation. On the other hand, there are no corresponding 2θ peaks for pure ZrN and CrN phases. The CrZrN peaks are shifted to lower 2θ angles as Zr content increased due to the replacement of larger atomic radius of Zr (0.161 nm) to the position of the smaller atomic radius of Cr (0.136 nm) [1], which result in a change in the lattice constants were identified for all as-deposited films. The result was agreed with the reference [6]. According to Bragg Law, the film's lattice constants are calculated from the CrZrN (111) plane. Figure 2 show the 2θ position and lattice constants of CrZrN(111) films as a result of Zr content. It can be seen that as the Zr content increases, the 2θ position and the lattice constant were both increases. When using Scherrer's equation from the FWHM of the XRD pattern, the crystal size of CrZrN thin films was calculated as a result of the Zr content performed in Table 1, were decreased from 8.5 to 6.4 nm for (111) plane and 8.3 to 6.9 nm for (200) plane as increasing of the Zr content.

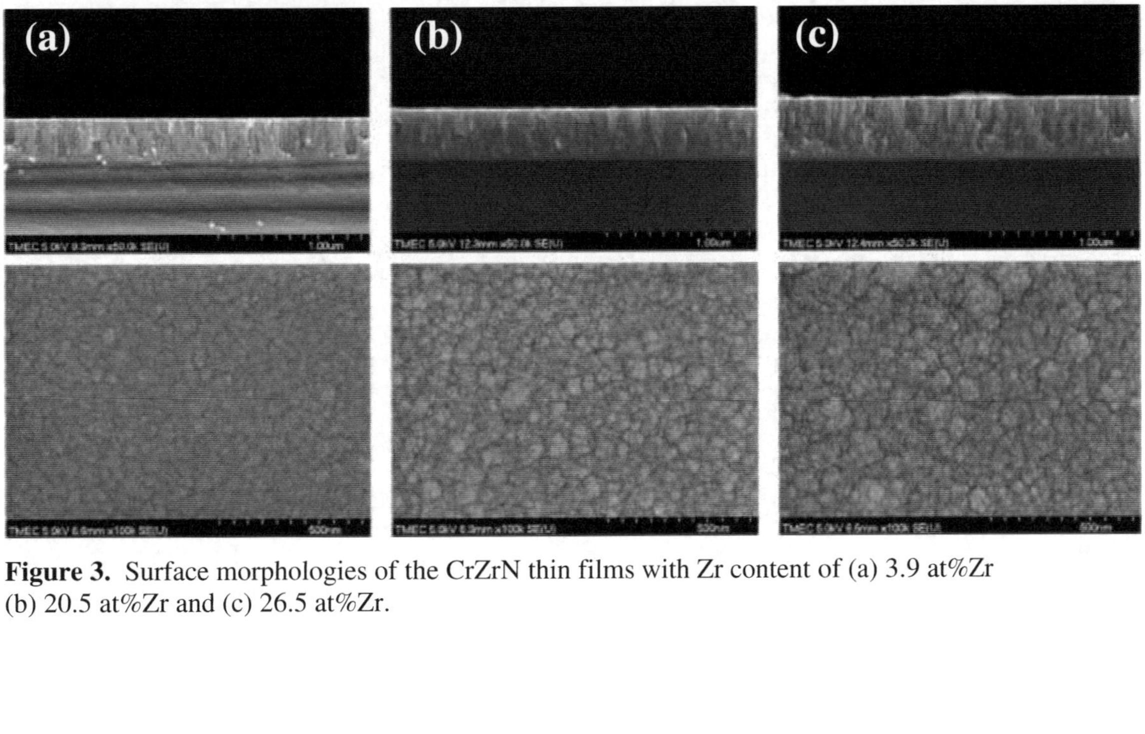

Figure 3. Surface morphologies of the CrZrN thin films with Zr content of (a) 3.9 at%Zr (b) 20.5 at%Zr and (c) 26.5 at%Zr.

The microstructure and morphologies of the CrZrN thin film with different Zr contents were analyzed by the FE-SEM technique (figure 3). At low Zr content (3.9 at%), it was found that the granule shape pattern with almost same size was indentified along the film surface but exhibited the grain boundaries between each grain. The grain aggregration across the surface resulting the enhancement of granule grain size were cleary indentifiled at higher Zr content (20.5 at%). More over, the grain boundaries were also investigated. The grains were kept the same shape but became bigger size with more void between each grain at highest Zr content (26.5 at%). The grain size increased can be explained that the larger the kinetic energy of sputtered deposited atom reaching to the surface of substrate [7].

The cross-sectional structure and thickness of FE-SEM micrograph were also listed and illustrated in figure 3 and table 1, respectively. At low Zr content (3.9 at%), the competitive grain growth with columnar structure were found. At highest Zr content of 26.5 at%, the porous microstructure which more open grain boundaries accompanied with fibrous grain along the coating were indentified. The cross-section pattern were well agreement with Zone T of Thornton 's zone model [8]. According to table 1, the thickness increased from 307 to 401 nm as a function of Zr content which increase by increasing the target sputtering current (I_{zr}) from 300 to 900 mA. The higher of sputtering current helps to increase the growth rate of deposited atom resulting the increase of film thickness [7].

4. Conclusion

The nanocrystalline CrZrN thin films were grown on Si(100) by reactive DC magnetron co-sputtering without heating and biasing to the substrate at a low sputtering current in order to analyse the effect of Zr incoperation on their structure and morphology. It was found that the films crystallographic orientation in (111) and (200) were found which implied that the solid solution with FCC structure of (Cr,Zr)N was successfully deposited even though Zr content less or more than Cr content. The 2θ diffraction peaks shifted to the lower angle wheareas the lattice constants were increase. An decrease in crystal size was observed as the Zr content in the films increase. The nanocrystalline structure of CrZrN film was performed due to a crystallite size of the film less than 10 nm. Higher composition and atomic ratio of Zr were investigated. The thickness of films increased when the Zr content increase. The microstructure showed columnar with more open grain boundaries as Zr content enhancement.

References

[1] Kim G S, Kim B S, Lee S Y and Hahn J H 2005 Structure and mechanical properties of Cr-Zr-N films synthesized by closed field unbalanced magnetron sputtering with vertical magnetron sources *Surf. Coat. Technol.* **200** 1669–75

[2] Cheng W L, Zhou Z F, Shum P W and Li K Y 2013 Effect of Ni addition on the structure and properties of Cr–Ni–N coatings deposited by closed-field unbalanced magnetron sputtering ion plating *Surf. Coat. Technol.* **229** 84–9

[3] Zeng X T, Zhang S, Sun C Q and Liu Y C 2003 Nanometric-layered CrN/TiN thin films: mechanical strength and thermal stability *Thin Solid Films* **424** 99–102

[4] Kim Y S, Kim G S and Lee S Y 2011 Influence of the N_2 partial pressure on the characteristics of CrZrN coatings synthesized using a segment CrZr target *J. Nanosci. Nanotechnol.* **11** 8792–7

[5] Yamamura Y and Tawara H 1996 Energy dependence of ion-induced sputtering yields from monatomic solids at normal incidence *At. Data Nucl. Data Tables* **62** 149–253

[6] Chantharangsi C, Denchitcharoen S, Chaiyakun S and Limsuwan P 2012 Structure and surface morphology of Cr-Zr-N thin films deposited by reactive DC magnetron sputtering *Procedia Eng.* **32** 868–74

[7] Bai T R, Liu C Q, Wang N, Liu S M, Wang H L, Jiang W W, Ding W Y, Fei W D and Chai W P 2015 Effects of sputtering voltage and current on the composition and microstructure of the CIGS films prepared by one-step pulsed DC magnetron sputtering *J. Alloys Compd.* **646** 532–40

[8] Berg S and Nyberg T 2005 Fundamental understanding and modeling of reactive sputtering processes *Thin Solid Films* **476** 215–30

Journal of Physics: Conference Series **2145** (2022) 012031 doi:10.1088/1742-6596/2145/1/012031

Effect of argon annealing method on structural and ferromagnetic properties in Fe-doped SnO_2 powders

S Nongkae[1], K Tangphanit[1], S Teeta[1], E Swatsitang[2] and K Wongsaprom[1,*]

[1]Department of Physics, Faculty of Science, Mahasarakham University, Mahasarakham, 44150, Thailand
[2]Institute of Nanomaterials Research and Innovation for Energy (IN-RIE), NANOTEC-KKU RNN on Nanomaterials Research and Innovation for Energy, Khon Kaen University, Khon Kaen, 40002, Thailand

*E-mail: kwanruthai.w@msu.ac.th

Abstract. Nanocrystalline powders of Fe-doped SnO_2 ($Sn_{1-x}Fe_xO_2$) (x = 0.00, 0.01, 0.03, 0.05) were prepared by a hydrothermal method. The powders were calcined in argon atmosphere at 600 ºC for 2 h, causing phase transition from diamagnetic and weak ferromagnetic behavior to a ferromagnetic state. No trace and other magnetic impurity phases was detected in the samples with Fe content up to 3%. The calcined samples of Fe-doped SnO_2 revealed the room temperature ferromagnetism with highest magnetization values of 434.07 memu/g at 15 kOe for x = 0.05. The room temperature ferromagnetism of samples originated from oxygen vacancies that occurred in the argon calcination process. In particular, oxygen vacancy shows a significant role in ferromagnetic coupling corresponding to F-center interaction.

1. Introduction

Nanostructures of tin dioxide (SnO_2), as a n-type semiconductor with a wide bandgap (E_g = 3.6 eV) has elicited much interested due to their potential for a wide range of applications [1-4]. Recently, diluted magnetic semiconductors (DMSs) have attracted much attention due to their novel magnetism and potential application in spintronics [5]. In addition, SnO_2 has an excellent electronic and magneto-optical properties. The performance of SnO_2 has been improved considerably by doping with Fe [6], Co [7], Ni [8] or Mn [9]. Some studies have reported that the calcination process had effects on the ferromagnetic behaviour of the transition metal doped SnO_2 system [10, 11]. Therefore, it is important to examine the effect of the calcination process on ferromagnetic behaviour of Fe-doped SnO_2 under an atmosphere that is lacking in oxygen.

In this study, Fe-doped SnO_2 ($Sn_{1-x}Fe_xO_2$) nanocrystalline powder was prepared with x = 0.00, 0.01, 0.03 and 0.05 by a hydrothermal method, and the prepared samples of powder were calcined at 600 ºC under argon atmosphere for 2 h. The calcined samples were then examined for crystal structure via X-ray diffraction (XRD). The particle size and shape were characterized by transmission electron microscopy (TEM). The ferromagnetic measurements were performed at 300 K using a vibrating sample magnetometer (VSM).

2. Experimental procedure

Fe-doped SnO_2 nanocrystalline powders were prepared via a hydrothermal method. $SnCl_2 \cdot H_2O$ (99.99%, Aldrich) and $FeN_3O_9 \cdot 9H_2O$ (99.99%, Aldrich) were used as initial materials. The precursor materials were dissolved in 15 ml deionized water (DI water) and 5 ml ethanol was added with continuous stirring at 50 °C for 3 h and then adjusted to pH 11 with sodium hydroxide (NaOH). The mixed solution was changed to a Teflon-lined stainless-steel autoclave with heating treatment at 180 °C for 24 h. The precipitates were picked by centrifugation and washed many times with DI water and ethanol separately, then dried the powders at 70 °C for overnight. The powders were calcined at 600 °C under argon atmosphere for 2 h. The crystal structure of the samples was examined by X-ray diffraction (XRD) using a Philips X-ray diffractometer (PW3710, The Netherlands). Size and shape of Fe-doped SnO_2 samples were performed by transmission electron microscopy (TEM) (Hitachi H8100 200 kV). The ferromagnetic behaviour was examined at 300 K by a vibrating sample magnetometer (Versa Lab VSM, Quantum Design).

3. Results and discussion

Figure 1(a) presents the XRD patterns of the SnO_2 and $Sn_{1-x}Fe_xO_2$ (x = 0.00, 0.01, 0.03, 0.05) samples. The diffraction peaks observed for the SnO_2 and $Sn_{1-x}Fe_xO_2$ samples can be indexed to the pure phase of tetragonal structure (JCPDS No. 41-1445). The XRD patterns of the SnO_2 and $Sn_{1-x}Fe_xO_2$ with x = 0.01 and 0.03 samples with no contaminated phase or magnetic impurity phase are detected. The exception is the sample of x = 0.05, which reveals an extra peak at $2\theta \sim 32.86$, 37.15 and 50.29 which can be an index as the Fe_2O_3 [12]. The detection of contaminated phase in the structure of this sample, reveals that of the Fe ions cannot completely insert into the position of the Sn^{4+} ions within the lattice. Therefore, the synthesis of $Sn_{1-x}Fe_xO_2$ by this hydrothermal method is limited by the solubility Fe ions instead of Sn^{4+} ions within the lattice being less than 5% [13]. After the samples were calcined in argon, the samples did not demonstrate a contaminated phase of Fe_2O_3 and other compounds in the structure (figure 1(b)). This indicates that calcining at 600 °C under argon provides reaction conditions that allows Fe^{3+} ions to be inserted in the structure of SnO_2. Moreover, after being calcined, the samples showed intensity increased and decrease peak width, which indicated more crystalline properties [14, 15].

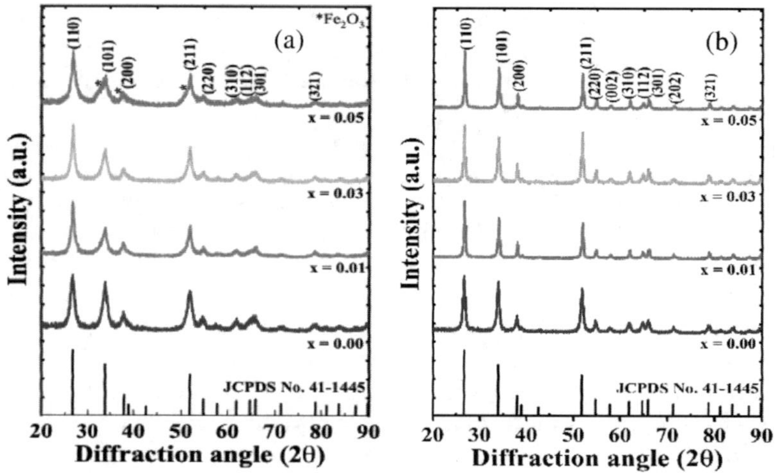

Figure 1. XRD patterns of (a) pre-calcined $Sn_{1-x}Fe_xO_2$ and (b) $Sn_{1-x}Fe_xO_2$ were calcined in argon.

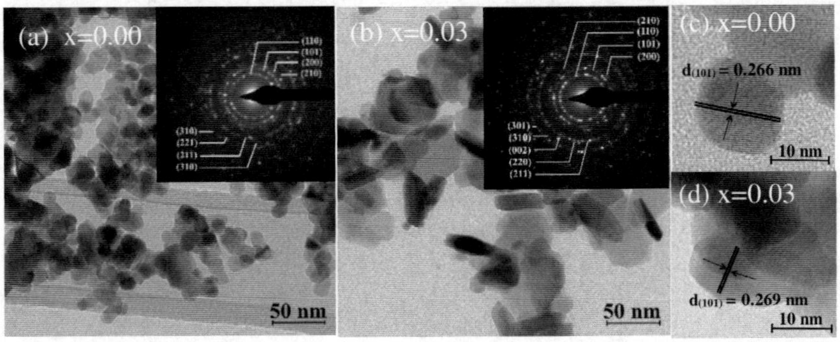

Figure 2. Images of Fe-doped SnO_2 samples: (a) TEM image with the corresponding SAED patterns (inset) for x=0.00, (b) TEM image with the corresponding SAED patterns (inset) for x=0.03, (c) and (d) HRTEM images of the samples.

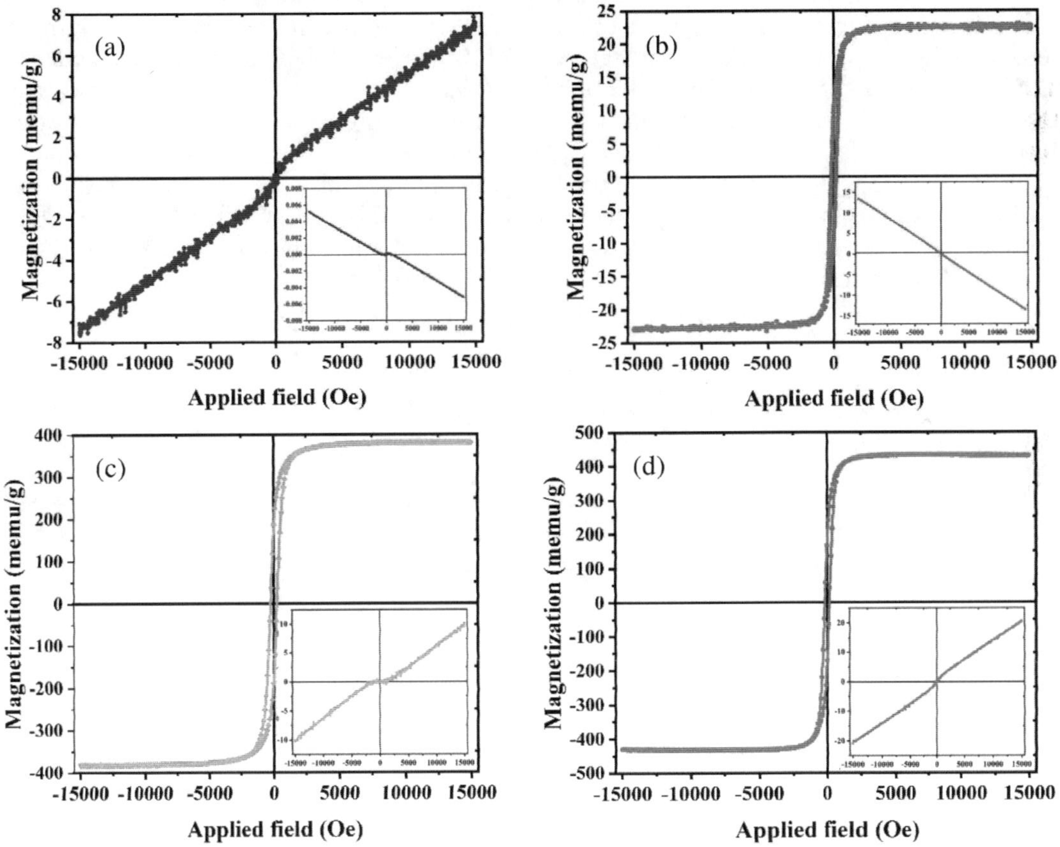

Figure 3. Magnetization vs magnetic field for $Sn_{1-x}Fe_xO_2$ samples were calcined in argon (a) x=0.00, (b) x=0.01, (c) x=0.03 and (d) x=0.05. Inset bottom right shows the magnetic behavior for pre-calcined samples.

The morphology and structure of the SnO_2 and 3% Fe-doped SnO_2 calcined in argon were examined by TEM with the corresponding selected area of electron diffraction (SAED) patterns as presented in figure 2. The particle size of the samples is ~15-23 nm. The SAED patterns of SnO_2 and 3% Fe-doped SnO_2 samples in the inset of figure 2 reveal pure phase of tetragonal structure without magnetic impurity phases of Fe, Fe_2O_3 or Fe_3O_4, which corresponds to XRD results. The HRTEM images indicate the presence of the lattice fringe in the samples of SnO_2 and 3% Fe doped SnO_2 (figure 2(c) and (d)). The lattice distance of 0.266 and 0.269 nm indicated in these figures corresponded to the (101) crystalline plane.

Figure 3 presents the VSM measurement of field dependence of magnetization for SnO_2 and $Sn_{1-x}Fe_xO_2$ samples at room temperature. The SnO_2 sample (figure 3(a)) exhibits weak ferromagnetic behavior after the sample was calcined under argon. The magnetization curves of SnO_2 doped with 1, 3 and 5% Fe are shown in figure 3(b), (c) and (d), respectively. All of the samples doped with Fe exhibited ferromagnetic behaviour. The magnetization of Fe-doped SnO_2 samples increased with increasing concentration of dopant. In addition, after doped samples were calcined in argon, the magnetization value are also increased. The doped samples showed ferromagnetic behaviour at room temperature with magnetization values of 22.5, 383.5 and 434.1 memu/g and coercivity of 183.8, 239.9 and 171.6 Oe for x = 1, 3 and 5%, respectively. These results suggest that the Fe content and calcination process play an important role on magnetism for the SnO_2 nanocrystalline system. However, pre-calcined samples doped with 1% Fe exhibited diamagnetic properties while the samples doped with 3 and 5% Fe show weak ferromagnetic properties with the magnetization of 0.6 and 2.2 memu/g. The observed difference between ferromagnetism for pre-calcined and calcined samples is directly related to the concentration of defects in the crystal structure. Mehraj *et al.* [10] observed the annealing temperature on magnetization for SnO_2 nanoparticles to be 0.75 emu/g and found that the concentration of oxygen vacancies was important for magnetic behaviour. Therefore, the difference of magnetization values in Fe-doped SnO_2 samples is possibly allocated to their difference in oxygen content. The plausible mechanism, which has been proposed to explain ferromagnetism in SnO_2 is an F-center exchange interaction [6]. An F-center, an electron in the d-shells trapped in oxygen vacancy, where the electron occupies and overlap with nearest neighbors. We expect that the ferromagnetic coupling of Fe^{3+} and Fe^{3+} ions by an electron trapped in oxygen vacancy, thus ferromagnetism is achieved for Fe-doped SnO_2 samples [6].

4. Conclusion

Nanocrystalline powders of SnO_2 and Fe-doped SnO_2 as diluted magnetic semiconductors have been prepared by a hydrothermal method. The calcination process in argon improved the ferromagnetism in the samples. All of doped samples showed ferromagnetic-like ordering with the highest magnetization of 434.1 memu/g at room temperature. The ferromagnetic-like ordering is due to the defect of oxygen vacancies in the structure. The magnetization value is directly related to the content of oxygen vacancies and crystallite size of SnO_2 nanocrystalline powder as a function of calcination process. This study showed that the ferromagnetism in SnO_2 compound can be improved by calcination process under an oxygen deficient atmosphere.

Acknowledgments

This research was financially supported by Faculty of Science, Mahasarakham University (Grant year 2021).

References

[1] Teeramongkonrasmee A and Sriyudthsak M 2000 *Sens. Actuators.* **66** 256
[2] Gu F, Wang S F, Lu M K, Qi Y X, Zhou G J, Xu D and Yuan D R 2004 *Opt. Mater.* **25** 59
[3] Ferrere S, Zaban A and Gsegg B A 1997 *J. Phys. Chem.* **101** 4490
[4] Subramanian V, Gnanasekar K I and Rambabu B 2004 *Solid State Ionics.* **175** 181
[5] Yakuphanoglu F 2009 *J. Alloys Compd.* **470** 55
[6] Coey J M D, Douvalis A P, Fitzgerald C B and Venkatesan M 2004 *Appl. Phys. Lett.* **84** 1332
[7] Ogale S B *et al.* 2003 *Phys. Rev. Lett.* **91** 077205
[8] Xiao W Z, Wang L L, Xu L, Wan W and Zou B S 2009 *Solid State Commun.* **149** 1304
[9] Tian Z M, Yuan S L, He J H, Li P, Zhang S Q, Wang C H, Wang Y Q, Yin S Y and Liu L 2008

J. Alloys Compd. **466** 26
[10] Mehraj S, Ansari M S, Al-Ghamdi A A and Alimuddin 2016 *Mater. Chem. Phys.* **171** 109
[11] Srinivas K, Vithal M, Sreedhar B, Manivel Raja M and Venugopal Reddy P 2009 *J. Phys. Chem.* C **113** 3543
[12] Li X, Wei W, Wang S, Kuai L and Geng B 2011 *Nanoscale* **3** 718
[13] Liu J F, Lu M F, Chai P, Fu L, Wang Z L, Cao X Q and Meng J 2007 *J. Magn. Magn. Mater.* **317** 1
[14] Peng C, Bai H, Xiang M, Su C, Liu G and Guo J 2014 *J. Electrochem. Sci.* **9** 1791
[15] Bhattacharjee A, Ahmaruzzaman M and Sinha T 2014 *Spectrochim. Acta* A *Mol. Biomol. Spectrosc.* **136** 751

Environment friendly concrete block made from Portland cement and aggregate replacement materials

B Kaewsai[1,2], P Torkittikul[3] and A Chaipanich[1,4]*

[1] Advanced Cement-based Materials Laboratory, Department of Physics and Materials Science, Faculty of Science, Chiang Mai University, Chiang Mai 50200, Thailand
[2]Master Degree's Program in Materials Science, Faculty of Science, Chiang Mai University, Chiang Mai 50200, Thailand
[3] Department of Civil Technology, Faculty of Industrial Technology, Lampang Rajabhat University, Lampang 52100, Thailand
[4] Center of Excellent in Materials Science and Technology, Chiang Mai University, Chiang Mai 50200, Thailand

*Corresponding author's E-mail address: arnon.chaipanich@cmu.ac.th

Abstract. This research work investigated the properties of concrete block made from Portland cement and aggregate replacement materials. Portland cement (PC) was replaced by fly ash (FA) at 10%, 20%, 30% and sand was replaced by bottom ash (BA) at 10% by weight. Water was used at 7% by weight of total solid mass. Binder : Sand : Stone dust ratio of 1 : 5 : 6, 1 : 4 : 5 and 1 : 3.5 : 4.5 were used. Compressive strength were tested after curing in air for 28 days. The results showed that compressive strength of 1 : 5 : 6 ratio was lower than others. Concrete block replaced PC by fly ash had lower compressive strength when amount of fly ash increased. Concrete block had lower compressive strength when replaced sand by bottom ash. As a result, the mixes with FA as PC replacement and BA as sand replacement at the ratio of 1 : 5 : 6 did not meet the requirement of Thai industrial standard. However, concrete block with PC replaced by fly ash at 10%, 20% and sand replaced by bottom ash at 10% of 1 : 3.5 : 4.5 ratio was higher than 1 : 5 : 6 ratio and this ratio meet the requirement of Thai industrial standard.

1. Introduction

Industrial by-products with pozzolanic properties have been used as a binder material to partly replace Portland cement. Today the Portland cement manufacturing industry emits 8% of the global carbon footprint [1]. Therefore, efforts have been made to reduce the use of Portland cement and replace it with other materials that can also improve the properties of various concrete. Reducing the use of cement by using pozzolans as an alternative binder is one of the accepted and effective methods. Pozzolans material according to ASTM C618 standard is defined as "pozzolans material is a silica-based material, or silica and alumina as the main component. Pozzolan by itself does not have the ability to set and bind like Portland cement. But if there is enough moisture, it can react with calcium hydroxide at normal temperature and resulting in a compound that has the ability to bind (cementitious)" [2]. In the electricity generation industry from the coal power plant, fly ash and bottom ash is obtained as a by-product. The storage or disposal is a necessary process in this industry [3]. Efforts to replace Portland cement with fly ash have been attempted as fly ash can then be pozzolanic reaction. The product is a binder. Silica or alumina in fly ash can react with calcium hydroxide under normal water and temperature conditions [4]. Then, it forms a binder that can be used as a replacement for cement, but the reaction takes place slowly. In general, the fly ash contains about 50% and 20% of the main components, namely silica and alumina, respectively [5]. While fly ash has been used more widely and is well known as a Portland cement replacement, bottom ash has not used in the same extent. In this work, a variety of mixes was used (FA was used to replace PC and BA was used to replace sand, different ratio liquid to solid was also used) to investigate the compressive strength compared to the industrial standard to determine the possibility of using the combination fly ash and bottom ash at a certain composition in order to meet the requirement of Thai Industrial Standard.

2. Materials and methods

2.1. Materials

Ordinary Portland cement from Siam Cement Group was used in this study. Fly ash and bottom ash are a by-product of electricity generation from Mae Moh coal Power Plant, Lampang Province, Thailand. Portland cement was replaced by FA at 10%, 20% and 30% by weight and sand was replaced by BA at 10% and 20% by weight.

2.2. Methods

2.2.1. Mixing. Binder to sand to stone dust ratio of 1 : 5 : 6, 1 : 4 : 5 and 1 : 3.5 : 4.5. Water was used at 7% by weight of total solid mass. Mixing for 60 s, then resting for 90 s and mixing again 60 s. All the ingredients are pressed into a concrete block mold, the size of which is width x height x length is 70 x 190 x 390 mm. according to Thai industrial standard.

2.2.2. Curing. Curing was the processed by which the hydration reaction of the cement continues. This will result in a hardened concrete block. In this study, concrete blocks were cured for 28 days in air.

2.2.3. Chemical technical methods. X-ray fluorescence (XRF) was used to determine the chemical composition of Portland cement and fly ash.

2.2.4. Physical technical methods. Compressive strength of concrete blocks were tested in accordance to Thai industrial standard 109-1974 [6].

3. Results and discussion

3.1. Chemical composition of Portland cement

Oxides in Portland cement are composed of two major groups the main oxides are CaO, SiO_2, Al_2O_3 and Fe_2O_3, which together comprise approximately 90% of the cement weight and the secondary oxide is MgO, Na_2O, SO_3, K_2O, TiO_2, P_2O_5 and gypsum. The various oxide content in Portland cement is expressed as a percentage by weight as shown in table 1.

3.2. Chemical composition of fly ash

The fly ash obtained from the Mae Moh coal power plant is class C, the compositions are shown in table 1. It was shown that fly ash from the Mae Moh coal Power Plant, Lampang Province, Thailand. contains silicon oxide (SiO_2), Alumina oxide (Al_2O_3), iron oxide (Fe_2O_3) and calcium oxide are 90.12%

Table 1. Chemical composition of Portland cement and fly ash by XRF.

Chemical composition	PC	FA
	%	%
SiO_2	19.13	37.26
CaO	64.47	19.46
Al_2O_3	4.98	20.64
Fe_2O_3	3.20	12.76
SO_3	3.23	2.35
MgO	1.88	2.65
K_2O	0.61	2.36
Na_2O	0.09	1.11
P_2O_5	0.06	0.21
TiO_2	-	0.44
MnO	0.01	0.11
Loss On ignition (LOI)	2.30	0.29
Specific gravity	3.15	2.1

3.3. Bottom ash as sand replacement

The results showed that the compressive strength of concrete block replaced sand by bottom ash lower compressive strength when amount of bottom ash increased. But the compressive strength of concrete block that used Binder : Sand : Stone dust ratio of 1 : 3.5 : 4.5 and replaced sand by bottom ash at 10% had the highest compressive strength for PC Mix as a result of increasing the ratio of cement to aggregates are shown in figure 1.

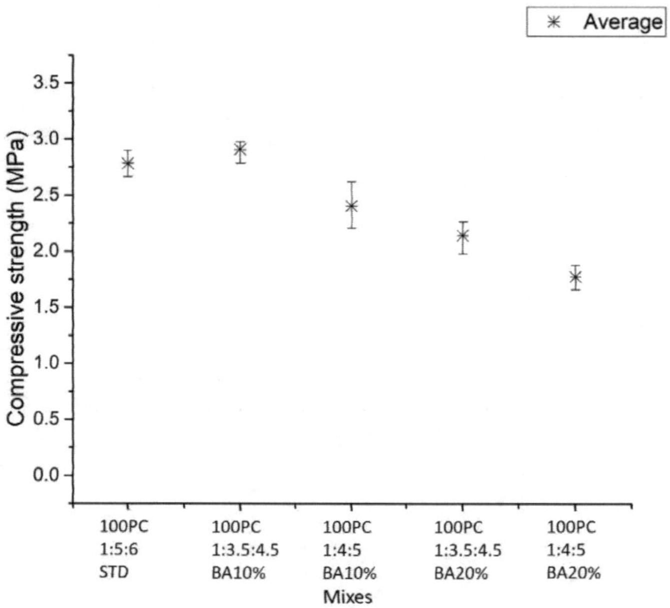

Figure 1. Compressive Strength of concrete blocks with Bottom ash as sand replacement.

3.4. *Fly ash as PC replacement*

As shown in figure 2, the results showed that the compressive strength of concrete block replaced by fly ash had lower compressive strength when amount of fly ash increased (70PC30FA) when compared to 100PC mix. However, concrete block with PC replaced by fly ash at 10% (90PC10FA) and 20% (80PC20FA) meet the requirement of the Thai industrial standard 58-1990 [7]. These mixes have compressive strength similar to 100 PC mix.

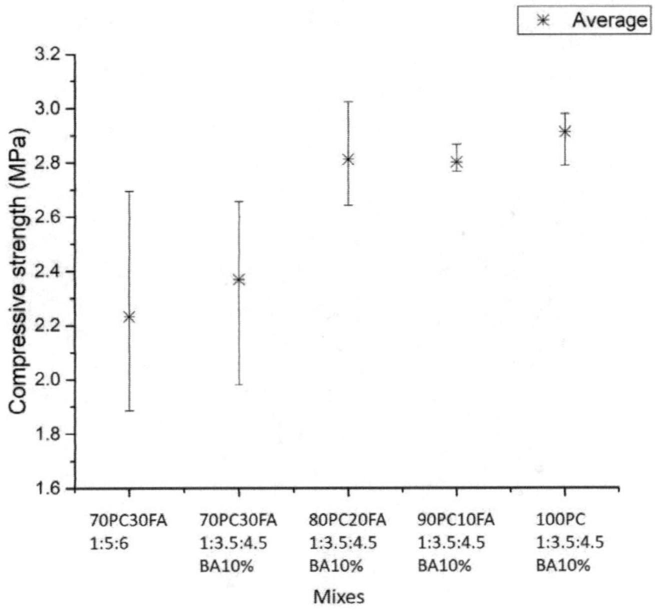

Figure 2. Compressive Strength of Concrete blocks with fly ash as PC replacement and bottom ash as sand replacement.

4. Conclusions

The results showed that concrete block that used Binder : Sand : Stone dust ratio of 1 : 3.5 : 4.5 and replaced sand by bottom ash at 10% by weight had the highest compressive strength for PC Mix and similar compressive strength were found in 10% and 20% fly ash mixes. From the results, fly ash can be used as a replacement for cement with fly ash up to 20% and bottom ash can be used as a replacement for sand at 10% by weight and meet the requirement of the Thai industrial standard 58-1990. For future work, the mixes will be analyzed using X-ray Diffraction (XRD) and Scanning Electron Microscope (SEM).

Acknowledgements

The authors are thankful to the members at the Advanced Cement- based Materials Laboratory (ACMRL). This research work was partially supported by Chiang Mai University.

References

[1] Robbie M A 2018 *Earth. Syst. Sci. Data* **10** 195
[2] ASTMC618-19 2019 *Standard Specification for Coal Fly Ash and Raw or Calcined Natural Pozzolan for Use in Concrete 100* (West Conshecken, PA: American Society for Testing and Materials)
[3] Kayali O 2008 *Const. Build. Mater.* **22** 2393
[4] Goodarzi F 2006 *Fuel* **85** 1418

[5] Zhang M H and Gjorv O E 1991 *ACI Mater. J.* **88** 240
[6] TIS109 1974 *Standard for sampling and testing concrete masonry units* (Bangkok,Thai Industrial Standard Institute, Ministry of Industry)
[7] TIS58 1990 *Standard for hollow non-load-bearing concrete masonry units* (Bangkok, Thai Industrial Standard Institute, Ministry of Industry)

Magnetic and electrochemical properties of nickel oxide microstructures prepared by hydrothermal method

N Boonraksa[1], S Maensiri[2], E Swatsitang[3] and K Wongsaprom[1,*]

[1]Department of Physics, Faculty of Science, Mahasarakham University, Mahasarakham, 44150, Thailand
[2]School of Physics, Institute of Science, Suranaree University of Technology, Nakhon Ratchasima, 30000, Thailand
[3] Institute of Nanomaterials Research and Innovation for Energy (IN-RIE), NANOTEC-KKU RNN on Nanomaterials Research and Innovation for Energy, Khon Kaen University, Khon Kaen, 40002, Thailand

*E-mail: kwanruthai.w@msu.ac.th

Abstract. Nickel oxide microstructures were succesfully synthesized by hydrothermal method. The structure, morphology, surface and porosity of the NiO hexagonal plates indicated the formation of NiO without appearing any secondary phases, occupying the typical cubic structure. The ferromagnetic behaviour was examined by vibrating sample magnetometer (VSM). The sample exhibited ferromagnetic behaviour at room-temperature with the magnetic moment value of ~ 160 memu/g. The cyclic voltammetry (CV) and galvanostatic charge-discharge (GCD) anylysis were used to examine the electrochemical capacitance of the sample. The specific capacitance of the sample at a current density of 1 A/g was obtained to be 174.14 F/g. The cycle stability was excellent usability 76.6% after 500 cycles at a current density of 5 A/g.

1. Introduction

Transition metal oxides (MnO_2, Co_3O_4, ZnO_2, NiO, etc.) are the most important active electrode materials for supercapacitors, where highly reversible redox processes are involved at the surface area [1-4]. Among these metal oxides, NiO is very popular due to its ferromagnetic behaviour above 523 K and it is an intrinsic p-type semiconductor ($E_g \approx 3.6$ - 4.0 eV) [5]. Apart from the magnetic properties, NiO has been considerable interest in the electrochemical properties. It has very good chemical, high thermal stability and a very high theoretical specific capacitance [6-10]. The electrochemical performance and magnetic properties of NiO have been studied by Ahmad *et al.* [11]. Electrochemical studies suggest the redox nature of this material with a strong pseudo capacitance with the specific capacitance of 210 F/g at scan rate of 100 mV/s. Vyas *et al.* [12] reported the magnetic properties of NiO with a value of magnetization of about 0.22 emu/g and a specific capacitance of 123.6 F/g at a scan rate of 10 mV/s.

In this paper, we study the magnetic and electrochemical performance of NiO microstructures synthesized by hydrothermal method. The structural analysis of NiO microstructures was employed X-ray diffraction (XRD). The morphology of the sample was observed by field emission scanning electron microscopy (FE-SEM). The chemical composition of the sample was investigated by energy dispersive X-ray spectroscopy (EDS). The porous nature of NiO microstructure was analysed using a Brunauer-Emmett-Teller (BET) analyser. The magnetization was performed using vibrating sample magnetometer (VSM) at 300 K. The electrochemical performance of materials was examined by cyclic voltammetry (CV) and galvanostatic charge-discharge (GCD).

2. Experimental procedure

In this work, nickel oxide microstructure was prepared by hydrothermal method. Nickel nitrate [Ni $(NO_3)_2.6H_2O$, 0.1 M] was used as a precursor. Nickel nitrate was dissolved and stirred in 50 mL of deionized water (DI water) at room-temperature, followed by addition to ammonia solution until its pH reached 9. The green solution was changed to a Teflon-lined stainless-steel autoclave with heating treatment at 180 °C for 24 h. The green precipitates were picked by centrifugation and washed many times with DI water and ethanol separately, then dried the green powder at 70 °C for 24 h. Finally, the dried sample was calcined at 300 °C for 3 h. The crystal structure of the NiO microstructure was analysed by XRD (Bruker-D8 advanced X-ray diffractometer). FE-SEM and EDS were used to observe the morphology and elemental composition. The pore size analysis of the sample were calculated from nitrogen adsorption-desorption isotherm by using a BET analyser. The magnetic behaviour was characterised by VSM (Versa Lab VSM, Quantum Design). The electrochemical properties were examined by CV and GCD.

3. Results and discussion

The crystal structure of the NiO sample was primary examined by XRD. Figure 1 reveals the XRD pattern of NiO sample prepared by hydrothermal method. The diffraction peaks observed for the sample can be indexed to the pure phase of cubic structure (JCPDS No.47-1049). In addition, these narrow and high intensity diffraction peaks indicate the good crystallinity of the sample. Figure 2 reveals the spectrum of EDS for NiO microstructures. The spectrum reveals peaks of Ni and O. No evidence of any impurity phase was identified by XRD and EDS. Moreover, the presence of C, Cu and Au came from carbon tape, copper stub and Au coating on the surface of sample, respectively. The detailed morphology of NiO microstructures was studied by FE-SEM (inset in figure 2). The FE-SEM image reveals the large hexagonal plates with some fractures due to the calcination at high temperature [13, 14]. The average crystallite size is about 787 ± 54 nm.

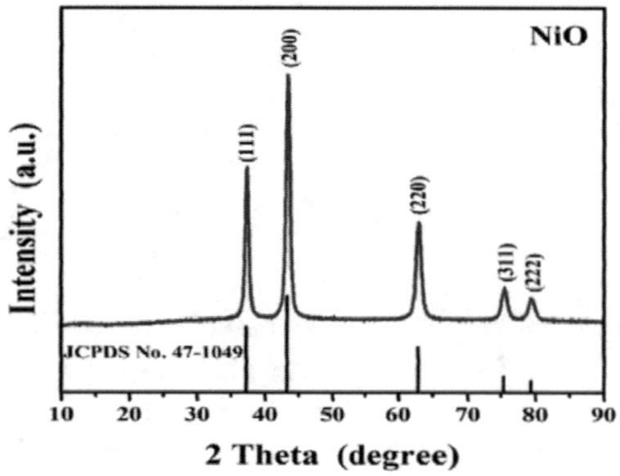

Figure 1. The pattern of XRD for NiO sample.

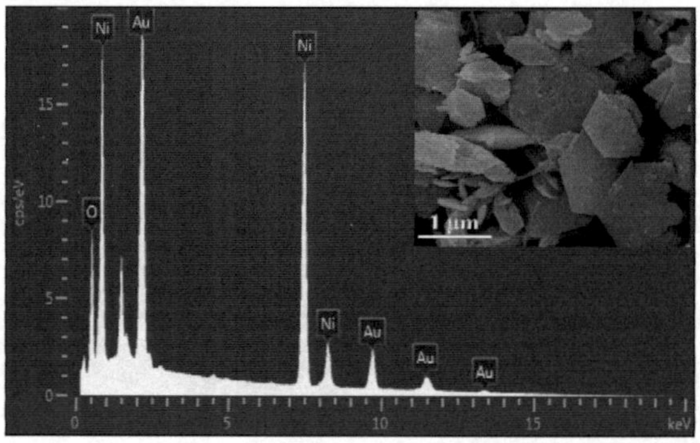

Figure 2. EDS spectrum of the sample, with inset showing its FE-SEM image.

The electrochemical properties of NiO microstructures were investigated by CV and GCD analysis. Figure 3(a) presents the cyclic voltammograms at 2, 5, 10, 20, 50, 100 and 200 mV/s in 6 M of KOH, revealing the interesting electrochemical behavior in the voltage range 0.0 – 0.55 V. The CV curves show pseudocapacitive behavior with the anodic peak appearing at 0.36 to 0.46 V, indicating the oxidation process of NiO to NiOOH. The reduction process of the reversible redox-reaction type are observed at 0.12 to 0.25 of the cathodic peak. The first process is suggested to be the adsorption and desorption of OH^- ions on the NiO surface structure, occuring near the interface of electrolyte and electrode [15], according to the equation (1). The second one is the intercalation/deintercalation of Ni ions due to the reversible redox-reaction of Ni^{2+}/Ni^{3+} as presented in equation (2).

$$(NiO)_{Surface} + OH^- \leftrightharpoons (NiOOH)_{Surface} + e^- \quad (1)$$

$$NiO + OH^- \leftrightharpoons NiOOH + e^- \quad (2)$$

Figure 3(b) reveals the GCD curves at different current densities (1, 2, 5, 10, 15, 20 A/g). The GCD curves exhibit the two levels of discharge, as a result of the change in the oxidation number of NiO/NiOOH due to the reversible redox-reaction, indicating the pseudocapacitive behavior [16]. The determined specific capacitance of the sample is a value of 174.14 F/g at the current density of 1 A/g. This value is less than theoretical capacitance [6], due to the low surface area with low active sites in our sample. The sample exhibits an excellent retention in discharge-charge test by 76.6% cycle after 500 cycles as presented in figure 3(c).

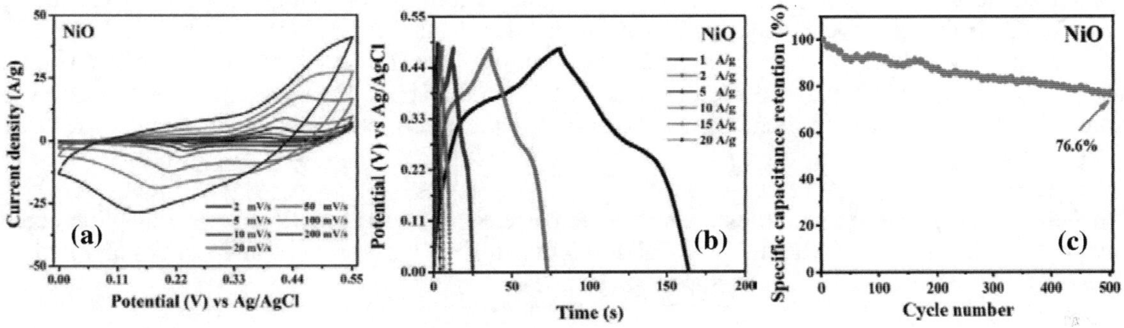

Figure 3. (a) CV curves of NiO microstructures, (b) GCD curves of NiO sample and (c) The capacitance retention of the sample.

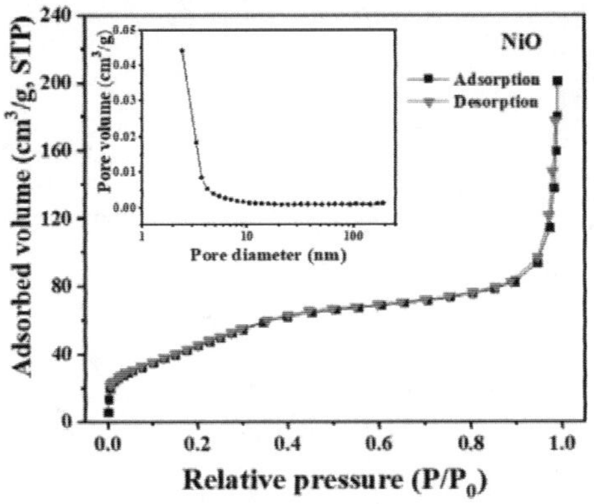

Figure 4. The nitrogen adsorption-desorption analysis with inset showing the pore size distribution.

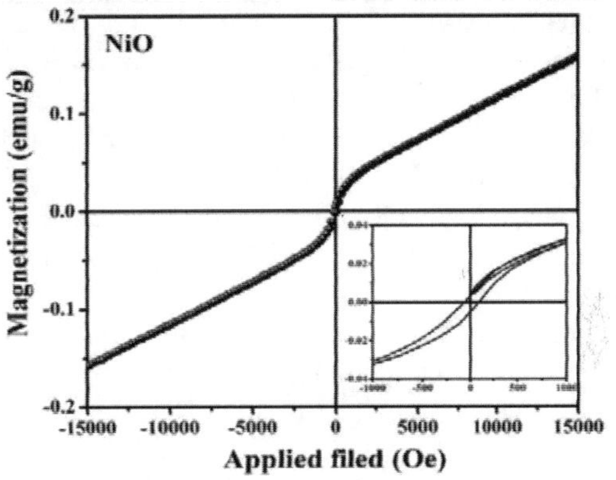

Figure 5. M-H curve of the NiO sample at room-temperature.

The porous nature is investigated by nitrogen adsorption-desorption. The nitrogen adsorption-desorption result, with revealed the Type-II of typical microporous material. Figure 4 reveals the specific surface area of the sample is 176.46 m^2/g, which calculated by BET. The calculation of pore size distribution was studied by the BJH (Barrett-Joyner-Halenda) method [inset in figure 4]. The NiO sample has the average pore size and pore volume of 2.43 nm and 0.11 cm^3/g, respectively. This result of the examining the surface area confirms the specific capacity of the material. Although the NiO has a high specific surface area caused by the calcination process, it yields a low specific capacitance. This might be result from the low average pore size and pore volume.

Figure 5 reveals the magnetization curve of the NiO sample at room-temperature. NiO microstructure shows weak ferromagnetic behaviour with a magnetization around 160 memu/g and coercivity of 80.6 Oe at the applied magnetic field of 15 kOe. A weak ferromagnetic behaviour of the sample, could result from the structural defects that formed an oxygen gap from thermal processes causing bound magnetic polaron (BMP) interactions [17, 18]. This result is consistent with the report of Bharathy and Raji [19], studying the effects of Co doping in NiO. The author found that the pure NiO and Co-doped NiO exhibited a weak ferromagnetic behaviour, causing by the oxygen gap defect with electron lone pair confined to it. Therefore, the occurred polarization could result in a weak ferromagnetism at room-temperature. Moreover, our work also consistent with the research of Suresh *et al.* [20] revealing the evidence of superparamagnetism and weak ferromagnetism in NiO nanoflakes.

4. Conclusion

Microstructures of NiO were synthesized by a hydrothermal method. The average crystalline size of the obtained product is about 787 $\pm$ 54 nm. The NiO microstructure exhibited a good specific capacitance (174.14 F/g) and a high cycle stability (76.6% capacitance retention after 500 cycles). The magnetic measurement showed the weak ferromagnetic behaviour of the NiO microstructure with magnetization of approximately 160 memu/g and coercivity of 80.6 Oe at the applied magnetic field of 15 kOe. The authors expect this study to further advance the basic knowledge for electrode and magnetic materials.

Acknowledgments

The authors would like to thank the center of excellence in advanced functional materials, Suranaree University of Technology, Thailand for CV and GCD facilities. This work was financially supported by Thailand Science Research and Innovation (TSRI) 2021.

References

[1] Li F, Xing Y, Huang M, Li K L, Yu T, Zhang Y X and Losic D 2015 *J. Mater. Chem.* A **3** 7855
[2] Wang Y, Lei Y, Li J, Gu L, Yuan H and Xia D 2014 *ASC Appl. Mater. Interfaces* **6** 6739
[3] Xiao X, Han B, Chen G, Wang L and Wang Y 2017 *Sci. Rep.* **7** 40167
[4] Khairy M and El-Safty S A 2013 *RSC Adv.* **3** 23801
[5] Din M I and Rani A 2016 *Int. J. Anal. Chem.* **2016** 3512145
[6] Chen J, Huang Y, Li C, Chen X and Zhang X 2016 *Appl. Surf. Sci.* **360** 543
[7] Li X, Xiong S, Li J, Bai J and Qian Y 2012 *J. Mater. Chem.* **22** 14276
[8] Ding S, Zhu T, Chen J S, Wang Z, Yuan C and Lou X W 2011 *J. Mater. Chem.* **21** 6602
[9] Yuan Y F, Xia X H, Wu J B, Yang J L, Chen Y B and Guo S Y 2010 *Electrochem. Commun.* **12** 890
[10] Song X and Gao L 2008 *J. Phys. Chem.* C **112** 15299
[11] Ahmad T, Ramanujachary K V, Lofland S E and Ganguli A K 2006 *Solid State Sci.* **8** 425
[12] Vyas R, Navin K, Tripathi G K and Kurchania R 2021 *Optik* **231** 166433
[13] Nathan T, Aziz A, Noor A F and Prabaharan S R S 2008 *J. Solid State Electrochem.* **12** 1003
[14] Liu X M, Zhang X G and Fu S Y 2006 *Mater. Res. Bull.* **41** 620
[15] Zhang L, Chen Q, Han X and Zhang Q 2018 *J. Clust. Sci.* **29** 1089
[16] Kong D S, Wang J M, Shao H B, Zhang J Q and Cao C N 2011 *J. Alloys Compd.* **509** 5611
[17] Schuele W J and Deetscreek V D 1962 *J. Appl. Phys.* **33** 1136
[18] Bi H, Li S, Zhang Y and Du Y 2004 *J. Magn. Magn. Mater.* **277** 363
[19] Bharathy G and Raji P 2018 *Physica B: Condens. Matter* **530** 75
[20] Suresh R, Ponnuswamy V, Sankar C, Maniekam M, Venkatesan S and Perumal S 2017 *J. Magn. Magn. Mater.* **411** 787

The facile one-step hydrothermal method to prepare MnO_2 nanoparticles: Structural and electrochemical properties

K Tangphanit[1], N Boonraksa[1], S Maensiri[2], E Swatsitang[3] and K Wongsaprom[1,*]

[1]Department of Physics, Faculty of Science, Mahasarakham University, Mahasarakham, 44150, Thailand
[2]School of Physics, Institute of Science, Suranaree University of Technology, Nakhon Ratchasima, 30000, Thailand
[3]Institute of Nanomaterials Research and Innovation for Energy (IN-RIE), NANOTEC-KKU RNN on Nanomaterials Research and Innovation for Energy, Khon Kaen University, Khon Kaen, 40002, Thailand

*E-mail: kwanruthai.w@msu.ac.th

Abstract. MnO_2 nanoparticles were successfully prepared via one-step hydrothermal method. The surface are properties of the MnO_2 nanoparticles were determined by BET nitrogen adsorption-desorption measurement. The XRD analyses confirm the pure phase of γ-MnO_2 and α-MnO_2, having orthorhombic crystal structure (JCPDS file no.14-0644 and 44-0141). FE-SEM analysis reveals the combination of massively small spherical particles with average particle size 54.8 nm. The electrochemical results revealed that the MnO_2 nanoparticles delivered the specific capacitance of 200.83 F/g at a current density of 1A/g. The cycle stability was usability 30% after 500 cycles at a current density of 5 A/g. The MnO_2 nanoparticles reveal a energy density of 3.62 Wh/kg under a power density of 43.11 W/kg.

1. Introduction

Current demand for energy is increasing, due to the energy crisis and global warming. This is to explore novel materials for advanced energy storage devices. Supercapacitors are the most promising candidates for advanced energy storage devices because of its excellent properties such as long cycle life, high power density and fast charge-discharge rate [1, 2]. One main objective for studying in this field is to explore novel material with high capacitance and fast charging [3]. The transition metal oxides (viz. NiO, Co_2O_3, TiO_2, RuO_2, MnO_2, Fe_2O_3, etc) [4-9] have been demonstrated as novel electrode materials for pseudocapacitors. Among these oxides, MnO_2 is prominent as it reveals excellent electrochemical behaviour which are widely used in applications [10-12]. It is well-known that the surface structure strongly influence the electrochemical behaviour. The different synthesis methods are enhanced the electrochemical properties of electrode materials. One of synthesis method to prepare electrode materials with various morphology and low cost is hydrothermal techniques. Xiao *et al.* [13] reported the sample of α-MnO_2 with a high capacitance 220 F/g at 5 mV/s in 0.1M Na_2SO_4 electrolyte solution prepared by the hydrothermal method. Moreover, different shape of MnO_2 reveal good capacitance of 317, 204 and 276 F/g at 5 mV/s for nanospheres, hollow urchins and smooth balls, respectively [14].

We herein present the structural and electrochemical properties of MnO_2 nanoparticles synthesized by the one-step hydrothermal method. X-ray diffraction (XRD) was used to determine the structure of the MnO_2 nanoparticles. The morphological analysis was investigated by field emission scanning electron microscopy (FE-SEM). Energy dispersive X-ray spectrometer (EDS) was used to characterize the chemical composition of the nanoparticles. The pore size analysis of the MnO_2 nanoparticles was studied using Brunauer-Emmett-Teller (BET) method. The electrochemical performance was examined by cyclic voltammetry (CV) and galvanostatic charge-discharge (GCD).

2. Experimental procedure

Nanoparticles of MnO_2 were prepared by the one-step hydrothermal method. Manganese nitrate [Mn $(NO_3)_2.6H_2O$] and potassium permanganate [$KMnO_4$] were dissolved and stirred in 100 mL of deionized water (DI water) at ambient temperature. The solution was changed to a Teflon-lined stainless-steel autoclave, sealed and maintained at 180 °C for 3 h. After the autoclave was cooled naturally to ambient temperature, the product was washed several times with DI water and ethanol separately. Then the product was dried in vacuum oven at 70 °C for 24 h. The structural analysis of the MnO_2 nanoparticles was investigated by XRD (Bruker-D8 advanced X-ray diffractometer). The shape and elements of the sample were examined by FE-SEM and EDS, respectively. The porous nature of the MnO_2 nanoparticles were performed using BET method. The electrochemical properties were examined by CV and GCD.

3. Results and discussion

Figure 1 presents the XRD pattern of the MnO_2 sample. The XRD pattern shows the diffraction corresponded to the γ-MnO_2 phase (JCPDS No.14-0644) with the orthorhombic structure. Moreover, the α-MnO_2 phase was also observed (JCPDS No.44-0141) and consistent with the research of Zhang *et al.* [15]. The peaks of impurity phases of Mn, Mn_2O_3 or Mn_3O_4 are not observed in the sample.

The elemental composition of MnO_2 nanoparticles was determined by EDS as presented in figure 2. The sample reveals peaks of Mn and O, no evidence of impurity phases, the presence of Au came from Au coated on the sample. The shape of the MnO_2 sample was studied by FE-SEM as shown in the inset of figure 2. FE-SEM image shows the small spherical particles of a uniform particle size. Moreover, these platelets are densely overlapped and aggregated. The average particle size is approximately 54.8 nm. The aggregation of particles was caused by the short reaction time of the hydrothermal process, consistent with the research of Wang *et al.* [16] synthesized a spherical α-MnO_2 using the hydrothermal method.

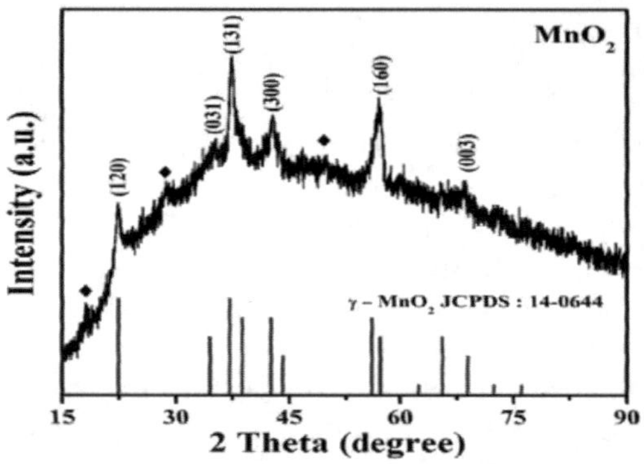

Figure 1. XRD pattern of MnO_2 nanoparticles, ♦ represents the phase of α-MnO_2.

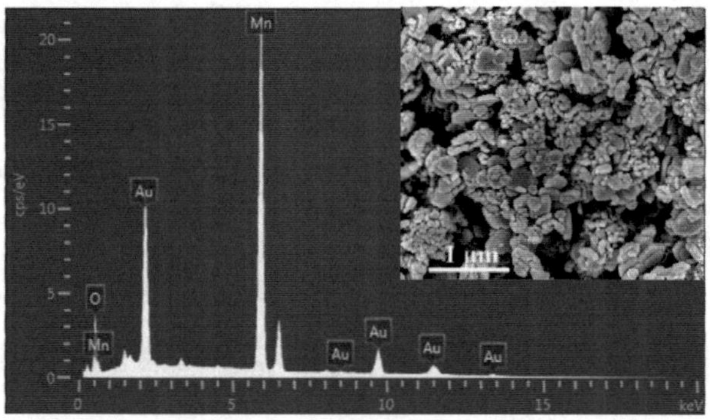

Figure 2. EDS spectrum and the inset shows the FE-SEM image of MnO_2 nanoparticles.

We investigated the specific surface area and pore size of the MnO_2 nanoparticles via nitrogen adsorption-desorption measurement. The result revealed the III Type isotherm, according to IUPAC classification as presented in figure 3. The calculation of pore size distribution was studied by the BJH (Barrett-Joyner-Halenda) method [inset in figure 3]. The pore volume and average pore diameter of the sample are 0.30 cm^3/g and 3.30 nm. The specific surface area is 7.01 m^2/g, which calculated by BET. According to the obtained structure of MnO_2 and those of low specific surface area, pore volume and pore diameter, this could result in a low specific capacity value.

The capacitive properties of the MnO_2 sample were examined by CV analysis. CV curves of the sample within the scan range 2 to 200 mV/s in a potential range -1 to 0.55 V are shown in figure 4. The CV curves show the pseudocapacitive behaviour of the sample, anodic peak appearing at 0.0 to 0.2 V and the cathodic peak shows two peaks at 0.2 and -0.3 to -0.4 V. The occurrence of more than one reaction in an intercalation / deintercalation process of the MnO_2 ion group follows the equation [17, 18].

$$MnO_2 + OH^- \rightarrow MnOOH + e^- \quad (1)$$
$$MnOOH + e^- \rightarrow HMnO_2 + e^- \quad (2)$$
$$MnOOH + HMnO_2^- \rightarrow Mn_3O_4 + H_2O + OH \quad (3)$$

From these reactions, it can be seen that the oxidation number of Mn has been changed from Mn^{4+} to Mn^{3+} and Mn^{2+}, during the reaction of the electrolyte and the electrode material. As a result, the electrode resistance and reaction resistance are increased. This result is agree with the study of Kaiyu *et al.* [18]. The increase in resistance could result in a lower specific capacitance value.

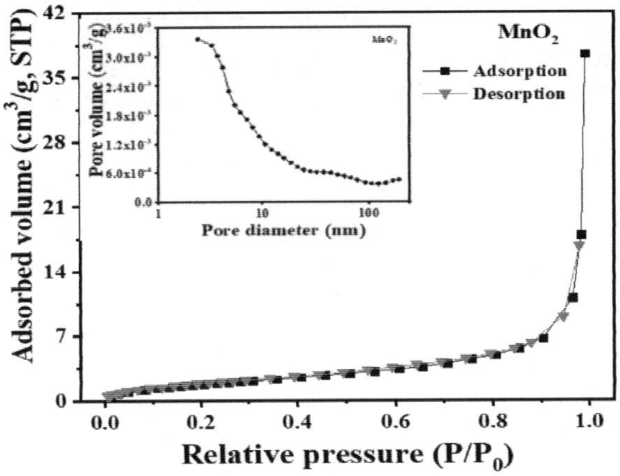

Figure 3. Nitrogen adsorption-desorption isotherms of MnO_2 nanoparticles.

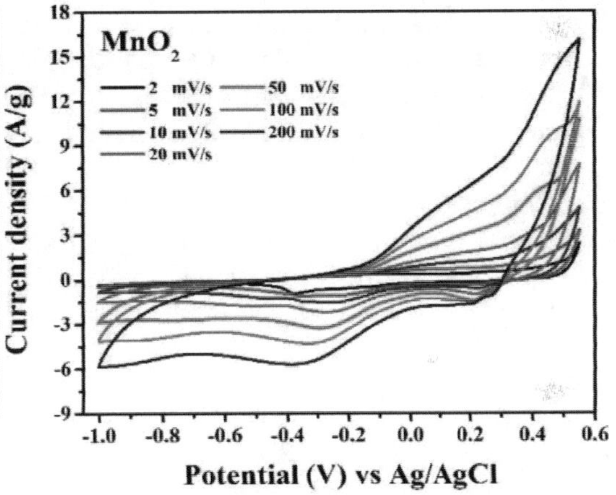

Figure 4. CV curves of the MnO_2 nanoparticles.

The electrochemical performance of the sample was further evaluated by GCD analysis as shown in figure 5. The sample electrode have three levels of discharge at different current densities. The first level is in a potential range -0.1 to 0.55V, the second level is in a potential range -0.5 to -0.1V and the third level is in a potential range 0.6 to -1V. The discharge is due to the change in the oxidation number of MnO_2, according to the reversible redox-reaction [19]. The specific capacitance is 200.83 F/g at a current density of 1 A/g. Figure 6 presents the cycle stability of sample with a usability of 30% after 500 cycles of charge and discharge at a current density of 5 A/g. The sample revealed an energy density of 3.62 Wh/kg under a power density of 43.11 W/kg. This low specific capacitance is due to the spherical shape of MnO_2, resulting in less contact area. In addition, it has a low pore diameter and pore volume.

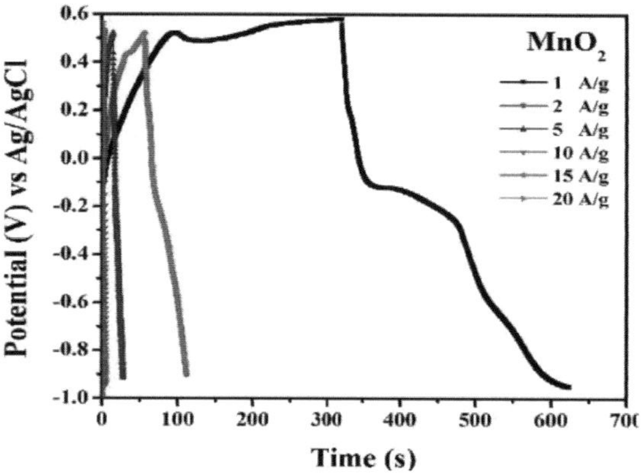

Figure 5. GCD curves of the MnO_2 nanoparticles.

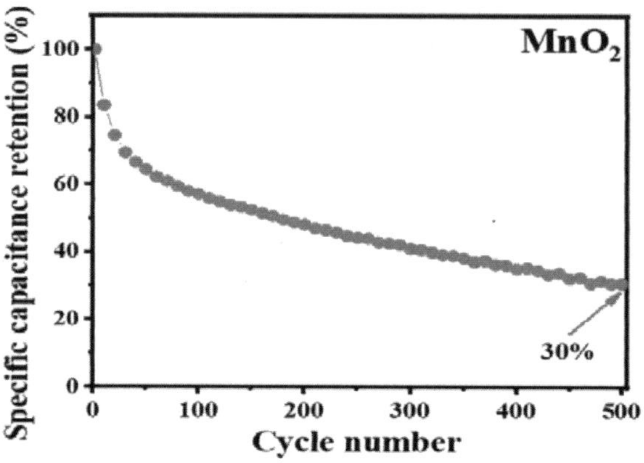

Figure 6. The cycle stability of MnO_2 sample at 5 A/g.

4. Conclusion

The MnO_2 sample was prepared by the one-step hydrothermal method. Pure phase of γ-MnO_2 and α-MnO_2 are detected in the structures. The average particle size is approximately 54.8 nm. More than one reaction occurred at the MnO_2 electrode resulting from a change in the oxidation numbers of Mn^{4+} to Mn^{3+} and Mn^{2+} ions. The MnO_2 nanoparticles showed a good specific capacitance of 200.83 F/g with 30% retention after 500 cycles. This research may be useful for further development of oxide synthesis techniques for fabrication of electrode materials.

Acknowledgments

The authors would like to thank the center of excellence in advanced functional materials, Suranaree University of Technology, Thailand for CV and GCD facilities. This work was financially supported by Thailand Science Research and Innovation (TSRI) 2021.

References

[1] Poonguzhali R, Shanmugam N, Gobi R, Senthikumar A, Viruthagiri G and Kannadasan N 2015 *J. Power Sources* **293** 790
[2] Ramirez-Castro C, Crosnier O, Athouël L, Retoux R, Bélanger D and Brousse T 2015 *J. Electrochem. Soc.* **162** A5179
[3] Wei C, Pang H, Zhang B, Lu Q, Liang S and Gao F 2019 *Sci. Rep.* **3** 2193
[4] Kim S-I, Lee J-S, Ahn H-J, Song H-K and Jang J-H 2013 *ACS Appl. Mater. Interfaces* **5** 1596
[5] Jayalakshmi M and Balasubramanian K 2008 *Int. J. Electrochem. Sci.* **3** 1196
[6] Lu X, Wang G, Zhai T, Yu M, Gan J, Tong Y and Li Y 2012 *Nano Lett.* **12** 1690
[7] Jiang Q, Kurra N, Alhabeb M, Gogotsi Y and Alshareef H N 2018 *Adv. Energy Mater.* **8** 1703043
[8] Kulal P M, Dubal D P, Lokhande C D and Fulari V J 2011 *J. Alloys Compd.* **509** 2567
[9] Xia A, Yu W, Yi J, Tan G, Ren H and Liu C 2019 *J. Electroanal. Chem.* **839** 25
[10] Ghodbane O, Pascal J L and Favier F 2009 *ACS Appl. Mater. Interfaces* **1** 1130
[11] Ray A, Roy A, Saha S, Ghosh M, Chowdhury S R, Maiyalagan T, Bhattacharya S K and Das S 2019 *Langmuir* **35** 8257
[12] Simon P and Gogotsi Y 2008 *Nat. Mater.* **7** 845
[13] Xiao W, Xia H, Fuh J Y H and Lu L 2009 *J. Power Sources* **193** 935
[14] Li N, Zhu X, Zhang C, Lai L, Jiang R and Zhu J 2017 *J. Alloys. Compd.* **692** 26
[15] Zhang Y, Li G Y, Lv Y, Wang L Z, Zhang A Q, Song Y H and Huang B L 2011 *Int. J. Hydrog Energy* **36** 11760
[16] Wang F, Zhou Q, Li G and Wang Q 2017 *J. Alloys. Compd.* **700** 185

[17] Jayachandran M, Rose A, Maiyalagan T and Vijayakumar T 2021 *Electrochim. Acta* **366** 137412
[18] Kaiyu L, Ying Z, Wei Z, He Z and Geng S 2007 *Trans. Nonferrous Met. Soc. China* **17** 649
[19] Kate R S, Khalate S A and Deokate R J 2018 *J. Alloys Compd.* **734** 89

Structural and electronic properties of $LiMnO_2$ doped with transition metals: A first-principles study

Nontawat Chaiyaocha[1] and Worasak Sukkabot[2]

Department of physics, Faculty of Science, Ubonratchatani University, Ubonratchatani 34130, Thailand

E-mail: [1]nontawat.ch.60@ubu.ac.th, [2]worasak.s@ubu.ac.th

Abstract. A spin density functional calculations of structural and electronic properties of $LiMnO_2$ doped with several transition metals (Sc, V and Tc) are reported. The physical properties of $LiMnO_2$ material are sensitive with the transition-metal dopants. Transition metal dopants enhance the lattice parameters and volumes, thus increasing the Li diffusion channel. The computations underscore that d orbitals of transition metals are located around the Fermi level. V doping in $LiMnO_2$ demonstrates the enhancement in the electronic conductivity due to the volumetric expansion. Finally, these results deliver a valuable information for the transition-metal doped $LiMnO_2$ cathode materials to improve the performance of lithium batteries.

1. Introduction

The environmental contamination and gradual exhaustion of oil resources have recently grown much attention. To overcome these concerns, the area of energy storage system has been substantially developed. The lithium-ion batteries have been utilized extensively in various application such as mobile phones, computers, cameras, electrical automobiles and hybrid electrical automobiles. The focus of this paper is on $LiMnO_2$ compound which constitutes the active component of the cathode. For the utilization of lithium-ion batteries, the performance of the cathode materials needs to be improved. To do so, several schemes have been implemented to solve this problem such as mixing carbon with the particles, formation of submicron or nanometer dimension and transition-metal doping to improve the electrochemical properties. [1-3] The $LiMnO_2$ has been a pronounced systematic consideration in the last years. [4, 5]

Here, this theoretical investigation reports the electronic properties of $LiMnO_2$ doped with Technitium (Tc), Vanadium (V) and Scandium (Sc) because there is no report of these transition metals doping in this cathode. Here, we utilize the generalized gradient approximation (GGA) with the Perdew, Burke, and Ernzerhof (PBE) [6] exchange-correlation functional as executed in CASTEP [7, 8] to convey the atomistic information. When doping with transition metals, the Broyden-Fletcher-Goldfarb-Shanno minimization algorithm [9-12] is used for the optimization. These computations can provide a better understanding and support the experimental works.

2. Theory

The spin polarized generalized gradient approximation (GGA) with PBE exchange correlation functional is used to determine the electronic structures in all systems of $LiMnO_2$. The magnetic states in all systems are antiferromagnetic (AFM). For the doping presentation, one of Mn atoms in the unit cell is replaced by Tc, V and Sc. The structural optimization is computationally achieved by the Broyden-Fletcher-Goldfarb-Shanno minimization algorithm. The criterion of the algorithm is given as energy of 2.0×10^{-6} eV/atom, force of 1.0×10^{-5} eV/Å, stress of 0.05 GPa and displacement of 0.001 Å. For all calculations, the plane-wave basis set with a cut-off energy of 520 eV is employed. The total energy within 2.0×10^{-6} eV/atom is calculated for the convergence. A 4×4×4 Monkhorst-Pack mesh is described for Brillouin zone integration.

3. Result

Here, $LiMnO_2$ structure is Pmmn. The structural parameters of $LiMnO_2$ doped with Scandium, Vanadium and Technetium atom are itemized in table 1. The calculations demonstrate that all transition metals doping in $LiMnO_2$ increase the lattice constants and their volumes. The changes of lattice constants and volumes are reported when doping with the transition metals due to the different ionic radius between Mn and dopants. These calculations highlight that the escalation of volume by transition-metal impurity can improve channels of lithium-ion diffusion.

Table 1. The computed lattice parameters, volumes of $LiMnO_2$ with Mn replacement by Sc, V and Tc.

	$LiMnO_2$	$Li(Mn, Sc)O_2$	$Li(Mn, V)O_2$	$Li(Mn, Tc)O_2$
a (Å)	2.926894	3.164833	3.124775	3.107963
b (Å)	9.268949	8.263267	8.307039	8.206066
c (Å)	5.832507	6.276250	6.456483	6.247384
Volume (Å^3)	155.089668	163.857134	167.264277	159.334211

The formation energies of $LiMnO_2$ doped with Scandium, Vanadium and Technetium atom are computed from this relation,

$$E^f(TM) = E_{dope}(TM) - E_{pure} - (\mu_{TM} - \mu_S) \quad (1)$$

Here, $E^f(TM)$ is the formation energy of the transition-metal doped cathodes. $E_{dope}(TM)$ is the total energy of the cathode with a transition-metal dopant (TM) and E_{pure} is the total energy of the pure cathode. μ_{TM} is the chemical potential of transition metals and μ_S (S=Mn or Co) is the chemical potential of Mn or Co atom in these cathodes. The chemical potentials (μ) are computed as the free atom energy. For the calculations, the free atom is placed in the center of a very huge empty cubic box with lattice constant of 30 atomic units. Table 2 lists the formation energies of $LiMnO_2$ doped with Sc, Tc and V atom. The calculations demonstrate that all transition metals doping in $LiMnO_2$ can be experimentally synthesized at the regular conditions. The highest value of the formation energies is observed in V doping in this cathode material.

Table 2. The calculated formation energies of $LiMnO_2$ with Mn substitution by Sc, Tc and V atom.

Material	Formation energy (eV)
$LiMnO_2$	-
$Li(Mn, Sc)O_2$	-5.6956
$Li(Mn, Tc)O_2$	-6.6503
$Li(Mn, V)O_2$	-7.8395

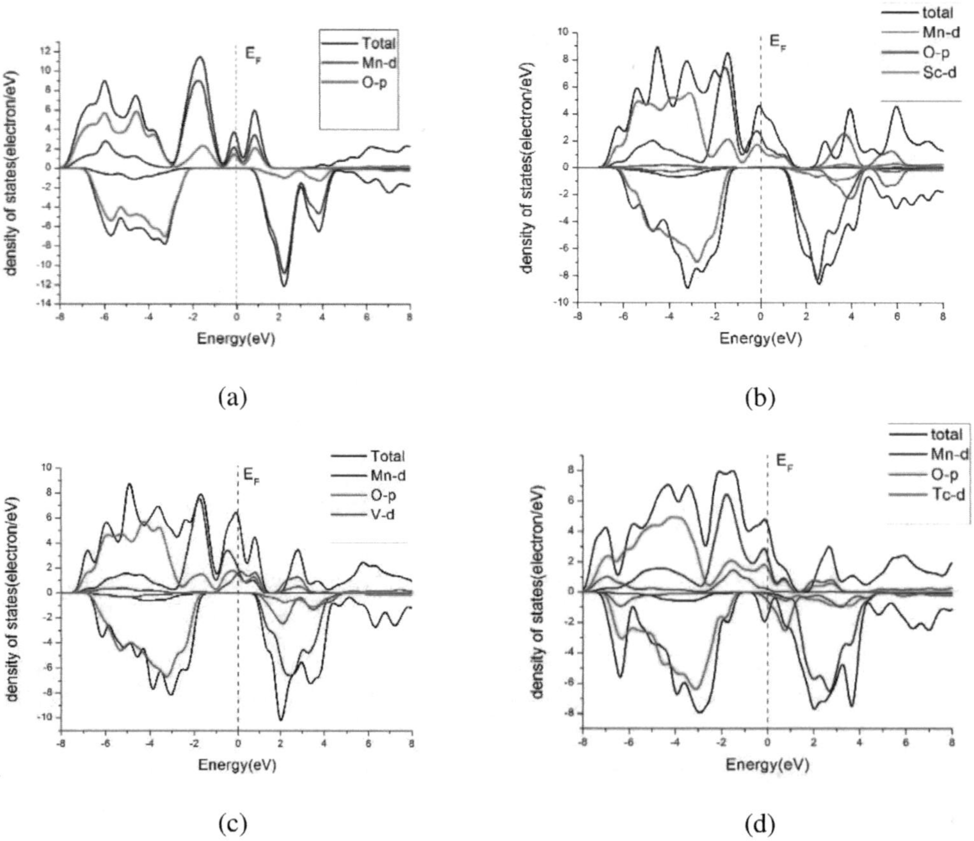

Figure 1. (a) Density of states of $LiMnO_2$ (b) $LiMnO_2$ doped with Sc (c) $LiMnO_2$ doped with V (d) $LiMnO_2$ doped with Tc.

Finally, figure 1 demonstrates the density of state of all studied materials. The dopants with d orbitals are mainly introduced around the Fermi level. Besides, all doped $LiMnO_2$ compounds are characterized as metal.

4. Summary and conclusion

In this work, the result shows that $LiMnO_2$ doped with Vanadium atom is the best material to synthesis a new battery because of the highest volume. This means Lithium ions can move easily in this cathode.

All transition metals doping in $LiMnO_2$ can be experimentally synthesized at the normal conditions. The highest value of the formation energies is observed in Li(Mn, V)O_2. All doped $LiMnO_2$ compounds are considered as metal. Finally, these computational results mainly provide the suggestion of transition-metal doped $LiMnO_2$ cathode for the novel lithium batteries.

Acknowledgement

We appreciate the financial support from the Science Achievement Scholarship of Thailand Project (SAST) and the Faculty of science, Department of physics, Ubonratchatani University.

References

[1] Wang G X, Yang L, Bewlay S L, Chen Y, Liu H K and Ahn J H 2005 *J. Power Sources* **146** 521
[2] Gaberscek M, Dominko R and Jamnik J 2007 *Electrochem. Commun* **9** 2778
[3] Chung S Y, Bloking J T and Chiang Y M 2002 *Nat. Mater.* **1** 123

[4] Hoang K 2017 *Phys. Rev. Mater.* **1** 075403
[5] Zhang Z *et al*. 2019 *Nano Lett.* **19** 8774
[6] Perdew J P, Ernzerhof K and Burke M 1996 *Phys. Rev. Lett.* **77** 3865
[7] Clark S J, Segall M D, Pickard C J, Hasnip P J, Probert M J, Refson K and Payne M C 2005 *Z. Kristallographie* **220** 567
[8] Segall M D, Lindan P J D, Probert M J, Pickard C J, Hasnip P J, Clark S J and Payne M C 2002 *J. Phys.: Condens. Matter* **14** 2717
[9] Broyden C G 1970 *J. Inst. Math.* Appl. **6** 76
[10] Fletcher R 1970 *Computer J.* **13** 317
[11] Goldfarb D 1970 *Math. Comput.* **24** 23
[12] Shanno D F 1970 *Math. Comput.* **24** 647

Mechanical and physical properties of water hyacinth and cogon grass fiber reinforced epoxy resin composites

P Kongkaew[1*], P Praneekrit[1], T Rudchapo[1] and K Khampui[2]
[1]Program of Physics, Faculty of Science and Technology, Rajabhat Maha Sarakham University, Muang Maha Sarakham, Maha Sarakham 44000 Thailand
[2]Program of Physics, Faculty of Education, Rajabhat Maha Sarakham University, Muang Maha Sarakham, Maha Sarakham 44000 Thailand

*Corresponding author's e-mail address : pongsathorn.ko@rmu.ac.th

Abstract. In this paper, the study investigates the mechanical and physical properties of water hyacinth and cogon grass fiber reinforced epoxy resin hybrid composites. Hand lay-up technique was used to fabricate the composites. Water absorption, microstructure, tensile properties, flexural properties, and impact strength tests for total fiber contents, 15 wt %, and different water hyacinth and cogon grass fiber ratios (10/0, 8/2, 6/4, 4/6, 2/8, and 0/10) were used to evaluate the investigation's effects. The addition of water hyacinth and cogon grass fiber into epoxy improves tensile, flexural, and impact properties while decreasing water absorption, according to the findings. Using a scanning electron microscope (SEM), the microstructure of the composites was analyzed, and surface fracture behavior and the void between the fiber and matrix were observed.

1. Introduction

Natural fiber composites are a new material trend in the automotive, construction, and packaging industries. [1]. Furthermore, they are environmentally friendly, renewable, and biodegradable resources [2]. However, a water hyacinth and cogon grass are used to reinforce a polymer matrix in epoxy hybrid composites in this study. Because of its appealing properties, water hyacinth (*Eichornia crassipe* (Mart.) Solms.) is a considered useful natural fiber to use as a reinforcement in composite materials. The cellulose structure is small in diameter but has a high concentration [3]. The water hyacinth possesses a high percentage of holocellulose in comparison to other fibers, which is advantageous in its use as a reinforcing material. Synthetic polymers could thus be reinforced with natural fillers such as water hyacinth to increase their physical and mechanical properties and achieve the desired features in specific applications [4]. Cogon grass (*Imperata cylindrica* (L.) Beauv.) is one of the worlds 's ten most aggressive weeds because of its ability to colonize, spread, and replace valuable plants [5]. This plant is used to make paper, thatch, and weave bags and mats, as well as for traditional medicine [6]. Apart from that, cogon grass is a promising material for use as a reinforcement in the development of polymer composites [5]. As a result, this study examined the physical, mechanical, and microstructure of water hyacinth hybridization with cogon grass composites.

Table 1. Physical and mechanical average values of tests performed.

Sample (Water hyacinth: cogon grass)	Water absorption (%)	Tensile strength (MPa)	Tensile modulus (MPa)	Flexural strength (MPa)	Flexural modulus (MPa)	Impact strength (kJ/mm^2)
Epoxy	3.65±0.07	1.99±0.67	30.83±14.88	32.71±3.38	267.33±0.57	0.35±0.07
10:0	3.36±0.83	20.35±1.88	373.16±2.37	35.49±9.65	1625.88±336.45	4.37±0.015
8:2	2.17±0.95	27.41±1.20	584.33±36.26	52.88±3.61	2053.36±287.12	3.46±0.14
6:4	2.13±0.39	23.55±0.41	382.66±154.39	58.13±6.62	2117.12±318.26	4.60±0.32
4:6	3.13±0.68	22.19±1.70	379.66±148.83	40.77±8.98	1760.85±219.22	4.09±0.46
2:8	2.54±0.41	24.79±0.50	421.33±60.33	50.72±6.34	1784.34±83.38	4.36±0.03
0:10	3.06±0.40	25.54±0.70	433.83±109.61	44.26±2.04	1769.98±425.38	4.95±0.47
p-value	0.08	0.00*	0.00*	0.02*	0.01*	0.00*

Where * indicates a significant difference at $p < 0.05$.

2. Materials and methods

2.1. Raw materials

The raw materials prepared in this research include water hyacinth and cogon grass fiber were collected locally from Kaonoi village Kalasin, Thailand. Both fibers had diameters in the range of 200-300 μm, which were obtained at an average moisture content of 4-13%. The matrix was prepared from epoxy resin and hardener, both of which had a density of 1.18 g/m^3, and were purchased at a local market in Bangkok, Thailand.

2.2. Fabrication of composite

To produce the hybrid composites, epoxy was used to reinforce fabric water hyacinth and cogon grass fibers. Firstly, a penknife was used to extract both fibers from the plants. The fibers are then soaked in water and sundried to eliminate any moisture that could cause poor fiber and matrix bonding. After that, the fibers were immersed in a NaOH solution for 3 hours (NaOH 97 percent Acilabscan Ar 1171-P1 KG, 5% by volume) [7]. Later, the fibers are cleaned with deionized water (DI) until the solution is neutral (pH7) and then dried in an oven for 120 h at 90 °C [8]. Thereafter, the particles were mixed and sieved until they were between 9 and 10 mesh. Fiber ratios of water hyacinth and cogon grass (10/0, 8/2, 6/4, 4/6, 2/8, and 0/10) were used to produce the fiber contents (15% by weight). Epoxy resin and hardener were mixed in a 10:1 weight ratio to create a matrix [9]. After that, the mixture is put into the acrylic mold. It was carried out in accordance with ASTM standards. (The water absorption, tensile test, flexural and impact test were determined in accordance with ASTM D570, ASTM D680 type 1, ASTM D790 and ASTM D256, respectively). The hand lay-up technique was used to fill the acrylic mold with a suitable amount of water hyacinth, cogon grass, and epoxy resin. The top and bottom of the specimen were then laminated with Mylar films to create a smooth surface. To compress the composite layer, two c-clamps were used to tighten the mold, which produced the compressive force. After 7 days of curing at room temperature, the composite samples were removed from the mold to allow for set-up. After that, scanning electron microscope (SEM) was used to analyze the microstructure of the composite samples, and five samples were prepared for each test to investigate the physical and mechanical properties. Finally, all data are measured standard deviation. The data was analyzed using a one-way ANOVA with a 95 percent confidence level for significant differences. The cut-off for significance is P-value, which is an alpha of 0.05.

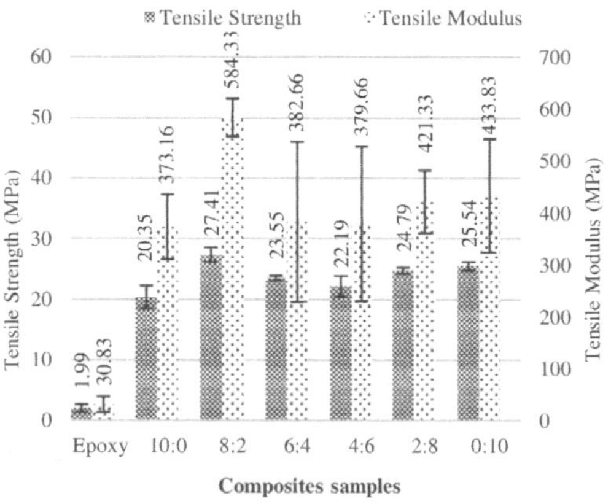

Figure 1. Tensile properties of composites.

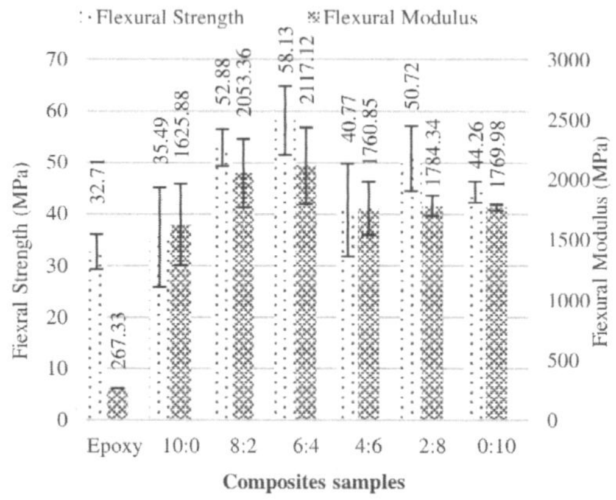

Figure 2. Flexural properties of composites.

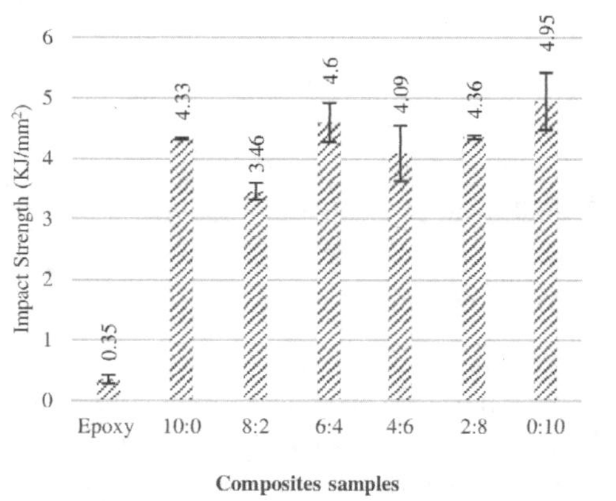

Figure 3. Impact strength of composites

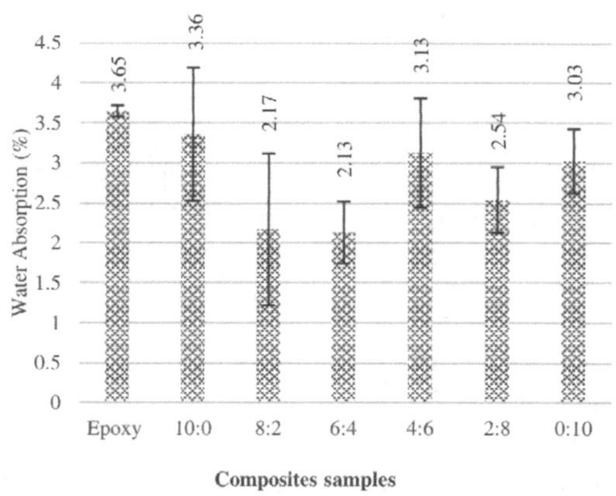

Figure 4. Water absorption of composites.

3. Results and discussion

3.1. Tensile test

Figure 1 illustrates the tensile strength and modulus of composites. In addition, table 1 shows a detailed statistical analysis. Statistically significant differences were found in the mean tensile strength and modulus ratios for both fiber types ($p < 0.05$). All-fiber ratios displayed higher modulus and tensile strength than pure epoxy resin, as shown in figure 1. The composites with 8/2 fiber ratios demonstrate high tensile strength and modulus values than the other fiber ratios, with the highest values of 20.41 MPa and 584.98 MPa, respectively. In addition, the high compatibility between both fibers in variable ratios may contribute to composites improved tensile properties. The strength of a composite can be increased or decreased by adding natural lignocellulose fibers to a polymer matrix. Fibers, such as lignocellulose fibers, can typically improve strength by absorbing stress transferred from the polymer [10]. The natural properties of the fiber and matrix, the adhesion between fiber and matrix, the aspect ratio of fiber and matrix length distribution, volume, and fiber orientation are all factors that Sathiamurthi *et al.* [11] reported influence tensile strength and modulus.

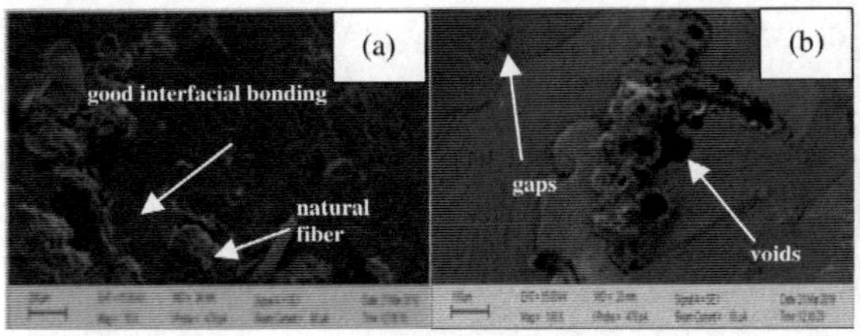

Figure 5. SEM micrographs of composites with fiber ratios of (a) 8/2 and (b) 10/0 after tensile testing.

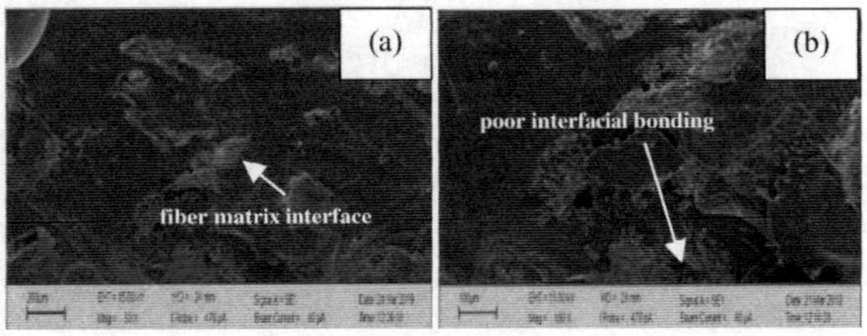

Figure 6. SEM micrographs of composites with fiber ratios of (a) 6/4 and (b) 10/0 after flexural testing.

3.2. Flexural test
Figure 2 illustrates that the flexural strength and modulus of composites are varied by fiber ratios. Table 1 shows the results of an analysis of variance (ANOVA) of the flexural properties. The analysis indicates that varying the hybrid fiber ratios resulted in significant differences ($p < 0.05$) in the composites mean flexural strength and modulus. As the fiber ratios increased from 10/0 to 6/4, the flexural strength and modulus increased, obtaining maximum values of 58.13 MPa and 2117.12 MPa, respectively. The improvement in flexural properties could be attributable to a similar cause as the tensile test results. However, flexural properties decreased at 4/6 fiber ratios, which could be attributed to poor interfacial bonding between the fibers and matrix due to lower fiber wetting. Poor flexural characteristics are caused by weak bonding between fiber and matrix, according to Khalil *et al.* [12]

3.3. Impact test
Figure 3 shows the results of altering the impact strength of the hybrid composites. Simultaneously, the detailed statistical investigation was directed, presented in table 1. Impact strength was not significantly different ($p > 0.05$) when the fiber ratios increased from 6/4 to 0/10 and 10/0. Nonetheless, further 8/2 fiber ratio brought about a significant change ($p < 0.05$) from each other, recording the highest value of 4.95 kJ/m^2 at 0/10 fiber, a ratio which was compared to other ratios. There is also a discontinuous trend in the impact strength from 10/0 to 0/10 fiber ratios, which can be observed. However, there is the lowest strength at 8/2 fiber ratio compared to other ratios except for epoxy resin. The decrease in the impact strength can be described as due to the "overcrowding effect" [13]. From these results, the filler was loaded further an over limit and agglomeration occurs, which affects the load transfer between the matrix and the filler. Inevitably, early failure reduced the mechanical strength [14].

3.4. Water absorption
Figure 4 illustrates the water absorption behavior from several composites after 30 minutes of immersion. Investigation indicated that the water absorption of the samples was not significantly different ($p > 0.05$) for all other fiber ratios as shown in table 1. From these results, it was observed that the 10/0 fiber ratio showed that maximum water uptake with an increase of 3.36%. Water absorption was lowest with 6/4 fiber ratio and increased by 2.13 percent, while epoxy resin has the highest absorption, which is compatible with the mechanical test results. The water absorption ability of the fiber depends on the greater adhesion hydrogen bonding between the network structure [15].

3.5. SEM morphology
The fracture morphology from the tensile and flexural test of samples of the highest and lowest tensile and flexural strength were investigated utilizing SEM a shown in figures 5 and 6. Figure 5 (a) observed good interfacial contact and high strength for 8/2 fiber ratios. At 10/0 fiber ratios, a generally high fiber pulls out and small gaps around the fibers were shown in figure 5 (b). In this micrograph, a fiber was pulled out, and it was observed that the sample had poor internal bonding between fiber and matrix.

Figure 6 (a) shows the image of 6/4 fiber ratios after the flexural test. From this micrograph, it was observed an ability high adhesion between fiber and matrix. As a result, if they have high strength and stiffness, reinforcement is strong adhesion to the matrix [16]. The void in the matrix in figure 6 (b) demonstrated that the porous structures were caused by trapped air bubbles within the matrix composites. The mechanical strength of the composites decreased as the concentration of water hyacinth fiber within the matrix increased. According to Fedulov *et al.* [17] void content was found to minimize the tensile and flexural failure load.

4. Conclusion

The effects of using water hyacinth/cogon grass fiber blends to reinforce epoxy resin composites were investigated. At water hyacinth to cogon grass blend ratio of 8/2, 6/4 and 0/10, epoxy resin hybrid composites produced at a constant fiber weight fraction of 15% achieved maximum tensile, flexural properties and impact strength respectively. The water absorption results showed the composite had a low water uptake rate which improved the mechanical properties. In a composite morphology investigation, fiber pull-out, interfacial characteristics and internal surface structures were all detected. Fiber reinforced epoxy composites could be used to replace synthetic fiber reinforced materials.

Acknowledgments

The author would like to acknowledge the funds provided by Research and Development Institute (RDI), Rajabhat Maha Sarakham University.

References

[1] Sapuan S M, Leenie A, Harimi M and Beng Y K 2006 *Mater. Des.* **27** 689
[2] Mukhopadhy S and Srikanta S 2008 *Polym. Degrad. Stab.* **93** 2048
[3] Goud G and Rao R N 2011 *Int. J. Polym. Anal. Charact.* **16** 239
[4] Rana A K, Mandal A and Bandyopadhyay S 2003 *Compos. Sci. Technol.* **63** 801
[5] Aiza Jaafar C N, Zainol I, Ishak N S, Ilyas R A and Sapuan S M 2021 *J. Mater. Res. Technol.* **146** 751
[6] Ruksakulpiwa C, Wanasut W, Singkum A and Ruksakulpiwat Y 2013 *Adv. Mater.* **747** 375
[7] Duigou A L, Davies P and Baley C 2009 *Polym. Degrad. Stab.* **94** 1151
[8] Kulawong S, Chanlek N and Osakoo N 2020 *J. Environ. Chem. Eng.* **8** 2
[9] Rao D N, Mukesh G, Mukesh G, Ramesh A and Anjaneyulu Y 2020 *Mater. Today Proc.* **27** 1703
[10] Lai C Y, Sapuan S M, Ahmad N, Yahya N and Dahlan K Z H M 2005 *Polym. Plast. Technol. Eng.* **44** 619
[11] Sachinkumar, Narendranath S and Chakradhar D 2021 *Mater. Today Proc.* **45** 1
[12] Khalil H P S A, Jawaid M, Hassan A, Paridah M T and Zaidon A 2012 *Composites and Their Applications* (London: Intech Open Limited) Chapter 8 p 187
[13] Sureshab B, Indushekhara N M, Varan CA, Sachin D and Pranao K 2021 *Mater. Today Proc.* **43** 1478
[14] Thostenson E T and Chou T W 2003 *J. Phys.* D *Appl. Phys.* **36** 573
[15] Mehra A K, Saini R and Kumar A 2021 *Compos. Commun.* **25** 6
[16] Biswas S, Kindo S and Patnaik A 2011 *Fibers Polym.* **12** 73
[17] Sapuan S M, Aulia H S, Ilyas R A, Atiqah A, Dele-Afolabi T T, Nurazzi M N, Supian A B M and Atikah M S N 2020 *Polymers* **12** 2211

Development and study on mechanical properties of small diameter artificial blood vessel by using electrospinning and 3d printing

A Sukchanta[1]*, P Kummanee[2] and W Nuansing[3]

[1]School of Biomedical Innovation Engineering, Institute of Engineering, Suranaree University of Technology, Muang Nakhon Rachasima, Nakhon Ratchasima 30000, Thailand
[2]Medical Physics Program, Faculty of Medicine and Public Health, HRH Princess Chulabhorn College of Medical Science, Lak Si, Bangkok 10210, Thailand
[3]School of Physics, Institute of Science, Suranaree University of Technology, Muang Nakhon Rachasima, Nakhon Ratchasima 30000, Thailand

*E-mail: m6200770@g.sut.ac.th

Abstract. The small diameter artificial blood vessel is synthesized with a diameter less than or equal to 6 millimetres. This technique has been used in coronary artery bypass grafting to treat coronary artery disease. Currently, the problem of coronary artery disease is still common, in addition to aortic aneurysm caused by the incompatibility of mechanical properties between the artificial blood vessel and the local blood vessel in the patient's body. This research aims to solve the aforementioned problems using electrospinning and 3D printing technologies, as many types of materials are supported, all parameters are easy to change, and the cost is low. In this report, we describe a design for a small diameter polylactic acid (PLA) vascular graft fabricated by electrospinning with solutions of PLA in AC/DMF (1:1) 10, 12 and 15% w/v at 4 mm. The electrospun PLA nanofibers are tested for their morphology, contact angle, and seam strength. As the results, the fibres are still no same direction alignment due to insufficient rotation speed. The filament holding force is in the range of 1.90-2.71 N and the contact angles are greater than 90° because the samples are not wettable and have hydrophobic property. Further on, we will investigate other required properties, such as cell culture and other mechanical properties. Furthermore, we will compare the results with 3D printed artificial blood vessels with small diameter.

1. Introduction

Cardiovascular diseases (CVDs) are one of the leading causes of death around the world. In 2016, 17.3 million people died from cardiovascular related reasons. Plaque formation in the arteries can affect the cardiovascular system, which is the cause of CVDs. Generally, bypass surgery is performed to allow peripheral or coronary revascularization. Although large diameter (inner diameter (ID) $>$ 6 mm) artificial vessels have been successfully used for bypass surgery, small diameter (inner diameter (ID) $\leq$ 6 mm) artificial vessels are still unsuccessful [1]. Most of the causes of poor performance of small diameter grafts are adverse biological responses, including delayed reendothelialization, surface thrombosis, and neointimal hyperplasia (IH) [2].

Tissue engineering is an interesting option for vascular grafts and for the production of small diameter vessels using biodegradable scaffolds Technologies for vascular grafts that resemble the structure of blood vessels. Electrospinning and 3D printing technology is widely used for this purpose, as this technique allows the fabrication of nano- to micro-scale structures and provides control over the composition and mechanical properties of the scaffolds [3].

The aim of this research is to develop and investigate the fabrication of scaffolds for vascular grafts using electrospinning and 3D printing, and to test the morphological and mechanical properties for the fabrication of small diameter vascular grafts. Moreover, the importance of 3D printing is able to determine the patterns of blood vessel as desire for the patient, who got damage in specific part of the blood vessel. The developed small-diameter vascular graft be mechanical testing which are young's modulus, tensile stress, elongation at break, compliance, burst pressure, suture retention, and contact angle for wetting capability that is an important property for cell culture of vascular graft. Finally, researchers expect to achieve new technical methods and conditions for synthesizing small-diameter artificial blood vessel by two early methods, including artificial blood vessel with proper mechanical properties for utilizing in coronary artery bypass grafting.

2. Materials and methods

2.1. Materials

Poly (lactic acid) (PLA) (Shenzhen eSUN Industrial, China). The PLA was dissolved in a 1:1 solvent mixture of 99.8% Dimethylformamide (DMF) (Sigma-Aldrich, Germany) and 99.5% Acetone (AC) (RCI Labscan, Thailand).

2.2. Preparation of PLA solutions

The polymer solutions for electrospinning were prepared by dissolving PLA in AC and DMF (50/50 v/v). PLA was dissolved to obtain solutions with concentrations 10, 12 and 15% w/v. To preparation the solutions concentration, the PLA were weighted and put in a glass bottle before added AC and DMF at a volume ratio of 1:1. All solutions were magnetically stirred at 120 °C for 4 h, until complete dissolution of the polymer.

2.3. Electrospinning process

In this work, a vertically rotating drum collector was used for electrospinning to fabricate nanofibers for small diameter artificial blood vessels. The electrospinning setup consisted of a high-voltage power supply, a syringe pump, and 10-mL syringe attached PTFE (Polytetrafluoroethylene) tube with an 18-G needle that cutting out the sharp tip of the needle on the other side of tube. The prepared solutions 4 ml were loaded into the syringe and injected at a constant flow rate of 6 mL/hr. The distance between the needle tip and the collector and the applied voltage were fixed at 15 cm and 15 kV, respectively. The speed of the collector was changed to 200, 350 and 450 rpm for the orientation of the nanofiber morphology studied. After the fabrication process, the electrospun PLA nanofibers were wrapped with longitudinally around a small diameter rod of 4 mm (see figure 1).

2.4. Characterization of vascular grafts

The surface morphology of the electrospun PLA nanofibers was characterized using a JEOL model JSM- 6010LV scanning electron microscope (SEM) with an accelerating voltage of 10 kV. Prior to observation, the entire sample was coated with gold sputter for 1 min (JEOL model JFC-1100E). The contact angle of electrospun PLA nanofibers were observed using a homely setup. The square (1 × 1 cm) electrospun PLA nanofibers were placed on a glass slice and the water was dropped on it. The diameters of the electrospun PLA nanofibers and the contact angle between water and the surface of the electrospun PLA nanofibers were measured manually using ImageJ analysis software.

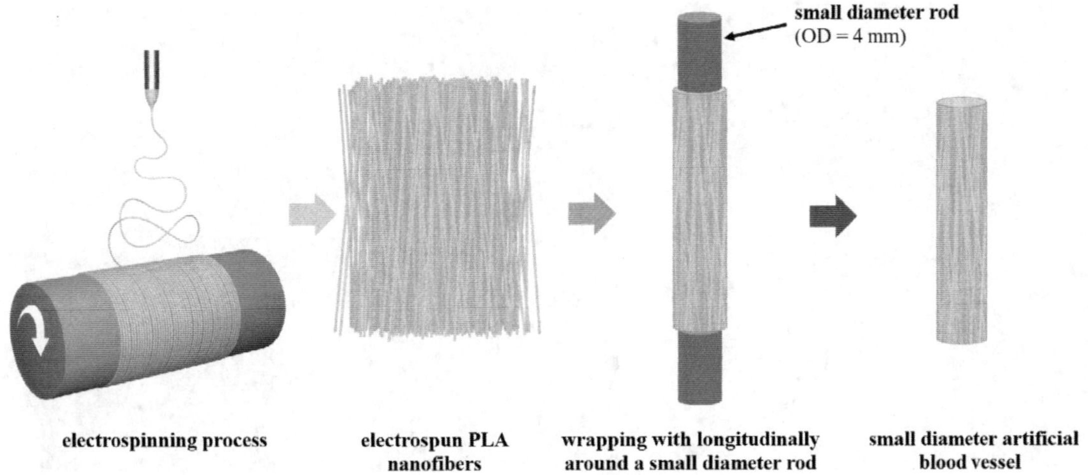

Figure 1. The process for fabricating small diameter artificial blood vessels with electrospun PLA nanofibers.

2.5. Suture retention testing

For the suture retention test, electrospun PLA nanofibers were cut into rectangles with 10 × 20 mm. The lower end was attached with a loading weight bag. The PTFE monofilament was passed through a 0.5 mm hole and 2 mm from the top end of the specimen. The suture loop was extended by adding the load to the bag until the membrane was broken (see figure 2). The SRS was defined as the maximum force measured during the test, which refers to the resistance of grafts to tearing out a suture. So, the maximum load weight can be converted into a force using Newton's second law ($\sum \vec{F} = m\vec{a}$).

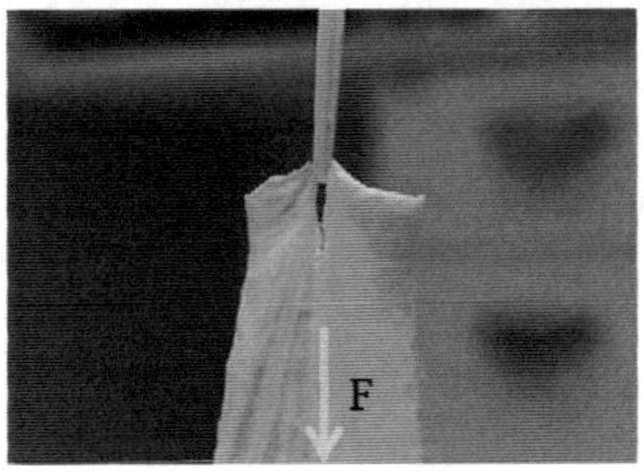

Figure 2. The rectangular specimen 20 × 10 mm with a single suture placed at a distance of 2 mm from the edge of the specimen for testing the suture retention testing.

3. Results and discussion

3.1. Surface morphology

The surface morphology of the electrospun PLA nanofibers of three concentrations was characterized using SEM, as shown in figure 3(B). As the result, all samples are inhomogeneous and do not show the same orientation direction. There are some beads and noticeable defects, as shown in SEM figure. Their average diameters are listed in table 1.

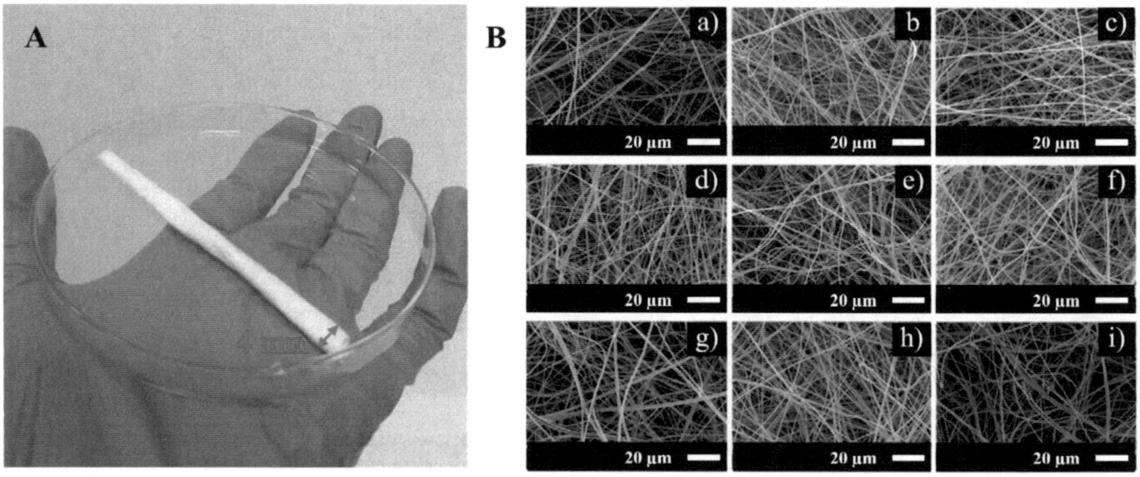

Figure 3. A) The electrospun PLA nanofibers wrapped with longitudinally around a small diameter rod of 4 mm and B) SEM image of PLA nanofibers (1000x) of (a-c) 10 w/v%, (d-f)12 w/v% and (g-i)15 w/v% concentration with 200, 350 and 450 rpm speed of collector, respectively.

Table 1. The average PLA fibres diameter with difference speed of collector, nm.

PLA Concentration (%w/v)	200 rpm	350 rpm	450 rpm
10%	722 ± 57	607 ± 35	740 ± 54
12%	736 ± 50	720 ± 42	766 ± 39
15%	1129 ± 84	965 ± 64	1009 ± 80

3.2. Contact angle

The PLA electrospun nanofibers were examined wetting capability by measuring the contact angle of water. The tested specimens are in rectangular shapeas, shown in figure 4. The contact angle of PLA nanofibers with rotation speed collector at 200 rpm are 124°, 89° and 139° for the PLA solutions with concentration of 10, 12 and 15%w/v, respectively.

Figure 4. Water droplets on PLA nanofibers from solution (a) 10, (b) 12 and (c) 15 % w/v.

3.3. Suture retention

The SRS of vascular grafts was measured by adding weight loading until the specimens ruptured, that is, by the maximum force measured during the test, which refers to the resistance of the grafts to tearing out a suture. The SRS of PLA nanofibers from solutions with concentration of 10, 12 and 15%w/v at higher collector speed (450 rpm) was much greater than at lower (200 rpm) and the SRS has also relative to concentration (10%w/v; 1.96, 0.75 N, 12%w/v; 1.90, 2.40 N and 15%w/v; 2.53, 2.71 N). In the generally required for successful implantation and long-term use of a vascular graft, the sufficient SRS must be above the threshold of 2.0 N [4].

4. Conclusion

This study is mainly focused on development and study on mechanical properties of small diameter artificial blood vessel by using electrospinning for cardiovascular tissue engineering.

PLA nanofibers diameter and fibre alignment of the vascular were optimized using different PLA concentration, PLA volume and rotation speed. As the results, the fibres are still none same direction alignment. That may be the insufficient rotation speed. There were somewhat typical results in the fibre diameter at the rotation speed of 350 rpm. This may be due to a malfunction of the instrument used. Suture retention strength was similar to previously measured human saphenous vein suture retention strength (1.92 ± 0.02 N), and human arteries suture retention strength (1.92 ± 1.16 N) [5]. The contact angles are more than 90° because the specimens have non wetting capability and present hydrophobic property. So, the PLA graft needs a surface modification such as plasma treatment [6] for increase wetting capability before cell culture process the results of this study show that the small diameter PLA artificial blood vessel can be an attractive substitute for a biocompatible and biodegradable artificial blood vessel [7].

In further we will study other required features i.e., cell culture, and other mechanical properties which are young's modulus, tensile stress, elongation at break, compliance, burst pressure. In addition, we will compare the results to 3D printed small diameter artificial blood vessel.

Acknowledgement

This research was supported by School of Physics, Suranaree University of Technology. The author would like to thank Dr. Wiwat Nuansing for the consultation, comments and recommendations on this research. We would also like grateful to Prof. Dr. Santi Mansiri (Advanced Materials Physics (AMP) Laboratory) to supported laboratory materials and instrument for this research.

References

[1] Marelli B, Alessandrino A, Farè S, Freddi G, Mantovani D and Tanzi M C 2010 *Acta Biomater.* **6** 4019–26
[2] Tan W, Bonani W and Madhavan K 2010 *Nanofibrous Materials for Vascular Tissue Engineering and Regeneration* (Boca Raton: CRC Press) chap 2 pp 1–26
[3] Li Z, Li X, Xu T and Zhang L 2019 *J. Appl. Sci.* **9** 2864
[4] Casasola R, Thomas N L, Trybala A and Georgiadou S 2014 *J. Polym.* **55** 4728–37
[5] Park S, Kim J, Lee M K, Park C, Jung H D, Kim H E and Jang T S 2019 *J. Mater. Des.* **181** 108079
[6] Ercolani E, Del Gaudio C and Bianco A 2013 *J. Tissue Eng. Regen. Med* **9** 861–88
[7] Wang D, Xu Y, Li Q and Turng L 2020 *J. Mater. Chem.* B **8** 1801–22

Synthesis of the platinum particle with the pH variation for the particle size control

P Thongnopkun* and W Kitprapot

Gems and Jewelry Research Unit, Faculty of Gems, Burapha University, Chanthaburi Campus, Chantaburi, 22170, Thailand

*E-mail: pimthong@go.buu.ac.th

Abstract. Platinum is a precious metal widely used in the jewelry industry due to its property and intrinsic value. The different particle sizes of platinum can be applied in various applications, especially for jewelry production. In the present article, submicron, and nano-sized platinum particle sizes were synthesized through simple chemical reduction methods and the effect of pH variation was revealed. The scanning electron microscope (SEM) images showed that the pH variations give rise to significant changes of the obtained particle size. The size of platinum particle was decreased from submicron to nanoscale while the pH was increased. The UV-Vis spectra indicated the maximum absorption at 220 nm confirming the spherical shape of the platinum particle. The FT-IR spectroscopy was used to analyze the residuals from the synthesis. The result showed that there is no indication of residual in the synthesized particle. Additionally, this synthesis can provide stability in terms of size and shape, as well as high production yield.

1. Introduction

Platinum is a noble metal largely applied in various applications i.e., a catalytic converter, dentistry equipment, platinum resistance thermometers, laboratory equipment, and jewelry. In jewelry manufacture, platinum is a problematic element in the production phase due to its high sintering and melting temperature. Nowadays, nanotechnology is a burgeoning field widely applied for the synthesis of platinum nanoparticles (Pt NPs) [1-2]. According to the size of a particle, the nanoparticle has a lower sintering and melting temperature than those of the microparticle. Moreover, both chemical and physical properties of Pt NPs were found to be productive and, in many cases, fascinating in the nanosized range. It can possess a wide range of properties that can be used for many practical applications, including the jewelry aspect. In recent years, articles on the synthesis of Pt NPs have been published and different features such as electrocatalyst, polymer membranes, cancer therapy, and coating have been discussed as well as the effect of synthesis parameter [1-2]. The size and shape of the synthesized particles have been affected by several processing parameters. Therefore, the purpose of this research is to synthesize platinum particle with the simplest chemical reduction and to control the obtained particle size by pH variation as well as to use it in further jewelry application. Moreover, this research concerns on developing this synthesis method for laboratory-scale to pilot-scale and to near the commercial scale. Percentage yield (%) was calculated to measure the efficiency of the chemical synthesis. The UV-Vis measurement was employed for investigating platinum nanoparticle in as-synthesized platinum solution. The particle sizes and surface morphology of the precipitated platinum particles were determined via scanning electron microscope (SEM). In addition, Fourier transform infrared (FTIR) spectroscopy was used to identify the residuals that could be occurred from chemical contamination.

2. Experiment

2.1. Materials

Simple chemical synthesis was performed to synthesize the platinum particle. To reduce the cost of the synthesis, in this research, platinum metal plates were dissolved with aqua regia for using as the platinum reactant. The aqua regia solution was prepared by mixing of concentrated hydrochloric acid (12M, Merck) and concentrated nitric acid (15.8M, Merck) in the ratio 3:1, respectively. L-ascorbic acid, of 99% purity Merck, was used as reducing agent. Sodium hydroxide, of 97% purity Merck, was performed to adjust the pH. All chemicals were used without further purification and deionized water was used throughout the reactions.

2.2. Synthesis of platinum particles

For the simple synthesis of the platinum particles, a certain volume of 0.1 M ascorbic solution was slowly added to the 0.5%w/v platinum solution and then continuously stirred with a stirring speed of 400 rpm and a stirring time of 15 min. Then, the solution was adjusted to various pH (pH 1, pH 4, pH 7, pH 10 and pH 13) by adding 0.1 M sodium hydroxide, and the solution was continuous stirred for 8 hrs. With increasing the pH values, the color of the solution changed from yellow to brown which indicated the formation of the Pt NPs. After 8 hrs, the result revealed colloidal platinum mixing with some precipitated platinum particles. The amount of platinum precipitates depended on a variation of pH value. The precipitates were removed from the solution and repeatedly washed with distilled water three times. The products were dried at 80 ^{o}C for 4 hrs and the percentage yield was calculated afterward. The platinum powder was stored before the SEM and FTIR analysis, while the as-obtained colloidal solution was separated for determining with UV-Vis measurement.

2.3. Instrumentation and data acquisition

After completing of the synthesis, the as-obtained colloidal solution was monitored by Cary 60 Agilent UV–Vis spectrophotometer to determine the remaining platinum particle in the sample solution. The particle sizes and morphology of the precipitated platinum powder under different pH values were obtained using LEO 1450 VP scanning electron microscope. FTIR spectra were recorded on an Alpha II Bruker FTIR spectrometer to analyze the residual from the synthesis method. For the FTIR measurements of capped platinum particles, a small amount of dried platinum was mixed with potassium bromide (KBr) to form a round disk suitable for transmission sampling technique. The samples were investigated in the wavenumber of 500 to 4000 cm^{-1} and reference against air.

3. Results and discussion

3.1. Percentage yield

Since the purpose of this article is not only to study the pH effect on the particle size of platinum but also to attempt developing the synthesis method from laboratory-scale to pilot-scale, the percentage yield is a parameter that should be concerned. The relationship between the pH value and the yields of platinum powder precipitated from the different pH conditions is shown in figure 1. As seen in figure 1, the results showed that pH plays a very important role in the yield of platinum particles. When increase pH, the percentage yield tends to be increased.

3.2. SEM analysis

The average particle sizes obtained from every pH condition are shown in figure 2. The results revealed that the average sizes of platinum particle decreased when the pH value was increased. The SEM images of the as-prepared platinum powder precipitated by pH variation are shown in figure 3. Considering at the pH 1, the sample consists of large agglomerates of spherical particles in the range of micron, average size about 2.15 microns. At the other pH conditions, the SEM images reveal submicron-sized to nano-sized platinum particle. The average sizes of the obtained particles at pH 4, pH 7, pH 10 and pH 13 were 490 nm, 466 nm, 372 nm, and 110 nm, respectively.

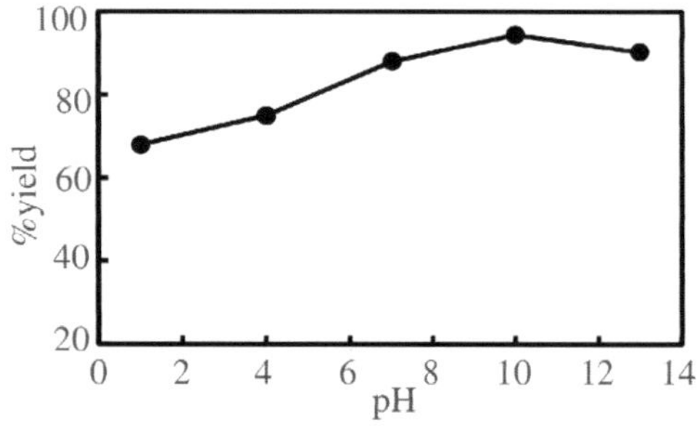

Figure 1. Relationship between increasing pH and percentage yield of precipitated platinum particle.

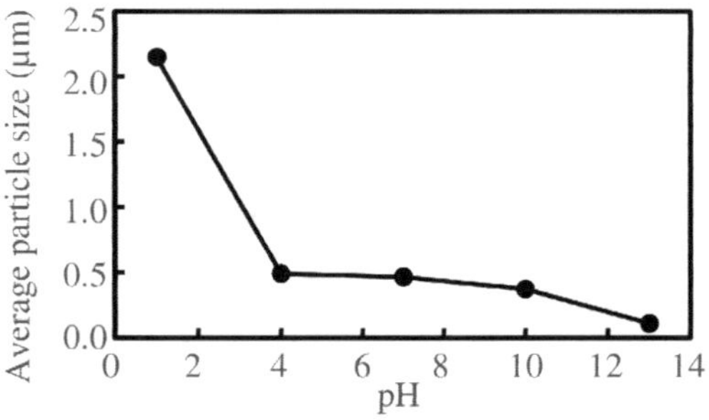

Figure 2. Relationship between increasing pH and particle size of precipitated platinum particles.

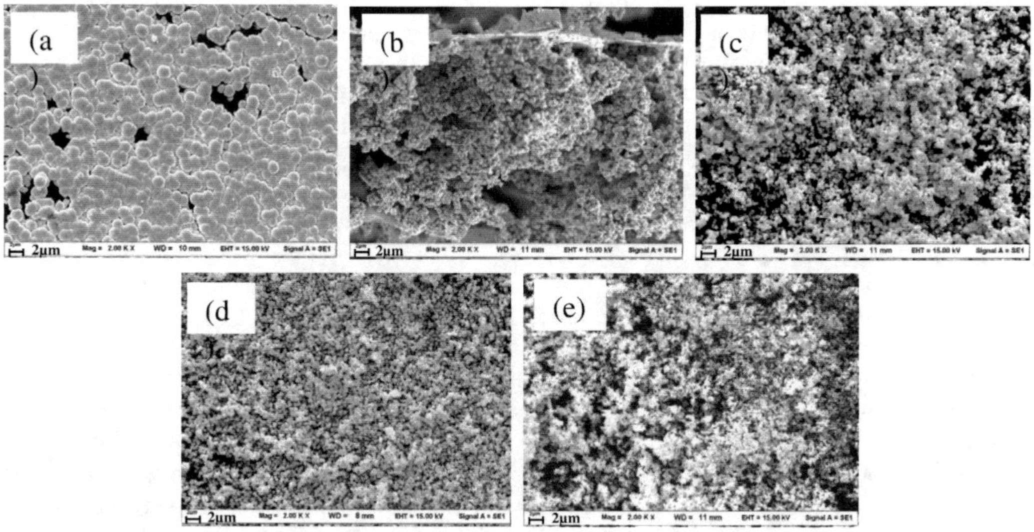

Figure 3. SEM images of precipitated platinum obtained from different pH values; (a) pH 1, (b) pH 4, (c) pH 7, (d) pH 10, and (e) pH 13.

3.3. UV-Vis spectra analysis

The UV-Vis spectra of as-obtained platinum colloids from different pH are presented in figure 4. Considering the spectrum of pH 1, the spectrum shows the absorption bands at 240 and 267 nm corresponding to $PtCl_4^{2-}$ and $PtCl_6^{2-}$ that remaining in solution [3-4]. Therefore, the results indicated that the reaction rate is very slow and the formation rate of Pt^0 is low. For this reason, a low percentage yield of platinum precipitation was achieved. Moreover, the size of platinum particles obtained from this condition is micro-sized particle which can be confirmed from the figure 3 (a). When the pH was increased, the band at 267 nm disappeared and the color of the colloidal solution changed. It can be assumed that the $PtCl_6^{2-}$ was reduced to $PtCl_4^{2-}$, and the formation of Pt^0 tend to completely be performed. In addition, the surface plasmon resonance peak shifts toward the short wavelength region as well as becomes narrower. This blue shift indicates a reduction in the mean diameter of the particles [5-6]. It can be implied that the particle size of the precipitated platinum particle is reduced to a smaller one as a nano-sized platinum particle and the percentage yield of high pH tends to be increased as well. Additionally, the UV-Vis spectra indicate the maximum absorption at 220 nm confirming the spherical shape of the platinum particle [3-5].

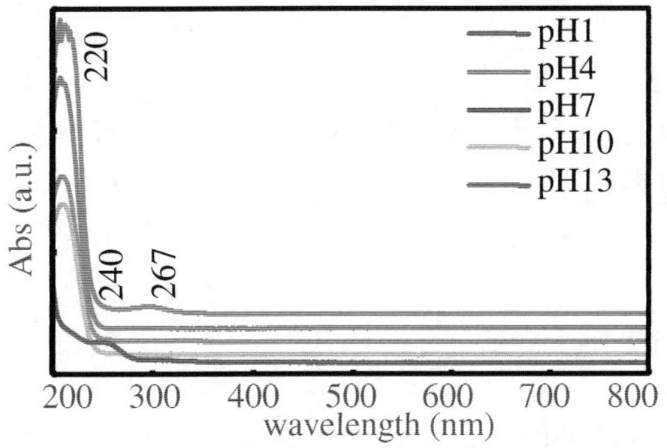

Figure 4. UV-Vis spectra of synthesized platinum obtained from different pH values.

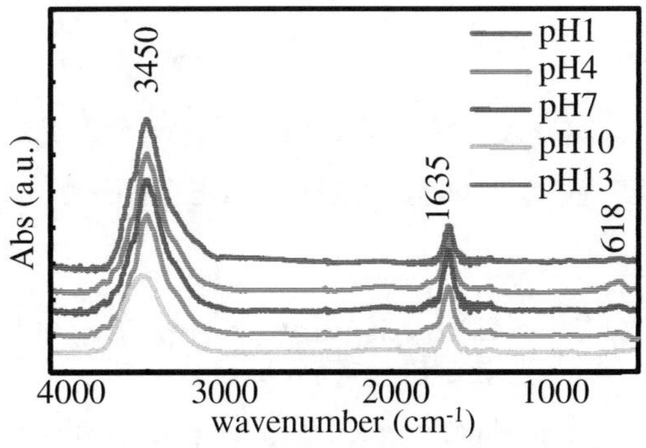

Figure 5. FTIR spectra of precipitated platinum obtained from different pH values.

3.4. FTIR analysis

Figure 5 displays the FTIR spectra of the dried platinum powder according to variation of pH values. All spectra present absorption bands at 3450, 1635, and 618 cm^{-1}. The bands relate to OH stretching, OH bending, and OH out of plane bending of absorbed water interference in KBr. The obtained spectra do not reveal any absorption bands corresponding to the other contaminants from the synthesis. It can be assumed that there is no residual in synthesized platinum powder.

4. Conclusion

In this research, the impact of pH variation on the simple chemical synthesis of the platinum particle was investigated. The results affirmed that the different pH affecting to the percentage yield and the size of platinum particle. By increasing pH value, the percentage yield of precipitated platinum was increased, and the particle size was decreased from submicron to the nano-sized, as seen in the SEM images. The UV-Vis spectra of prepared samples showed characteristic properties of Pt NPs and platinum ion remaining in colloid solution. The FTIR spectra confirmed that there is no indication of residual contamination in the synthesized particles. Therefore, this simple synthesis attributing to pH control can produce the purity platinum powder with high production yield. It can be applied for further jewelry application and other industries.

Acknowledgments

The authors gratefully acknowledge Faculty of Gems, Burapha University, Research Grant No. 01/2562 for financial support and Christy Gems Co., Ltd. (Thailand) for supporting the platinum metal plate. In addition, the authors sincerely thank Miroscopic center, Faculty of Science, Burapha University, Thailand, for SEM analysis.

References

[1] Stepanov A L, Golubev A N, Nikitin S I and Osin Y N 2014 *Rev. Adv. Mater. Sci.* **38** 160
[2] Islam M A, Bhuiya A K and Islam M S 2014 *Asia. Pac. J. Energy Environ.* **1** 107
[3] Tajerian T, Monsefi M and Rowan A 2019 *J. Nanostructure Chem.* **9** 197
[4] Herricks T, Chen J and Xia Y 2004 *Nano. Lett.* **12** 2367
[5] Gharibshahi E and Saion E 2012 *Int. J. Mol. Sci.* **13** 14723
[6] Alqadi M K, Abo Noqtah O A, Alzoubi F Y, Alzouby J and Aljarrah K 2014 *Mater. Sci-Poland.* **32** 107

Fabrication of encapsulated graphene-based heterostructure using molybdenum as edge-contacts

Illias Klanurak[1], Kenji Watanabe[2], Takashi Taniguchi[3], Sojiphong Chatraphorn[1] and Thiti Taychatanapat[1*]

[1] Department of Physics, Faculty of Science, Chulalongkorn University, Bangkok 10330, Thailand
[2] Research Center for Functional Materials, National Institute for Materials Science, Tsukuba, 305-0044 Japan
[3] International Center for Materials Nanoarchitectonics, National Institute for Materials Science, Tsukuba, 305-0044 Japan

[*]Corresponding author's e-mail: thiti.t@chula.ac.th

Abstract. Graphene is an intriguing platform to study exotic quantum transport phenomena due to its intrinsically high mobility and remarkable electronic properties. To achieve high-performance device, graphene is usually encapsulated between thin sheets of hexagonal boron nitride (hBN) to protect graphene layer from extrinsic impurities. Cr/Au is typically employed to make contacts with the edges of the heterostructure. In this research, Mo is used as an alternative electrode for graphene without adhesion layer to simplify the fabrication process. hBN-graphene-hBN heterostructures were fabricated by a pick-up technique and etched in O_2/CHF_3 gases to expose graphene edges. Mo contacts were deposited onto the substrates by sputtering. We achieved ohmic contacts between graphene and Mo. The contact resistance reaches the maximum of around 1,300 $\Omega\cdot\mu m$ at charge neutrality point and decreases to 975 $\Omega\cdot\mu m$ at the density of 4×10^{12} cm^{-2}. We observed that the contact resistance increases over time likely due to the oxidation of Mo but remained ohmic after 2 months. The intrinsic transport characteristics of graphene can still be obtained by using four-probe measurement. Here, we realized a high-quality twisted bilayer graphene device with a room-temperature mobility of 27,000 $cm^2/V\cdot s$ indicating that Mo can be used as edge-contacts to probe the transport properties of graphene.

1. Introduction

Two dimensional (2D) materials have gained an enormous interest in the past decade due to their remarkable electrical and mechanical properties. Graphene, a 2D layer of carbon atoms arranged in a honeycomb lattice, is the first 2D material that was extensively studied. In recent years, the observation of superconducting state in twisted bilayer graphene [1] sparks an interest to use a twist angle between layers of graphene or other 2D materials to modify their electronic properties. Nevertheless, to observe exotic phenomena associated with the twist angle, a high-quality graphene device, i.e., a device with low-defect and high-mobility, is a crucial requirement.

hBN is typically used as an insulated substrate for graphene to improve device quality. It can also be used to encapsulate graphene layer, ensuring that extrinsic impurities have neglectable effects on transport properties of graphene. However, the encapsulated structure imposes new limitation for making electrical contacts between graphene and metal electrodes. One way to overcome this limitation is to expose the edge of graphene by using O_2/CHF_3 plasma to etch the heterostructure. Then, metal electrodes are deposited onto the heterostructure, making one dimensional contact to the edge of graphene [2]. Typically, Au is employed to make electrical contacts with the edges of graphene using Cr as an adhesion layer to improve adhesion between graphene and the metal.

Here, we present Mo as an alternative electrode to make edge-contact with graphene. Mo is selected in this study due to its ability to form covalent bond with carbon atom [3] enabling the realization of the contact without adhesion layer. The carbide formation between Mo and graphene can also enhances orbital overlap at the Mo-C interface which is an important factor for making high quality contact [2].

2. Methods

In this study, we fabricated field-effect transistor (FET) using twisted bilayer graphene (tBLG) as a transport channel and Mo as source-drain electrodes. The tBLG layer was encapsulated between two thin sheets of hBN. Then electrical transport measurement was performed on the device to characterize the quality of graphene and metal contacts.

2.1. hBN-graphene-hBN assembly

Graphene and hBN flakes were obtained by mechanical exfoliation and deposited onto highly doped silicon wafer with thermally-grown SiO_2. Optical microscope was used to identify usable flakes. Raman spectroscopy was employed to determine the number of graphene layers, and atomic force microscopy was used to determine thickness of hBN flakes. The hBN flakes with thickness of around 10-30 nm were selected to encapsulate graphene layer. The twisted bilayer graphene was created form a single flake of monolayer graphene using tear-and-stack technique [4]. The assembly process was performed under a microscope with a temperature-controlled stage. Here, we used poly(bisphenol A carbonate) film (PC) on a block of Polydimethylsiloxane (PDMS) as a stamp to pick-up top-hBN flake from Si/SiO_2 substrate. The top-hBN was then used to tear a monolayer graphene flake into two parts: one part remained on Si/SiO_2 and the other part was picked up with top-hBN. After that, the part of graphene on top hBN was aligned to pick up the part of graphene on Si/SiO_2 which had been rotated by an angle of $\sim 1.1°$ to form tBLG. The top hBN-tBLG structure was used to pick up bottom-hBN flake, creating hBN-tBLG-hBN heterostructure (BNGBN). All the pick-up process was conducted at 100°C. In the final step, the heterostructure was slowly released onto another Si/SiO_2 substrate at stage temperature of 180°C to remove contaminants that trapped between interfaces during the assembly process.

2.2. BNGBN characterization

Raman spectroscopy with a laser excitation of 532 nm was employed to characterize the BNGBN. A normalized Raman spectrum of the stacked graphene region is shown in figure 1. We observed a few prominent peaks in the spectral range of 1,100-3,000 cm^{-1}. Raman peaks centered at 1,365 cm^{-1} and 1,582 cm^{-1} are characteristic of in-plane vibration of hBN and graphene respectively [5]. The lack of D-peak in the Raman spectrum indicates that the graphene layers are defect-free [6]. To confirm small twist angle between two graphene layers, we investigate 2D peak of graphene at 2,700 cm^{-1} since this peak is sensitive to electronic band structure of graphene. We found that the shape of 2D peak is significantly different than that of Bernal-stacked bilayer graphene as shown in figure 1(b). This implies that our tBLG still retains its twisted angle even after heated at 180°C in the layer assembly process.

2.3. Mo contact

To make electrical contacts with the edges of graphene, a rectangular shaped Poly(methyl methacrylate) (PMMA) etch-mask was defined on the BNGBN using electron-beam lithography (EBL). The region of the BNGBN outside of the mask was then etched away by O_2/CHF_3 plasma to expose the edge of graphene. Contacts were defined by EBL followed by DC sputtering of 80 nm-thick Mo film onto the substrate. The deposition process was performed in a high vacuum chamber with a base pressure of 9.6×10^{-7} mbar. Ar gas at a pressure of 6.1×10^{-3} mbar was used as sputtering gas at 550 W. After lift-off process, we obtained tBLG FET device with Mo electrodes as shown in figure 2(a, b).

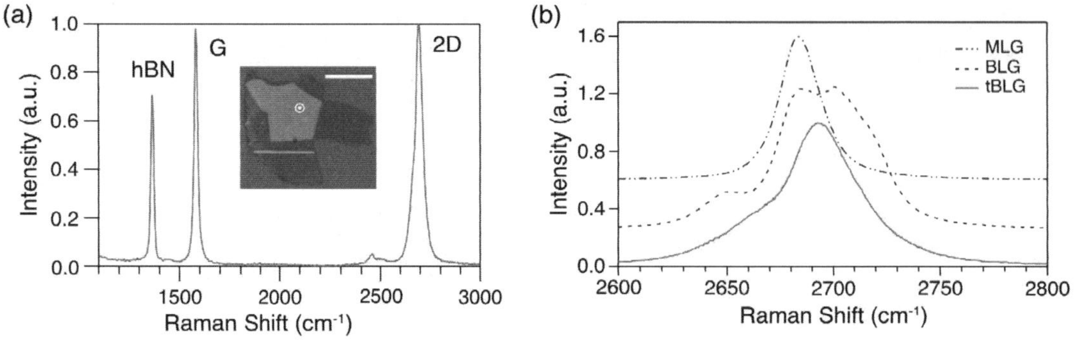

Figure 1. Raman spectrum of the BNGBN using 532 nm laser excitation. (a) Raman spectrum in the spectral range of 1,100-3,000 cm^{-1} and (b) Comparison between Raman 2D peak of graphene for monolayer graphene (MLG), Bernal-stacked bilayer graphene (BLG) and tBLG. The Raman spectra are vertically offset for clarity. The inset shows optical image of the BNGBN on Si/SiO_2 substrate with a circle indicating the position of the laser illuminated on the BNGBN (scale bar is 20 μm).

3. Results and discussion

Electrical transport measurement was performed on the BNGBN at room temperature to characterize the quality of contacts. Inset of figure 2(c) shows the linear behavior between current, I_{SD}, and source-drain voltage, V_{SD}, between contacts 3 and 4 indicating the ohmic contacts between Mo and the edges of graphene. The standard lock-in technique with excitation frequency of 17 Hz was used to measure the transport properties of the BNGBN. Carrier density, n, in graphene channel can be electrically tuned by applying voltage, V_{BG}, to a silicon back-gate. Using parallel plate capacitor model, carrier density in tBLG can be determined by the relation $n = \left(\frac{\varepsilon_{r1}\varepsilon_{r2}\varepsilon_0}{e\varepsilon_{r2}d_1+e\varepsilon_{r1}d_2}\right) V_{BG}$ where e, d_1, d_2 and ε_0 are electron charge, thickness of SiO_2, thickness of hBN and vacuum permittivity, respectively. Here, we used ε_{r1}=3.7 and ε_{r2}=3.2 for the dielectric constant of SiO_2 and hBN [7].

Figure 2(b) illustrates the measurement setup for measuring contact resistance. First, a bias current, I_{34}, of 100 nA was applied between contacts 3 and 4 and the voltage, V_{34}, across contacts 3 and 4 was measured as a function of V_{BG}. The measured resistance $R_{34}=V_{34}/I_{34}$ can be expressed as $R_{34}=R_{C3}+R_{tBLG}+R_{C4}$ where R_{tBLG} is a resistance of graphene channel and R_{Ci} is the contact resistance between contact i and graphene. Then, the voltage, V_{14}, across contacts 1 and 4 was measured using the same configuration for current bias. The resistance $R_{14}=V_{14}/I_{34}=R_{tBLG}+R_{C4}$ was obtained. The contact resistance between contact 3 and graphene as a function of V_{BG} can be extracted by subtracting R_{34} with R_{14} as shown in figure 2(c). We observed that the contact resistance depends on carrier density in tBLG channel. The contact resistance reaches the maximum value of 1,300 Ω·μm at charge neutrality point (CNP) and decreases to 975 Ω·μm at 60 V away from CNP which corresponds to carrier density of ~ 4×10^{12} cm^{-2}. This result is comparable to BN/monolayer graphene/BN devices fabricated by using Cr/Au as edge contacts [2] whose contact resistance varies between 200-2,000 Ω·μm over similar range of carrier density. To confirm the reproducibility of ohmic contact between Mo and graphene, we fabricated another FET device with Mo edge contacts. We observed that all of 6 contacts in our experiment exhibit ohmic behavior. This indicates that Mo can be used to make reliable contact to the edge of graphene.

We also noticed the change in contact resistance when measuring it again after 57 days as can be seen in figure 2(d). We observed two-fold increase in maximum contact resistance near CNP and asymmetry in the contact resistance around CNP when the charge carrier changes between electron and hole. The contact resistance for electron is larger than hole at the same carrier density. This asymmetry imitates behavior of contact resistance of high work function metal like Pd or Au which is inconsistent with the low work function of Mo [8]. We argue that this asymmetry may be the result of the oxidization of Mo contacts into MoO_x. The large difference in work function between graphene (~4.5 eV) and MoO_x

(~6.8 eV for MoO_3) causes charge transfer at the contact interfaces resulting in hole doped region at the vicinity of the contacts [3]. The p-doping of tBLG near the contacts acts as p-n junction leading to high contact resistance when the charge carrier is electron. Nevertheless, the contacts between Mo and tBLG still exhibit ohmic behavior as can be seen in the inset of figure 2(d).

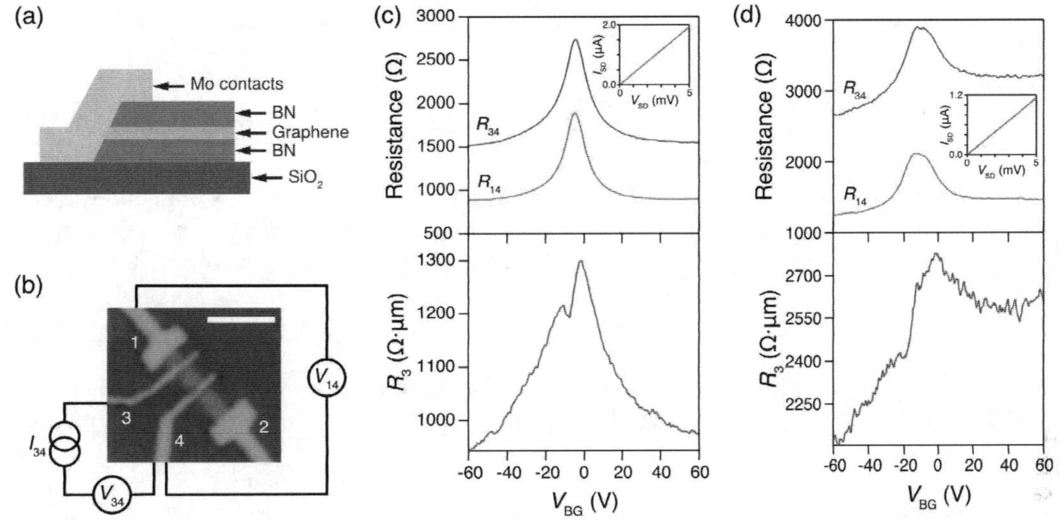

Figure 2. (a) Schematic diagram illustrating cross-section view of the tBLG FET device with Mo edge contacts, (b) diagram of measurement setup for probing contact resistance between Mo and tBLG where each contact is labelled with number 1-4 (scale bar is 5 μm), (c) measured resistance R_{14} and R_{34} as a function of V_{BG} obtained after the device fabrication along with R_3 calculated by subtracting R_{34} with R_{14} and multiplying the result with the length of 1D contact and (d) contact resistance R_3 obtained after 57 days from the first measurement. Insets in (c) and (d) show ohmic behavior between contacts 3 and 4 of the device measured without applying voltage to the back-gate.

Despite the increase in contact resistance of Mo, the intrinsic transport properties of BNGBN can still be obtained by using four-probe measurement to eliminate the effects of contacts. The CNP was located near the zero back-gate voltage when measure immediately after the fabrication process. However, after 3 months, the CNP was shifted to V_{BG} ~ -16 V (see figure 3(a)). Even though the tBLG is encapsulated by hBN, it is possible that dangling σ-bonds at the exposed edges of graphene may react with ambient contamination in air leading to H-terminated edge. The H-terminated edges of tBLG induce n-doping to graphene sheet resulting in the negative shift of CNP [9]. Even after such doping, the device still exhibits a very small hysteresis during sweeping V_{BG} forward and backward as a result of encapsulated structure as shown in figure 3(a). Mobility, μ, of the device is calculated from the slope of conductivity, σ, using Drude model, $\mu=\frac{1}{e}\frac{\partial\sigma}{\partial n}$, as shown in figure 3(c). The room-temperature mobilities of electron and hole reach the values as high as 27,000 and 21,000 $cm^2/V{\cdot}s$ at electron and hole density of 5×10^{11} cm^{-2} and -4×10^{11} cm^{-2}, respectively. The mobilities in our device are comparable to the state-of-the-art, encapsulated monolayer graphene devices which have room-temperature mobility of ~40,000 $cm^2/V{\cdot}s$ [2]. This indicates that Mo contacts do not hinder the quality of the graphene and can be employed to study its intrinsic electronic transport properties. Contrary to Bernal-stacked bilayer graphene, we observed non-monotonic behavior in conductivity when V_{BG} was higher than 20 V, shown in figure 3(c). We argued that this may result from the emergence of a new Dirac cone in the electronic band structure at high energy, which is a phenomenon that typically observed in small twist angle bilayer graphene [1].

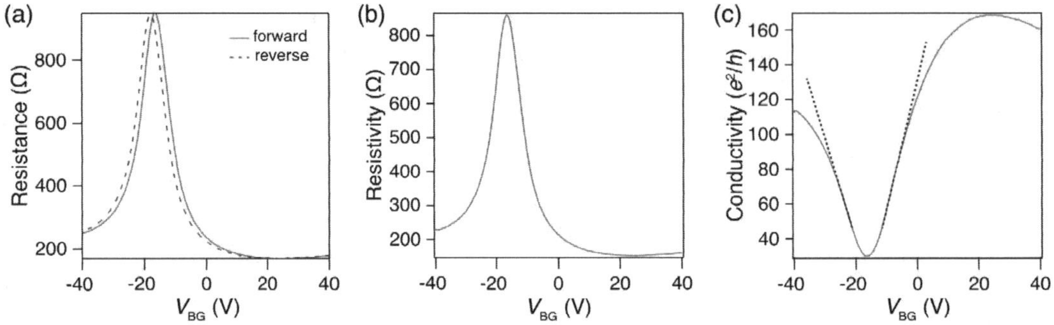

Figure 3. (a) Dirac peak of tBLG obtained from four-probe measurement after 3 months by sweeping V_{BG} forward (solid line) and backward (dash line), (b) resistivity and (c) conductivity extracted from the Dirac peak. The dash lines in figure 3(c) show the slope of conductivity used for calculating electron and hole mobilities.

4. Conclusion

In this work, FET based on hBN-encapsulated tBLG was fabricated using Mo as edge contacts. We realized Ohmic contacts between Mo and tBLG. Despite the increase in contact resistance overtime, it still remains Ohmic and four-probe measurement can still be used to obtain intrinsic electrical properties of graphene. The high mobility at room-temperature indicates that Mo has negligible effect to carrier transport in FET channel. These results demonstrate that Mo can be used as an alternative metal to probe intrinsic characteristics of graphene.

Acknowledgments

This work was supported by the Research Fund for DPST graduate with First Placement (Grant no. 002/2015) and Ratchadaphiseksomphot Endowment Fund of Chulalongkorn University, Grant for Research. I.K. would like to acknowledge the Development and Promotion of Science and Technology Talents Project (DPST) scholarship. The authors would also like to thank K. Jaruwongrungsee (NECTEC, Thailand), B. Namnuan, and P. Panchawirat (ThEP Center, Thailand) and for help with wire bonder, sputtering and E-beam lithography respectively.

References

[1] Cao Y, Fatemi V, Fang S, Watanabe K, Taniguchi T, Kaxiras E and Jarillo-Herrero P 2018 *Nature* **556** 43–50
[2] Wang L *et al.* 2013 *Science* **342** 614–7
[3] Zhang Y and Qu X-P 2019 *AIP Adv.* **9** 055221
[4] Kim K *et al.* 2016 *Nano Lett.* **16** 1989–95
[5] Cuscó R, Gil B, Cassabois G and Artús L 2016 *Phys. Rev.* B **94** 155435
[6] Ferrari A C and Basko D M 2013 *Nat. Nanotechnol.* **8** 235–46
[7] Campos L C, Young A F, Surakitbovorn K, Watanabe K, Taniguchi T and Jarillo-Herrero P 2012 *Nat. Commun.* **3** 1239
[8] Xia F, Perebeinos V, Lin Y M, Wu Y and Avouris P 2011 *Nat. Nanotechnol.* **6** 179–84
[9] Brenner K, Yang Y and Murali R 2012 *Carbon* **50** 637–45

Angle-dependent spectrum measurement of polystyrene opal-like structure described by Bragg-Snell diffraction and perturbed photonic band structure

N Sitpathom[1*], T Muangnapoh[2], P Kumnorkaew[2], S Suwanna[1], A Sinsarp[1] and T Osotchan[1]

[1] Department of Physics, Faculty of Science, Mahidol University, Bangkok 10400, Thailand
[2] National Nanotechnology Center (NANOTEC), National Science and Technology Development Agency, Pathum Thani 12120, Thailand

*E-mail: nonthanan.sit@student.mahidol.ac.th

Abstract. Optical diffraction of opal structure, a colloidal photonic crystal, can be predicted by Bragg-Snell diffraction and photonic band structure. Theoretical prediction and optical measurement are frequently slightly different due to distance variation of particle packing. In this research, opal of 310 nm polystyrene beads was fabricated by self-assembly process and optically investigated in transmission spectra at varied angles. The measured spectra had less agreement to the Bragg-Snell prediction at large angle of detection. To explore influence of packing distance on optical response, photonic band structures were numerically simulated via plane-wave expansion method at presence of perturbed length in primitive lattice vectors. Extending each primitive vector with fixing others provided a different eigen-frequency of the first photonic band, although they had a symmetrical perturbation on (111) face-centered cubic. Perturbation on lattice length became much strong when the disturbing direction was out of eigenstate orientation plane.

1. Introduction

Optical diffraction of colloidal photonic crystals, frequently called as opals, can be well predicted by Bragg-Snell prediction [1] in some regions. The Bragg-Snell equation combines between diffraction of parallel crystal planes (Bragg's diffraction) and refraction at interface of two media (Snell's law) with various incidence angle. The Bragg-Snell equation from (111) surface can be written in general form for (hkl) plane diffraction as

$$\lambda_{(hkl)}(\theta) = 2d_{(111)}n_{eff}\left(\frac{3}{h^2+k^2+l^2}\right)^{1/2}\cos(\theta-\Psi_{(hkl)})\,. \tag{1}$$

The diffracted wavelength $\lambda_{(hkl)}$ from (hkl) plane at incidence angle θ is calculated from the product of (111) d-spacing, effective refractive index n_{eff} and cosine of relative angle of (111) and (hkl) planes $\Psi_{(hkl)}$. There were several reports that the measured Bragg diffraction angles can be differ from the theoretical prediction values [2-3]. Moreover, the diffraction intensity becomes weaker in disordered opal and its peak slightly shifts to longer wavelength [4-5]. The main cause of disordered opal was assumed by variation of particle diameters, thermal fluctuation in self-assembly process [6] and solution evaporation rate [7]. For face-centered cubic (fcc) structure of opal, anti-band crossing of (111) to (200) and $(\bar{1}11)$ was observed at 30° for U and K symmetry points, respectively, and at 40° in the orthogonal axis for W symmetry point [8].

Fabrication of opals is easily made by self-assembly process of colloidal packing with natural forces such as capillary and Van der Waals forces [9]. For solution process, a convective deposition is a well-alternative approach among the others for example spin coting, dip coating or vertical deposition [9]. In order to obtain better crystal quality, the convective deposition can achieve by using substrate vibration during coating [10].

In this work, the convective deposition with vibration assistance was used for opal film fabrication. Opal microstructure was observed by scanning electron microscope (SEM) and the self-assembled opals were optically characterized by transmission spectra at varied angles. Moreover, an imperfection of opal structures was computationally studied via plane-wave expansion method (PWEM).

2. Experiment and simulation methods

The fcc structures of polystyrene particles (purchased from Thermo Fisher Scientific) with diameter of 310±9 nm (standard deviation < 3%) were prepared by convective deposition [11]. The glass substrate was moved with speed of 10 μm/s by linear motor and vibrated at frequency of 40 Hz during self-assemble process. The microstructure of fabricated opal films were observed by SEM with conductive ultrathin film coating.

For optical characterization, the transmission spectra were examined for investing Bragg diffraction and analyzing opal crystal quality. The transmission measurement was performed in Shimadzu UV-Vis 2600 spectrometer with varying the sample angle from -50° to 0° and 0° to +50° with respect to normal axis of the substrate.

The photonic band structures and their related electric fields were studied on varied perturbation axes by PWEM which is having eigenfrequencies and eigenstates by solving Maxwell's eigen-equation. The simulation was performed by MIT Photonic Bands (MPB), open-source software. The primitive vectors and high symmetry points used in this PWEM simulation were taken from reference [12]. Three primitive vectors of fcc were defined as $\mathbf{a}_1$, $\mathbf{a}_2$ and $\mathbf{a}_3$ for $(0,1/2,1/2)$, $(1/2,0,1/2)$ and $(1/2,1/2,0)$, respectively. To study photonic bandgap change under perturbed lattice, the translation vector along each primitive vector was extended by 10%. Thus, the sets of imposed perturbed vectors were $(1.1\mathbf{a}_1,1.0\mathbf{a}_2,1.0\mathbf{a}_3)$, $(1.0\mathbf{a}_1,1.1\mathbf{a}_2,1.0\mathbf{a}_3)$ and $(1.0\mathbf{a}_1,1.0\mathbf{a}_2,1.1\mathbf{a}_3)$.

3. Results and discussion

The opal film of 310 nm polystyrene particles presented violet color as shown in figure 1 (a). The opal structure showed hexagonal layer packing as observed by SEM (figure 1 (b)). There are multi-domain orientations of the order particles and small areas of disorder particle' assembly at grain boundaries. Multiple domains in self-assembly can be caused by fast evaporation rate of solution and simultaneous crystal formation in different places [7]. Moreover, the variation in thermal gradient in solution and hydrophilicity of substrate surface [6] would induce point and line defects in opal film.

Transmission spectra of the 310 nm opal films were measured at -50° to 0° and 0° to 50° with 5° increment and plotted in figure 1 (c). Predicted-diffracted wavelengths estimated by Bragg- Snell equation were labelled on the spectra by black, red and green arrows for (111), (200) and $(\bar{1}11)$ plan of fcc, respectively. The measured spectra and the estimated diffraction peaks had well agreement in the region where the angles varied between -20° to 0° to +20° and became different for the wider angles. This may occur due to a higher chance of multi-domain measurement at large angle with larger projection area. Additionally, the diffraction peaks merging of (111) and $(\bar{1}11)$ planes at angle of about 25° can be another factor of the shift [2]. The misaligned stacking sequence of particle layers can disturb the opal crystal structure and can result in variation of packing period.

Journal of Physics: Conference Series **2145** (2022) 012040 doi:10.1088/1742-6596/2145/1/012040

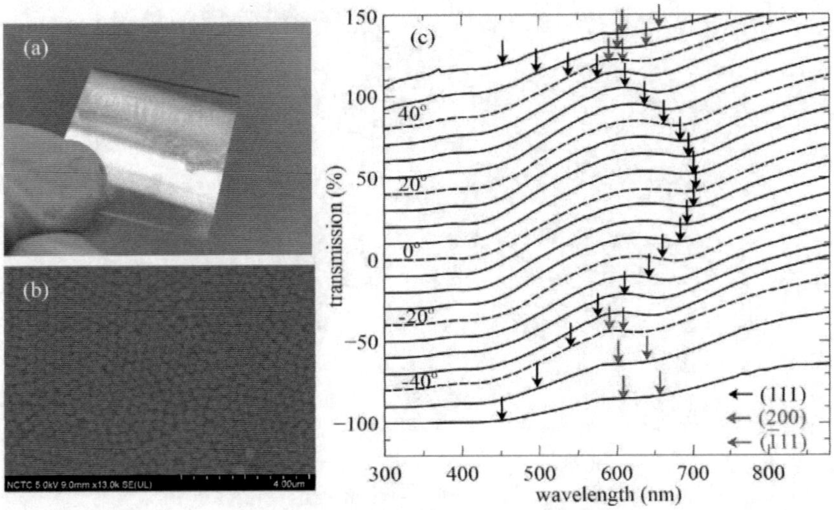

Figure 1. (a) Photograph of fabricated opal film presenting violet color, (b) SEM image of order 310 nm nanoparticle opal film and (c) transmission spectra measured at angles from -50° to 0° to +50° with black, red, and green arrows labelled of predicted diffraction wavelength from (111), (200) and ($\bar{1}$11) fcc planes, respectively.

Simulation of eigen electric fields on photonic band structures was applied to explore the effect of perturbed lattice distance on photonic band gap. The lower band-edge of the first bandgap in ΓL direction, corresponding to [111] fcc, indicated strong effect for the case of perturbation vector $(1.0\mathbf{a}_1, 1.1\mathbf{a}_2, 1.0\mathbf{a}_3)$, see the circle in figure 2 (b). For electric-field energy density of this perturbation axis, the mode field was spatially spanned out of plane, while those in the other cases were extended in plane. Although three primitive vectors had the same projected angle to vector [111], only the perturbed $\mathbf{a}_2$ direction case had the most influence on the first photonic bandgap due to pointing out to the plane of field localization direction. Additionally, the spatial shape of electrical energy localization become more spherical at the $\mathbf{a}_2$ perturbation, see color contours figure 3. Having optical localization in different geometrical shape would be a reason of shifted eigenfrequency.

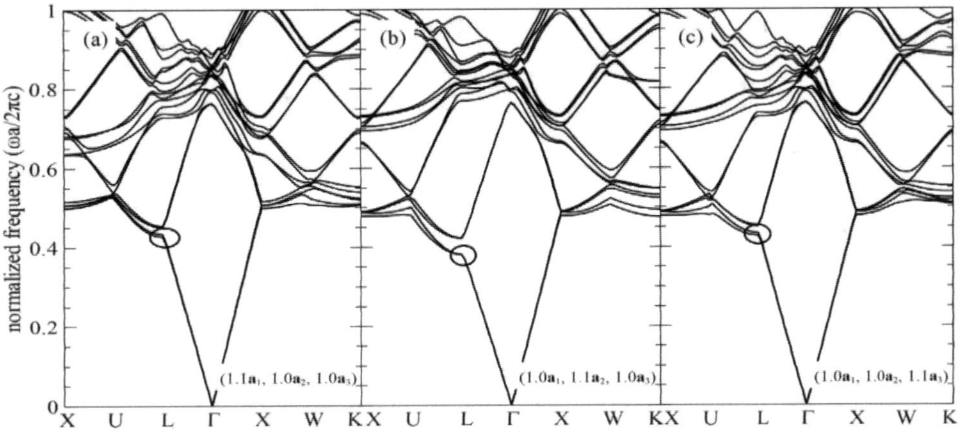

Figure 2. Photonic band structures of perturbed opal crystals with primitive vectors of (a) $(1.1\mathbf{a}_1, 1.0\mathbf{a}_2, 1.0\mathbf{a}_3)$, (b) $(1.0\mathbf{a}_1, 1.1\mathbf{a}_2, 1.0\mathbf{a}_3)$ and (c) $(1.0\mathbf{a}_1, 1.0\mathbf{a}_2, 1.1\mathbf{a}_3)$.

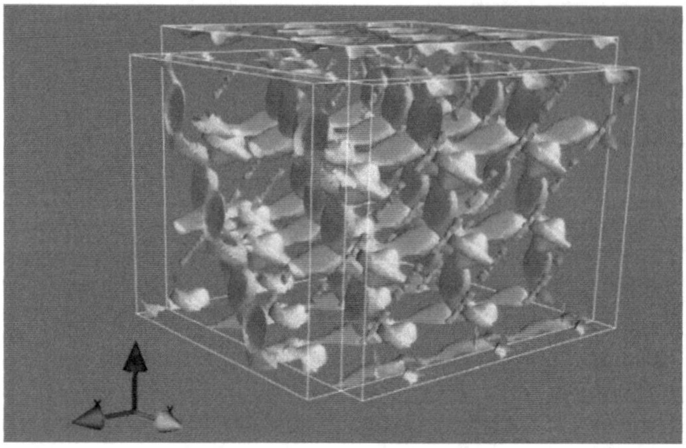

Figure 3. Contours of electric-field energy density fixing the value at 3.5 shown in blue, green and red colors for perturbation on $\mathbf{a_1}$, $\mathbf{a_2}$ and $\mathbf{a_3}$, respectively. The x-, y- and z-axis presented in red, green and blue arrows.

4. Conclusion
Colloidal photonic crystals or opal structures was fabricated by convective deposition with vibration assistance and exhibits optical diffraction predicted by Bragg-Snell equation. The measured diffracted wavelengths have some agreement to the theoretical prediction in region of low angle of incidence but the shift occurred at large angle probably due to several grains involving at this large projection area of incidence and non-ideal face-centered cubic packing. For simulation by plane-wave expansion method, the perturbation effect on photonic bandgap became strong for the perturbed crystal with Eigen electric field mode extending in out of localization plane.

Acknowledgments
The first author would like to thanks the Thailand Development and Promotion of Science and Technology Talent Program for a scholarship. This project was partially supported by the Research Nanotechnology Network (RNN), National Nanotechnology Center.

References
[1] Baryshev A V, Kaplyanskiĭ A A, Kosobukin V A, Limonov M F and Skvortsov A P 2004 *Phys. Solid State* **46** 1331–9
[2] Baryshev A V, Khanikaev A B, Uchida H and Inoue M 2006 *Phys. Rev.* B **73** 033103
[3] Omshankar, Priya and Nair R V 2019 *J. Opt. Soc. Am.* B **36** 2338–45
[4] Palacios-Lidón E, Juárez B H, Castillo-Martínez E and López C 2005 *J. Appl. Phys.* **97** 063502
[5] Rengarajan R, Mittleman D, Rich C and Colvin V 2005 *Phys. Rev.* E **71** 016615
[6] Boettcher J M, Joy M, Joshi K, Muangnapoh T and Gilchrist J F 2015 *Langmuir* **31** 10935–8
[7] Cheng S and Grest G S 2013 *J. Chem. Phys.* **138** 064701
[8] Rybin M V, Samusev K B and Limonov M F 2007 *Photonics Nanostructures*: *Fundam. Appl.* **5** 119–24
[9] Vogel N, Retsch M, Fustin C-A, del Campo A and Jonas U 2015 *Chem. Rev.* **115** 6265–311
[10] Joy M, Muangnapoh T, Snyder M A and Gilchrist J F 2015 *Soft Matter* **11** 7092
[11] Sitpathom N, Kumnorkaew P, Muangnapoh T, Rutirawut T and Osotchan T 2017 *Materials Today: Proc.* **4** 6485–90
[12] Setyawan W and Curtarolo S 2010 *Comput. Mater. Sci.* **49** 299–312

Corrigendum: Efficiency of supercapacitor with $CaTiO_3$-filled polysulfone separators (2021 *J. Phys.: Conf. Ser.* **2145** 012041)

S Suwanwong*, C Yuenyao and A Hutem
Physics Division, Faculty of Science and Technology, Phetchabun Rajabhat University, Phetchabun 67000, Thailand

Page 3:
In the Results and discussion section, the following text appears:
"The efficiency of supercapacitor using $CaTiO_3$- filled PSF separators in the weight percentage of 0.5, 1.0 and 2.0 wt% to analyze specific energy (as shown in figure 5) and specific capacitance (as shown in figure 6) using Battery Testers system,"

This should read:
"The efficiency of supercapacitor using $CaTiO_3$- filled PSF separators in the weight percentage of 0.5, 1.0 and 2.0 wt% to analyze specific energy and specific capacitance (as shown in figure 6) using Battery Testers system,"

Page 3:
In the Results and discussion section, the following ***Figure*** appears:

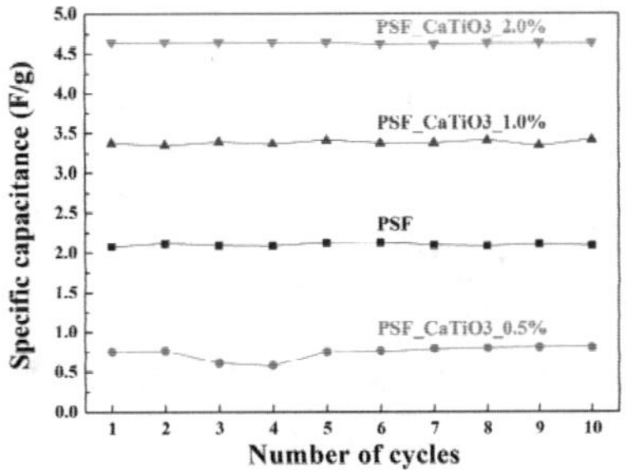

Figure 4. *FT-IR spectra of $CaTiO_3$-filled PSF separators.*

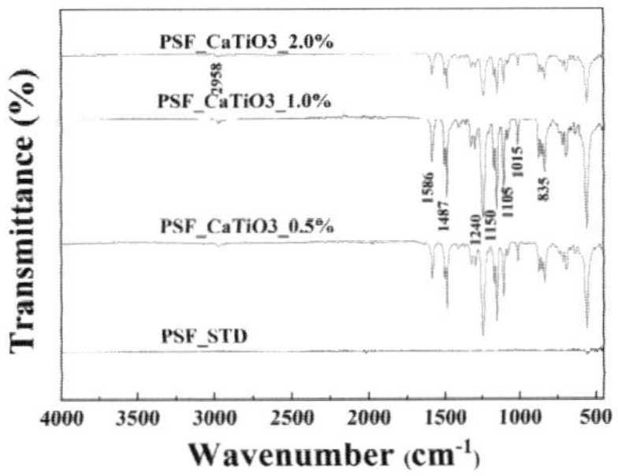

Figure 6. *Specific capacitance of supercapacitors with $CaTiO_3$-filled polysulfone separators.*

This should ***Figure***:

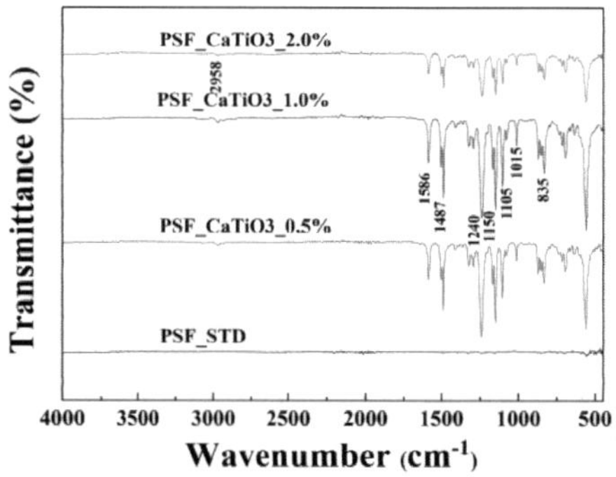

Figure 4. *FT-IR spectra of $CaTiO_3$-filled PSF separators.*

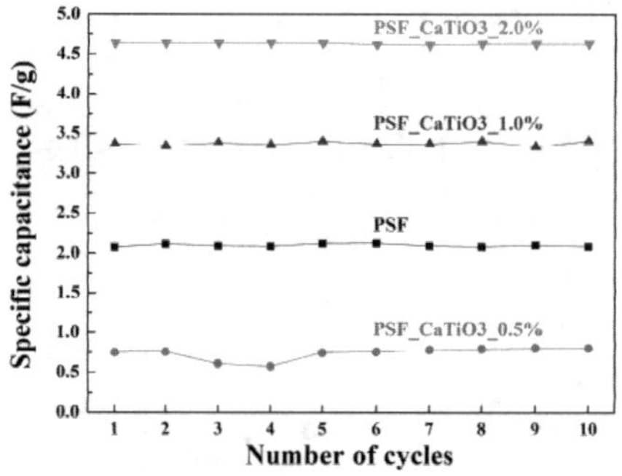

Figure 6. *Specific capacitance of supercapacitors with $CaTiO_3$-filled polysulfone separators.*

Journal of Physics: Conference Series **2145** (2022) 012041 doi:10.1088/1742-6596/2145/1/012041

Efficiency of supercapacitor with $CaTiO_3$-filled polysulfone separators

S Suwanwong*, C Yuenyao and A Hutem

Physics Division, Faculty of Science and Technology, Phetchabun Rajabhat University, Phetchabun 67000, Thailand

*Corresponding author E-mail: sanit.suw@pcru.ac.th

Abstract. Supercapacitors are one of the energy storages designed to serve the increasing demand for electricity nowadays. They are durable and can charge electricity faster and keep electric charge longer. In this study, electrodes for supercapacitors were made from aluminum foils coated with carbon nanotube film and separated by electrolyte solution and a separator. A separator could prevent short circuit but allow ions to pass through, and consequently increased storage layers of electric charge. The separators used in this study were made from polysulfone containing $CaTiO_3$ 0.5, 1.0 and 2.0 wt% with perovskite properties, high dielectric constant, electrical resistivity and energy density. After that, they were built in coin-cell form. It was found from the study that the addition of 2.0 wt% $CaTiO_3$, the largest proportion of all samples, provided a maximum specific energy at 4.03 mWh/g and a maximum specific capacitance at 4.64 F/g. accounting for 2.17-time higher than that of polysulfone without $CaTiO_3$. The functional group analysis of PSF and $CaTiO_3$ separators showed that the increasing of $CaTiO_3$ in the PSF separators, the 2958 cm^{-1} in C-H stretch peaks reduced, and consequently specific energy and specific capacitance were higher. Thus, supercapacitors with $CaTiO_3$-filled polysulfone separators are suitable for improving efficiency of supercapacitors in energy storage from electrical supply.

1. Introduction

The constantly increasing demand for electric energy in the present world results in the need for energy storage such as batteries, capacitors and supercapacitors. The problems are that batteries require long charging time while providing short cycle time and capacitors, though can be fast charged, have low storage capacity and discharge quickly. Thus, supercapacitors with durability, faster electric charge and longer electric storage are developed. Since electrodes for supercapacitors were made from aluminum foils coated with carbon film [1,2], and separated by an electrolyte solution and a separator to prevent short circuit but allow ions to pass through, storage layers of electric charge increase when dielectric material is added to separator sheets. The added material possesses perovskite properties including catalyst, sensor and high dielectric constant. Calcium titanate ($CaTiO_3$) is a high-dielectric-constant substance with high electrical resistivity and high energy density [3]. Because of its wide space between the particles, $CaTiO_3$ allows greater amount of energy to be stored, which is called internal-barrier-layer-capacitor model. It is suitable for inventing ceramic capacitors and polymer capacitors [4] by using polysulfone-based polymers as a polymer sheet [5,6] which has low dielectric constant and porosity [7].

In this study, the researcher was interested in developing supercapacitors using $CaTiO_3$-filled polysulfone separators by analyzing functional groups in the separators using FT-IT techniques and oxidation numbers of Ca, Ti and O ion using XPS technique in order to increase the efficiency of supercapacitors in energy storage from electricity sources such as wind turbines and solar cells.

2. Materials and methods

In this section, the preparation of electrodes, electrolyte solution, $CaTiO_3$ and separators is elaborated. To prepare nano-carbon electrodes, LA 133 (Linyi Gelon LIB Co., Ltd.) was dissolved in Di water and mixed well. Then, nano-carbon was added to the LA 133 adhesive solution and stirred at 60°C for 3 hours. After that, it was poured into a 5 x 10 cm aluminum foil sheets, scrapped longitudinally and at 120°C for 30 minutes to obtain films on an aluminum sheets used as electrodes. Electrolyte solution was made of the mixture of Tetraethylammonium tetrafluoroborate with solution density at 1 M and Acetonitrile. $CaTiO_3$ was prepared by ball milling 1 mol calcium oxide powder with 1 mol titanate dioxide power for 6 hours and sintering at 1,000°C for 8 hours. $CaTiO_3$ crystal structure, then, was investigated using X-ray diffraction (XRD) (Bruker, model D2 PHASER) in the range of 10 - 80°. For separators, 15 wt% Polysulfone (PSF) was dissolved in Dimethylacetamide (DMAc) and stirred well at 60°C for 4 hours. Then, the solution was poured onto glass, scrapped using doctor blade and immersed in DI water to allow the PSF separators to have phrase transformation before they took out from the glass. The separators with 15 wt% PSF solution were filled with $CaTiO_3$ at 0.5, 1.0 and 2.0 wt%, respectively. Through the steps above, 4 separators were obtained. The surfaces of the separators were examined through scanned images of electrodes using Field Emission Scanning Electron Microscope (FE-SEM) (Carl Zeiss, model AURIGA), and functional group analysis was used to analyze the $CaTiO_3$ - filled separators using FT-IR spectrometer (PerkinElmer, model Spectrum Two). After that the oxidation state of Ca, Ti and O ion of $CaTiO_3$-filled separators was investigated using X-rays Photoelectron spectroscopy (XPS) technique from BL3.2a beamline at the Synchrotron Light Research Institute (SLRI), Thailand. To assemble supercapacitor, nano-carbon electrodes on aluminum foil and the separators were cut into circles with 1.6 cm and 1.8 cm diameters, respectively (see figure 1), then coin-cell were made with a hydraulic press. Their efficiency was measured with the Battery Testers System (NEWARE, model BTS 3000) by charging and discharging 10 mA current under electric potential difference (V) of 2.5 V in order to analyze specific energy (W) and calculate specific capacitance (C) as shown in equation (1).

$$W = 1/2\, CV^2 \tag{1}$$

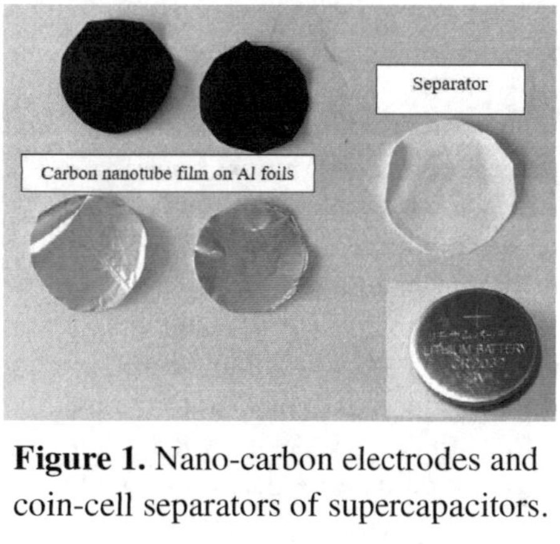

Figure 1. Nano-carbon electrodes and coin-cell separators of supercapacitors.

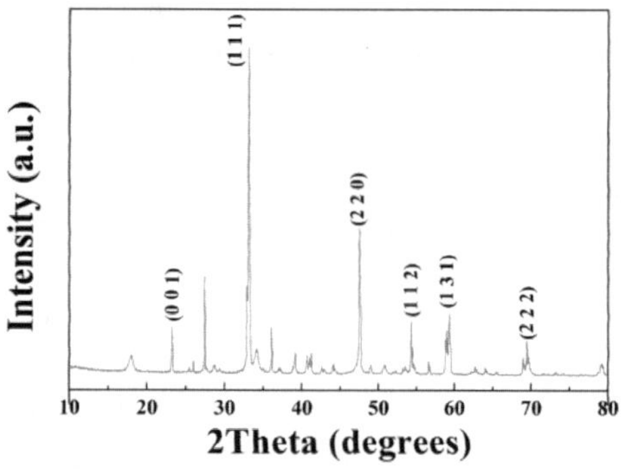

Figure 2. Crystal structure of $CaTiO_3$ from XRD technique.

3. Results and discussion

Calcium oxide and titanium dioxide in a 1:1 mole ratio was mixed and sintered at 1,000°C for 8 hours to generate $CaTiO_3$, then its crystal structure was examined using XRD technique as shown in figure 2. After measuring the diffraction angle 2-theta range of 10-80°, it was found that the crystal structure corresponds to $CaTiO_3$ (ICSD code: 031864) with planes 001, 110, 220, 112, 131 and 222. The examination of polysulfone (PSF) surface in DMAc solution as shown in figure 3(a) and polysulfone containing $CaTiO_3$ 0.5, 1.0 and 2.0 wt% as shown in figure 3(b)–(d) revealed that the increasing amount of $CaTiO_3$ in PSF separators results in the replacement of $CaTiO_3$ particles in the porosity.

The functional group analysis of PSF and $CaTiO_3$ separators using FT-IR spectrometer as shown in figure 4 was conducted. The results revealed no interesting peak in the PSF separators. However, for $CaTiO_3$-filled PSF separators, the peak at 2958 cm^{-1} in C-H stretch, at 1589 and 1487 cm^{-1} in C-C stretch, at 1240, 1150, 1105 and 1015 cm^{-1} in C-N stretch and at 835 cm^{-1} in C-H stretch were observed. Additionally, the greater amount of $CaTiO_3$ made all of the aforementioned peaks weaker. The analysis of XPS spectra of 2wt% $CaTiO_3$-filled separators as in figure 5 showed that the oxidation numbers of Ca ions were Ca2s, Ca3s Ca2p and Ca3p; those of Ti ions were Ti2p and Ti3p and those of O ions were O1s and O2s.

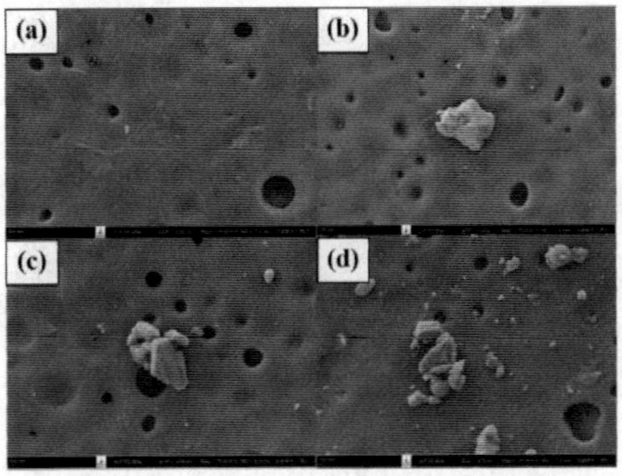

Figure 3. FE-SEM image of separators of (a) PSF (b) PSF with 0.5% $CaTiO_3$, (c) PSF with 1.0% $CaTiO_3$ and (d) PSF with 2.0% $CaTiO_3$.

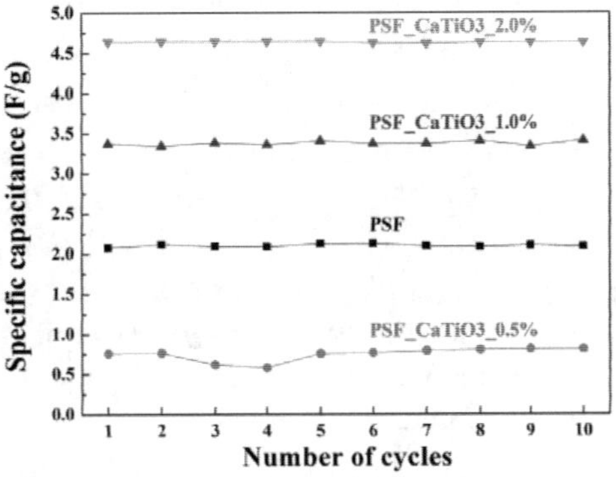

Figure 4. FT-IR spectra of $CaTiO_3$-filled PSF separators.

The efficiency of supercapacitor using $CaTiO_3$- filled PSF separators in the weight percentage of 0.5, 1.0 and 2.0 wt% to analyze specific energy (as shown in figure 5) and specific capacitance (as shown in figure 6) using Battery Testers system, brand NEWARE model BTS 3000 (as shown in table 1) was examined. The results showed that supercapacitors with PSF separators held specific energy at 1.85 mWh/g and specific capacitance at 2.13 F/g. Supercapacitors with $CaTiO_3$-filled separators in the weight percentage of 2.0% (PSF_CaTiO3_2.0%) generated specific energy and specific capacitance at 4.03 mWh/g and 4.64 F/g, respectively. Supercapacitors with $CaTiO_3$-filled separators in the weight percentage of 1.0% (PSF_CaTiO3_1.0%) generated 2.96 mWh/g of specific energy and 3.41 F/g of specific capacitance, respectively. The specific energy and specific capacitance produced by $CaTiO_3$-filled separators in the weight percentage of 0.5% (PSF_CaTiO3_0.5%) were the least as they were marked at 0.70 mWh/g and 0.81 F/g, respectively.

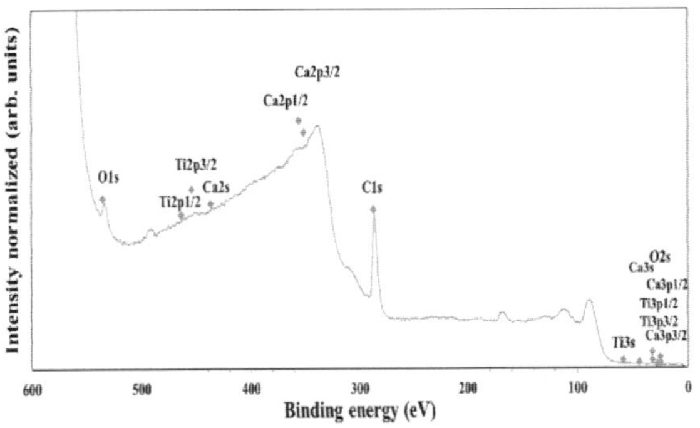

Figure 5. XPS spectra of supercapacitors with 2 wt% $CaTiO_3$-filled polysulfone separators.

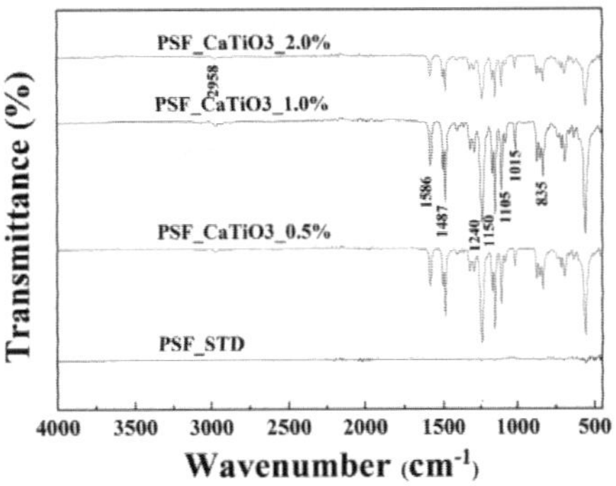

Figure 6. Specific capacitance of supercapacitors with $CaTiO_3$-filled polysulfone separators.

Table 1. Maximum specific energy and specific capacitance of supercapacitors with $CaTiO_3$-filled polysulfone separators.

Sample	Specific energy (mWh/g)	Specific capacitance (F/g)
PSF	1.85	2.13
PSF_CaTiO3_0.5%	0.70	0.81
PSF_CaTiO3_1.0%	2.96	3.41
PSF_CaTiO3_2.0%	4.03	4.64

It was found from the study that the mixture of CaO and TiO_2 in a 1:1 mole ratio with sintering at 1,000°C for 8 hours to generate $CaTiO_3$ caused phase transformation of crystal structure. This corresponds to the crystal structure of $CaTiO_3$. The analysis of FE-SEM images of PSF separators from XRD technique revealed that the greater amount of $CaTiO_3$ replaced porosity in separators which lowered porosity and consequently increased storage layers of electric charge.

The functional group analysis of PSF and $CaTiO_3$ separators showed that the increasing of $CaTiO_3$ in the PSF separators, the 2958 cm^{-1} in C-H stretch peaks reduced, and consequently specific energy and specific capacitance were higher. The analysis of XPS spectra of $CaTiO_3$-filled separators revealed the presence of Ca, Ti and O ion. It can be seen when the amount of $CaTiO_3$ in PSF separators increases, the amount of Ca, Ti and O increases accordingly, and this is in line with the decrease of C-H stretch peaks (2958 cm^{-1}) making energy storage more effective. Thus, supercapacitors with 2.0 wt% $CaTiO_3$ provided a maximum specific energy at 4.03 mWh/g and a maximum specific capacitance at 4.64 F/g accounting for 2.17-time higher than that of PSF separators without $CaTiO_3$, and followed by the addition of 1.0% $CaTiO_3$ while the addition of 0.5% $CaTiO_3$ could produce the least.

4. Conclusion

The addition of $CaTiO_3$ in PSF separators can increase storage layers of electric charge. Supercapacitors filled with 2.0% $CaTiO_3$ provided 2.17-time higher than the polysulfone without $CaTiO_3$. Thus, supercapacitors with $CaTiO_3$-filled polysulfone separators are suitable for improving efficiency of supercapacitors in energy storage from electrical supply and electronic devices.

Acknowledgements

This study was supported by Physics Division, Science Center, Faculty of Science and Technology, and Research and Development Institute, Phetchabun Rajabhat University which received funding from Thailand Science Research and Innovation (TSRI) under Thailand research strategy and innovation plan in the fiscal year 2020.

References

[1] Deng M, Yang B and Hu Y 2005 *J. Mater. Sci.* **40** 5021–23
[2] Xiong W *et al* 2011 *J. Power Sources* **196** 10461–64
[3] Zhou H Y, Liu X Q, Zhu X L and Chen X M 2017 *J. Am. Ceram. Soc.* **101**(5) 1999–2008
[4] Hu Y, Zhang Y, Liu H and Zhou D 2011 *Ceram. Int.* **37**(5) 1609–13
[5] Chittrakarn T, Tirawanichakul Y, Sirijarukul S and Yuenyao C 2016 *Surf. Coat. Tech.* **296** 157–63
[6] Ruangdit S, Chittrakarn T, Anuchit S, Tirawanichakul Y and Yuenyao C 2017 *Malaysian J. Anal. Sci.* **21**(2) 372–80
[7] Zheng X, Pu Z, Hu L, Tian Y, Xia J, Cheng J and Zhong J 2019 *J. Mater. Sci. Mater. Electron.* **30**, 18168–76

Energy conversion of electrostrictive poly(vinylidene fluoride-co-hexafluoropropylene)/Graphene composites

R Ruadroew[1,a], P Thainiramit[1,2,b] and C Putson[1,2,c]

[1]Department of Physics, Faculty of Science, Prince of Songkla University (PSU), Songkhla, 90112, Thailand
[2]Center of Excellence in Nanotechnology for Energy (CENE), Songkhla, 90112, Thailand

E-mail : [a]6010210368@psu.ac.th; [b]panu.t@psu.ac.th; [c]chatchai.p@psu.ac.th

Abstract. This study investigates energy-conversion properties of the electrostrictive polymer, poly(vinylidene fluoride-co-hexafluoropropylene), P(VDF-HFP), filled with graphene nanosheets (GNPs). The composites (i.e., P(VDF-HFP) and GNPs) were fabricated by using the solution casting method. The dielectric constant of these electrostrictive materials was measured to observe the energy conversion property with different frequencies using an LCR meter. Their mechanical properties were measured using a photonic sensor with varying various input vibrations and electric fields to calculate their electrostrictive coefficients. These characterized results revealed that dielectric constants and electrostrictive coefficients were significantly increased when GNPs fillers were filled higher. For the electrical property, the generating current, which was measured across these polymer films, increased proportionally with respect to the adding GNPs. In this obtained result, the main finding of P(VDF-HFP)/GNPs composites is a promising electrostrictive material for applications of electromechanical energy conversions in many smart-material systems.

1. Introduction

Energy harvesting is the process, which can converse small amounts of ambient energy into electrical energy for using suitable transducers in many smart systems. For portable and self-powered electronic devices, energy conversion using electrostrictive materials has the potential applications to be electromechanical transduction as a promising technique. In fact, electrostrictive is the property of materials that can compress itself that induced strain (S) in the regions of high electric field strength according to equation (1)

$$S = ME^2 \tag{1}$$

Where M is an electrostrictive coefficient [1-2]. Moreover, when this material is received stress (T) from vibration under an electric field, it can perform a psudopiezoelectric property. Then, it can generate electricity from the stress (T), and this specific property of this promising material for energy-harvesting devices. In addition, the key factor to be developed for increasing the capability of electrostrictive material is a dielectric property. This factor relates directly to a polarization (P) [3]. The effective process to manipulate the dielectric permittivity is to add a filler into a polymer matrix. As this expected result, the filler can lead to the increasing of charges between the boundary of the material matrix; furthermore, it can affect the increase of interfacial polarization as well as the dielectric constant to be increased too. In this work, Poly(vinylidene fluoride-co-hexafluoropropylene) polymer, P(VDF-HFP), is selected because it has unique properties, such as flexibility, low processing temperature, and low dielectric constant. P(VDF-HFP) can be enhanced its electrostrictive property by adding graphene nanoplatelets (GNPs). Its unique properties are high electrical conductivity, high modulus, and high specific surface area [4]. Thus, the composite P(VDF-HFP)/GNPs film is used to study the electrical, electrostrictive properties, and capability of energy conversions.

2. Experimental

2.1. Preparation of P(VDF-HFP) with GNPs composite films

The GNPs 0 (for PVDF-HFP pure), 1, 2, 3, and 4 wt% were added into the DMF and dispersed by using ultrasonic for 20 minutes. Then, the P(VDF-HFP) granules 20 g were added in solution and mixed continuously at 80 °C for 3 hours to gain a homogeneous solution. The final solution was cast onto a glass surface by a solution casting method and dried in the oven at 80 °C for 14 hours to remove the solvent.

2.2. Electrical properties and electrostrictive properties measurement

The electrical properties and electrostrictive properties of all samples were characterized. The electrical properties especially the dielectric constant was assessed across 1-10^5 Hz by an LCR meter (Model IM 3533, HIOKI)[3]. Then, the electrostrictive behavior was assessed by measuring the electric field (E), induced strain (S) with a photonic displacement sensor (Model MTI-2100, Photonic sensor, sensitivity 0.0058 mm/mV) at a low frequency and low electric field strength (f = 1 Hz, $E \leq 2$ MV/m)[2]. The sample was placed between two electrodes, and the electric field (E) was applied to the sample across the thickness direction from high voltage (Model 610E, Trek). The Log-in amplifier (Model SR510, SRS) was applied to measure the induced strain in the following direction.

2.3. Energy conversion measurement using the energy-harvesting experiment setup

These samples were sputter-coated with gold electrodes (20×20 mm). They were then fixed with a mechanical shaker providing sinusoidal 20 Hz from a function generator and the signal amplified using a power amplifier. The dc-bias electric field was applied across both electrodes using a high-voltage power supply. The laser displacement sensor (Model IA030, KEYENCE) [1] was employed to observe the sample displacement. The induced output voltage from the polymer was amplified by a current preamplifier (sensitivity 1mA/V), then the signal was filtered to 20 Hz by a lock-in amplifier, and thus sending all output signals to the oscilloscope display. The measured voltage output can calculate the induced current of the electrostrictive material by varying the sensitivity of the current preamplifier.

3. Results and discussion

Figure 1 shows that the dielectric constant decreased while the frequency increased because of the effect of oriented polarization. It occurred when molecules of material, having a dipole moment, were induced to rotate under an AC-electric field. At the low frequency, dipole moment could follow the direction of the electric field, but it could not reorient at the high frequency. In addition, the dielectric constant also proportionally increased when the GNPs filler was more added because the GNPs filler conducted the interfacial polarization occurring from the induced charges in the boundary of the two-phase. The charge induced dipole moment between two boundaries. The increase of this specific interfacial surface area affected by the conductive filler (i.e., GNPs) was able to influence the dielectric constant to significantly increase as well. The electrostrictive coefficient (M), which could be calculated from the slope of the graph between induced strain (S) and square magnitude of electric field (E^2) increased by adding the GNPs content, as shown in figure 2. In this case, the GNPs content was conducted to increase the total polarizations related to the dielectric constant. In addition, the electrostrictive coefficient (*M*) was proportional to the dielectric constant according to equation (2) [2].

$$M \ \alpha \ \varepsilon_0(\varepsilon_r - 1)^2(Y \times \varepsilon_r)^{-1} \tag{2}$$

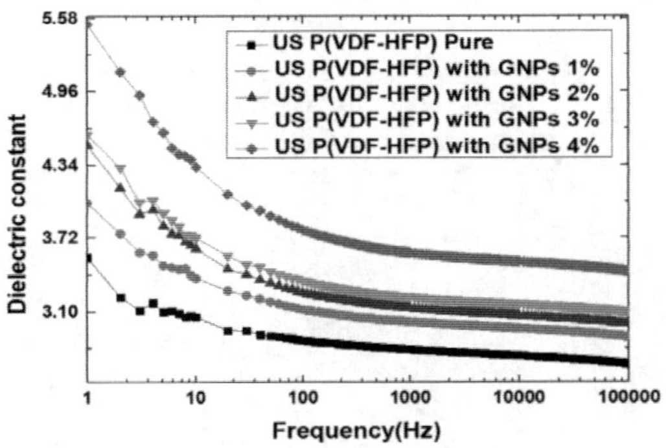

Figure 1. Dielectric constant as a function frequencies of all samples.

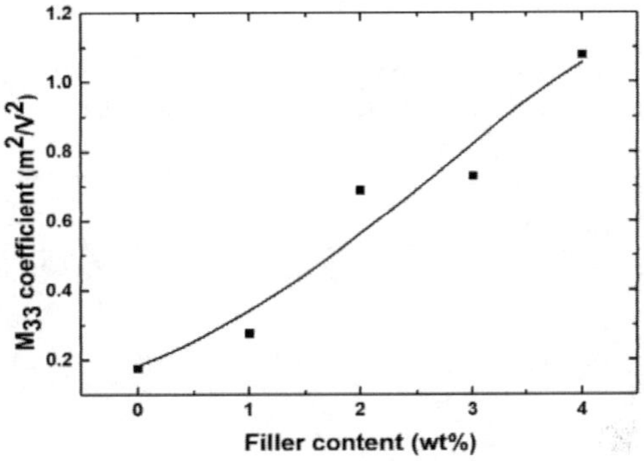

Figure 2. Electrostrictive coefficient versus the GNPs content.

The voltage calculated from the displacement on polymer using a photonic sensor is shown in figure 3. In the neat P(VDF-HFP) film, the voltage output at the operating frequency of 20 Hz was increased due to the increased displacement on the film from the mechanical shaker, as shown in figure 4. In addition, voltage output also increased with GNPs content, and the highest induced current from the sample was obtained with the maximum GNPs content, as shown in figure 5. In the theoretical modeling, the stress (*T*) on the samples could induce strain (*S*), by mechanical shaker under electric field, and electrical displacement (*D*) was also undergone, as shown in equation (3) and equation (4),

$$S_1 = s_{11}T_1 + M_{31}E_3^2 \tag{3}$$

$$D_3 = \varepsilon_{33}E_3 + 2M_{31}E_3T_1 \tag{4}$$

Hence, these two parameters led to the polymer generating output current [1] according to equation (5),

$$I = 2M_{31}YE_{dc}\int\left(\frac{\partial s}{\partial t}\right) dA \tag{5}$$

The increase of electrostrictive coefficient (*M*) with GNPs content was also enhanced polymer abilities to compress itself, and led to the change of more strain under the same electric field. For this reason, the P(VDF-HFP) with more GNPs content adding could be induced more strain, led to the increase of the output voltage. Then the polymer composite could generate more output current under the stable dc-electric field, as shown in figure 6.

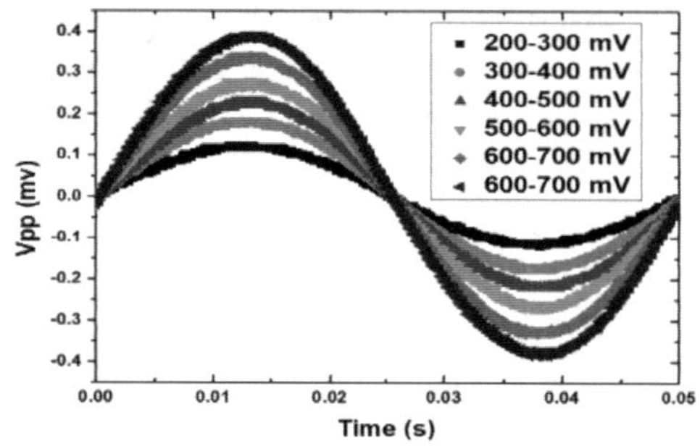

Figure 3. Input voltage signal as a function of time with various displacement.

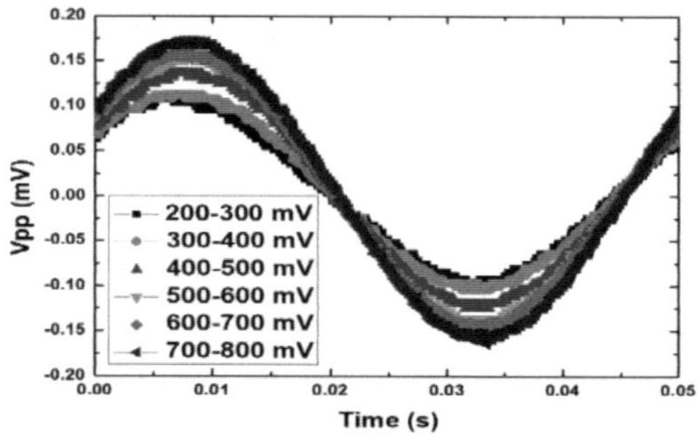

Figure 4. Output voltage from neat P(VDF-HFP) film measured as a function of time with various displacemen.

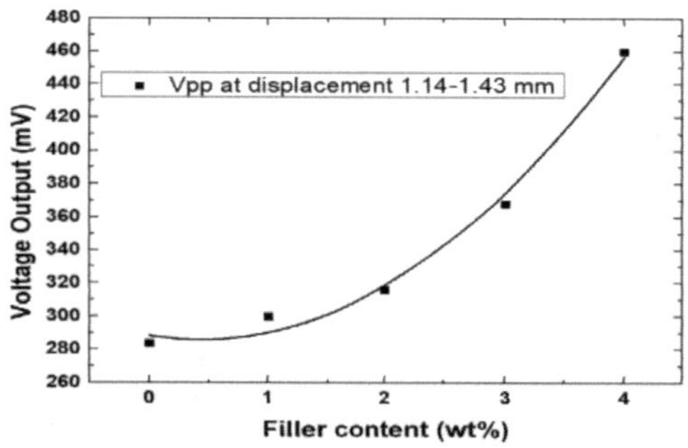

Figure 5. Voltage output versus GNPs contents.

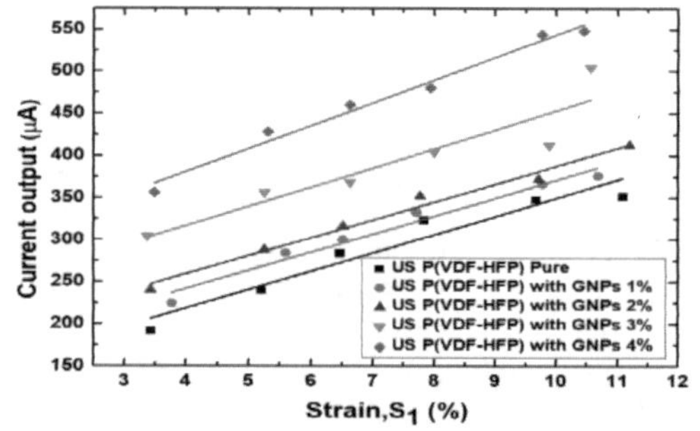

Figure 6. Induced current versus strain of all samples.

4. Conclusion

The P(VDF- HFP) filled with GNPs composites was fabricated to study energy conversions of electrostrictive polymers. The electromechanical conversions of these polymers were improved their electrostrictive properties by proportional increases of the GNPs. Their dielectric constants and electrostrictive coefficient (*M*) were also increased by the GNPs due to the increment of interfacial polarization at the boundary between GNPs particles and P(VDF- HFP) matrix. This experiment can enhance the electrostrictive property that was presented by measured results with various input vibrations and electric fields. Then, the polymers can provide a higher electrical- output voltage and current than neat polymers, depended on GNPs content adding. Hence, the potential applications of this electrostrictive material may be better smart actuators as well as energy harvester, based on energy-conversion capabilities of the electrostrictive materials.

Acknowledgements

This work was supported by Development and Promotion of Science and Technology Talents Project (DPST), Institute for the promotion of teaching science and technology Thailand (IPST) and Thailand Center of Excellence in Physics (ThEP-61-PIP-PSU3). We also acknowledge the Department of Physics Faculty of Science, Prince of Songkla University (PSU), Thailand.

References

[1] Nawaka K and Putson C 2020 *Compos. Sci. Technol.* **198** 108293
[2] Jaaoh D, Putson C and Muensit N 2016 *Compos. Sci. Technol.* **122** 97–103
[3] Ardimas and Putson C 2018 *Mater. Sci. Forum* **917** 117–21
[4] Tohluebaji N, Putson C and Muensit N 2019 *Polymers* **11** 1817

Ferroelectric properties and breakdown strength of layer-by-layer poly(vinylidene fluoride-co-hexafluoropropylene) (P(VDF-HFP)) and polyurethane (PU) for energy storage application

C Chooseng, S Chaipo and C Putson*

Department of Physics, Faculty of Science, Prince of Songkla University, Hat Yai, Songkhla 90112, Thailand

*Corresponding author's e-mail: chatchai.p@psu.ac.th

Abstract. Ferroelectric polymers are one of the next-generation pulsed capacitor materials for the potential application in capacitive energy storage. This polymer with higher saturated polarization, smaller remnant polarization, and higher electrical breakdown are the most promising candidates. In this work, the dielectric properties and energy storage capacity of the bilayer polymer films of Poly(vinylidene fluoride-co-hexafluoropropylene) (P(VDF-HFP)) and polyurethane (PU) were studied. These bilayer polymers were prepared by layer-by-layer method at the condition of variable layer thickness. The results show that the dielectric constants and the saturated polarization of the bilayer films increased, and bilayer films with P70/PU30 exhibit electrical high breakdown strength up to 379 V/μm. Moreover, enhanced energy storage density and the energy efficiency of the bilayer constrictors will be discussed for the capacitive energy storage polymers.

1. Introduction

Dielectric polymer materials are principal for capacitors due to their attributed lightweight, flexibility, scalability, and high dielectric strength. Dielectric constant (ε_r) and electrical breakdown strength (E_b) were the two main keywords to enhance energy storage density of dielectric materials, which depend on the volume of interfacial polarization and structure [1]. Polyurethane (PU) is one of the electroactive polymers that have a high ε_r and easily respond to an external electric field [2]. However, the increasing ε_r of material reduced their E_b and energy efficiency causes an increment of energy loss within material [3]. For instance, the PU/BT50 polymer composites presented the ε_r and E_b were 13.6 and 59.8 MV/m, compared to pure PU ε_r and E_b were 7.4 and 88.6 MV/m [4]. Therefore, it is important to enhance both the energy storage density and the energy efficiency of dielectric materials. The consequence of layered films can enhance dielectric properties and energy capacity. The layer-structured 15 vol% BT/P(VDF-HFP) composite exhibits simultaneously increasing dielectric constant and breakdown strength (ε_r~ 20 @10 kHz, E_b 241 MV/m), and comparison with their single-layer (ε_r ~ 11 @10 kHz, E_b 97 MV/m) [5]. Ferroelectric copolymer, Poly(vinylidene fluoride- co-hexafluoropropylene) P(VDF-HFP) is a semi-crystalline polymer that provided high energy density and high E_b (317 MV/m) [6]. In this research, we are interested in the bilayer film of P(VDF-HFP) and PU using the layer-by-layer solution casting method, and studied on the electrical properties and energy storage capacity.

2. Experimental

2.1. Fabrication of polymer films

The solution of single-layer film, 5 g of P(VDF-HFP) was dissolved in 20 mL of N, N-dimethylformamide (DMF) by stirring with 500 rpm at 50°C for 3 h and 5 g of PU was dissolved in 25 mL of DMF by stirring with 600 rpm at 80°C for 3 h. Then, single-layer film was cast onto the glass substrate by the casting method. After that, the films were dried in an oven at 80 °C for 12 h of P(VDF-HFP) and 14 h of PU. The thickness of the films was controlled around 30±5 μm and measured via the thickness gauge (PEACOCK, G-7C). Fabrication of bilayer film, with the volume fractions of P(VDF-HFP) to PU was 100: 0, 70: 30, 50: 50, 30: 70, and 0: 100. The solution was prepared with the same ratio of single-layer film. First, the P(VDF-HFP) was cast onto glass substrate as the first layer and dried for 1 h in the oven at 80°C and then the PU was cast onto P(VDF-HFP) layer. Finally, bilayer films were dried at 80°C for 12 h in the oven.

2.2. Characterization

The dielectric properties were measured by LCR meter (IM 3533 HIOKI) in the frequency range of 1-10^5 Hz at room temperature. The dielectric constant (ε_r) and conductivity (σ) were calculated by equations (1) and (2), respectively.

$$\varepsilon_r = \frac{Cd}{\varepsilon_0 A} \tag{1}$$

$$\sigma = \frac{Gd}{A} \tag{2}$$

where C is the capacitance, d is the sample thickness, ε_0 is the permittivity of air (8.853×10^{-12} Fm^{-1}), A is the electrode area, and G is the conductance. The Energy storage density (U_e) and the energy efficiency (η) can be estimated in equations (3) and (4), respectively.

$$U_e = \int_{P_r}^{P_s} EdP \tag{3}$$

$$\eta = \frac{U_e}{U_e + U_l} \times 100\ \% \tag{4}$$

where P is the polarization, P_s is the saturation polarization, P_r is the remnant polarization, E is the electric field, and U_l is energy loss was calculated in the close area of the hysteresis loop.

The breakdown strength (E_b) was measured by Dielectric Breakdown Test System (PolyK Technologies State College) at the applied external electric field (E) 0-10 kV, and the Weibull model as equation (5) was used to analyze the breakdown probability of all samples.

$$P(E) = 1 - \exp\left[-(E/\lambda)^k\right] \tag{5}$$

where $P(E)$ is the breakdown probability of the material when the electric field was applied, k is the parameter related to the reliability of the sample and the shape parameter k shows the distribution of E_b, λ is the breakdown probability of dielectric breakdown [7].

3. Results and discussion

3.1. Electrical properties

Figure 1(a) shows the dielectric constant of bilayer films is higher than the single-layer film of P(VDF-HFP). The values of dielectric constant and dielectric loss of PU increase when the frequency was decreased (figure 1(a) and 1(b)). Also, the conductivity increased with frequency (figure 1(c)). The dielectric constant and dielectric loss depend on the frequency with polarization in the dielectric material. At low frequencies, the motion and rotation of dipoles in polymer films easily respond to the electric field. In case of neat PU, the dielectric constant and dielectric loss have the highest when compared with other conditions due to the interfacial polarization between soft and hard segments. The increment of the dielectric constant of bilayer films is also with two matrix and interfacial polarization at the boundary region. The conductivity of all conditions increased with frequency because of the charge carrier mobility.

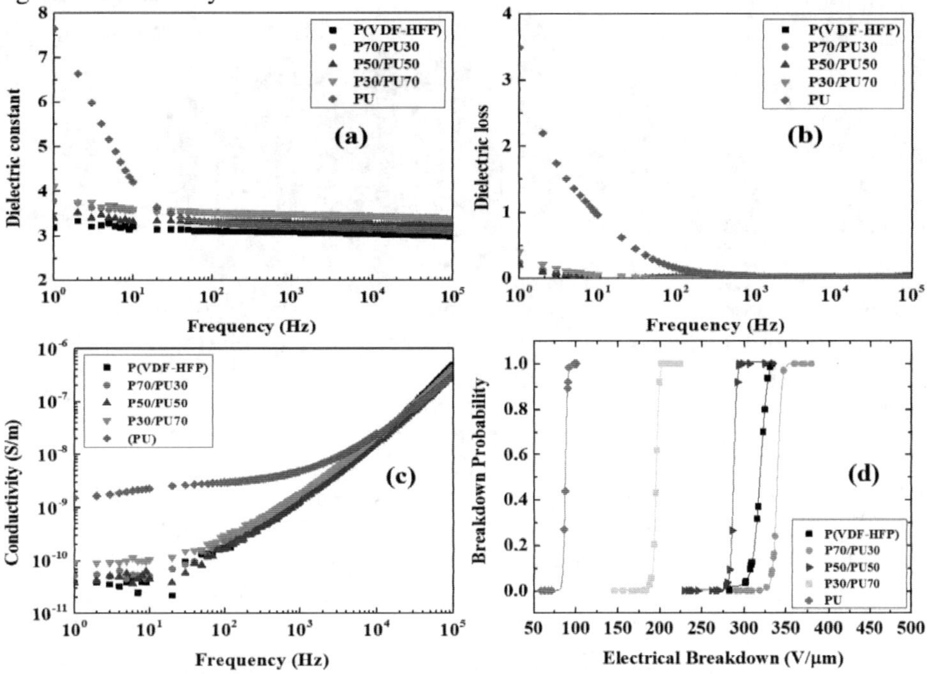

Figure 1. Frequency dependence of (a) dielectric constant, (b) dielectric loss, (c) conductivity and (d) breakdown probability of all polymer films with different volume fractions.

Figure 1(d) shows the probability of electrical breakdown ($P(E)$) of PU, P30/PU70, P50/PU50, P(VDF-HFP) and P70/PU30, respectively. The breakdown strength of polymers depends on the external electric field. The influences such as structure, grain size, the porosity of materials, and environmental contamination affected to E_b. The bilayer P70/PU30 exhibited the highest E_b at 379 V/μm compared with pure P(VDF-HFP) at 333 V/μm that had more conductive pathway interface in the neat film. The intrinsic electric breakdown is a material characteristic and is electronic in nature [8]. The charge with interfacial polarization of heterogeneous polymer causes PU has lowest breakdown strength.

3.2. Energy storage capacity

The polarization of film increased with electric field due to the motion and alignment in domain film. For instance, the saturation polarization (P_s) of P70/PU30 was 0.60 μC/cm² at the electric field of 40

MV/m and increased to 1.75 μC/cm² at the electric field of 100 MV/m. The remant polarization (P_r) of pure P(VDF-HFP) was 0.32 μC/cm² under the electric field of 100 MV/m, while the remnant polarization of P70/PU30 is 0.43 μC/cm² under the same electric field. The single-layer of P(VDF-HFP) and bilayer of P70/PU30 show that the energy storage density is higher than bilayer P50/PU50 and P30/PU70 (figure 2(a)) due to the result of saturation polarization from ferroelectric properties. The energy efficiencies of all films are drop when electric field was increased (figure 2(b)) due to an increase of the energy loss and the remnant polarization. The highest energy efficiency of bilayer film P70/PU30 at 40, 60, and 80 MV/m, is 77%, 70%, and 61%, respectively. The energy loss of polymer films importantly consists of the dielectric loss and the conduction loss from leakage current at the high electric field. The resulting of efficiency at 100 MV/m, the bilayer film P70/PU30 has lower than pure P(VDF-HFP) due to it has energy loss more than pure P(VDF-HFP).

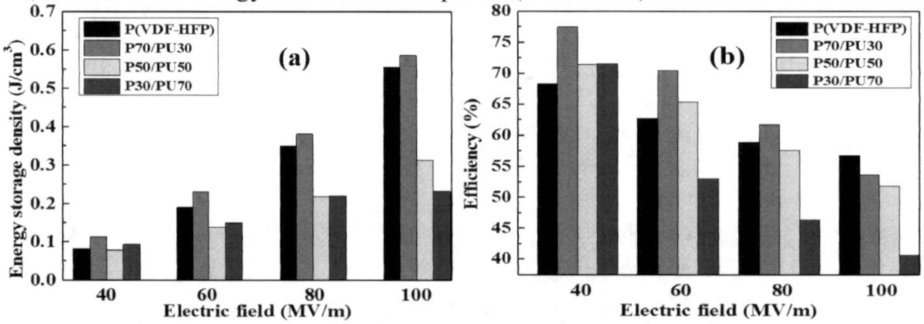

Figure 2. Energy storage density (a) and energy efficiency (b) of polymer films with different volume fractions as a function of the electric field.

4. Conclusions

In summary, the bilayer films provided high dielectric constant, energy efficiency, and the breakdown strength of pristine P(VDF-HFP) film. The result showed that the energy storage density of P70/PU30 is higher than other films. The energy efficiency of P70/PU30 increases with increasing interfacial polarization of PU, and decreasing dielectric loss of P(VDF-HFP) segment. Moreover, the electrical breakdown strength of bilayer P70/PU30 provided the highest E_b.

Acknowledgements

This work was supported by Development and Promotion of Science and Technology Talents Project (DPST), Institute for the promotion of teaching science and technology Thailand (IPST) and Thailand Center of Excellence in Physics (ThEP-61-PIP-PSU3). We also acknowledge the Department of Physics Faculty of Science, Prince of Songkla University (PSU), Thailand.

References

[1] Sun C, Deng H, Ji W, Zhou H, Zhao G, Du R and Fu Q 2019 *Nanocomposites* **5** 36–48
[2] Wongtimnoi K, Guiffard B, Bogner-Van de Moortèle A, Seveyrat L, Gauthier C and Cavaillé J-Y 2011 *Compos. Sci. Technol.* **71** 885–92
[3] Chen J, Wang Y, Yuan Q, Xu X, Niu Y, Wang Q and Wang H 2018 *Nano Energy* **54** 288–96
[4] Yin G, Yang Y, Song F, Renard C, Dang Z M, Shi C Y and Wang D 2017 *ACS Appl. Mater. Interfaces* **9** 5237–43
[5] Sun L, Shi Z, Liang L, Wei S, Wang H, Davoud D, Sun K and Fan R 2020 *J. Mater. Chem.* C **8** 10257–65
[6] Luo B, Wang X, Wang H, Cai Z and Li L 2017 *Compos. Sci. Technol.* **151** 94–103
[7] Pedroli F, Marrani A, Le M Q, Froidefond C, Cottinet P J and Capsal J F 2018 *J. Polym. Sci.* B *Polym. Phys.* **56** 1164–73
[8] Palneedi H, Peddigari M, Hwang G T, Jeong D Y and Ryu J 2018 *Adv. Funct. Mater.* **28** 1803665

Preparation and electrostrictive properties of polyurethane filled with polypyrrole-carbon black for the energy harvesting application

K Hirunchulha, K Jehlaeh and C Putson*

Department of Physics, Faculty of Science, Prince of Songkla University, Hat Yai, Songkhla 90112, Thailand

*Corresponding author's e-mail: chatchai.p@psu.ac.th

Abstract. The polymer composites based on electrostrictive materials with high polarization are attracting scientists because of the prospect of application in energy conversion and electromechanical performance. Polyurethane is one of a good flexible dielectric polymers which is more interesting in electromechanical applications due to its easy fabrication, high dielectric constant, and high electrostrictive coefficient. This work focuses on the dielectric properties and electrostrictive properties of Polyurethane (PU) filled with Polypyrrole-Carbon Black (PPy-CB) various the different concentrations. All polymer composites were prepared by using a film-casting process in solution with DMF solvent. The dielectric properties and conductivity of thin films were investigated by using LCR meter at 1-10^5 Hz. The results showed that the dielectric constant of composites significantly increases when the concentration of PPy-CB was increased. Moreover, the electrostrictive coefficient of composites also increases when the concentration of PPy-CB was increased. Furthermore, the electrical breakdown stretches were analyzed by using the Weibull model, this result will discuss the relationship when the PPy-CB fillers were used.

1. Introduction

Nowadays, smart material technology advancement has widely evolved in sensors and actuators, which can be applied to the fields of medicine, the electronic industry, and robots. Electrostrictive polymer is one of the smart materials that can change shape or size under an electric field. It has lightweight, high flexibility, quick response, and high specific energy. Electrostrictive materials can cover the mechanical properties to electrical properties. In fact, an increase of permittivity is needed to enhance energy harvesting. Polyurethane is one of the electrostrictive polymers which provide the large electrostrictive coefficient [1]. Moreover, Polyurethane can be considered as an heterogeneous material with soft and hard segments. The electric properties are controlled with hard segments, while mechanical properties relate to the soft segment. The hard/soft ratio of Polyuethane can be adjusted with the conductive fillers [2]. Polypyrrole is one of a conductive polymer that is widely used due to its good electrical conductivity and thermally stability. Moreover, dropping carbon black in the polymers has improved electrical conductivity. Among the various ways to enhance the dielectric constant, the development of polymer nanocomposites introduced with conductive fillers is of great interest. Owing to it is more flexible to tune their electrical and mechanical properties by choosing a suitable condition. The electrostrictive properties are significantly related to the dielectric constant and modulus, which strongly depend on the dispersion of the fillers in the polymer matrix, and the microstructure state of the composite. This work proposed the dielectric and electrostrictive properties of polyurethane/polypyrrole-carbon black thin films. Furthermore, the electrostrictive performance with the electrical breakdown stretches will be discussed.

2. Experimental

2.1. *Preparation of thin films filled with polypyrrole-carbon black*

Polyurethane (PU) pellets, DP 9370A, supplied from Covestro AG (Germany), N, N-dimethylformamide (DMF, 99% purity) purchased from RCI Labscan Limited, Thailand. All polymer composites were dissolved in DMF by magnetic stirring at 80°C for 3 hours. Polypyrrole-carbon black (PPy-CB) particles were then added to DMF and dissolved with sonication for 20 minutes. The PPy-CB solution was dropped in the PU solution. Then, the PU and PPy-CB solution was stirred at 80°C for 1 hour to ensure homogeneity. PU composite films were obtained by using a film-casting process of polyurethane/polypyrrole-carbon black (PU/PPy-CB) composited with 0.25, 0.50, 0.75, or 1.00 %wt. The solution was dried in the oven for 16 hours to remove the solvent. Finally, the PU/PPy-CB films were taken off from the glass surface.

2.2. *Characterization methods*

2.2.1. Measuring the dielectric properties by using LCR Meter. Samples thickness is 100±5 μm in all cases. The dielectric constant, loss tangent, and AC conductivity of all polymers composites were investigated by LCR meter (IM 3533 HIOKI) at room temperature with a frequency range of 1-10^5 Hz.

2.2.2. Measuring the electrostrictive properties by using an electrostrictive setup. The electrostriction is generally defined as a quadratic coupling between strain (S), and electric field (E). The electrostrictive coefficient (M) depends on the dielectric properties (ε_r) and Young's modulus (Y) [3], as shown in equation (1).

$$M = \frac{\varepsilon_0 \left(\varepsilon_r - 1\right)^2}{Y} \tag{1}$$

where ε_0 is the permittivity of free space. The electrostrictive coefficient (M) relates strain (S) to the square of the applied electric field [4].

$$S = ME^2 \tag{2}$$

The electric field-induced strain was investigated with the electrostrictive setup with help of a photonic sensor (MTI-2100) and high voltage supply (Trek Model 610E).

2.2.3. Measuring the dielectric breakdown properties by using polarization loop & dielectric breakdown test system. The dielectric breakdown (E_b) of obtained polymer composites were measure with the electric field between 0 and 10kV by Dielectric Breakdown Test System (PolyK Technologies State College). In this case, 12 breakdown tests were used to evaluate the E_b which was analysed by using the Weibull model, as shown in equation (3).

$$P(E) = 1 - \exp\left[-\left(\frac{E}{\lambda}\right)^k\right] \tag{3}$$

Journal of Physics: Conference Series **2145** (2022) 012044 doi:10.1088/1742-6596/2145/1/012044

where P is the breakdown probability of the material when the electric field E was increased, λ is the breakdown probability of dielectric breakdown at 63.2%, and k is the parameter related to the reliability of the sample and the shape parameter k shows the distribution of E_b.

3. Results and discussion

3.1. Dielectric properties

The dielectric constant, loss tangent and conductivity from 1 Hz to 10^5 Hz of PU/PPy-CB composites is shown in figure 1 (a), (b) and (c), respectively. It is found that the dielectric constant of PU and composites decreases when the progressive increase of frequency. This behavior occurs due to the loss of one of the polarization contributions (interfacial polarisation, orientation polarisation, etc.). Thus, this decrease can be unambiguously attributed to the loss of the space charges induced interfacial polarization contribution. Furthermore, it is shown that the dielectric constant, loss tangent, and conductivity of PU/PPy-CB film composites increase when the concentration of PPy-CB was increased, which is related to the increasing of the charge density on the surface of the dopant, induced the interfacial polarization. It may be attributed that an addition contribution to the quantity of accumulated charge when the PPy-CB was used. The loss tangent of PU/PPy-CB 0.75%wt is higher than PU/PPy-CB 1.00%wt because of the degree of filler dispersion in PU matrix. It may be attributed to some agglomerates of PPy-CB when increasing fillers. Thus, the conductivity of PU/PPy-CB 0.75%wt and PU/PPy-CB 1.00%wt at very low frequency is not different.

3.2. Electrostrictive properties

The electrostrictive property of PU and PU/PPy-CB film composites was observed in figure *2*, the electrostrictive coefficient can be evaluated from the slope of the curve, following equation (2). Figure *3* shows the electrostrictive coefficient of PU and PU/PPy-CB film composites. The results show that the electrostrictive coefficient slightly increases when the concentration of PPy-CB was increased. In fact, the electrostrictive properties are directly related to the dielectric constant and modulus. Consequently, the increment of the dielectric constant can be explained by the dipole moment, the interfacial polarization, and conductive charge distribution. However, the modulus of composite increases with increasing filler content. It may be attributed that the elastic properties of PU composites decrease due to the polymer chains was stocked when fillers were used. Therefore, the slight enhancement of the electrostrictive coefficient versus PPy-CB content can be explained by the additive of conductive charges in PU matrix compare with the increasing the dielectric constant as compared to the increasing modulus.

3.3. Dielectric breakdown properties

Figures 4 and 5 present the electrical breakdown probability and the electrical breakdown strength of PU and PU/PPy-CB film composites, respectively. It showed that a decrease of the electrical breakdown strength was observed when PPy-CB loading was increased. It can explain that the decreased electrical breakdown strength of PU with PPy-CB loading is related to (i) enhanced mobility of charge carriers (ii) the high conductivity of fillers and (iii) increment of heterogeneousness of polymer matrix when the filler was used [5].

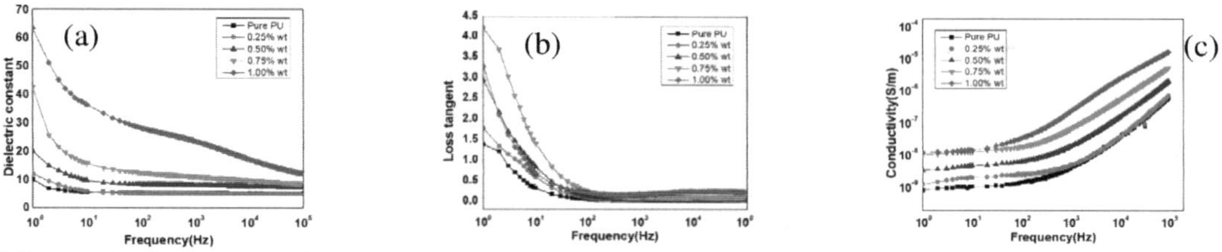

Figure 1. Frequency dependences of (a) the dielectric constant (b) the loss tangent and (c) the conductivity for composites with various PU/PPy-CB contents.

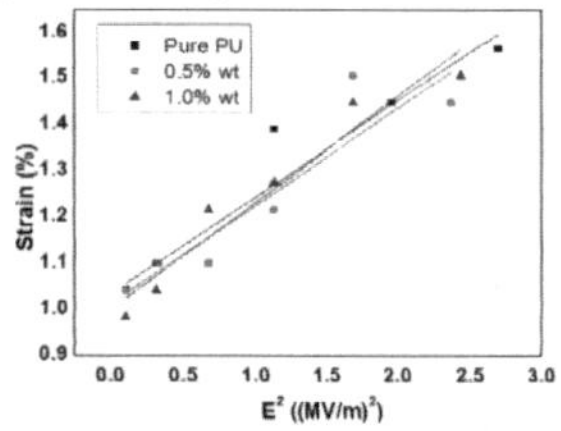

Figure 2. Strain response of PU/PPy-CB composite as a function of squared electric field.

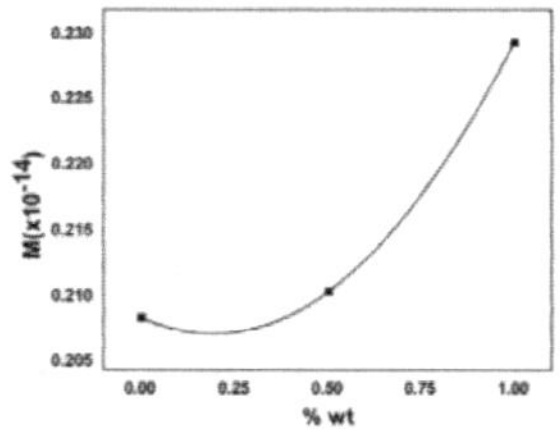

Figure 3. The electrostrictive coefficient versus the PPy-CB loading.

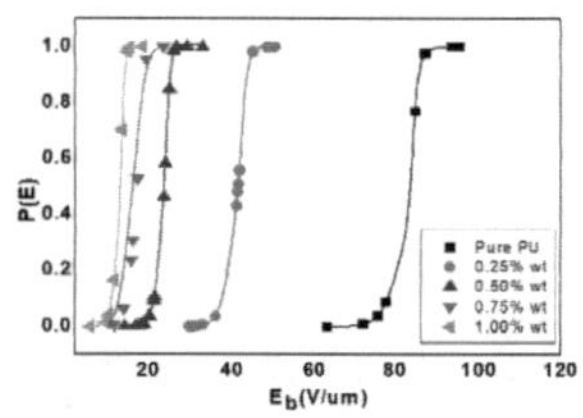

Figure 4. Breakdown probability versus electrical breakdown strength.

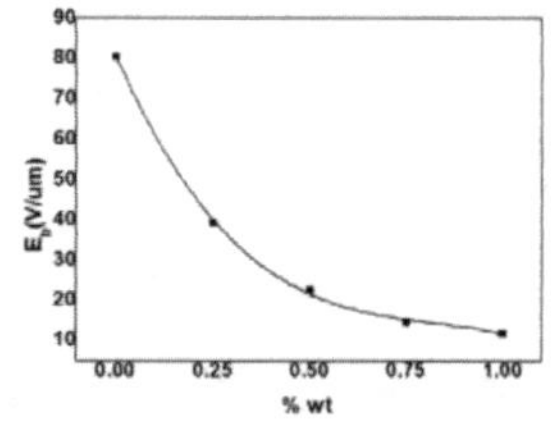

Figure 5. Electrical breakdown strength versus the PPy-CB loading.

4. Conclusion

This work studied the dielectric and electrostrictive properties of PU/PPy-CB film composites. All samples were prepared by using a film-casting process. The results showed that the dielectric constant and the electrostrictive efficiency of PU/PPy-CB film composites increase when the PPy-CB loading was used. While a decrease of the electrical breakdown strength was observed with the PPy-CB loading. They can be candidate materials for electromechanical application which is related to the high conductivity of fillers, charge mobility, and polarization distribution based on the heterostructure materials.

Acknowledgements

This work was supported by Development and Promotion of Science and Technology Talents Project (DPST), Institute for the promotion of teaching science and technology Thailand (IPST) and Thailand Center of Excellence in Physics (ThEP-61-PIP-PSU3). We also acknowledge the Department of Physics Faculty of Science, Prince of Songkla University (PSU), Thailand.

References

[1] Ardimas, Putson C and Muensit N 2018 *Compos. Sci. Technol.* **158** 164–74
[2] Jaaoh D, Putson C and Muensit N 2016 *Compos. Sci. Technol.* **122** 97–103
[3] Nawaka K and Putson C 2020 *Compos. Sci. Technol.* **198** 108293
[4] Jaaoh D, Jarawae R, Intan M and Chaidana H 2018 *Appl. Mech. Mater.* **879** 32–40
[5] Yin X, Liu Q, Galineau J, Cottinet P J, Guyomar D and Capsal J F 2016 *Eur. Polym. J.* **76** 88–98

Correlation between PM2.5 and meteorological variables in Chiang Mai, Thailand

P Anusasananan*, D Morasum, S Suwanarat and N Thangprasert

Department of Physics, Faculty of Science, Ramkhamhaeng University, Bangkok 10240, Thailand

*E-mail: panatcha.a@rumail.ru.ac.th

Abstract. Air pollution is a major concern for the population in Chiang Mai, northern Thailand, as it is for most people in other large cities around the world. Hazy skies and pollution alert have become normal during late winter and entire summer almost every year. Prolonged expose to PM2.5 can have acute and chronic effects on the respiratory and cardiovascular systems. This research aims to study the correlations between PM2.5 and meteorological variables (rainfall and temperature) in Chiang Mai during 2017 and 2020. The cross wavelet transform (XWT) and wavelet coherence (WTC) have been used to examine these relations by assessing the presence of common power and the relative phase in the time-frequency space. The XWT between PM2.5 and rainfall shows a significant common power in two dominant period bands, one in the period between 10-14 months and the other one between 5-7 months. The first common power occurs during all observed time intervals, so it is obviously related to natural annual periodicities of PM2.5 and rainfall. The second band, which occurs only in the year 2019 may be connected with the beginning of the monsoon season which starts in May and brings a stream of warm moist air to Chiang Mai. Our data shows that PM2.5 typically begins to rise starting in November, and it remains high until March of the next year. The PM2.5 is low in rainy season since rain has a wet scavenging effect on PM2.5. The WTC, which is a measure of the correlation between two time series, indicate that there is a significant correlation between PM2.5 and rainfall in the 10-14 month band. The phase difference between these two time series is defined by arrows. The phase arrows pointing to the left indicated the anti-phase relation, when rainfall increases, PM2.5 decreases and vice versa. The correlation coefficient (r) between PM2.5 and rainfall in rainy season is equal to 0.8504. Our study also finds that there is a proven correlation between PM2.5 and temperature in a day time scale with the correlation coefficient equal to 0.9249. In a one-day period, PM2.5 is low in the day time and high at night. An understanding of how climate variability may impact PM2.5 concentration in Chiang Mai will help the government with better planning and preparation to prevent environmental hazard from PM2.5 pollution.

1. Introduction

In 2020 Chiang Mai not only had to fight the Covid19 pandemic as the rest of the world had to do, but also had to fight a different battle: air pollution. For several months, Chiang Mai was blanketed by haze and smoke from ravaging forest fires, making its air quality rank as one of the worst cities for pollution for many straight days with Air Quality Index (AQI) of more than 200. The AQI is a report on a broad spectrum of air pollutants for general public such as PM2.5, PM10, carbon dioxide, and so on. PM2.5 refers to atmospheric particulate matter (PM), which is a diameter of less than 2.5 micrometers. There are many different sources of PM2.5, including natural and anthropogenic sources. PM2.5 is hazardous to human health because it can get the residual in the lungs and into blood vessels, causing acute and chronic effects for the respiratory and cardiovascular systems [1-7]. In figure 1 smoke and haze blanket large portions of Southeast Asia on 12 April 2019, using the visible infrared imaging radiometer suite instrument (VIIRS) on the Suomi National Polar-Orbiting Partnership spacecraft.

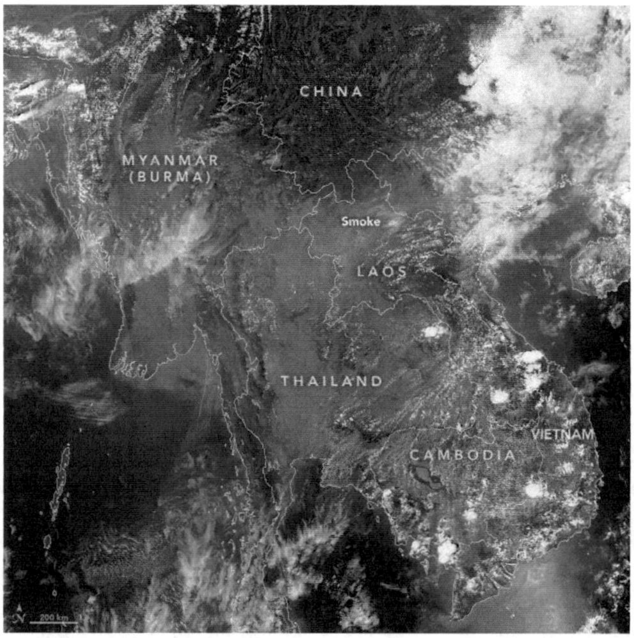

Figure 1. Unhealthy smoke over Thailand. Photo courtesy of the Earth Observatory – NASA.

There are several works that have studied the characteristic of PM2.5 and the impact of meteorological parameters on the mechanism of pollution creation. In this research, we will study the relation between PM2.5 and meteorological variables (rainfall and temperature) at Chiang Mai during the period 2017-2020 using the wavelet transform, which is developed and widely used as a time-frequency tool due to its ability to detect variations of power within the non-station data for information at different time scales. Mathematically continuous wavelet transform (CWT) of signal $d(t)$ is defined as follows [8, 9]:

$$W_c(a,b) = \int_{-\infty}^{\infty} d(t)\psi^*_{a,b}(t) \tag{1}$$

and

$$\psi_{a,b}(t) = \frac{1}{\sqrt{a}}\psi\left(\frac{t-b}{a}\right) \tag{2}$$

where $\psi_{a,b}(t)$ is the mother wavelet and a and b are scale and translational position parameters, respectively. The location parameter of a defines how the scale is stretched and of b gives the exact position for the wavelet. The cross wavelet transform (XWT) is a common tool that examine the relationship between two individual time series. It can be extended to the analysis of the correlation in the form of a time-frequency space between two time series with a power spectrum and a phase difference with 95% confidence level. The XWT power spectrum is produced absolute magnitude $|W_{xy}|$ by using the cross wavelet spectrum W_{xy} between two time series *x(t)* and *y(t)* with the wavelet transforms W_x and the complex conjugate of W_y. These wavelet transforms also can use to construct the continuous relative phase to examine a consistent phase relationship causing between the times [10].

2. Methodology

2.1. Study area and data

The monthly PM2. 5, rainfall and temperature data from the year 2017 to 2020 were used for this examination, taken from a meteorological station in Chiang Mai (~18.8°N, 98.8°E), Thailand. This area founded in 1296 and used to be the capital of the ancient Lanna Kingdom, is located 700 km north of the capital of Bangkok (~13°N, 101°E). Its terrain is in a verdant valley on the banks of the Ping River, surrounded by the mountain ranges of the highlands with an area of approximately 20,107 km^2. Its climate has a tropical wet and dry with warm to hot weather year-round and the nighttime conditions during the dry season are cool and much lower than daytime highs by recording the maximum temperature 42. 2 °C (108. 3 °F) in May 2005. The amounts of PM2. 5, rainfall, and temperature were collected and recorded every hour (except rainfall for every three hours), comprising a total of more than 100,000 data points, which were analyzed. The PM25 data was collected by a tapered element oscillating microbalance (TEOM) with the Thermo Scientific 1405 TEOM Continuous Ambient Particulate Monitor.

2.2. Wavelet analysis

The daily and monthly mean of PM2.5, rainfall and temperature data were calculated. Then, the wavelet analysis were used to identify such time series with significant multi-temporal scales. The XWT power spectrum were taken to investigate the correlation between the PM2.5 and rainfall and between PM2.5 and temperature. The XWT spectrum is defined by:

$$W_{xy} = W_x(a,b)W_y^*(a,b) \tag{3}$$

where * denotes the complex conjugate of $W_y(u,s)$. The XWT spectrum uncovers areas in the common power of a time-frequency space which represent the local covariance between both time series.

The wavelet coherence (WTC) between two W can be perceived significant coherence of local correlation of the two time series in a time-frequency space with confidence levels against red noise. The WTC is defined as the squared absolute value of the smoothed XWT spectra normalized by the smoothed individual wavelet common power spectrum of each series with a smoothing operator S, given by

$$R^2(a,b) = \frac{\left|S(s^{-1}W_{xy}(a,b))\right|^2}{S\left(s^{-1}\left|W_x(a,b)\right|^2\right)S\left(s^{-1}\left|W_y(a,b)\right|^2\right)}, \tag{4}$$

The WTC phase difference is used to characterize the correlation between two time series, which gives details on the time-frequency space about delays of oscillation of the two time series, defined as follow:

$$\phi_{xy}(a,b) = \tan^{-1}\left(\frac{\Im\left\{S\left(s^{-1}W_{xy}(a,b)\right)\right\}}{\Re\left\{S\left(S^{-1}W_{xy}(a,b)\right)\right\}}\right). \tag{5}$$

The consistent phase relationship is indicated by arrows on the WTC plots. This work and wavelet plots use R programming language.

 2145 (2022) 012045 doi:10.1088/1742-6596/2145/1/012045

3. Results and discussion

The PM2.5 concentration, rainfall and temperature data were measured in Chiang Mai from 2017-2020 as shown in figure 2. All of the time series obviously show the periodicity in a one-year period. As can be seen in the distribution of the measured data, we can observe three clearly evident principles: 1) the higher the amount of rainfall, the smaller the average values of PM2.5, especially in the rainy season 2) the higher the temperature in summer, the higher the PM2.5 concentration, and 3) at other times of the year the relation between PM2.5 and temperature does not reveal any unique dependence and represents rather complex relations. The maximum daily PM2.5 in Chiang Mai from January 2017 to December 2020 is 126.44 μg/m^3, the minimum is 5.9 μg/m^3, and the average is 28.35 μg/m^3. During this period, the maximum daily rainfall is 124.8 mm. and the average is 7.76 mm. And the temperature is highest in summer (34.7°C) and lowest in winter (15.8°C), with the average at 26.77°C. In figure 2, the amount of the PM2.5 concentration and rainfall move in opposite directions. However the rainfall decreases, the PM2.5 increases.

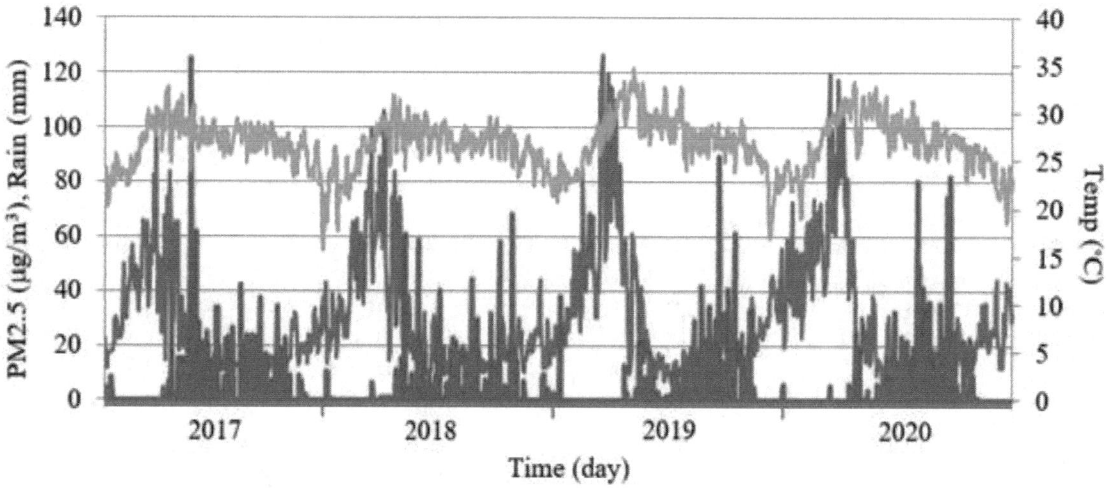

Figure 2. Time series of daily PM2.5 (blue), rainfall (red) and temperature (green) in Chiang Mai between January 2017 and December 2020. The maximum daily PM2.5 is 126.44 μg/m^3, which is higher than the recommended standard. PM2.5 is high in summer and low in rainy season opposite the amount of rainfall. On the other hand the temperature is high in summer and low in winter.

3.1. Relation between PM2.5 and rainfall

3.1.1. Wavelet power spectrum. The wavelet power spectrum (WPS) of the monthly PM2.5 concentration in Chiang Mai over a period of four years from Jan. 2.17 to Dec. 2020 is shown in figure 3. The horizontal axis is time in months and the vertical axis is the period in months. The colours with the power ranging from weak (blue shades) to strong (red shades) in the figure stand for the structure of the PM2.5 concentration variety. The relative power by the absolute value square gives the information at a certain period and in a certain month. The WPS of the PM2.5 is evaluated using four years of the data over from 2 to 6 months. The WPS shown in figure 3, analogous to temporally evolving spectrum, for PM2.5 shows two bands of high variance near a wavelength of 6 months from 2018 to 2019 and an annual seasonal trend of one year through the complete observed time series. The WPS of monthly rainfall in Chiang Mai between Jan. 2017 and Dec. 2020 is also shown in figure 3. It shows two high variance bands, one near a wavelength of 3 months for very short time in the beginning of 2017 and another one at one year period from 2017 to 2020 at the same time as PM2.5. Another significant cycle in the rainfall records occurs between 5 and 7 months with less power in the middle of 2019 to 2020.

Overall, the modes of PM2.5 and rainfall variability in Chiang Mai are wavelength varying from 11 to 13 months. As we can see, the patterns look similar, suggesting that there is a correlation between rainfall and PM2.5 in one year periods.

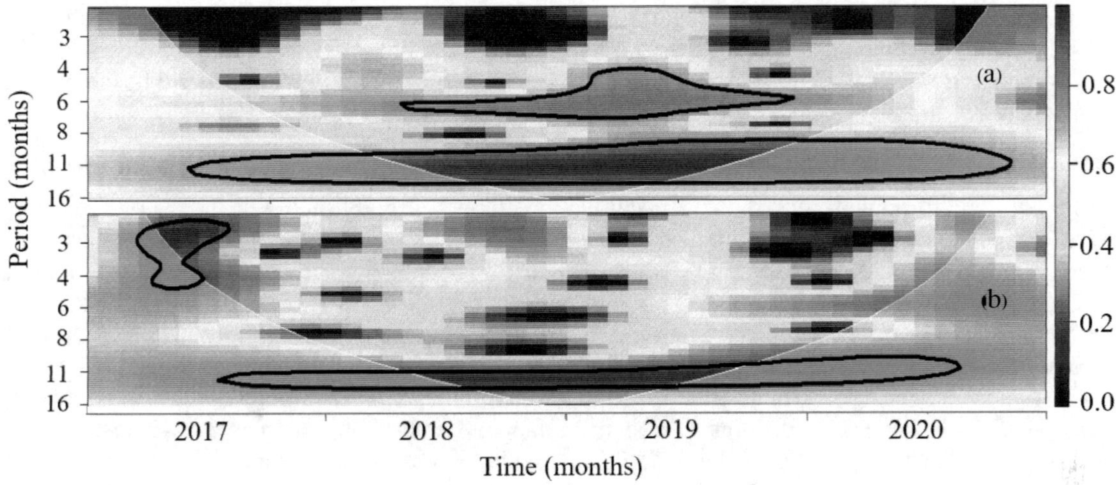

Figure 3. WPC of the PM2.5 (a) and rainfall (b).

3.1.2. Cross wavelet transform and wavelet coherence. The XWT between the PM2.5 concentration and rainfall in figure 4 shows at what frequencies and times the two series both have high variance, regardless of whether variations in one series are synchronous with those in the other. As we can observe from figure 4 PM2.5 and rainfall demonstrate significant correlation at an annual frequency between 10-14 months from 2017 to 2020 and between 5-7 months in 2019.

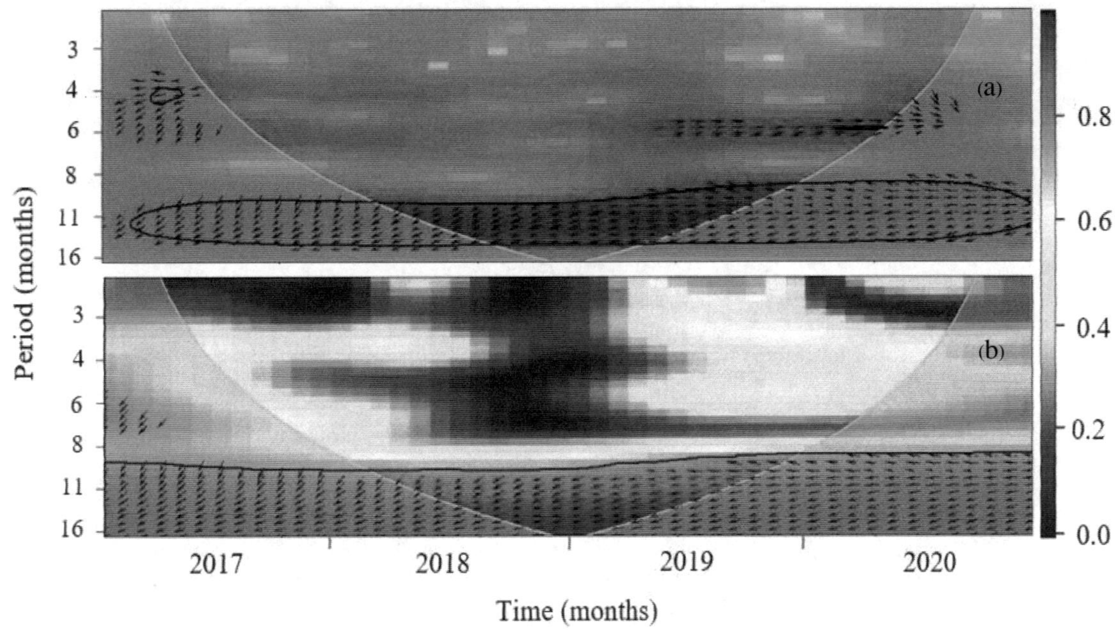

Figure 4. XWT (a) and WTC (b) between PM2.5 and rainfall time series.

The WTC shows whether variations in the two series are coherent (red corresponding to highest coherence) and in-phase or anti-phase (arrows to right, perfectly in-phase; arrows to left, anti-phase). As we can see from the figure, the red colours and right-pointing arrows through-out the time axis (near a wavelength of one year for WTC) indicate that variations near the one year period in PM2.5 and rainfall are approximately anti-phase and coherent through out the observed record, 2017-2020. The PM2.5 and rainfall are anti-phase with the pointing arrows to the left. Such phase different indicate that when the rainfall increases, the PM2.5 decreases. The high power spectrum are shown in the period band of 9-16 months between the two ends of the entire observation period.

3.1.3. Annual PM2.5 and rainfall. Since we can observe from WTC that there is a relation between PM2.5 and rainfall annually, as shown in figure 5, we plotted a graph between the average monthly PM2.5 and the average monthly rainfall over the period of four years from 2017 to 2020. As we can observe from this figure, rainfall has an effect on the PM2.5. Whenever there is a change in rainfall, there is a change in the PM2.5 because a raindrop in the atmosphere can attract tens to hundreds of tiny aerosol particles and coagulation before hitting the ground, which is a natural phenomenon to clear the air of pollutants. There is a significant negative correlation between the PM2.5 and rainfall. The higher the rainfall, the lower the concentration of PM2.5, which indicates that rainfall has a wet scavenging effect on PM2.5. The PM2.5 begins to rise from January to the maximum in March and linearly decreases until April. From March, the PM2.5 drops again to the lowest point in June since there is rain from the monsoon season. The PM2.5 is nearly stable from June until November in the rainy season Then its concentration increases linearly again to reach the high level in December. However, the rainfall is lowest in January, and February and then continues to increase to more than 200 mm in May. The rainfall is maximum in September at around 250 mm. From December to April, the amount of fine-particle dust in Chiang Mai is much worse than recommended standards reaching two times and March is the worst month.

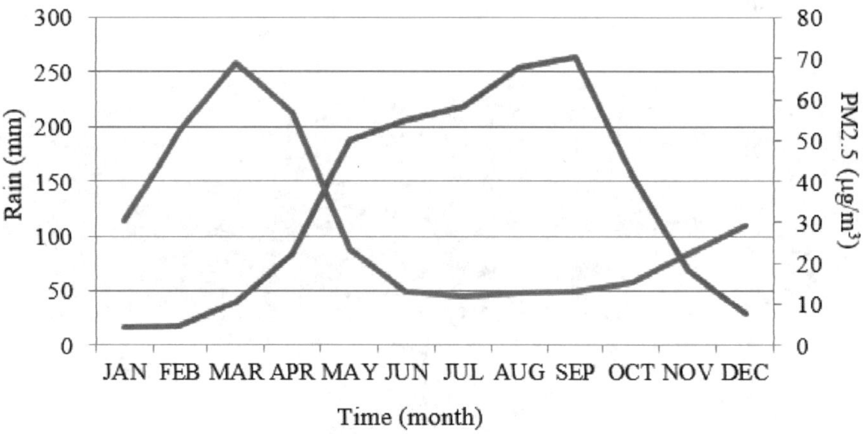

Figure 5. Average PM2.5 (red) and rainfall (blue) during 1 year period.

3.2. Relation between PM2.5 and temperature

3.2.1. Wavelet power spectrum. The WPS of the average hourly PM2.5 concentration and the average hourly temperature from 15-22 March 2020 is shown in figure 6 by the horizontal and vertical axis represents the time in hours and the period in hours, respectively. The colours with the power ranging from weak (blue shades) to strong (red shades) in the figure stand for the structure of the PM2.5 concentration variety. The relative power by the absolute value square gives the information at a certain period and in a certain month.

The WPS of the PM2.5 is evaluated over a period from 2 to 64 hours. The first oscillation is seen clearly between the entire observation period with the high power and the second oscillation occurs only 20 March 2020. Such periods are connected with natural annual periodicities. The WPS of the temperature is shown in figure 6 (b). The pattern looks similar to the WPS of PM25 in the diurnal cycle, which indicates a correlation between the temperature and PM2.5 in one day period.

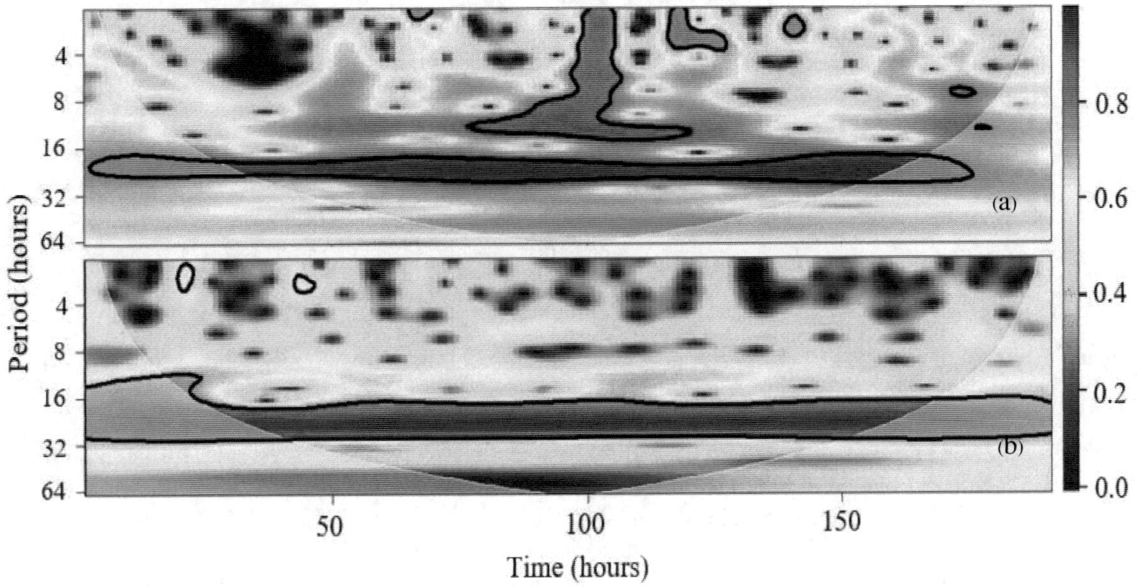

Figure 6. WPS of the average hourly PM2.5 concentration (a) and the average hourly temperature (b) on 15-22 March 2020. On the x-axis each unit means one hour.

3.2.2. Cross wavelet transform and wavelet coherence. The XWT between the PM2.5 concentration and temperature is shown in figure 7(a). Its significant common power spectrum appears clearly the first band covering 20-28 hours the entire observed period and the second band between 10-12 hours around 17-20 March 2020. The arrows of the phase difference in the significant band indicate the anti-phase (pointing to the left) correlation. The WTC is used to explore co-movement and the phase difference between such time series, as shown in figure 7 (b). The colored shading and the thick black line represent the squared coherence and the 95% significant level, respectively. Its high power spectrum is obviously with the period band of 20-28 hours between the entire observation periods with the temperature is leading, and 10-14 hours from 17 to 21 March 2020. The arrow of the phase relationship between are anti-phase. When the temperature increase, the PM2.5 decreases, and vice versa.

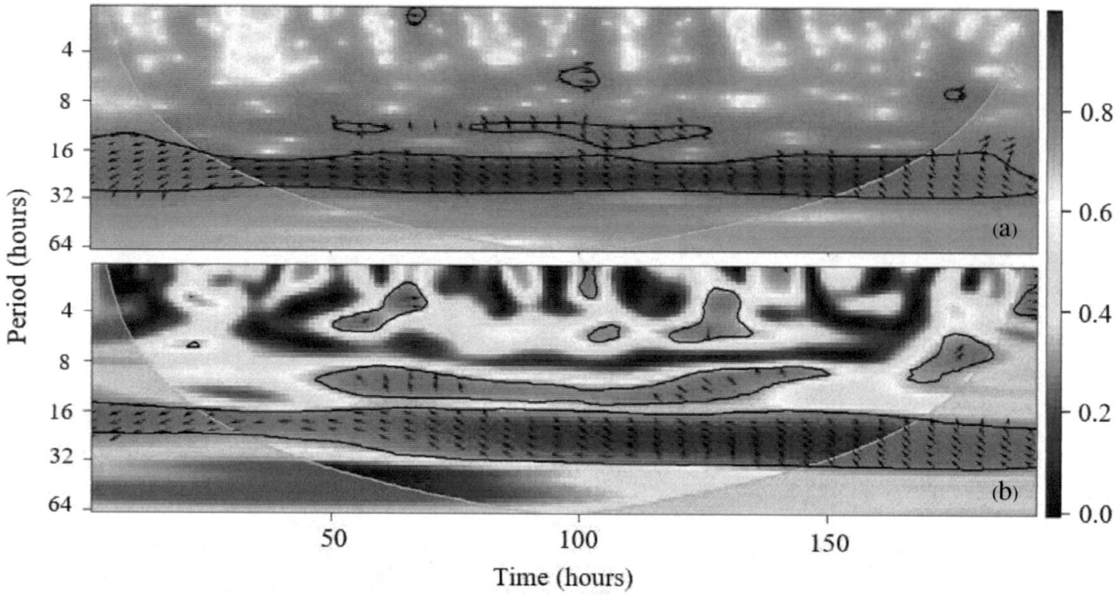

Figure 7. XWT (a) and WCT (b) between PM2.5 and temperature time series from 15-22 March 2020. On the x-axis each unit equals one hour.

3.2.3. Diurnal PM2.5 and temperature. To study the PM2.5 concentration and temperature in a one-day period, we calculated the average hourly the PM2.5 and temperature in each day over a period of 192 days in 2020, as shown in figure 8. The graph is seen clearly the trends of the PM2.5 and temperature move in opposite directions. The PM2.5 is high between 5 to 6 a.m. and linearly decreases to reach the lowest between 13-15 p.m. But the temperature is minimum between 5 to 6 a.m. and linearly increases to reach the highest between 1 to 3 p.m. As hot air from the earth's surface rises, it also takes the PM2.5 to higher altitudes. It is noticeable that at about 8 to 9 a.m. the PM2.5 is constant instead of decreasing since at these hours the activity outdoors is increasing, causing the pollution from tuk-tuks, cars, and trucks to add up.

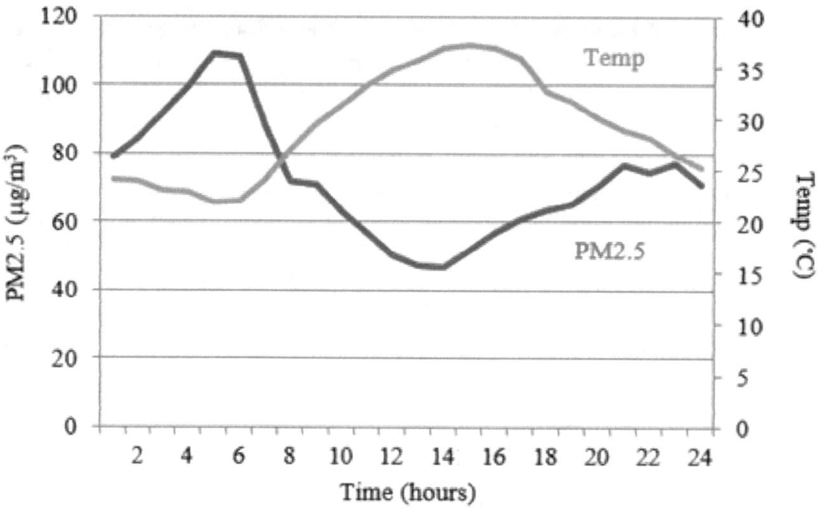

Figure 8. Average hourly PM2.5 (red curve) and temperature (green curve)

4. Conclusion

In this work, the correlation between the PM2.5 and meteorological parameters (rainfall and temperature were studied by using the wavelets. The results show that the PM2.5 and rainfall correlated annually with the correlation coefficient of 0.8504 in the rainy season. The XWT indicates that the PM2.5 and rainfall co-vary anti-phase. The PM2.5 is high between January and April, while the rainfall is low during these months. Rains conclude in Chiang Mai by the end of October, so there are no showers to clean the air until late April. In these four months in northern Thailand, the frames burn their fields to prepare for planting in the following year. Although it is illegal due to harm to the environment and people's health, their farmers continue this practice due to the lack of cheaper alternatives. With Chiang Mai is surrounded by mountains, namely Doi Saket, Doi Suthep, Doi Khun Tan, and Doi Inthanon, this area traps the smog over such months. The XWT analysis indicates that the average hourly PM2.5 and the average hourly temperature are co-varied out of phase. Since the correlation coefficient between these two variables is negative and close to 1, it also indicates that PM2.5 and the temperature tend to move in opposite directions i.e., when the temperature increases, PM2.5 decreases and vice versa. The PM2.5 is high around 5-6 a.m. and linearly decreases to reach the lowest level between 1 and 3 p.m. But the temperature is low between 5 and 6 a.m. and linearly increases to reach its highest points between 1 and 3 p.m. In order to avoid being affected by PM2.5 pollution, the key for the populace is to stay put indoors and use fans and filters to purify the poor indoor air quality. In case that it is necessary to do some activities out door, people should wear a pollution filtering mask and choose the time when the PM2.5 concentration is low. N95-grade masks protect person from 95% of the harmful particles the size of 0.3 microns or larger. As for the authorities, the Ministry of Natural Resources and Environment should be aware of the problem and try to control and prevent the burning of agricultural materials and waste in forests on plantations and in preserved areas before and during the burning season from January until April as shown by our data.

Acknowledgments
The authors are grateful to Department of Physics, Faculty of Science, Ramkhamhaeng University, Bangkok, the Pollution Control Department, and the Department of Meteorology for the data, for their support.

References

[1] Du Y, Xu X, Chu M, Guo Y and Wang J 2016 *J. Thorac Dis.* **8(1)** E8
[2] Bell M L, Ebisu K and Belanger K 2007 *Environ. Health Perspect.* **115(7)** 1118
[3] Liang Y, Fang L H, Zhang K, Kan H, Brook J R and Sun Q 2014 *Environ. Health* **13(1)** 102
[4] Brunckreef B and Holgate S T 2002 *Lancet* **360** 1233
[5] Kloog I, Ridgway B, Koutrakis P, Coull B A and Schwartz J D 2013 *Epidemiology* **24** 555
[6] Young D E, Kim H, Parworth C, Zhou S, Zhang X, Cappa C D, Seco R, Kim S and Zhang Q 2016 *Atmos. Chem. Phys.* **16** 5427
[7] Zanonetti A, Franklin M, Koutrakis P and Schwartz J 2009 *Environ. Health* **8** 58
[8] Morlet J, Arens G, Fourgeau E and Glard D 1982 *Geophysics* **47** 203
[9] Torrence C and Compo G P 1998 *Bull. Am. Meteorolo. Soc.* **79** 61
[10] Bloomfield D S, McAteer R T J, Lites B W, Judge P G, Mathioudakis M and Keenan F P 2004 *Astrophys. J.* **617** 623

Sensitivity of different physics schemes using WRF model in Typhoon Damrey (2017) over the Indochina region

Kulaya Keawsang-in[1], Sujittra Ratjiranukool[2,*] and Pakpoom Ratjiranukool[3]

[1] PhD's Degree Program in Applied Physics, Faculty of Science, Chiang Mai University, Muang District, Chiang Mai, 50200, Thailand
[2] Department of Physics and Materials, Faculty of Science, Chiang Mai University, Muang District, Chiang Mai, 50200, Thailand
[3] Department of Physics, Faculty of Science and Technology, Chiang Mai Rajabhat University, Muang District, Chiang Mai, 50300, Thailand

*E-mail: sujittra.r@cmu.ac.th

Abstract. To reduce the tropical cyclone impact of lives and economics, the precise forecast of the event is required. The typhoon Damrey (2017), which caused ravaging of the strong wind, heavy precipitation, flash flood and storm surge over the Indochina region, was simulated by Weather Research and forecasting (WRF) model. The simulated duration was 8 days starting from 31 October 2017 00 00 UTC to 8 November 2017 00 00 UTC. The NCEP 6-hour global FNL (final analysis) data at 1-degree resolution is provided for initial condition. The WRF model was run in a single domain of 20 km horizontal resolution bounded 0 to 20 N and 96°E to 124°E. The different physics schemes, which are the microphysics schemes, the planetary boundary layer schemes and cumulus parameterization schemes, were emphasized to examine the suitable schemes in tropical cyclone simulation over the Indochina region. To evaluate the reliability of the simulation of tropical cyclone, the track-position is correlated with the Japan Meteorological Agency (JMA) observation. The results show that the typhoon simulation forced by Belts-Millers-Janjic cumulus, WSM6 microphysics was suitable for simulating of typhoon Damrey.

1. Introduction

One of the most changeable weather events that effects to human lives, constructions, economy system, etc. is tropical cyclone event. The characteristic of tropical cyclone is lowest pressure at center over tropics. Tropical cyclone brings a heavy rain, storm surge and flash flood [1]. Reducing the social and economic impacts and the cost of forecasting, the precise tropical cyclone research is required. Typhoon Damrey (2017) was chosen in this work because it was the second deadliest storm of the 2017 Pacific typhoon season. It developed in the South China Sea and moved to Indochina and Thailand.

The Weather Research and Forecasting (WRF) Model is a mesoscale numerical weather prediction system including a non-hydrostatic process [2]. In this study, the options of parameterization schemes are recognized which the physics parameterization is a strong influence in model including track position and intensities, i.e., wind speed, central pressure, precipitation [3].

2. Analysis

In this study, the WRF model has been configured one-way nested domain bounded by 0 -24 N and 96-124 E with 20-km resolutions. The initial and boundary conditions were forced by NCEP 6 hourly final analysis (FNL). The NCEP FNL is reliable as it includes all of accessible observed data. There are 160 x 120 grid points overall selected domain. The simulation was applied with 32 vertical levels.

The main aim in Typhoon Damrey simulation was to recognize the physics parameterization impacts on tropical cyclone simulation. Thus, the combinations of three physics parameterizations, i.e., microphysics, planetary boundary layers, and cumulus parameterization schemes, are employed to perform the sensitivity of physics options. The experiment design was showed in table 1. The Kain-Frinsth cumulus, Kessler microphysics and Yonsei university are frequency appeared since there are many worldwide tropical cyclone researches especially over Asia as [4–7].

Table 1. The experiment design for typhoon Damrey simulations.

Simulation no.	Cumulus scheme	Microphysics scheme	PBL scheme
1	Kain/Fristch	Kessler	YSU
2	Belts-Millers-Janjic	Kessler	YSU
3	Grell-Freitas	Kessler	YSU
4	New Grell	Kessler	YSU
5	Kain/Fristch	WSM3	YSU
6	Kain/Fristch	Ferrier	YSU
7	Kain/Fristch	WSM6	YSU
8	Kain/Fristch	Kessler	MYNN

3. Results and discussion

The WRF model is simulated the 12-hour track position from 31 October 2017 06.00 UTC to 4 November 2017 18.00 UTC using different physics parameterizations, the follow results are presented.

3.1. Cumulus schemes

Cumulus parameterization including the shallow and deep cumulus cloud that effects to the vertical heat, moisture and momentum. Cumulus schemes involve tendencies of moisture and momentum and shallow convection. Each schemes have different options as shown in table 2. Moreover, parameterization of cumulus scheme significantly impacts on tropical cyclone track position and intensities with coarser resolution.

Table 2. Cumulus schemes [2].

Scheme no.	Scheme name	Moisture tendencies	Momentum tendencies	Shallow Convection
1	Kain/Fristch	Q_c Q_r Q_i Q_s	no	yes
2	Belts-Millers-Janjic	-	no	yes
3	Grell-Freitas	Q_c Q_i	no	no
5	New Grell	Q_c Q_i	no	yes

The location of low surface pressure indicated the track of typhoon Damrey. It shown in figure 1 that the typhoon originally located over the South China Sea and moved toward to Vietnam,

Cambodia, Laos and Thailand also. Difference of cumulus scheme sensitivity shows 2-4° northward compared to the observed track of Japan meteorological agency. It implied that cumulus scheme effects to tropical cyclone track position. In this experiment, the simulation with Belts-Millers-Janjic cumulus scheme nearly shows the closet track to the observation, approximately 0.2° southward for first 48 hours and 0.2-0.4 degree southward later.

3.2. Microphysics schemes

Microphysics parameterizations provide the atmospheric heat and moisture tendencies which are major parameters in regional weather. There are different number of hydrometeors and interactions for the processes in microphysics schemes [3] as shown in table 3.

Table 3. Microphysics schemes [2, 3].

Scheme name	Kessler (mp=1)	WSM3 (mp=3)	Ferrier (mp=5)	WSM6 (mp=6)
hydrometeors and interactions	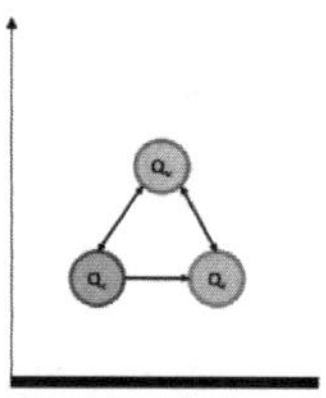	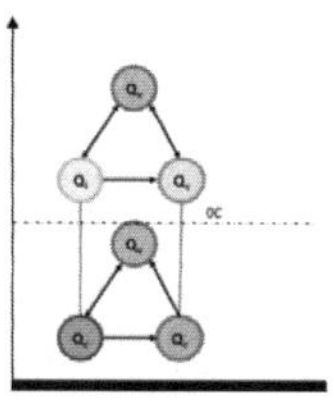	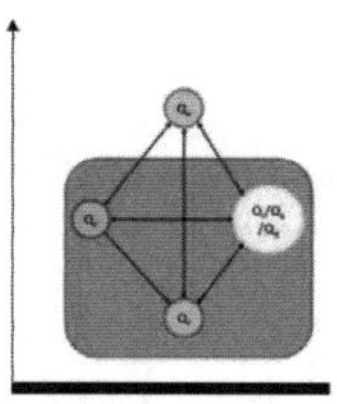	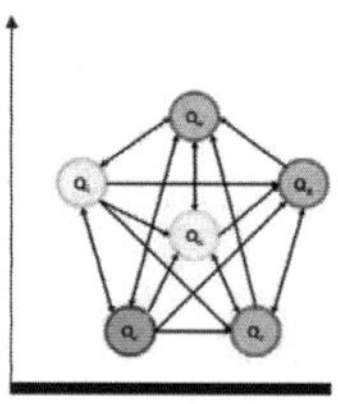

The KF cumulus and YSU planetary boundary layer schemes are selected because these schemes are applied for WRF model in various tropical cyclone studies [4]. From figure 2, the sensitivity of microphysics shows approximately 0.2 -1° differences from the observation. After 48 hours, the simulation with Kessler and Ferrier moved northwestward meanwhile its WSM3 moved southwestward. The selection of WSM6 microphysics performs tracking position of the tropical cyclone closer than Kessler, WSM3 and Ferrier microphysics schemes, with approximately 0.2° differences. Furthermore, several studies indicate the WSM6 microphysics scheme is perceptive to tropical cyclone track forecasting over Asia and western north Pacific [5-6].

3.3. Planetary boundary layer schemes

The Planetary boundary layer parameterization provides the profiles of vertically temperature and moisture and horizontally momentum in the atmosphere. The upper air swirling, the precipitation and the transport of energy and moisture are influenced by these Planetary boundary layer schemes. The YSU planetary boundary layer scheme is a typical nonlocal scheme meanwhile MYNN planetary boundary layer scheme is a local turbulent kinetic energy schemes [7,8].

The track position of early simulation with YSU planetary boundary layer scheme is similar to its MYNN PBL scheme and there is approximately 0.2° difference from the observation as shown in figure 3. After 48 hours, the track position of the simulations moved northward about 1-2°from observation. The track position analysis reveals that the simulation with YSU scheme similar to its MYNN 2.5 level planetary boundary layer schemes. Although the 5 tropical cyclones simulation over the Bay of Bengal showed that the YSU made a minimum track position error [7].

4. Conclusion

The sensitivity of different physics parameterizations was studied in typhoon Damrey simulation using WRF model. The simulated track position is compared to Japan Meteorological Agency observed data. The results shown that the simulation with Belts-Miller cumulus scheme and WSM6 microphysics scheme perform the best ability in tracking position of typhoon Damrey. Meanwhile, both planetary boundary layer schemes similarly effect to the tropical cyclone simulation. Furthermore, the statistical tests, i.e., track position error, root mean square error, etc., is required for validation of tropical cyclone simulation in future work.

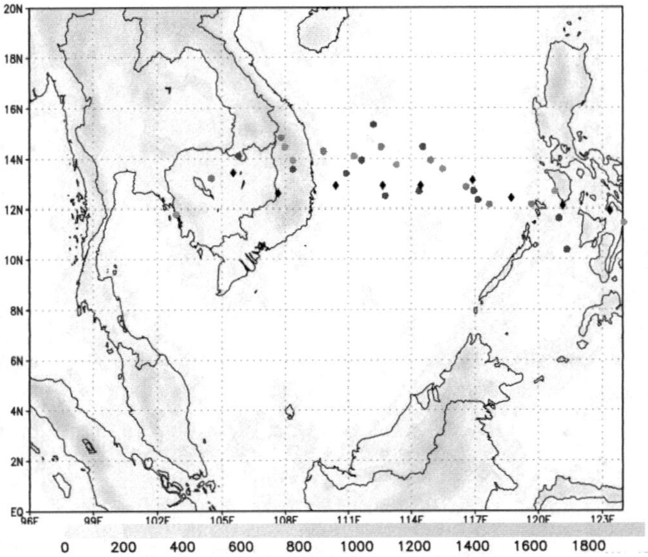

Figure 1. 12-hourly track position of typhoon Damrey including observation (black), Kain-Frinsth (red), Belts-Millers-Janjic (blue), Grell (orange) and new Grell (purple) cumulus scheme.

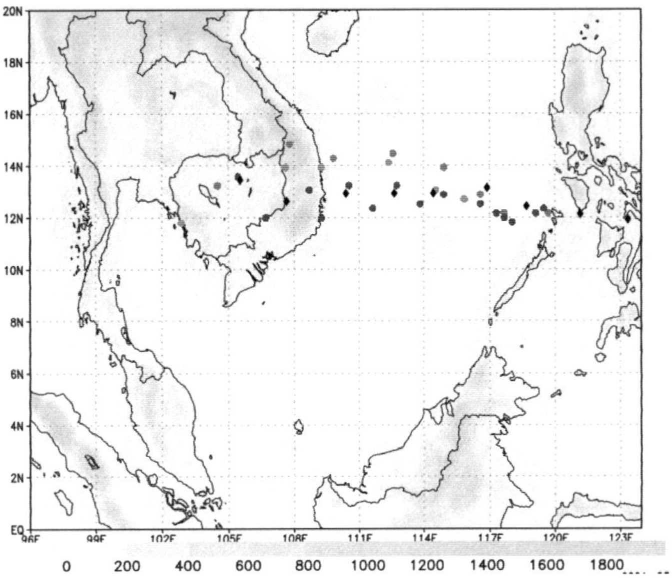

Figure 2. 12-hourly track position of typhoon Damrey including observation (black), Kessler (red), WSM3 (blue), Ferrier (orange) and WSM6 (purple) microphysics scheme.

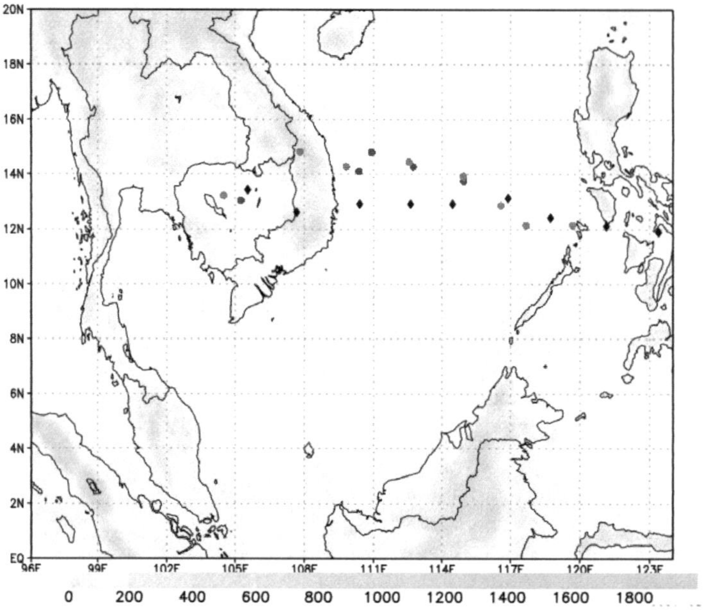

Figure 3. 12-hourly track position of typhoon Damrey including observation (black), Yonsei university (red) and MYNN 2.5 level (blue) PBL scheme.

Acknowledgement
This research would not have been possible without the contribution of Science Achievement Scholarship of Thailand.

References

[1] Yanai M 1964 Formation of tropical cyclones *Rev. Geophys.* **2** 367–414
[2] Skamarock W C, Klemp J B, Dudhia J, Gill D O, Barker D, Duda M G, Huang X Y, Wang W and Powers J G 2008 *A Description of the Advanced Research WRF Version 3* (Boulder, CO: National Center for Atmos pheric Research)
[3] Stensrud D J 2011 *Parameterization Shemes: Keys to Understanding Numerical Weather Prediction Models* (Cambridge: Cambridge University Press)
[4] Islam T, Srivastava P K, Rico-Ramirez M A, Dai Q, Gupta M and Singh S K 2015 Tracking a tropical cyclone through WRF–ARW simulation and sensitivity of model physics *Nat. Hazards* **76** 1473–95
[5] Maw K W and Min J 2017 Impacts of microphysics schemes and topography on the prediction of the heavy rainfall in Western Myanmar associated with tropical cyclone ROANU (2016) *Adv. Meteorol.* **2017** 3252503
[6] Park J, Cha D H, Lee M K, Moon J, Hahm S J, Noh K, Chan J C L and Bell M 2020 Impact of cloud microphysics schemes on tropical cyclone forecast over the Western North Pacific *J. Geophys. Res. Atmos.* **125** e2019JD032288
[7] Rajeswari J R, Srinivas C V, Reshmi Mohan P and Venkatraman B 2020 Impact of boundary layer Physics on tropical cyclone simulations in the bay of Bengal using the WRF model *Pure Appl. Geophys.* **177** 5523–50
[8] Cohen A E, Cavallo S M, Coniglio M C and Brooks H E 2015 A review of planetary boundary layer parameterization schemes and their sensitivity in simulating southeastern U.S. cold season severe weather environments *Weather Forecast.* **30** 591–612

Drought analysis over northern Thailand

Pakpoom Ratjiranukool[1] and Sujittra Ratjiranukool[2,*]

[1] Department of Physics, Faculty of Science and Technology, Chiang Mai Rajabhat University, Muang District, Chiang Mai, 50300, Thailand
[2] Department of Physics and Materials, Faculty of Science, Chiang Mai University, Muang District, Chiang Mai, 50200, Thailand

*Corresponding author's email address : sujittra.r@cmu.ac.th

Abstract. In this research, the Kalman filter method was applied for correcting precipitations simulated by a high-resolution regional climate model named Non-hydrostatic Regional Climate Model (NHRCM) during the period of 1980-1999. The improved average monthly precipitations were close to the stational observations. To reduce systematic error, the Kalman filter method was also applied to simulated monthly precipitations during the future period of 2080-2099. They were analysed to evaluate drought conditions during March-April (out rainy season) and June-July (in rainy season) by using Standardized Precipitation Index, SPI. Preliminary Analysis shows that drought conditions during both periods slightly mitigate. Furthermore, the drought over upper northern Thailand was found in the wettest month during the southwest monsoon period, September. The other months during the monsoon active are wetter than the period of 1980-1999.

1. Introduction
The frequency of occurrence and duration of drought has recently increased over many parts of the world associating with higher greenhouse gas emissions. The economic and environmental impacts of drought become more severe, particularly over the agricultural region. Precipitations below normal conditions over Thailand were found in the middle of the southwest monsoon period due to weakening of the monsoon during mid-June to mid-July, reduction in precipitation amount and wet day also the delay of southwest monsoon onset in dry season during March-April. There are many drought indices to monitor drought and its variation, for example, the precipitation anomaly percentage (PA), effective drought index (EDI), composite drought index (CDI), the Palmer drought severity index (PDSI), standardized precipitation index (SPI). In this work, projected precipitation from a regional climate model under the high emission scenario, Representative Concentration Pathway (RCP) 8.5 was analyzed to evaluate drought associating with climate change using SPI.

2. Materials and methods
2.1. Model simulation
The Non-Hydrostatic Regional Climate Model (NHRCM) has been developed by Meteorological Research Institute (MRI)/Japan Meteorological Agency (JMA). NHRCM was applied to simulate daily precipitation over Thailand during the baseline period of 1980-1999 and the future period of 2080-2099. The model was forced by the 20-km resolution MRI-AGCM3.2 model under RCP 8.5 scenario. The simulation was conducted over the area bound 4.6 °N - 22.1 °N and longitude of 93.7 °E - 107.7 °E at

Journal of Physics: Conference Series **2145** (2022) 012047 doi:10.1088/1742-6596/2145/1/012047

5-km resolution with 50 vertical levels. Parameterization schemes play main role in climate simulation [1]. These are examples in this work, Mellor-Yamada-Nakanishi-Niino level 3 scheme for planetary boundary layer scheme, Kain-Fritsch convective scheme for cumulus convection, and MRI/JMA Simple Biosphere (iSiB) model for land surface model.

2.2. Kalman filter

Kalman filter technique requires linear or non-linear discrete datasets which are not necessary to be a series. Kalman filter generally needs a set of equations to examine an unknown process by combining observation datasets which are related to temporal evolution process [2]. The technique minimises the estimations of mean square error. Considering time t, let M_t be a model simulated parameter that is related to an unknown process and O_t an observed parameter at the same time. The process M changes from time $t-1$ to t and unknown parameter can be examined by the following equation:

$$M_t = A_t \times M_{t-1} + K_t \tag{1}$$

the observation changes from time $t-1$ to t and unknown parameter is examined by the following equation:

$$O_t = B_t \times O_{t-1} + L_t \tag{2}$$

whereas the system and observation parameters which vary with time are coefficient parameters A_t and B_t, respectively which are needed to calculate before applying the filter. K_t and L_t are random parameters which are independent according to the normal distribution.

2.3. The standardized precipitation index (SPI)

The standardized precipitation index (SPI) is used widely to classify drought or wetness for any timescale up to 36 months, i.e., monthly, seasonally, annually period [3]. It requires monthly rainfall as input data for computing accumulated rainfall which is fitted to a gamma probability density function defined by

$$g(x) = \frac{1}{\beta^{\alpha}\Gamma(\alpha)} x^{\alpha-1} e^{-x/\beta} \tag{3}$$

where $\alpha > 0$, α is shape parameter, $\beta > 0$, β is scale parameter and $x > 0$, x is precipitation. The gamma function, $\Gamma(\alpha)$, is defined as:

$$\Gamma(\alpha) = \int_0^{\infty} y^{\alpha-1} e^{-y} dy \tag{4}$$

where the parameters can be calculated as: $\hat{\alpha} = \frac{1}{4A}\left(1 + \sqrt{1 + \frac{4A}{3}}\right)$, $\hat{\beta} = \frac{\bar{x}}{\hat{\alpha}}$ and $A = \ln(\bar{x}) - \frac{\sum \ln(x)}{n}$

n is number of precipitations. The gamma function is undefined for $x = 0$, however, monthly cumulative simulated precipitations are over 0 mm.

The cumulative density function is defined as:

$$G(x) = \int_0^{x} g(x)dx = \frac{1}{\hat{\beta}^{\hat{\alpha}}\Gamma(\hat{\alpha})} \int_0^{x} x^{\hat{\alpha}-1} e^{-x/\hat{\beta}} dx \tag{5}$$

where $t = \frac{x}{\hat{\beta}}$ so that $G(x) = \frac{1}{\Gamma(\hat{\alpha})} \int_0^{t} t^{\hat{\alpha}-1} e^{-t} dt$

The cumulative density function refers to the probability of precipitation event is less than or equal to the specified precipitation. Cumulative density function then subsequently transformed to a normal distribution which is SPI [4]. The drought classifications of SPI were shown in table 1.

Table 1. SPI classifications from extremely wet to extremely drought.

SPI Value	Drought Category
2.00 and above	Extremely wet
1.50 to 1.99	Severely wet
1.00 to 1.49	Moderately wet
0 to 0.99	Mild wet
-0.99 to 0	Mild drought
-1.00 to -1.49	Moderately drought
-1.50 to -1.99	Severely drought
-2.00 and less	Extremely drought

3. Results and discussion

It was found that the average 20-year simulated precipitation at some station points was up to 50% lower than the observation during the southwest monsoon period. By correcting the precipitation with Kalman filter, the root mean square error of the corrected precipitation was reduced by 80%. The annual cycle of precipitation at Chiang Mai and Chiang Rai was shown in figure 1. Monthly projected SPI analysis reveals that there was drought differentiate between an individual station and the average of all stations (figure 2). The probability of occurrence of drought at an individual station during March-April and June-July indicates that mild wet and mild drought were very likely found in the future (figure 3). The spatial SPI classifications of the 13 meteorological stations for the period of 2080-2099 were interpolated using Inverse Distance Weighting (IDW) interpolation technique in ArcGIS to cover northern Thailand. The higher risk of drought over the region was found during June-July (figure 4).

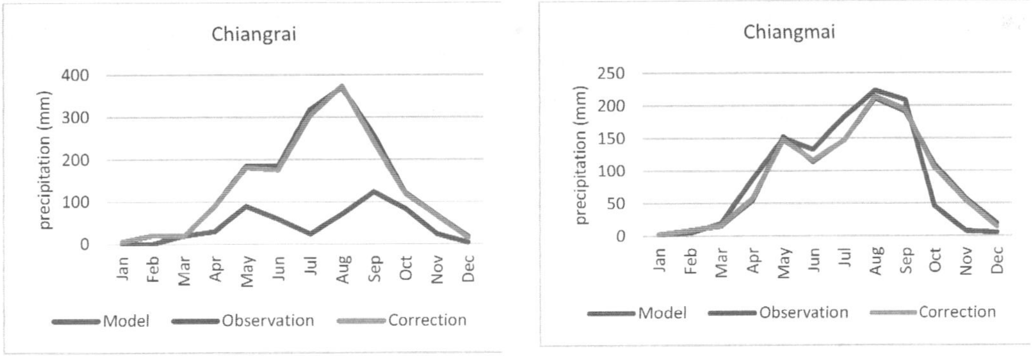

(a) (b)

Figure 1. Averaged monthly simulated, observed, and corrected precipitation at (a) Chiang Rai and (b) Chiang Mai station during 1980-1999.

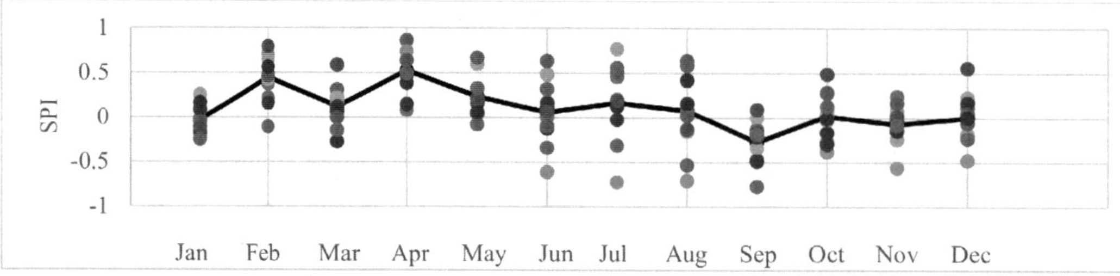

Figure 2. Average monthly SPI at 12 meteorological stations over northern Thailand (2080-2099). The black line is average SPI of all station.

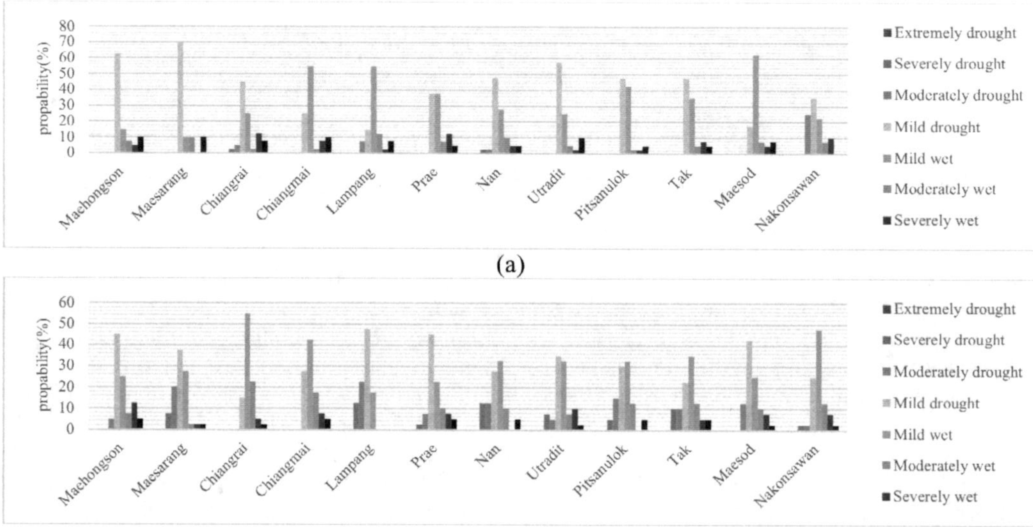

(a)

(b)

Figure 3. Drought probability at 12 meteorological stations over northern Thailand (2080-2099) during (a) March-April and (b) June-July.

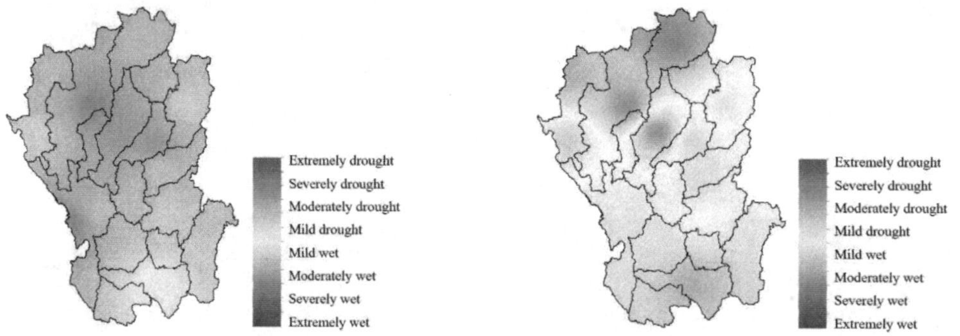

(a) (b)

Figure 4. Spatial wet and dry condition classified by SPI over northern Thailand (2080-2099) during (a) March-April and (b) June-July.

Drought depends on various environmental and meteorological parameters such as precipitation, surface runoff, groundwater, soil moisture, and evapotranspiration. Suggestion for further study would be Palmer Drought Severity Index using temperature, precipitation, and soil moisture.

4. Conclusion

It was found that drought condition in the future (2080-2099) remains the same as baseline period (1980-1999) over most of northern Thailand. Mild wetness was found from March to April which implied more frequently tropical storms or early southwest monsoon onset. Mild drought conditions across the region were found during southwest monsoon active, June to July, which implied weaker monsoon.

Acknowledgement

The calculations were conducted by Meteorological Research Institute (MRI) supercomputer system support by the TOUGOU Program of the Ministry of Education, Culture, Sports, Science and Technology (MEXT) of Japan. This work was also supported by Faculty of Science, Chiang Mai University and Faculty of Science and Technology, Chiang Mai Rajabhat University.

References

[1] Cruz F T, Sasaki H and Narisma G T 2016 Assessing the sensitivity of the non-hydrostatic regional climate model to boundary conditions and convective schemes over the Philippines *J. Meteorolo. Soc. Japan* **94** 165–79

[2] Houtekamer P L and Mitchell H L 2001 A sequential ensemble Kalman filter for atmopheric data assimilation *Mon. Weather Rev.* **129** 123–37

[3] Guenang G M and Kamga F M 2014 Computation of the Standardized Precipitation Index (SPI) and its use to assess drought occurrences in Cameroon over recent decades *J. Appl. Meteorol. Climatol.* **53** 2310–24

[4] Shah R, Bharadiya N and Manekar V 2015 Drought index computation using Standardized Precipitation Index (SPI) method for Surat District, Gujarat *Aquat. Procedia* **4** 1243–9

Evaluation of projected decadal wind energy potential in Chiang Mai

Pakpoom Ratjiranukool[1] and Sujittra Ratjiranukool[2,*]

[1] Department of Physics, Faculty of Science and Technology, Chiang Mai Rajabhat University, Muang District, Chiang Mai, 50300, Thailand
[2] Department of Physics and Materials, Faculty of Science, Chiang Mai University, Muang District, Chiang Mai, 50200, Thailand

*Corresponding author's email address : sujittra.r@cmu.ac.th

Abstract. One of the alternative energy sources which have globally pay more attention is wind power. In this research, the Weibull wind speed distribution was applied to evaluate wind energy potential during the period of 2081-2090. The projected wind datasets in this study were simulated by the Non hydrostatic Regional Climate Model (NHRCM). The boundary condition was the 20 km resolution MRI Atmospheric General Circulation Model (AGCM20) under the RCP8.5 scenario. The surface wind datasets in Chiang Mai were analyzed to 3-time intervals, i.e., May-August, September-December and January-April. Two averaged Weibull distribution parameters, i.e., k, shape parameter, and c, scale parameter were determined. The shape and scale parameters fluctuated 1.15 to 1.34 and 0.38 m/s to 0.74 m/s, respectively. It was found that the surface mean wind speed from May through August was stronger than from September through December.

1. Introduction

Wind energy is one of the sustainable energy sources which is dynamic and depends on weather conditions. It can be applied for small-scale usages such as battery charging, water pumping to industrial scales such as wind power plants. Wind speeds and directions in a local area may differ from the general circulation of the atmosphere. Wind speed plays an important role in wind energy potential. To evaluate wind energy potential on a regional scale, the wind speed probability distribution must be determined accurately. History record of observed wind speed may be evaluated wind potential for the current climate at a certain stationary location.

A regional climate model is a tool for simulating wind speed at any point over the region. It can be done either over the current climate to investigate the new points where possibly provide high wind energy potential or evaluate the potential of wind energy resources in the future concerned with climate change effect. There are regional and global climate models provided by various institutes for simulating climate parameters, including wind speed. Those models have been developing to improve model performances to understand past climate and project future climate with higher degrees of confidence. A higher resolution climate model is commonly run with more detailed processes to provide surface climate parameters in probability density function and extreme study. Using realistic RCM to project surface wind resource is limited.

Some regions, for example, Thailand, are affected by the prevailing winds. During the southwest monsoon period, mid-May to mid-October, the wind blows in from the Bay of Bengal and reaches the western part of the mountainous area in Thailand. During the northeast monsoon period, November to January, the wind comes from cold and dry higher latitudes. In this work, the surface wind projected by high-resolution RCM energy was statistically analyzed using the Weibull distribution to evaluate wind energy potential in Chiang Mai, Thailand. The finding of this study possibly provides more information to quantitative wind speed distribution at the end of this century.

2. Materials and methods

2.1. Simulation of wind data

The decadal wind datasets in Chiang Mai were collected from the projected climate simulation of the Non-hydrostatic Regional Climate Model (NHRCM), a high-resolution climate model developed by the Japan Meteorological Agency. The regional model was driven by the 20 km resolution MRI Atmospheric General Circulation Model (AGCM20) under the high emission scenario, Representative Concentration Pathway (RCP) 8.5. The model simulation was set in 5 km resolution which the mountainous and complex topography was applied in the simulation. The selected domain covers Indochina, part of the South China Sea and Bay of Bengal which has land and sea interaction impact. The Initial time condition was selected at the beginning of April when it was in the period of the dry season to avoid the atmosphere and land moisture effects. Parameterization schemes are Kain-Fritsch scheme for cumulus convection and Mellor-Yamada for Planetary Boundary Layer [1]. One month simulation was abandoned as model spin-up. Daily wind simulations for the whole period of 2081-2090 were manipulated by using the nearest grid point method to collect wind datasets in Chiang Mai for wind potential energy analysis.

2.2. Weibull distribution

The probabilistic model of wind speed Weibull distribution was commonly analyzed to study wind potential at different locations. Two functions, i.e., the probability density function (PDF) and the cumulative distribution function (CDF), were also determined in this study [2]. The Weibull density probability function, $f(v)$ expresses the probability of the wind or the frequency of occurrence which has a certain speed (v) in m/s during the selected period [3]. It can be determined by equation (1).

$$f(v) = \frac{k}{c}\left(\frac{v}{c}\right)^{k-1} e^{-\left(\frac{v}{c}\right)^k} \tag{1}$$

where k is the nondimensional shape parameter and c is the scale parameter [4, 5].

The cumulative distribution function of the velocity v, $F(v)$, implied the probability of the wind which below a certain speed. It can be determined by integrating PDF as shown in equation (2).

$$F(v) = \int_0^{\infty} f(v)dv = 1 - e^{-\left(\frac{v}{c}\right)^k} \tag{2}$$

The k and c parameters can be determined by the graphical method. Firstly, The CDF can be rewritten as the following equation.

$$1 - F(v) = e^{-\left(\frac{v}{c}\right)^k}$$

Using logarithm transformation, it can be rewritten as a linear equation.

$$\ln\{-\ln[1 - F(v)]\} = k\ln(v_i) - k\ln c \tag{3}$$

By plotting the linear regression between $\ln(v_i)$ as x-axis and $\ln\{-\ln[1 - F(v)]\}$ as y-axis. It is clearly seen that k is the slope and $-k\ln c$ is the y-intercept. The small value of k parameter indicates that the behavior distribution of wind speed reduces to an exponential distribution. On the other hand, the increased shape parameter gradually transforms wind speed distribution to normal distribution. The scale parameter, c , corresponds to wind speed distribution around the average. Increasing the value of c parameter affects increasing wind speed variability while maintaining its shape.

3. Results and discussion

The surface wind speeds and directions in Chiang Mai during 2081-2090 were collected from the NHRCM simulation under RCP 8.5 scenario. The data were analyzed in 3-time intervals, i.e., May-August, September-December, and January-April. The Weibull shape and scale parameter were determined by the graphical method shown in table 1. The minimum value of k implies that the smallest wind speed distribution around the average during the southwest monsoon active period was found in the future.

The calculation of k and c were applied to display the Weibull distribution and cumulative Weibull distributions. The Weibull probability density functions indicated the frequency of surface wind speed were shown in figure 1 (a). The graph displays the most prevalent wind speed during three-time intervals. The most prevalent wind speed was 0.1 m/s during September-December and January-April, while the range of 0.2-0.4 m/s was found as the most prevalent wind speed during May-August. It is suggested that the falling frequency indicates less probability at a certain wind speed. In the future, the probability of occurrence of the wind speed beyond 2.4 m/s was less in Chiang Mai.

The mean wind speed variations for the future period ranged from 0.4 m/s during September-December to 0.8 m/s during May-August (figure 2). The strongest in monthly mean wind speeds were found in June, July, February, and March, respectively, while the weakest wind speeds were found in November and December, respectively (data not shown here). It implies that the wind speed during the middle of the monsoon active period is stronger than the other period. On the other hand, the periods excluding the wet season have the least maximum mean wind speed.

The Weibull cumulative distribution functions were shown in figure 1 (b). CDF analysis revealed that at Chiang Mai, 50% of the data ranged from 0.25 to 0.6 m/s and below, while 80% of the data ranged from less than 0.6-1.2 m/s, across the period considered. The wider variation of wind speeds was also found above 50% of the data. From this analysis, it was found that wind speed below 1.7 m/s (1.45 m/s) was more prevailing at the probability of 95% occurrence during the southwest (northeast) monsoon active.

Table 1. The shape and scale parameter at Chiang Mai.

time interval	**k**	**c**
May-August	1.34	0.74
September-December	1.15	0.38
January-April	1.20	0.55

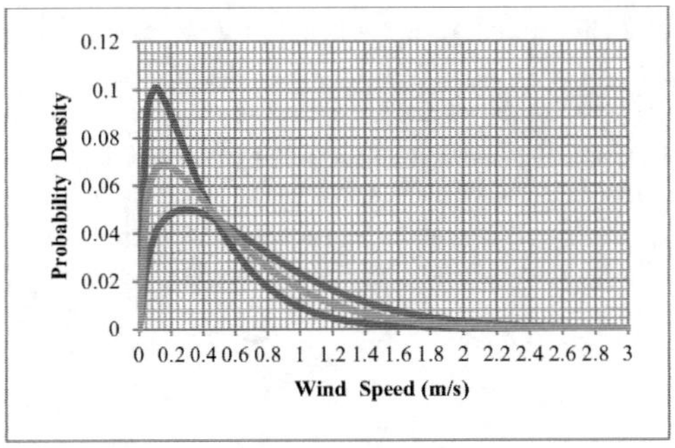

(a)

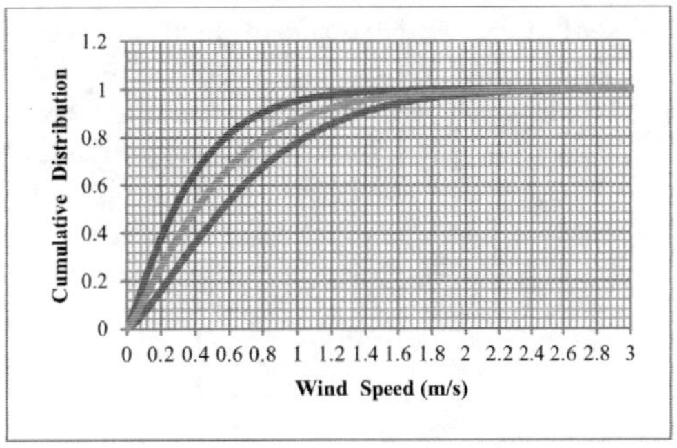

(b)

Figure 1. (a) Weibull probability distribution functions (b) cumulative distribution functions at Chiang Mai during May-August (blue), September-December (red) and January-April (green).

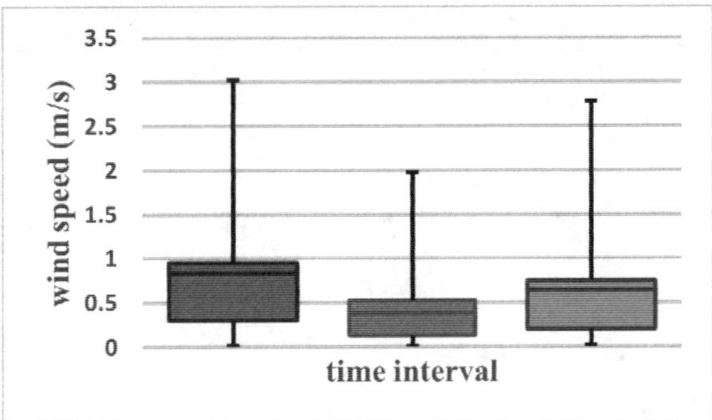

Figure 2. Box plot of wind speed during May-August (blue), September-December (red) and January-April (green). The line within the box plots indicates the median value, while the box bottom and top show the 25th and 75th percentile, respectively. Whiskers indicate the maximum and minimum.

4. Conclusion

The Weibull probability density functions of wind speeds were the right-skewed distributions during all three intervals. It was found that 90% of total wind speeds were less than 1.3, 0.5, and 1.1 m/s during May-August, September-December, and January-April, respectively. It could be noticed that the wind speeds at Chiang Mai in the future during May to August were higher than other periods.

Acknowledgement

The model simulation was support by the TOUGOU Program of the Ministry of Education, Culture, Sports, Science and Technology (MEXT) of Japan. The calculations were conducted by The Meteorological Research Institute (MRI) supercomputer system. This work was also supported by Faculty of Science, Chiang Mai University and Faculty of Science and Technology, Chiang Mai Rajabhat University.

References

[1] Cruz F T and Sasaki H 2017 Simulation of present climate over Southeast Asia using the non-hydrostatic regional climate model *Sci. Online Lett. Atmosphere* **13** 13–8

[2] Mekalathur B H, Balasubramanian S, Sanjeevikumar P and Holm-Nielsen J 2019 Wind potential assessment by Weibull parameter estimation using multiverse optimization method: A case study of region in India *Energies* **12** 2158

[3] Gul M, Tai N, Huang W, Nadeem H and Yu M 2020 Evaluation of wind energy potential using an optimum approach based on maximum distance metric *Sustainability* **12** 1999

[4] Mouangue R M, Kazet M Y, Kuitche A and Ndjaka J M 2014 Influence of the determination methods of k and c parameters on the ability of Weibull distribution to suitably estimate wind potential and electric energy *Int. J. of Renew. Energy Dev.* **3** 145–54

[5] Carrillo C, Cidrás J, Díaz-Dorado E and Obando-Montaño A F 2014 An approach to determine the Weibull parameters for wind energy analysis: The case of Galicia (Spain) *Energies* **7** 2676–700

The archaeomagnetic field recorded in ancient kiln walls in Si Satchanalai, Sukhothai

R Supakulopas*

Division of Physical Science, Faculty of Science, Prince of Songkla University, Hat Yai, Songkhla, 90110, Thailand

*Corresponding author's email address: radchagrit.s@psu.ac.th

Abstract. Archaeological dating is crucial in archaeology as it is a key to understand human history. However, traditional dating methods used by archaeologists such as potassium-argon dating and luminescence dating can provide ambiguous age results, e.g., argon loss during the dating returns young apparent ages. Therefore, I plan to establish an archaeomagnetic secular variation (ASV) curve to resolve this problem and use the ASV curve as an alternative tool to date archaeological artefacts. However, archaeomagnetic data in Thailand are absent from literature. Therefore, the ASV curve cannot be constructed from the archaeomagnetic data for this locality. To provide archaeomagnetic data to construct the ASV curve, the directions of the Earth's magnetic field recorded in kiln walls from Ban Ko Noi (KN123, age 1,370 ± 100 A.D.), Si Satchanalai were measured. The mean declination and inclination of 49.6° and 32.6° with 95% confidence limit of 5.4° were determined from 10 samples from kiln KN123. Mean directions from this study were also compared with the directions of the Earth's magnetic field in Thailand during 1,370 A.D. from the global archaeomagnetic field model ARCH3k.1. Declination and inclination from this study show significant departure from the field predicted by the ARCH3k.1 model.

1. Introduction

Archaeological dating methods used by archaeologists in Thailand include radiometric dating [1, 2] and luminescence dating [3, 4]. Despite their robustness, these methods still have some limitations. For example, a large amount of historically materials is used in the radiocarbon dating process [5]. The direct determination of ages cannot be made directly from artefacts such as baked clay or ceramic, but instead peat buried in the same soil layer with the artifacts. Moreover, as we are in a carbon-rich environment, archaeological artefacts can be contaminated by current-day carbon. This leads to a significant error of the samples' ages. With regards to the luminescence dating, the advantage of this method is that the determination of ages can be measure directly from archaeological artefacts. However, this method is considerably time consuming due to sample preparation processes. Furthermore, the data can be disturbed by noises during the dating procedure [3].

Archaeomagnetic dating is considered as an alternative dating of historical artefacts. There are two types of archaeomagnetic dating methods: secular variation (SV) dating [6] and magnetic reversal dating [7]. The first method relies on the directional and intensity changes of the Earth's magnetic field over time. The directional and intensity dating are the local pattern-matching method with the accuracy of ±25 years. This method is suitable for dating of 10-ka samples or younger. The magnetic reversal dating, known as magnetostratigraphy, is based on the global pattern matching of the normal and reverse polarity of the Earth's magnetic field. As the polarity of the magnetic field reverses every ~0.25 Ma, and the current Matuyama reversal chron is ~0.78-2.6 Ma [8, 9], therefore, this method is used to date samples older than 0.78 Ma with the accuracy of ±0.01 Ma [7].

In this proposed research, I focus on the archaeomagnetic secular variation (ASV) dating, which can be considered as a derivative dating method [10]. The technique relies on the Earth's magnetic field trapped in archaeological samples during the first firing [11]. As the geomagnetic field is a vector, both the direction and the intensity of the field are recorded in the artefacts and can be measured directly from the samples [12]. The variations of the direction and intensity, hereafter palaeodirection and palaeointensity, that change over time recorded in the artefacts can be used to construct and calibrate archaeomagnetic secular variation (ASV) curves [6]. The archaeological dating using the ASV curves is a suitable method for dating artefacts such as baked clay, slag, kilns and potteries [13-16]. This method reduces the age error gaps and considerably less time consuming, e.g., measurements of palaeodirection of two samples can be done in 40 minutes. However, no ASV curve in Southeast Asia has been calibrated and constructed, rendering the technique not applicable to date archaeological artefacts in the region. Therefore, the aim of this research is to collect palaeomagnetic data to construct the ASV curve, beginning from the 13th – 17th century samples in Si Satchanalai and Sukhothai, Thailand.

2. Fieldwork in Si Satchanalai, sample preparation and palaeomagnetic measurement

The fieldwork campaign to collect pottery and kiln wall samples was held in October 2020 at Ban Ko Noi and Ban Pa Yang in Si Satchanalai (17.476284°N, 99.755171°E), and Sukhothai historical park in Mueang Sukhothai (17.0326621°N, 99.697276°E). Samples were drilled using a portable electric drilling machine equipped with a diamond drill bit and a water cool system. Samples were orientated and measured azimuths and dips using an orientator before taking the cores out of the drilled holes. Approximately 1-cm diameter cores were drilled from the kiln walls. In case the kiln walls are fragile, brick samples were taken by an axe and a hammer. Before taking brick samples, samples were orientated using a Brunton compass. The Sun compass was not measured in this study as all sites are under the roofs. A total of 33 cores and 18 bricks were collected covering the time spanning ~13th-17th centuries. Then, samples were shipped to the palaeomagnetic laboratory, Prince of Songkla University for preparations and palaeomagnetic measurements. In sample preparation processes, brick samples were cut into 2-cubic centimeter cube using a hand saw, yielding ~6-10 specimen per brick. The natural remanent magnetisation (NRM) directions including declination and inclination were measured using an AGICO JR6 spinner magnetometer. Note that the standard demagnetisations: alternating field (AF) and thermal demagnetisations were not performed here.

3. Results and discussions

The palaeomagnetic directional data obtained from a kiln (KN123, age 1,370 ± 100 A.D. [17]) in Ban Ko Noi, Si Satchanalai are presented here. Declination and inclination data from 10 specimens from kiln KN123 were plotted on an equal area projection (figure 1(a)). The sample mean directions were calculated using the Fisher statistics [18], yielding the mean declination and inclination of 49.6° and 32.6° with the 95% confidence limit (α_{95}) of 5.4°. Overall, the data are tightly cluster on the sphere.

The mean directions were compared with the global archaeomagnetic field model during the past 3 ka, ARCH3k.1 [19]. The ARCH3k.1 predicts the inclination of ~18° with the confidence limit of 3.4° for the site locality during 1,370 A.D. (figure 1(b)). The inclination data deviate from the ARCH3k.1 model with no overlap between the confidence error. With regards to the declination data, the declination data show higher deviation from the ARCH3k.1 model during 1,370 (dec = -1.9° with the confidence error of 3.2°) than the inclination data do (figure 1(a)).

As the mean directions are tightly cluster with the precision parameter $k \geq 50$, and were calculated from 10 specimen per site (the number of specimen per site $N = 10$), it is clear that the data meet the modern site selection criteria of the palaeomagnetic community ($k \geq 50$ and $N \geq 5$) [20]. High departure of the data from the model may be from these reasons. (1) The ARCH3k.1 model was constructed using the palaeomagnetic data during the past 3 ka. Most of the data are from Europe. Minor data are from North America, South America and East Asia. Therefore, the model may have the bias toward Europe. (2) The samples may acquire secondary overprint such as isothermal remanent magnetisation (IRM) induced by lightning strike which occurs commonly in tropical region. Therefore, samples may be partially remagnetised or fully demagnetised by the IRM and the secondary overprint deviates the NRM directions that samples acquired during the time of the kiln production.

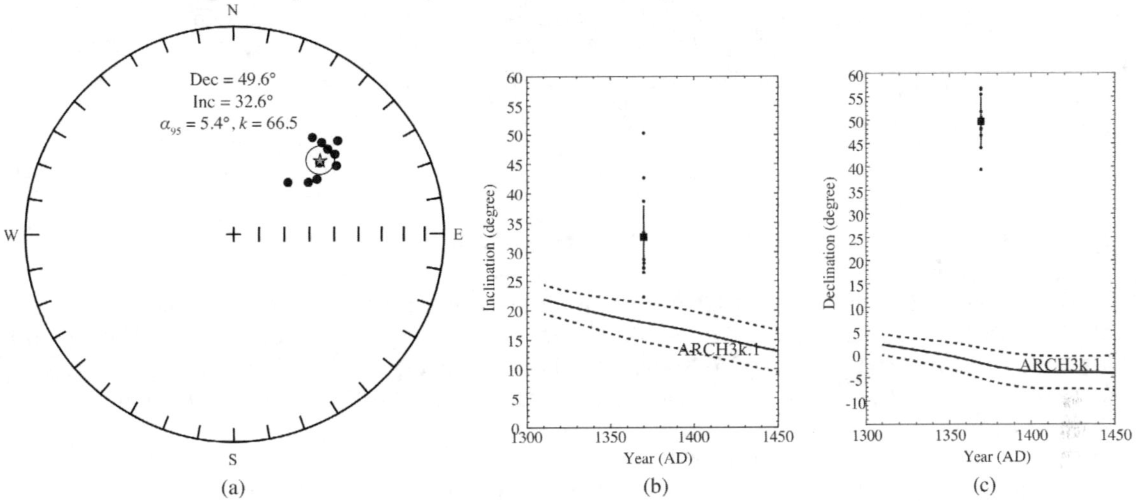

Figure 1. (a) Equal area projection of the NRM directions (black circles) with the site mean direction (yellow star) and 95% confidence cone (circle). (b) and (c) represent the plots of mean inclination and declination (big squares) with the 95% confidence boundaries against the ARCH3k.1 model (black line) with the upper and lower boundaries (dash lines). Black circles show directions from individual specimen.

4. Conclusion

The palaeomagnetic directions from the kiln KN123 show large deviations from the archaeomagnetic field model ARCH3k.1. To examine whether samples acquire secondary overprint after the production time, samples should be fully demagnetised using the standard demagnetisation protocols including stepwise AF and thermal demagnetisations to isolate the characteristic remanent magnetisation (ChRM) acquired during the kiln production time. Rock magnetic properties including hysteresis loops, backfield curves and thermomagnetic curves should also be measured to understand magnetic domain grains, types of magnetic minerals and its Curie temperature. The domain grain information provides coercivity data of magnetic minerals in kilns and confirms whether samples easily acquire secondary overprint. Besides, the palaeomagnetic intensity measurements will be performed to understand the behaviour of the palaeomagnetic field during the $13^{th} - 17^{th}$ in Southeast Asia.

Acknowledgement

Radchagrit Supakulopas would like to express appreciation to the Faculty of Science, Prince of Songkla University for funding this research under the project grant number SCI6304004S, and archaeologists and staff at Si Satchanalai historical park and Sukhothai historical park for fieldwork support.

References

[1] Sako T 2019 *Ancient Maritime Cross-culture Exchanges: Archaeological Research in Thailand* (Bangkok: Rung Silp Printing Co. Ltd.) pp 167–91

[2] Thongaram B 2019 *Ancient Maritime Cross-culture Exchanges: Archaeological Research in Thailand* (Bangkok: Rung Silp Printing Co. Ltd.) pp 193–207

[3] Puttagan T, Chawchai S, Surakiatchai P, Chalorsantisakul S and Preusser F 2019 *Archaeol. Anthropol. Sci.* **11**(10) 5393–403
[4] Pailoplee S, Won-in K, Chaisuwan B and Charusiri P 2016 *Songklanakarin J. Sci. Technol.* **38**(6) 699–705
[5] Cersoy S, Zazzo A, Rofes J, Tresset A, Zirah S, Gauthier C, Kaltnecker E, Thil F and Tisnerat-Laborde N 2017 *Sci. Rep.* **7**(1) 7141
[6] Sternberg R S 1997 *Chronometric Dating in Archaeology* (Boston: Springer) pp 323–56
[7] Barr S M, Macdonald A S, Haile N S and Raynolds P H 1976 *J. Geol. Soc. Thailand* **2**(1) 1–10
[8] Ogg J G 2012 *The Geologic Time Scale* (Boston: Elsevier) pp 85–113
[9] Haneda Y, Okada M, Suganuma Y and Kitamura T 2020 *Prog. Earth. Planet. Sci.* **7** 1–22
[10] Aitken M J 1990 *Science-Based Dating in Archaeology* (London: Longman)
[11] Cai S, Jin G, Tauxe L, Deng C, Qin H, Pan Y and Zhu R 2017 *Proc. Natl. Acad. Sci. U.S.A.* **114**(1) 39–44
[12] Tauxe L, Butler R F, Van der Voo R and Banerjee S K 2010 *Essentials of Paleomagnetism* (Berkeley: University of California Press)
[13] Donadini F, Motschi A, Rösch C and Hajdas I 2012 *J. Archaeol. Sci.* **39**(7) 2153–66
[14] Ech-Chakrouni S, Hus J and Marchal J P 2015 *Studia Geophysica et Geodaetica* **59**(4) 578–93
[15] Shin S, Park Y H, Cheong D and Shin S C 2018 *Geosciences Journal* **22**(1) 1–9
[16] Stillinger M D, Hardin J W, Feinberg J M and Blakely J A 2016 *Near Eastern Archaeology* **79**(2) 90–106
[17] Barbetti M and Hein D 1989 *World Archaeololgy* **21**(1) 51–70
[18] Fisher R 1953 *Proceedings of the Royal Society of London Series* A **217**(1130) 295–305
[19] Korte M, Donadini F and Constable C G 2009 *Geochem. Geophys. Geosyst.* **10**(6) Q06008
[20] Johnson C L *et al.* 2008 *Geochem. Geophys. Geosyst.* **9**(4) Q04032

Integrated studies to investigate paleochannel aquifer in Dan Chang District, Suphan Buri Province, Western Thailand

S Muangnoi[1], N Chaimanee[2] and P Pananont[1,*]

[1] SEIS-SCOPE, Department of Earth Sciences, Faculty of Science, Kasetsart University, 10900, Thailand
[2] Department of Marine and Coastal Resources, 10210, Thailand

*Corresponding author's email address: fscipkp@ku.ac.th

Abstract. Climate change is currently causing droughts in many parts of Thailand, especially in Dan Chang District, Suphan Buri Province where many areas are outside irrigation coverage and deep groundwater explorations are less successful due to groundwater shortages. Therefore, there is a need to explore other shallow groundwater (i.e. less than 15 m deep) resources to relieve the drought problems in the communities. This work uses integrated studies of remote sensing, geology and geophysics to identify the location of a paleochannel that contains shallow groundwater resources in the Nong Makha Mong Subdistrict, Dan Chang District, Suphan Buri Province. Potential sites were selected using preliminary remote sensing analysis along with field surveys. A high-resolution topography map from aerial photos was generated and 2D resistivity imaging surveys were conducted at the selected sites to delineate the location of the paleochannel. The drilled wells prove the success of the integrated study to identify the shallow groundwater in the paleochannel at a depth of 3 to 15 m with a groundwater yield of 4 m^3/h. The sediments from the borehole suggest that a combination of alluvial deposits and fluvial sediments. Due to the relative lower cost of drilling shallow wells, this work could be used as a pilot project for local communities to explore shallow groundwater aquifers in paleochannels in areas that face a severe drought crisis and have very little deep groundwater potential.

1. Introduction

Climate change is presently causing droughts in many parts of Thailand especially in the Western Thailand where irrigation systems are inaccessible and the decrease of rainfall. As a result, the local communities, especially in the agricultural areas have been suffered by water shortages. Groundwater is considered one of the important choices to alleviate water shortage problems, furthermore during the dry season which is the beginning of the cultivation season for farmers.

A buried paleochannel is an old water passage such as a channel or river that existed in the past that has been covered by the current sediments. Paleochannel has the potential to store shallow groundwater that can be used as an alternative water resource. A typical paleochannel can be distinguished by its low width to length ratio and can be tortuous in plain view [1]. Although paleochannels can be obscured on the subsurface due to covert by sediments or being transformed by human activities [2], they can be identified using various techniques. Aerial photos, digital elevation model (DEM) and other remote sensing data can be used to determine the preliminary locations of paleochannels [3]. Geophysical methods can then be applied to pinpoint the exact location of the paleochannels and to delineate their geometries such as width, length and depth. Several geophysics methods can be applied to determine the location and define the geometries of the paleochannels such as 1D and 2D electrical resistivity surveys, electromagnetic surveys [3], ground penetrating radar surveys (GPR) [4] or seismic refraction surveys [5].

As the study area has a geomorphology that indicates abundant paleochannels, we are interested in exploring the shallow groundwater resources from these buried paleochannels in Nong Makha Mong Subdistrict, Dan Chang District, Suphan Buri Province, Western Thailand. We integrate various techniques such as geology and remote sensing to preliminarily determine the locations of possible buried paleochannels and then we conduct a geophysical survey such as the two-dimensional resistivity imaging survey in the selected sites to delineate the precise locations of the paleochannels. Therefore, the shallow groundwater can be extracted.

2. Geological setting

Dan Chang District is located to the east of Suphan Buri Province and is portrayed as a gradually sloping basin from the west to the east. The geology of Dan Chang District consists of the Ordovician (O) limestone with partly marble and calc-silicate. There are also phyllite and slate (SD) and colluvial deposits of the Quaternary (Qc) and rock fragments and clay ([6]: figure 1). This area is considered as the upstream area of the Mae Klong river basin and the Tha Chin river basin with the Tanaosri mountain range to the west and there is considered one of the rain shadow areas of Thailand. As a result, the average rainfall in Dan Chang District is less than the average rainfall in Suphan Buri Province.

The Krasiao reservoir is one of the major water resources in the region (figure 1) but most communities in Dan Chang District have very little benefit from the Krasiao reservoir because the streams from the reservoir flow out of Dan Chang District into other Districts of Suphan Buri. Therefore, most communities in Dan Chang District have had ongoing problems with water shortages and they have to use deep groundwater for their needs instead.

The groundwater resources are widely developed in Dan Chang District, the confined aquifers are averagely more than 40 m deep with a limited groundwater yield of 1 to 10 m^3/h. In the Nong Makha Mong Subdistrict in Dan Chang District, there are many small streams and paleochannels that could have been buried due to the changes in sedimentary deposition environment and land uses in the past and can be potentially good shallow groundwater resources.

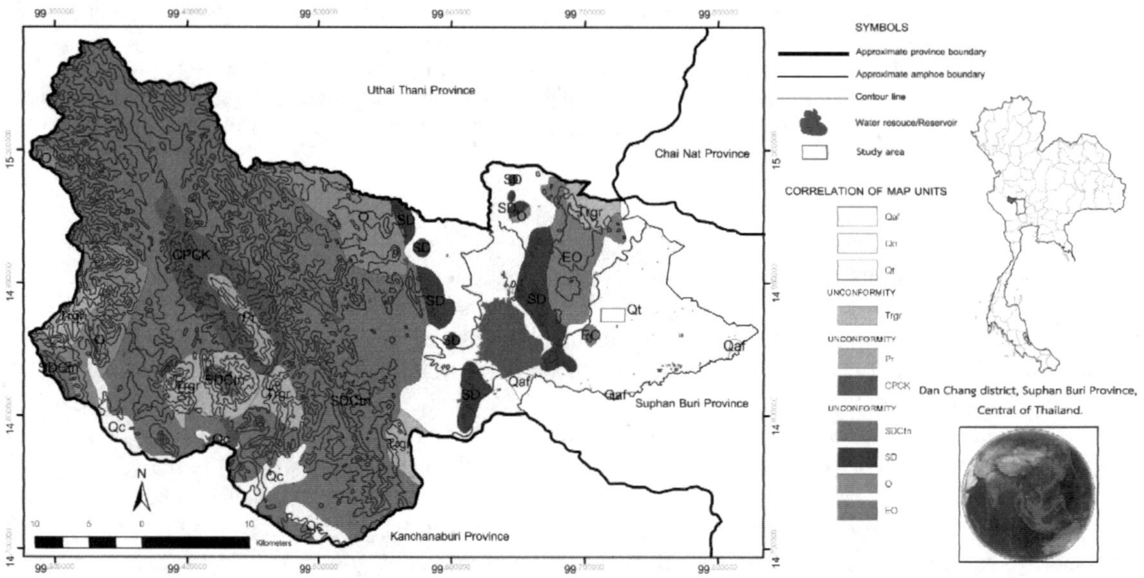

Figure 1. Geology of Dan Chang District, Suphan Buri Province [6] and location of study area.

3. Methodology

3.1. Identification of Potential Paleochannels

A paleochannel is an old channel that consists of unconsolidated sediments deposited in the past or they could be inactive rivers or channel systems [2]. These infilled deposits consist of clay and silt on the top with interbedded sand and clayey sand. The sequence of infilled sediments expresses the geologic history [7].

This work used aerial photos from Google Earth ® time-lapse to identify gross locations, directions and orientations of paleochannels in Nong Makha Mong Subdistrict, Dan Chang District, Suphan Buri Province, Western Thailand. The high-resolution satellite imageries show the different colors of various vegetations that can be used to preliminarily identify the locations of potential paleochannels in the study area (figure 2).

The example high-resolution aerial photos from Google Earth ® in figure 2 show the different colors of various vegetations which indicate the possible locations and directions of paleochannel at various times. Over time, the study area has been transformed by human activities to the agricultural areas. These aerial photos interpretations can be used to identify the preliminary locations of potential paleochannels in the study area.

Figure 2. Examples of interpretations of time-lapse aerial photos from Google Earth ® for the selected site to locate the preliminary potential paleochannels for further investigations. The yellow and blue dashed lines show the boundaries of paleochannel.

3.2. 2D resistivity imaging surveys

2D resistivity imaging survey is based on the principle that the distribution of electrical potential in the ground around a current-carrying electrode depends on the subsurface electrical resistivity and distribution of the surrounding soils and rocks [8].

2D resistivity imaging surveys with the Wenner-Schlumberger electrode configuration (48 to 72 electrodes with 5 m electrode spacing) for a total length of 1,415 m (figure 3) were performed and the apparent resistivity can be calculated by using equation (1) given in ([9]: figure 4).

Journal of Physics: Conference Series **2145** (2022) 012050 doi:10.1088/1742-6596/2145/1/012050

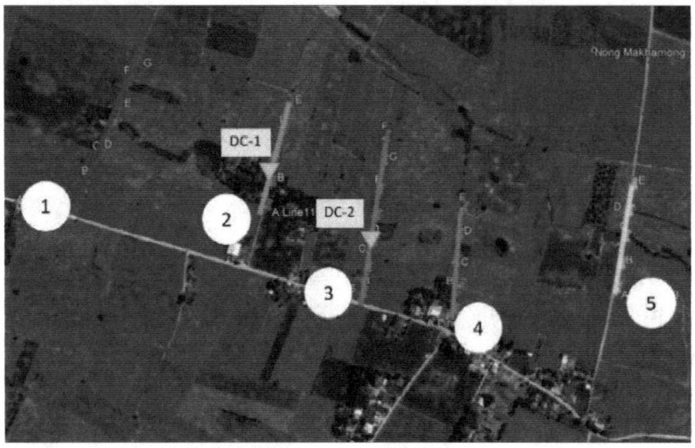

Figure 3. 2D-resistivity survey lines consist of 5 lines with total length of 1,415 m and locations of drilled wells (DC-1 and DC-2) on survey lines.

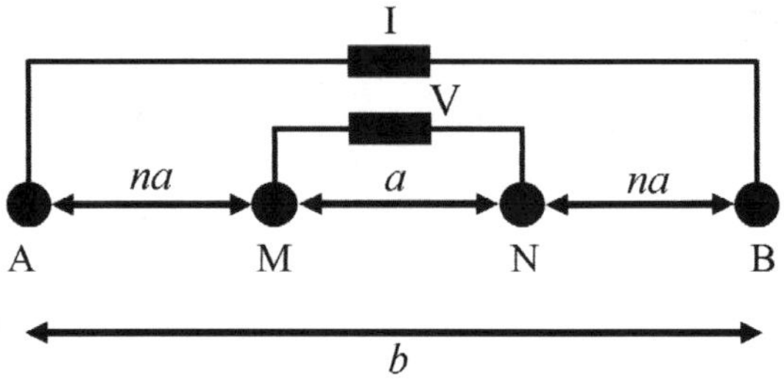

$$\rho_a = \frac{2\pi as}{(s - a)} R \quad (1)$$

Figure 4. The Wenner-Schlumberger electrode configuration used in 2D resistivity imaging survey in this work [9].

When $s=(a^2+b^2)^{0.5}$ and R is the measured resistance. a, is the distance between the current electrodes pair A-M (or N-B) and the potential pair M-N. The "n" is the ratio of distance between the A-M (or N-B) electrodes and when the "n" factor is equal to 1 that is similar to the Wenner electrode configuration (distance between electrode = a). b, is the spacing between the A and B electrodes. ρ_a is the apparent resistivity value for Wenner-Schlumberger array [9,10]. The acquired data were inverted to obtain the 2D cross sections of subsurface resistivity values by the Res2Dinv® software.

4. Results and discussion

4.1. 2D-resistivity imaging profiles

The results of 2D resistivity imaging survey indicate a potential paleochannel with a width from about 50 to 160 m (could extend up to a few hundred meters) and about 10 to 20 m deep which is clearly shown on survey lines 1, 2 and 3 (figure 5).

Resistivity values from survey lines 1, 2 and 3 can be separated into 3 zones. The top zones of each profile (at depth from 0 to 5 m) represent low resistivity about 5 to 10 ohm.m. The resistivities then increase to moderate values of 10 to 50 ohm.m in the middle zone underlying the top zone at depths of 5 to 15 m. The bottom of profiles show high resistivity values of more than 100 ohm.m.

In addition, the resistivities of survey lines 4 and 5 show high values on the top zone (in the center to the north end) and low resistivity (2 to 10 ohm.m) below the top zone which are different than those of survey lines 1, 2 and 3.

The sediments collected lithology log from boreholes on survey lines 2 and 3 correspond well with the 2D resistivity imaging survey results (figure 6).

4.2. Comparing interpreted 2D-resistivity profiles and lithological log of boreholes

The results of interpreted 2D resistivity imaging surveys help identify the paleochannel and lead to precise locations of drilled wells. Depositional cross-section models were interpreted based on both grain size of sediments and electrical resistivity values.

The survey line 2 cross-sectional model displays the deposition of paleochannel in the SN direction. The results of electrical resistivity values and grain sizes from the DC-1 well can be distinguished into three zones. Low to moderate resistivity of the top layer from 1 to 3 m associated with silty-clay and clay of the topsoil layer. The middle layer of the cross-sectional model at about 3 to 15 m deep represents sand and silty clay of paleochannel deposits which are related to low to middle resistivity values. The bottom of the profile is interpreted as granite weathered bedrock.

The survey line 3 cross-section model shows several paleochannels in the SN direction. The average thickness of paleochannels is less than that of the survey line 2. Silty clay and clay are the main compositions in paleochannel deposits here. The bottom of the profile is interpreted as granite weathered bedrock at about 2 to 15 m deep from the surface.

The discovery wells detect shallow groundwater in this paleochannel at a depth of 5 to 9 m in the DC-1 well and 3 to 12 m in the DC-2 well with a groundwater yield of 4 to 4.2 m^3/h. The sediments from boreholes suggest that alluvial-fluvial deposits (figure 6).

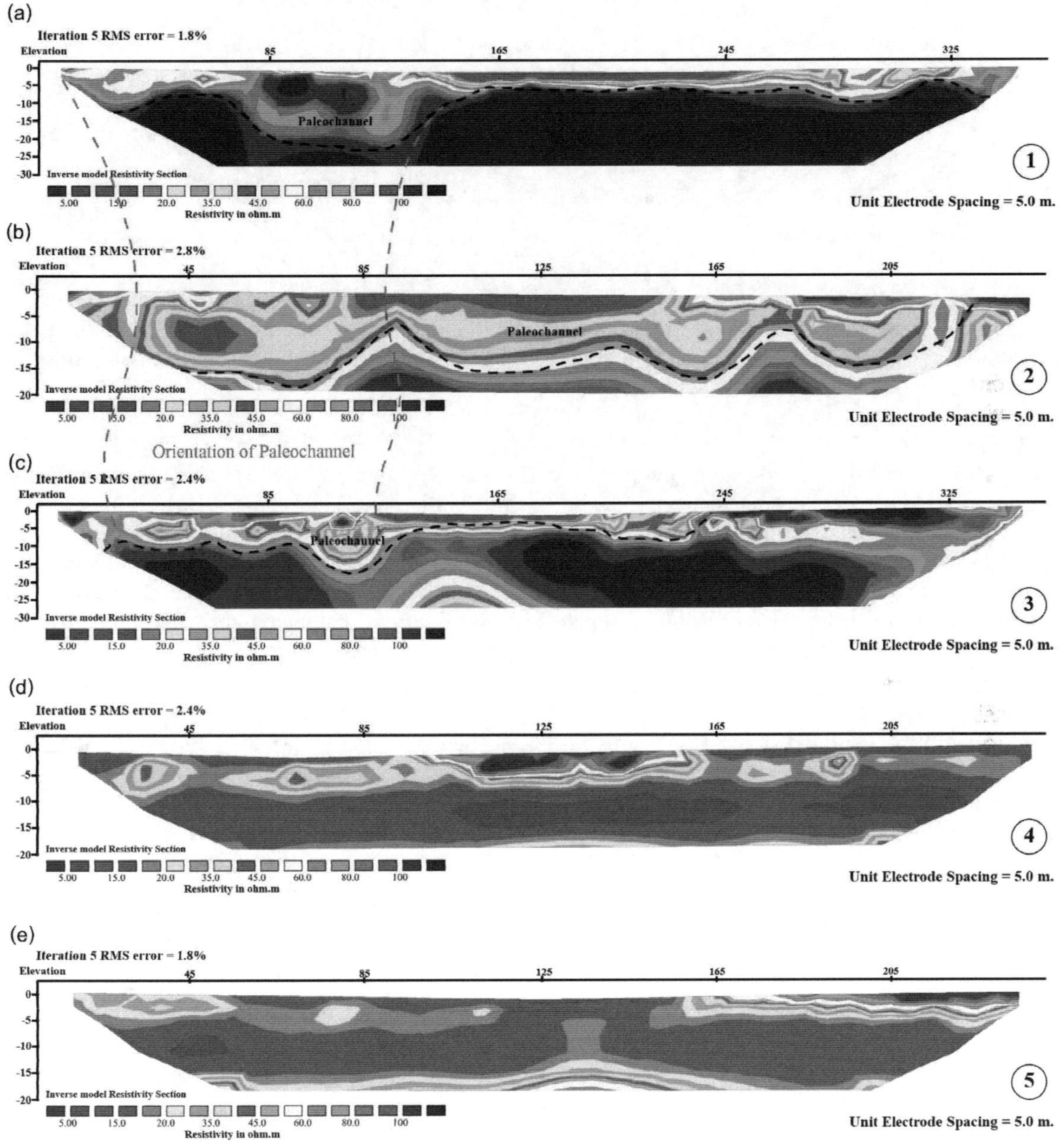

Figure 5. A result of 2D resistivity imaging surveys. (a) the survey line 1, (b) the survey line 2, (c) the survey line 3, (d) the survey line 4 and (e) the survey line 5. The black and red dashed lines indicate the boundaries of potential paleochannel in the study area.

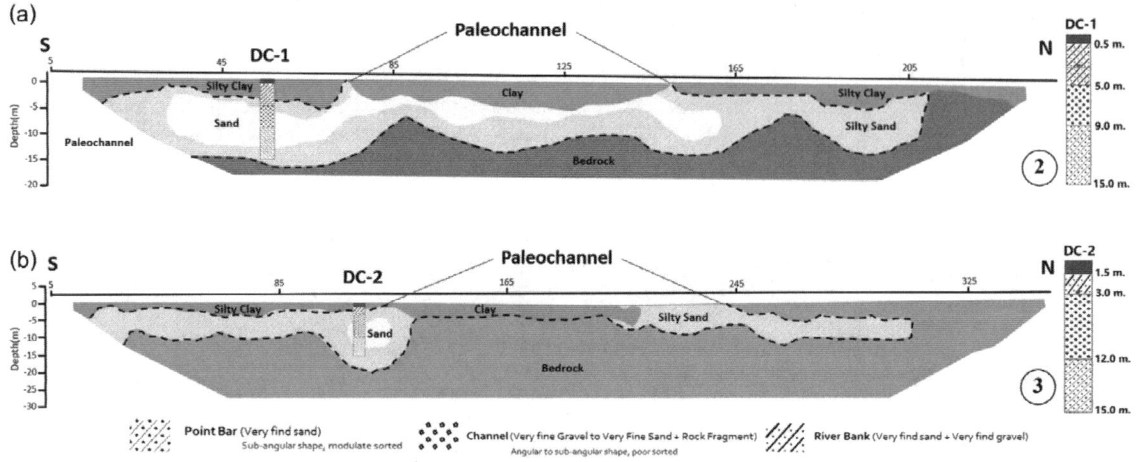

Figure 6. Integrated interpretation of paleochannel from 2D resistivity imaging survey based on subsurface resistivity and lithology log from borehole data. (a) the survey line 2 interpretation of the profile with DC-1 lithology log from the borehole and (b) the survey line 3 interpretation of the profile with DC-2 lithology log from the borehole.

5. Conclusion

Integrated studies of remote sensing, geology and 2D resistivity imaging surveys were used to identify the location of a paleochannel in Nong Makha Mong Subdistrict, Dan Chang District, Suphan Buri Province. The results of the 2D resistivity imaging survey indicate a clear paleochannel. The discovery groundwater wells detect shallow groundwater from this paleochannel depth of 3 to 15 m with a groundwater yield of about 4 m^3/h. This work can be used a pilot project for local communities to search for shallow groundwater resources in paleochannels in areas that encounter very severe drought crisis and have very little deep groundwater potential or surface water resources.

References

[1] Chandra S, Choudhury J, Maurya P K, Ahmed S, Auken E and Verma S K 2019 *Explor. Geophys.* **51** 74–83
[2] Bates and Robert L 1980 *Glossary of Geology* (Virginia: American Geological Institute)
[3] Signanini P, Torrese P, Rainone M L and Colantonio F P 2013 *Proc. Symp. Appl. Geophyics Eng. Environ. Probl.* **24** 316–8
[4] Nimnate P, Thitimakorn T, Choowong M and Hisada K 2017 *Open Geosci.* **9** 675–88
[5] Holzschuh J 2002 *Geophys.* **67** 690–700
[6] Department of Mineral Resource 2014 *Geological and Mineral Resources Management, Suphanburi Province* (Bangkok: Ministry of Natural Resources and Environment Thailand)
[7] Kumar V 2011 *Encyclopedia of Snow, Ice and Glaciers* ed Singh V P, Singh P and Haritashya U K (Amsterdam: Springer) p 803
[8] Jamaluddin and Emi P U 2018 *IOP Conf. Ser.: Earth Environ. Sci.* **118** 012006
[9] Pazdirek O and Blaha V 1996 *58th EAGE Conf. Exhib.* **48** 00228
[10] Loke M H 2002 *Tutorial : 2-D and 3-D Electrical Imaging Surveys* (Penang: Geoelectrical)

The remanent magnetisation recorded in the Chesapeake Bay impact crater, Virginia

R Supakulopas[1,3]* and S M Tikoo[2,3]

[1] Division of Physical Science, Faculty of Science, Prince of Songkla University, Hat Yai, Songkhla, 90110, Thailand
[2] Department of Geophysics, Stanford University, Stanford, California, 94305, United States
[3] Department of Earth and Planetary Science, Rutgers University, Piscataway, New Jersey, 08854, United States

*Corresponding author's email address: radchagrit.s@psu.ac.th

Abstract. During impact events, planetary crusts experience high pressures that can impart rocks with shock remanent magnetisation (SRM) if an ambient magnetic field or demagnetise rocks if a field is absent. If rocks experience substantial impact heating or are pressurised above ~40 GPa (inducing melting and recrystallisation) they may instead record a thermo-viscous remanent magnetisation (TVRM) as they cool below their Curie temperatures. Understanding impact re-magnetisation is crucial for studying terrestrial impact craters, but also unraveling the history of long-lived core dynamo fields on other planetary bodies. In this research we studied impact-related re-magnetisation recorded in natural rock samples from the Chesapeake Bay impact crater, Virginia. As a case study, here we discuss the natural remanent magnetisation (NRM) of two samples of different rock types: a suevite (sample I9-UI, depth 1.40 km beneath the ground) and a schist (sample S32, depth 1.67 km beneath the ground) using thermal and alternating field demagnetisation. The suevite represents a sample that contains material that experience impact remelting, whereas the schist represents an unmelted rock. From the NRM spectra, we found that the sample ITH9-UI was remagnetised by TVRM due to impact-related heating, while the sample STH32 shows the indication of shock deformation of magnetic minerals.

1. Introduction

Hypervelocity impact events on the planets drive Earth scientists to explore the effects of impacts on rocks. During impact events, shock-induced remanent magnetisation (SRM) can record the ambient field such as the Earth's magnetic field throughout geological time when the field is presence [1]. If an ambient field is absent, the shock pressure rearranges the magnetic moments in rocks randomly, in a process known as shock demagnetisation [1]. The SRM is crucial in palaeomagnetism as it is a key to understand long-lived core dynamo magnetic field of the planets, and also transient magnetic fields generated and amplified by impact plasmas.

There are several experiments designed in laboratories to understand SRM (e.g., [1, 2]). However, many of these studies are conducted using hydrostatic analogs of SRM. Laboratory experiments show that SRM is preferentially recorded in low-coercivity ferromagnetic grains [i.e., SRM can easily be removed by low alternating magnetic field (e.g., 10 mT)]. In a natural setting, SRM may be overprinted by thermoviscous remanent magnetisation from impact-related heating (TVRM), viscous remanent magnetisation (VRM) from long-term exposure to a planetary magnetic field after an impact and isothermal remanent magnetisation (IRM) from effects such as lightning strikes. Therefore, SRM may not survive in nature after the impact events. With regards to the relationship between SRM and magnetic unblocking temperatures, SRM seems to occupy grains spanning nearly the entire unblocking temperature spectrum of rocks up to the Curie temperature. As natural craters on both terrestrial and extraterrestrial crusts experience shock level pressures in the order of ~1-50 GPa, samples from impact craters are ideal to study the SRM. Therefore, in this research we sought to explore the nature of impact remagnetisation at the Chesapeake Bay impact crater, Virginia and determine whether SRM was recorded in rock samples there.

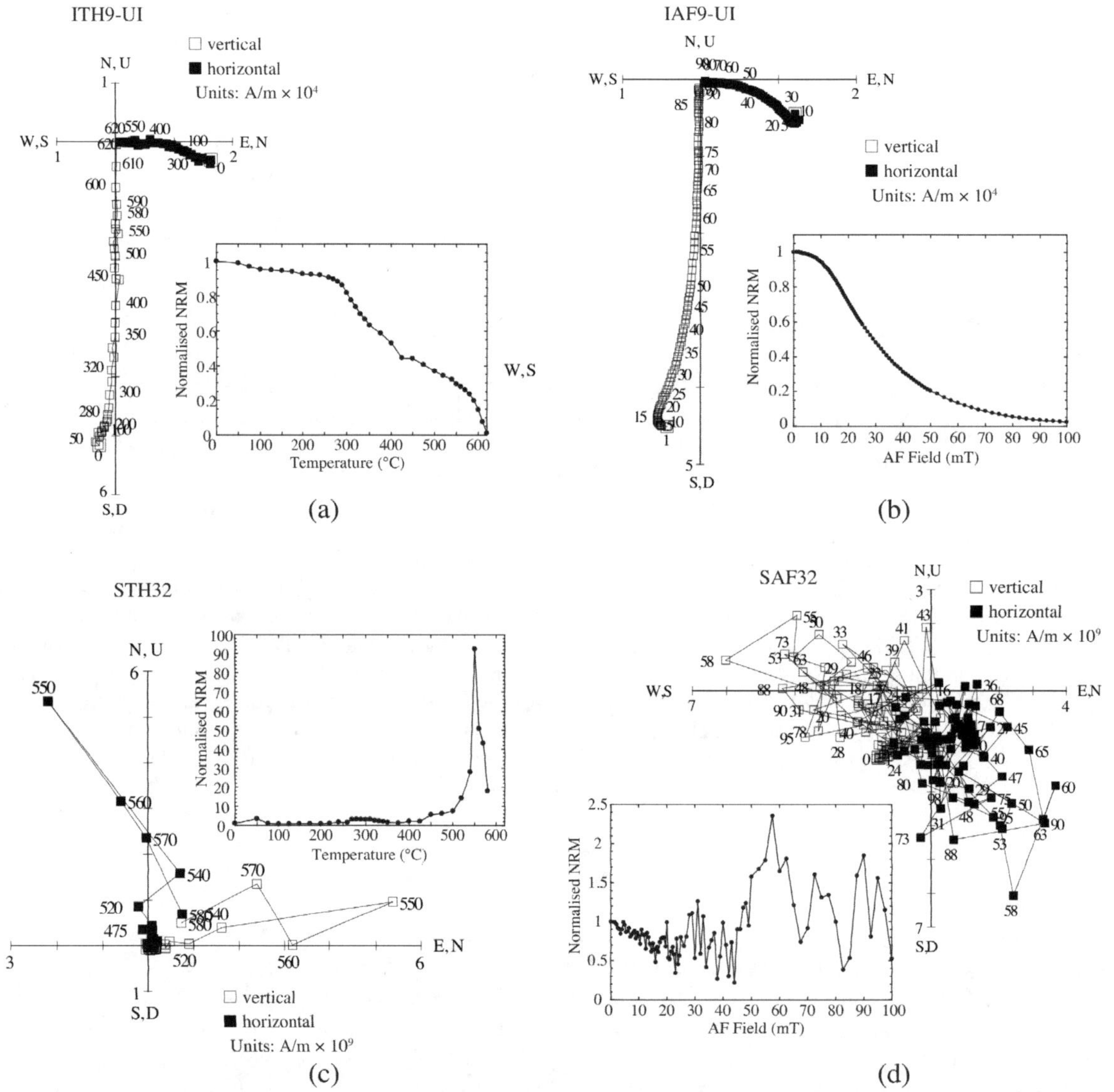

Figure 1. Zijderveld diagrams showing the declination (black square, vertical field) and inclination (open square, horizontal field) of the palaeomagnetic field. Inset diagrams show the normalised NRM spectrum. (a) and (b) show thermal and AF demagnetisation of sister specimens from a suevite sample from the impact layer. (c) and (d) represent the thermal and AF demagnetisation of a schist sample from the schist-pegmatite layer.

2. Sample collection and palaeomagnetic measurements

2.1. Chesapeake Bay impact crater and sample collections

The Chesapeake Bay impact crater was formed during the late Eocene (~35.5 Ma) by a meteorite crashing onto Earth [3]. The International Continental Scientific Drilling Program (ICDP) held a scientific drilling campaign to sample this impact structure in 2005 at the Eyreville farm, Virginia and successfully obtained three cores: core A (depth 124.6-940.9 m), core B (depth 737.6-1766.3 m) and core C (depth 0-140.2 m). These cores are stored at the ICDP core repository at Rutgers University. In our effort to determine the nature of impact remagnetisation at the crater, we focus on the Eyreville core B as this core contains an impact-melt-containing breccia (suevite) layer (depth ~1,393-1,550 m) containing materials that experienced a shock pressure in the order of 35-50 GPa during the impact [4], and a schist-pegmatite layer (~1,550-1,766 m depth) which represents the impact target rocks. We collected 19 samples from the impact breccia layer and 30 samples from the schist-pegmatite layer. Samples were cut into a 2×2 cm squares with a thickness of ~0.5 cm, yielding ~2-6 specimens for each sample.

2.2. Palaeomagnetic measurements

Palaeomagnetic measurements were performed in the Palaeomagnetism Laboratory at Rutgers University. Standard demagnetisation experiments including thermal demagnetisation and alternating field (AF) demagnetisation were performed on 19 specimens from the impact layer and 30 specimens from the schist-pegmatite layer. For the thermal demagnetisation, the specimens were stepwise heated from 50 °C to 620 °C using an ASC TD48-SC thermal demagnetiser to isolate natural remanent magnetisation (NRM) directions. The remaining NRM and directions were measured using a 2G Cryogenic SQUID magnetometer after each heating step. During AF demagnetisation experiments, peak ac fields of 1-100 mT were applied to gradually demagnetise the NRM, and the remaining NRM and magnetisation directions was measured after each AF demagnetising field. The palaeomagnetic data were illustrated and interpreted on Zijderveld diagrams [5] and NRM decay plots (figure 1).

3. Results and discussion

Here we present palaeomagnetic data from a suevite sample (sample I9-UI, depth 1.40 km) from the impact layer and a schist sample from the schist-pegmatite layer (sample S32, depth 1.67 km). Samples with the AF and TH codes in front of the numbers are referred to as AF-demagnetised and thermal demagnetised samples.

3.1. Impact layer

The thermal demagnetisation data show that two magnetisation components are recorded in the suevite: A high-temperature (HT) component likely attributable to primary cooling after the impact is blocked between 620 °C and 350 °C (figure 1(a)). Curvature of the Zijderveld diagram connects the HT component to a second low-temperature (LT) magnetisation component with a somewhat different direction blocked between 350 °C and room temperature. It is possible that the LT component was acquired during prolonged cooling accompanied by block rotation within the crater but further work is necessary to explore that hypothesis. To determine the magnetic mineralogy of our samples, we consider the NRM spectrum (figure 1(a) inset). Our data show two phases of magnetic minerals: pyrrhotite (Curie temperature ~320 °C) and oxidised titanomagnetite (Curie temperature ~580 °C). It is clear that the impact layer acquires the TVRM.

AF demagnetisation revealed the same two magnetisation components observed in the thermal data: a high coercivity (HC) component with a direction matching the HT direction from the sister specimen was present in grains with coercivities between 50-100 mT. A low coercivity (LC) component matching the LT direction was blocked between 1-50 mT. The NRM spectrum (figure 1(b) inset) shows the median destructive field (MDF, the field required to remove half of the NRM) was ~29.1 mT.

3.2. Schist-pegmatite layer
Thermal demagnetisation of the schist sample (figure 1(c)) reveals significant scatter on the Zijderveld diagram during the heating steps between 50-400 °C. The sample acquires a spurious magnetisation from heating in the thermomagnetic oven above 400 °C due to thermochemical alteration of the sample (figure 1(c) inset). Therefore, data points above 400 °C are not considered here. Unlike the impact layer, the characteristic component of the NRM for low temperature steps (50-400 °C) cannot be determined from the Zijderveld diagram. AF demagnetisation data from the sister specimen also exhibit scattered magnetisation directions (figure 1(d)). The NRM spectrum is not smooth, precluding determination of the MDF (figure 1(d) inset). The data from both experiments indicate that the schist-pegmatite layer apparently did not acquire any stable SRM as a result of the impact but may in fact have experienced a net demagnetisation effect.

A full shock demagnetisation of intact magnetic minerals may be excluded because there is an ambient magnetic field on Earth. Pyrrhotite is the primary magnetic mineral in the schist sample while magnetite is the minor magnetic mineral [6]. The pyrrhotite grains are single domain grain with medium coercivity [6]. Experimental studies on shock pressure effecting on pyrrhotite grains show that pressures exceeding the Hugoniot elastic limit of 3.5 GPa of pyrrhotite causes the fracturing and deformation of pyrrhotite grains, forming amorphous domains [7]. The high shock pressures may also reduce magnetic grain sizes to the superparamagnetic range that cannot preserve stable magnetisation over geological time [7]. The amorphous domains reduce the ability of remanent acquisition in pyrrhotite [7]. We tentatively conclude that the schist sample may have lost its ability to acquire a stable magnetisation due to fracturing and deformation of the pyrrhotite grains. However, further studies to measure the magnetic domain grains and mineral phases, e.g., measurements of hysteresis, backfield curve and Curie temperature should be performed to provide better conclusions.

4. Conclusions
Paleomagnetic study of samples from the Chesapeake Bay impact crater indicates that different forms of magnetisation are recorded by different rock types within the crate. Melt-bearing breccias (suevites) containing pyrrhotite and titanomagnetite recorded TVRM as they cooled following the impact event. The schist target rocks show signs of impact-related demagnetisation due to the fracturing and deformation of pyrrhotite. We did not detect any evidence of SRM in our rocks, supporting prior observations that this form of magnetisation is not well-preserved in nature.

References
[1] Nagata T 1971 *Pure Appl. Geophys.* **89**(1) 159–77
[2] Tikoo S M, Gattacceca J, Swanson-Hysell N L, Weiss B P, Suavet C and Cournède C 2015 *J. Geophys. Res. Planets.* **120**(9) 1461–75
[3] Gohn G S, Koeberl C, Miller K G, Reimold W U, Cockell C S, Horton Jr. J W, Sanford W E and Voytek M A 2006 *Eos* **87**(35) 349–55
[4] Jackson J C, Horton Jr. J W, Chou I M and Belkin H E 2016 *Meteorit. Planet. Sci.* **51**(5) 946–65
[5] Zijderveld J D A 1967 *Tectonophysics* **4**(2) 121–53
[6] Elbra T, Kontny A and Pesonen L J 2009 *Geol. Soc. Am. Spec. Pap.* **458** 119–35
[7] Mang C, Kontny A, Fritz J and Schneider R 2013 *Geochem. Geophys. Geosyst.* **14**(1) 64–85

Single-axis solar tracking system referring to date and time

N Chaijum*, L Cheunchantawong and T Siriram

Program of Electromechanic Manufacturing Engineering, Faculty of Engineering and Industrial Technology, Bansomdejchaopraya Rajabhat University, Bangkok, 10600, Thailand

*Corresponding author's e-mail: natawee.ch@bsru.ac.th

Abstract. This article is about designing and building a single-axis solar tracking system referring to the sun position database. The objectives are as follows: 1. to design and build a solar tracking system, and, 2. to compare the power produced from the solar tracking system with that from the stationary solar panel. The angle of the solar panel from the solar tracking system is positioned at a constant altitude angle, 15 degrees, facing south, and the moving part was the azimuth, which follows the position of the sun. Latitude and longitude coordinates are identified by an Arduino UNO R3 microcontroller board for processing data, reading coordinates of the sun's angle degrees from the SD card module, and commanding the servo motor to rotate to adjust the angle of the solar panel in a position perpendicular to the sun. Results from the experiment were collected in October 2020 from 9 AM to 4 PM. The system changed the angle degree every 30 minutes. It is found that the solar tracking system can easily be created and controlled, and can also accurately follow the sun's position all day long. Moreover, it can generate more electric power than that generated by the stationary solar panel by up to 15%. The system is applicable and can generate more electric power than other tracking systems, although the results were collected during the rainy season when the weather was generally cloudy and rainy throughout the month.

1. Introduction

Generating electricity from solar energy continues to increase in Thailand every year. In 2018, cumulative installed capacity was 2,962.44 megawatts. Solar power generation system can be divided into 2 groups as follows: 1. on-grid solar power generation system, and 2. off-grid solar power generation system. From the conclusion of the measurement of solar radiation intensity data by the Department of Alternative Energy Development and Efficiency in 2020, the mean was 4.807 kWh/m^2-day. Moreover, according to the solar radiation intensity data statistics in Thailand during 2016-2020, it is found that the daily solar radiation intensity per month is low (especially in October 2020) in the rainy season (from mid-May to mid-October). The mean of daily solar radiation intensity per month is 13.364, which is extremely low [1-3]. However, solar radiation intensity is the important variable for making the electricity generating with solar cells more efficient.

Implementing solar cells to generate electricity to get the most efficiency is based on various factors, for example, the most suitable solar panel installation in Thailand with angle of inclination in Bangkok area is 15 degrees facing south [4]. However, the popular controllable factor to enable the solar panel to generate more electric power in each day is controlling solar panels to obtain the highest solar intensity at different times of the day, which is mostly use light dependent resistors (LDR) instead of a light intensity sensor. The researchers studied the light intensity measurement system to compare the power generation efficiency of the day and found that light dependent resistors (LDR) work well when the sky is clear as it is able to measure the light intensity effectively. One to four light dependent resistors or more will be installed in the different positions and angles of the solar panel [5-9]. However, using the light dependent resistors (LDR) when the sky is cloudy is not efficient since the intensity of the sun's rays when hitting the clouds is reflected back into the atmosphere and pass through the clouds before hitting the ground. There are different proportions for different periods and the time that the weather changes. In the rainy season, most of the sky is covered by the clouds, especially cumulus and cumulonimbus clouds, which are low level clouds and are important factors for the relationship of clouds and the sun's rays [10-12]. Because of this reason, the light intensity received by the light dependent resistors in the cloudy sky will cause errors and will directly lead to the system control unit. When the system receives the data from the light dependent resistors, the motor will adjust its direction in that position. For the image processing method, which is used instead of the light dependent resistors to detect the brightest point to track the position of the sun, it works well when the sky is clear. However, when it is cloudy, the tracking system will have less efficiency [13]. Therefore, we will have to use a suitable sun tracking system in designing and building a solar tracking system to be the most efficient at different times of the year [14-15].

From the problems stated above, the researchers, therefore, designed and constructed a single-axis solar tracking system by referring to www.timeanddate.com and www.suncalc.org for the position of the rising and setting of the sun at different times in a day. This is to enhance the sun tracking system to generate electric power in the rainy season, or on cloudy days, so that the solar panels can generate maximum power. Arduino board is used in the circuit control system and all the equipment to process the data derived from SD card module of the sun's degree of arc in each period. Moreover, the Arduino board is also used to activate the servo motor to rotate the solar panel at specified time.

2. The structure and algorithms design

2.1. The structure of the single-axis solar tracking system referring to date and time

Structural design of the single-axis solar tracking system to be used with 40-watt and 2.7-kilogram monocrystalline solar panel can rotate 180-degree in horizontal axis. The solar panel specifications are shown in table 1. Moreover, it is designed with a rectangular base to make it durable in windy weather. The servo motor, which controls the direction of solar cells, is positioned in the middle so that the solar panel can move freely and accurately at the specified angle. Figure 1 show the single-axis solar tracking system whose main components are: 1) solar panel, 2) LCD display, 3) servo motor for controlling the rotation of the spur gear, 4) Arduino board and 5) battery.

Table 1. Specification of solar panel.

Specification	Values
Maximum power (P_{max})	40 W
Open circuit voltage (V_{oc})	21.5 V
Short circuit current (I_{sc})	2.38 A
Maximum power voltage (V_{mp})	18 V
Maximum power current (I_{mp})	2.22 A
Dimensions	664 x 408 x 18 mm

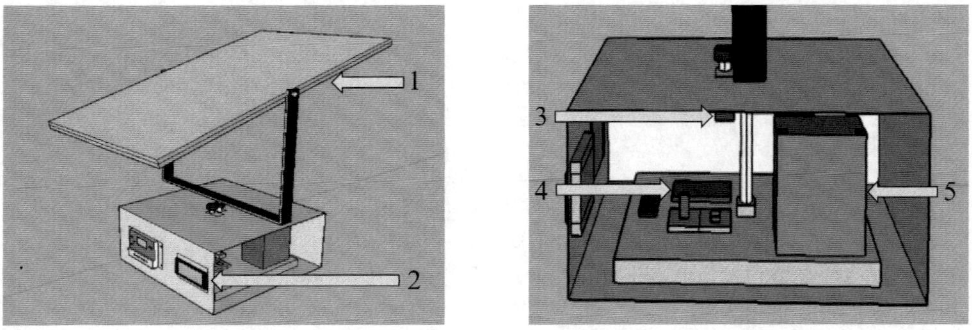

(a) (b)

Figure 1. The design (a) Structure of a single-axis solar tracking system. (b) Control box.

2.2. Algorithms of the single-axis solar tracking system referring to date and time

Algorithms of the single-axis solar tracking system can be explained as follows: Arduino board reads the date; (d), time; (t), and the azimuth angle data; (ψ) from an .csv file in the SD card module, but azimuth angle in the file must be minus 90 degrees because the servo motor can rotate 180 degrees. After that, the Arduino board commands the servo motor to rotate to adjust the angle of the solar panel towards the sun in the set direction for 30 minutes. When it reached the specified time, the Arduino board would read the azimuth angle data of the next period and command the servo motor to run continually. When it finished rotating in the set period, the system commands the servo motor to move towards the default arcdegree at 9 AM of the following day until Day 31 based on the system program details shown in figure 2 and circuit's connection of single-axis solar tracking system shown in figure 3.

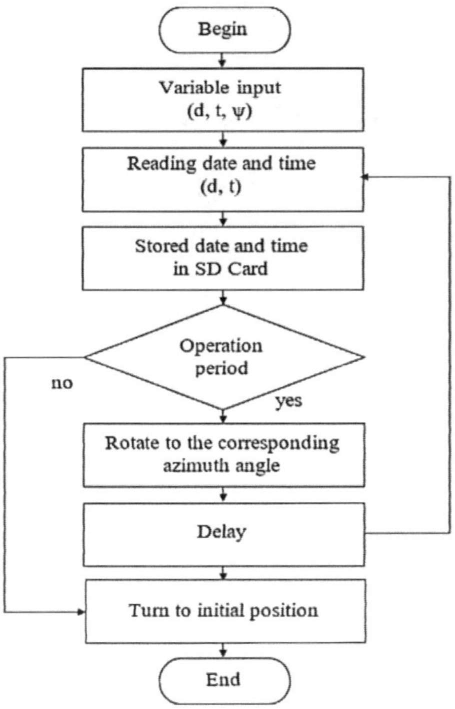

Figure 2. Algorithm for a single-axis solar tracking system.

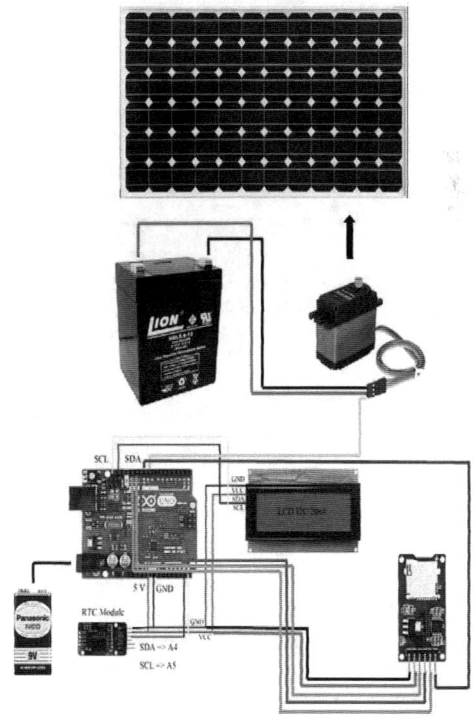

Figure 3. Device and circuit's connection of single-axis solar tracking system.

3. Experiments

To evaluate the electric power generated between single-axis solar tracking system referring to date, time and stationary solar panel, the researchers stored the azimuth angle of the date and time used in the experiment from www.timeanddate.com and www.suncalc.org in October 2020 as shown in figure 4 in the SD card module. In this way, the Arduino board could read and command the servo motor to rotate in the set direction, which was to rotate every 30 minutes from 9 AM to 4 PM. The researchers also measured the current and voltage with a multimeter every 30 minutes. The area used for the experiment was 13 Floor of Building 24 in Bansomdejchaopraya Rajabhat University, which is an open space and does not have any shadows that obscure the position of the solar panels at all times. In order to find the latitude and longitude of the experiment area, the researchers used Google Maps program to find the geographic coordinates to compare with the coordinates of the website and location data of the sun to receive the right coordinates.

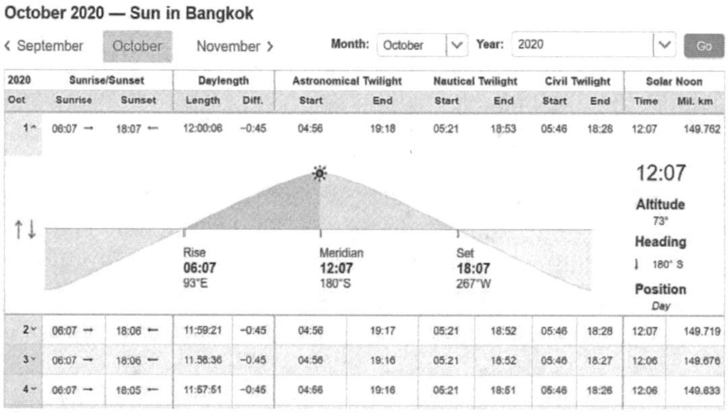

(a)

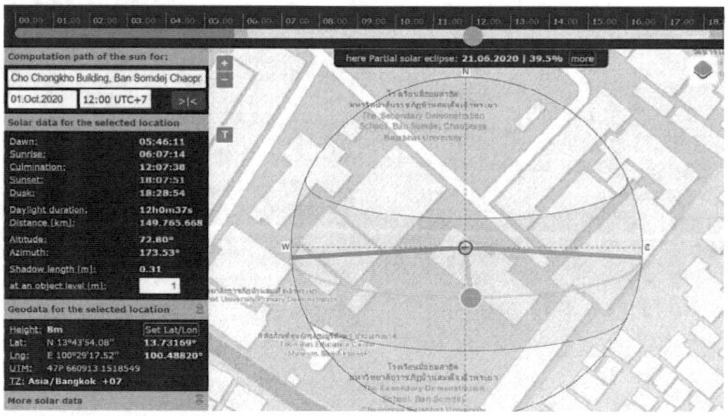

(b)

Figure 4. Date and time, and degree of arc data from (a) www.timeanddate.com [16] and (b) www.suncalc.org [17].

4. Results and discussion

The results are divided into 2 parts: the percentage of arcdegree accuracy when moved by solar cell system's servo motor, and the amount of electric power generated by the single-axis solar tracking system referring to date, time and stationary solar panel.

4.1. The percentage of arcdegree accuracy when moved by solar cell system's servo motor

The solar cell structure is durable and can take the weight load of the solar cell well. The operation control system is kept in an acrylic box, which is lightweight and easy to be moved as shown in figure 5(a). The test to find the percentage of arcdegree accuracy when moved by the solar cell system's servo motor was done 10 times and the azimuth angle was chosen on 1st October 2020. The experiment was done continuously from the first arcdegree (at 9 AM) to the last (at 4 PM); this would count as one round. The test simulation was done on-site and real-time prior to the collection date as shown in figure 5(b). The researchers find that using the servo motor to move the solar panels of the system has an error at 1.07% for the whole system as shown in table 2.

(a) (b)

Figure 5. The field experiment (a) Position of single-axis solar tracking system referring to date and time. (b) The arcdegree that the servo motor can rotate at 2.30 PM.

Table 2. The test result of finding the azimuth angle accuracy as of 1st October 2020.

Time (Hr)	System's angle data (°)	Rotatable angle data (°)	Error percentage (%)
9.00	17	17.20	1.18
9.30	21	21.10	0.48
10.00	26	26.40	1.54
10.30	34	34.20	0.59
11.00	45	45.40	0.89
11.30	61	62.00	1.64
12.00	84	84.90	1.07
12.30	109	110.20	1.10
13.00	129	130.40	1.09
13.30	142	144.00	1.41
14.00	151	152.00	0.66
14.30	157	159.00	1.27
15.00	161	163.60	1.61
15.30	165	166.30	0.79
16.00	168	169.20	0.71
	Total mean		1.07

This system uses more components than that in the stationary solar panel. The components and their cost are as follows: one 40-watt monocrystalline solar panel at $36, one Arduino board at $5, one SD card module at $1, one SD card at $2, one LCD display at $2, one RTC module at $1, one servo motor at $8, gear and rod at $8, and two bearings at $4, which make the overall cost of the solar tracking system at $67. The installation and maintenance are also small due to the system being only the prototype.

4.2. The amount of electric power generated by single-axis solar tracking system referring to date, time and stationary solar panel

It is found that the average amount of electric power of single-axis solar tracking system was 12.72 watts and that of the stationary solar panel was 11.05 watts. It can be said that the single-axis solar tracking system has average electric power 15% higher than the other. For sunny days and cloudy days, the single-axis solar tracking system generated an average amount electric power more than that of the stationary solar panels at 11% and 10%, respectively as shown in figure 6. This system can generate more electric power even on cloudy days.

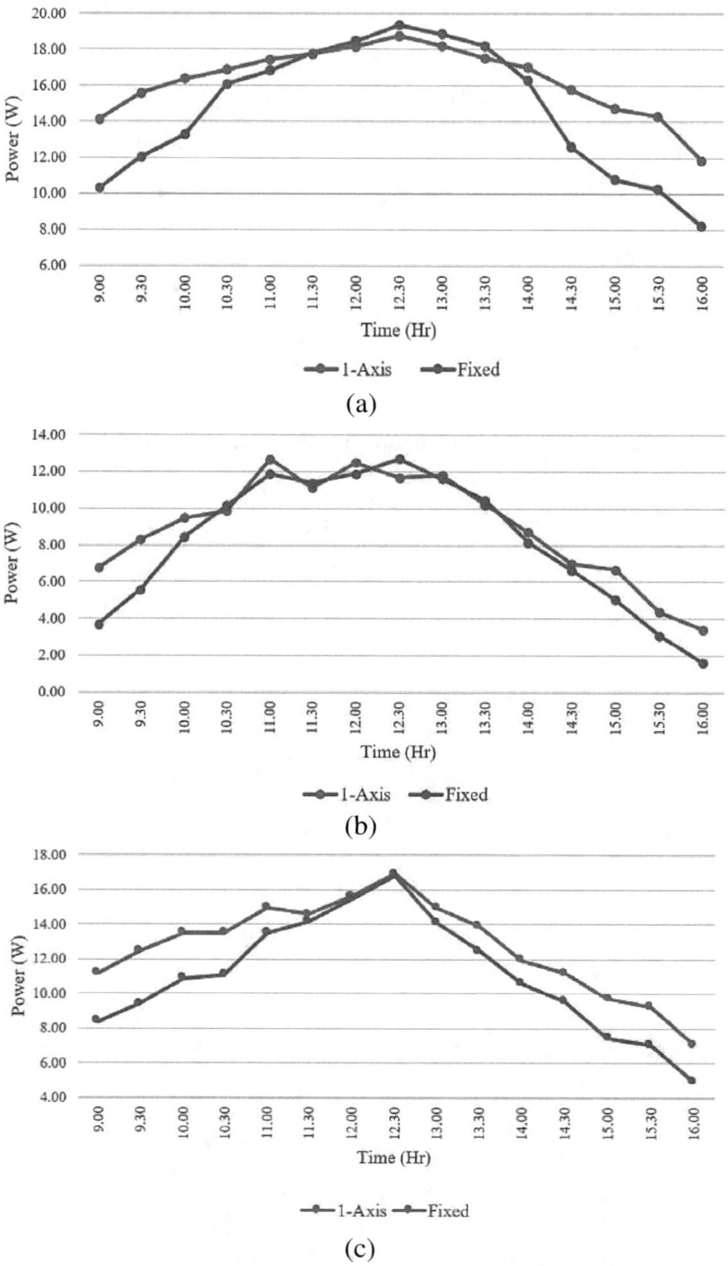

Figure 6. Electric power generated for (a) sunny days, (b) cloudy days and (c) October 2020.

The single-axis solar tracking system is a small-size prototype so it is suitable for domestic use or low-voltage equipment in remote areas. This system is also suitable to install on flat surfaces as opposed to inclined surfaces. Due to its lightweight structure, the system is easily portable. However, it is not advised to use the system during natural disasters, especially those related to storms due to its low stability.

The advantages of this system are its low cost and that it is easy to use. The components used in the system can easily be found in the market and the cost is low. In addition, it can also be used as a study approach and developed into a larger system. However, to design and build a larger system, its main parameters must be studied further, for instance, the motor that controls the overall system structure. Since these parameters depend on the size and number of solar panels, it requires more cost as a result.

Besides building a larger system to generate more electric power which requires more expense, the size and number of solar cells can be increased instead to increase the surface area receiving the sunlight, thus the intensity, as the price of solar panels tend to decrease in the future.

5. Conclusion

Single-axis solar tracking system referring to date and time is durable and lightweight, which is easily portable. Moreover, the servo motor that forces the solar panel to move at different times is accurately implemented throughout the experiment. The error is 1.07%, which is extremely low. Moreover, the servo motor consumes very little energy and is cheap in price [18]. It is also found that the single-axis solar tracking system generates 15% more electric power because of the solar tracking system referring to date and time works well on a cloudy day or in the rainy season [19]. The overall cost for the solar tracking system is at $67 as it is a small prototype. In conclusion, the solar cell system can adjust the direction of the solar panel to get more sunray almost all day long, except for 11.30 AM – 1 PM, the single-axis solar tracking system referring to date and time can generate electric power at almost the same amount as that of the stationary solar panel. The solar tracking system referring to date and time can collect the date; (d), time; (t), and the azimuth angle data; (ψ) for many years depending on memory card storage size. Using the arcdegree of the sun at different times from www.timeanddate.com and www.suncalc.org as a solar panel direction indicator is practical, even in the rainy season. Therefore, in the rainy season or on cloudy days, the single-axis solar tracking system referring to date and time should be used, or the sun tracking system should be adjusted to rotate twice a day to reduce motor power consumption during unnecessary periods.

The researchers intend for future studies to focus on developing a larger solar tracking system with adjustable solar panels to receive the highest light intensity for whole day in all weather conditions. Furthermore, the researchers intend to re-design the system to use lower-cost components and to improve the tracking system to the highest efficiency to connect with internet of things (IoT) to display important information and to remotely control the system.

Acknowledgements

The author would like to express sincere thanks to the Research and Development Institute, Bansomdejchaopraya Rajabhat University and the Faculty of Engineering and Industrial Technology, Bansomdejchaopraya Rajabhat University, Bangkok, Thailand.

References

[1] Department of Alternative Energy Development and Efficiency 2019 *Thailand Alternative Energy Situation 2019* (Bangkok: Ministry of Energy)
[2] Wattana S and Wattana B 2020 *Int. Energy J.* **20** 101–14
[3] Climatological Center 2020 *Monthly Weather Summary in Thailand October 2020* (Bangkok: Thai Meteorological Department) pp 1–16
[4] Jacobson M Z and Jadhav V 2018 *Sol. Energy* **169** 55–66
[5] Tharamuttam J K and Ng A K 2017 *Energ. Proc.* **143** 629–34
[6] Munanga P, Chinguwa S, Nyemba W R and Mbohwa C 2020 *Proc. CIRP* **91** 571–76

[7] Farhan A J 2020 *IOP Conf. Ser.: Mater. Sci. Eng.* **757** 012042
[8] Abadi I, Setyawan E H and Ardiansyah M 2020 *IOP Conf. Ser.: Earth Environ. Sci.* **520** 012001
[9] Ndubuisi A O, Imolorhe E and Akinyemi M L 2018 *IOP Conf. Ser.: Earth Environ. Sci.* **173** 012044
[10] Yongsiriwith P, Khoonphunnarai P, Intarachumnum R and Kongmeetong S 2018 *J. Phys.: Conf. Ser.* **1144** 012034
[11] Mohanapriya V, Manimegalai V, Praveenkumar V and Sakthivel P 2021 *IOP Conf. Ser.: Mater. Sci. Eng.* **1084** 012073
[12] Imthiyas A, Prakash S, Vijay N, Abraham A A and Kumar B G 2020 *IOP Conf. Ser.: Mater. Sci. Eng.* **993** 012124
[13] Stinia H, Kukliński K, Jaiswal A, Fadnavis R, Meshram S, Dhote N, Gadge G, Dudek M and Raźniak A 2021 *IOP Conf. Ser.: Earth Environ. Sci.* **642** 012009
[14] Amelia A R, Irwan Y M, Safwati I, Leow W Z, Mat M H, Rahim M S A and Esa S M 2020 *IOP Conf. Ser.: Mater. Sci. Eng.* **767** 012052
[15] AL-Rousan N, Isa N A M and Desa M K M 2018 *Renew. Sust. Energ. Rev.* **82** 2548–69
[16] Suncal 2020 *Bansomdejchaopraya Rajabhat University* Retrieved 25 Sep. 2020, from https://www.suncalc.org/#/13.7317,100.4882,19/2020.10.01/21:59/1/3
[17] Time and Date 2020 *Time and date Sunrise and sunset times in Bangkok* Retrieved 25 Sep. 2020, from https://www.timeanddate.com/sun/thailand/bangkok?month=10&year=2020
[18] Hammoumi A E, Motahhir S, Ghzizal A E, Chalh A and Derouich A 2018 *Energy Sci. Eng.* **6**(5) 607–20
[19] Kuttybay N, Saymbetov A, Mekhilef S, Nurgaliyev M, Tukymbekov D, Dosymbetova G, Meiirkhanov A and Svanbayev Y 2020 *Energies* **13**(19) 5226

Improved mixing height estimates from atmospheric LiDAR measurements

Ronald Macatangay[1*], Worapop Thongsame[2], Raman Solanki[3], Ying-Jen Wu[4], Sheng-Hsiang Wang[4], Titaporn Supasri[1] and Jirasak Noisapung[1]

[1]Astronomical Research Unit, National Astronomical Research Institute of Thailand, Mae Rim, Chiang Mai, 50180, Thailand
[2]University of Colorado, Boulder, Colorado, 80309, USA
[3]State Key Laboratory of Severe Weather, Chinese Academy of Meteorological Sciences, Haidian District, Beijing, 100081, China
[4]Cloud and Aerosol Laboratory, Department of Atmospheric Sciences, National Central University, Taoyuan City, 320, Taiwan

*Corresponding author's Email: ronald@narit.or.th

Abstract. In this study, an improvement in the estimation of the mixing height is carried out by introducing a time-dependent maximum and minimum analysis altitude (TDMMAA) in the Haar wavelet covariance transform (WCT) technique applied to atmospheric light detection and ranging (LiDAR) measurements generally used in mixing height estimations. Results showed that the standard method usually overestimates the mixing height and that the proposed algorithm is more robust against clouds and residual layers in the boundary layer that generally occur in the nighttime and early morning. The TDMMAA method does have a bit of subjectivity especially in defining the analysis periods as well as the top and bottom of the analysis altitudes as it needs user experience and guidance. Moreover, the algorithm needs to be further objectively refined for automation and operational use, validated with in-situ profile measurements, and tested during different atmospheric conditions.

1. Introduction

The exchange between the surface of the Earth and its atmosphere is primarily driven via turbulence, and this turbulent domain of the atmosphere adjacent to the ground surface is specifically referred to as the mixing layer. Within this layer, air pollutants, such as particulate matter, are assumed to be well-mixed [1]. The top of the mixing layer is often referred to as the mixing height (MH), which generally undergoes distinct diurnal variations with higher daytime altitudes and lower nighttime altitudes. Assuming constant emissions throughout the day, as well as calm and cloud-free weather conditions, the mixing height significantly controls the diurnal variation of air pollutants, especially particulate matter. This causes the concentrations of particulate matter to be generally high during the nighttime when the mixing height is low, and low during the daytime when the mixing height is high [2]. Its accurate estimation is therefore necessary.

In this study, an improved algorithm for estimating the mixing height is investigated and compared to the standard method using atmospheric light detection and ranging (LiDAR) measurements performed over the Fang district of Chiang Mai province in northern Thailand (near the border with

Myanmar observed to have very high air pollution episodes during the dry season) from March 4-31, 2019 as part of the Profiling with LiDAR and UAV Multi-scale Experiment (PLUME) of the Seven Southeast Asian Studies (7SEAS) Spring 2019 measurement campaign.

2. Methodology

The mixing height can be estimated using atmospheric light detection and ranging (LiDAR) measurements. This technique uses an Nd:YAG laser, usually having a visible wavelength (532 nm) and energies of 4 μJ, that is normally shot vertically into the atmosphere reaching altitudes up to 20 km. As the laser encounters particulates, the radiation is scattered back towards the receiver (telescope and detector) of the system. The measured signal is called the LiDAR backscatter profile and after removing the effect of the system characteristics, the resulting signal is called the normalized relative backscatter profile. This signal is proportional to the number concentration of particulates in the atmosphere and is used to estimate the mixing height.

The mixing height is estimated by finding the maximum change or gradient in the normalized relative backscatter profile. This is done by using a basis function, in this case the Haar wavelet, to perform a pattern search of sudden signal jumps. The pattern search in done by using the wavelet covariance transform (WCT) with translated and dilated versions of the basis function analyzed through a certain altitude range (analysis altitudes). Using various dilation values of the Haar wavelet, the noise in the signal is reduced at higher dilation values. From this, the mixing height is estimated from the altitude of the peak wavelet covariance transform coefficient [3].

The standard algorithm normally has a fixed analysis altitude range, usually from 0 to 4 km from the ground. In this study, we introduce a time-dependent maximum and minimum analysis altitude (TDMMAA) to make the mixing height estimate more robust against atmospheric phenomena (e.g. clouds, nighttime and early morning residual layers) and instrumentation features (e.g. noise, condensation due to enclosure air conditioning). These features also help identify analysis time periods when the standard algorithm is sensitive to. These sensitive analysis periods are the maximum mixing height period, the morning transition period, the afternoon transition period, and the minimum mixing height period.

3. Discussion

Putting all the normalized backscatter profiles together in a time-height-intensity plot as shown in figure 1, atmospheric (clouds on March 8, 2019; residual layers in the boundary layer every early morning and during the nighttime) and instrumentation (noise and condensation in the early morning of March 9 and 10, 2019) features caused an overestimate in the mixing height calculated using the standard algorithm as shown in the red line in figure 1. The black line in figure 1 is the estimated mixing height calculated using a time-dependent minimum and maximum analysis altitude (TDMMAA) as outlined in table 1. As can be seen in figure 1, the TDMMAA did not overestimate the effect of clouds, residual layers, noise and condensation. This algorithm was then applied to the entire dataset from March 4 – 31, 2019 as shown in figure 2.

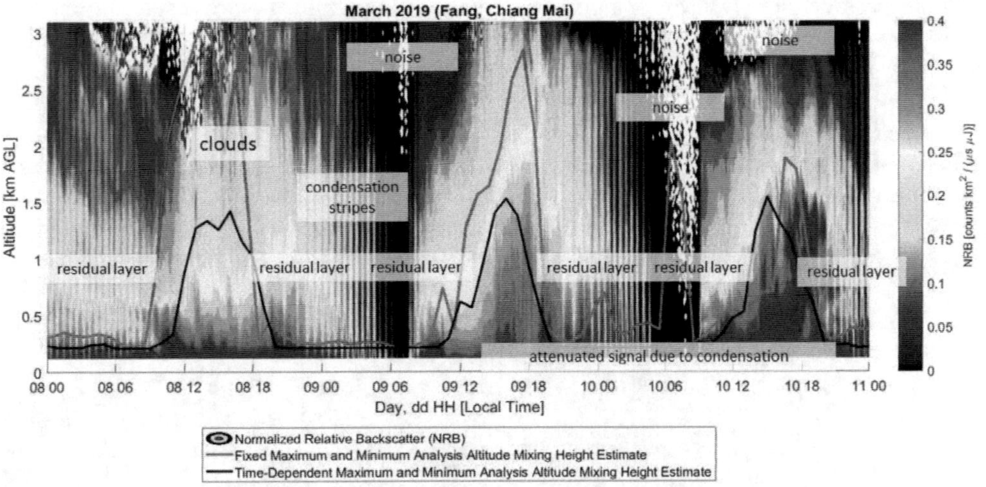

Figure 1. Atmospheric and instrumentation features detection and algorithm testing from March 8 – 10, 2019 over Fang district in Chiang Mai province in northern Thailand.

Table 1. TDMMA Parameters. Defined analysis periods (in local time, LT) and time-dependent maximum (top) and minimum (bottom) analysis altitudes (in meters above ground level, mAGL). The red font depicts changes from the parameters of the previous day.

Date	Period	Start Time (LT)	End Time (LT)	Top (mAGL)	Bottom (mAGL)
March 8, 2019	Maximum MH	1300	1800	2000	1000
	Morning Transition	1200	1300	2000	200
	Afternoon Transition	1800	2000	2000	500
	Minimum MH			400	119.9
March 9, 2019	Maximum MH	1500	1800	2000	1000
	Morning Transition	1200	1500	1500	200
	Afternoon Transition	1800	2000	2000	500
	Minimum MH			400	119.9
March 10, 2019	Maximum MH	1500	1800	2000	1000
	Morning Transition	1200	1500	1500	200
	Afternoon Transition	1800	2000	1500	500
	Minimum MH			400	119.9

The success of the TDMMAA algorithm greatly depends in the proper definition of the analysis periods as well as the maximum (top) and minimum (bottom) analysis altitudes. This introduces subjectivity in the algorithm and needs an experienced or a well-guided user. Future work entails making the algorithm more objective by coming up with monthly TDMMAA parameters or by using machine learning in defining these variables. Validation with in-situ profile measurements using unmanned aerial vehicles that were also deployed nearby the atmospheric LiDAR site during the same study period needs to be also performed to gain better confidence in the algorithm. Further testing of the algorithm in various atmospheric conditions and deployment locations of the atmospheric LiDAR in Chiang Mai province will also be done as a next step.

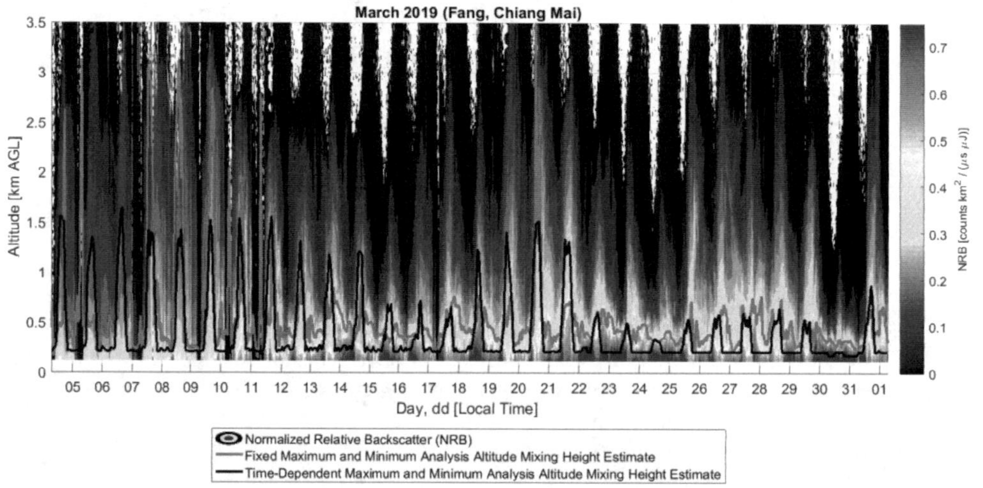

Figure 2. Applying the algorithm for the entire dataset from March 4 – 31, 2019.

4. Conclusion

A time-dependent maximum and minimum analysis altitude (TDMMAA) has been incorporated to the standard Haar wavelet covariance transform (WCT) method used in estimating the mixing height from atmospheric LiDAR observations over Fang, Chiang Mai in northern Thailand during the 2019 dry season. This method proved more robust against atmospheric and instrumentation features that the standard Haar WCT method tend to overestimate the mixing height estimate. However, there is some subjectivity in the TDMMAA method that needs to be further investigated in order to produce a more objective algorithm.

Acknowledgements

The authors would like to thank the Thailand Science Research and Innovation office and the Asia Pacific Network (APN) for Global Change Research (CBA2018-01MY-Wanthongchai - Integrated Highland Wildfire, Smoke and Haze Management in the Upper Indochina Region project) for partially supporting this work.

References

[1] Seibert P, Beyrichm F, Gryningm S E, Joffrem S, Rasmussenm A and Tercierm P 2000 Review and intercomparison of operational methods for the determination of the mixing height *Atmos. Environ.* **34** 1001–27
[2] Solanki R, Macatangay R, Sakulsupich V, Sonkaew T and Mahapatra P S 2019 Mixing layer height retrievals from miniMPL measurements in the Chiang Mai valley: implications for particulate matter pollution *Front. Earth Sci.* **7** **308** 1–11
[3] Brooks I 2003 Finding boundary layer top: application of a wavelet covariance transform to lidar backscatter profiles *J. Atmos. Ocean. Technol.* **20** 1092–105

An application of reflective holographic gratings for measurement of cylindrical curvature

Chanikan Inneam, Keerayoot Srinuanjan*and Witoon Yindeesuk

Department of Physics, Faculty of Science, King Mongkut's Institute of Technology Ladkrabang, Bangkok, 10520, Thailand.

*E-mail: keerayoot.sr@kmitl.ac.th

Abstract. This paper presented an application of reflective holographic gratings for the measurement of cylindrical curvature. The surface of the fabricated holographic grating was coated with gold by the sputtering method, where it became a reflective holographic grating. The grating was attached to the surface of various radius cylindrical objects. The diffraction pattern produced by the bent grating with different radius was observed by illuminating a laser beam normal to the grating surface. The gratings constant were calculated from the observed diffraction pattern. The relationship between the grating constants and the radius of cylindrical objects was obtained. The grating constant and the reciprocal of the radius of cylindrical objects was a linear relationship, with the least R-square between 0.85-0.97. Moreover, the y-intercept of the relationship between the grating constants and the reciprocal radius was consistent with the grating constant of the non-bended grating. As the radius of the grating approach is infinite, the reciprocal of the radius approaches zero, which is a non-bend grating. We can apply this method to measure the radius of cylindrical objects.

1. Introduction

Curvature is an important physical parameter that needs to be monitored in engineering, bridge, machinery, architectural structure, health monitoring, biomechanics, robotics, and aerospace. Various optical fiber-based curvature sensors have been widely studied due to their distinct advantages of simple structure and high sensitivity. Researchers have proposed methods for the physical parameter based on a different structure, such as holographic optical elements (HOEs) [1], long-period grating (LPG) [2–3], fiber Bragg grating (FBG) [4–7], photonic crystal fiber (PCF) [8–9], a multicore fiber (MCF) [10–11], and in-fiber Mach–Zehnder interferometer (MZI) [12–14]. They could be well applied to the physical parameter measurement. However, the methods mentioned above have some accuracy problems and complication for setting up.

This research proposes a method for measuring the radius of cylindrical objects by applying reflective holographic gratings. The reflective holographic grating was fabricated by the method in Ref. [15]. Then, the holographic grating was coated with gold by the sputtering method. The grating was attached to the surface of various radius cylindrical objects. A diffraction pattern appears on the screen observed by illuminating a laser beam normal to the bent grating surface. The grating constant is calculated from the diffraction pattern where the grating constant corresponds to the radius. Then grating constant is confirmed by Scanning Electron Microscope (SEM). The relationship between the gratings constant and the reciprocal radius of the objects and discussed. Using this method, the radius of a cylindrical object could easily be measured, uncomplicated experiment setup and able to measure the radius in a narrow space. In the future, we may have more devices to support better measurement.

2. Experiment

2.1. Fabrication of reflective holographic grating

Transmission grating fabricated from an alternative setting of Michelson interferometer, Inneam *et al.* [15], were fabricated. By adjusting the angles between the two arms of the Michelson interferometer, cross-beam angle, we can obtain a fringe pattern that has a linear shape. The grating sample 1 to 5 have cross-beam angles (β) 0.2°, 0.4°, 0.6°, 0.8°, and 1.0° respectively. The surface of the fabricated holographic grating was coated with gold by the sputtering method. The sputtering method used gold as a target material bombarded with the Argon plasma, the plasma current of 18 mA in 120 s, and the coating thickness is 408 Å. The transmission grating became the reflective holographic grating.

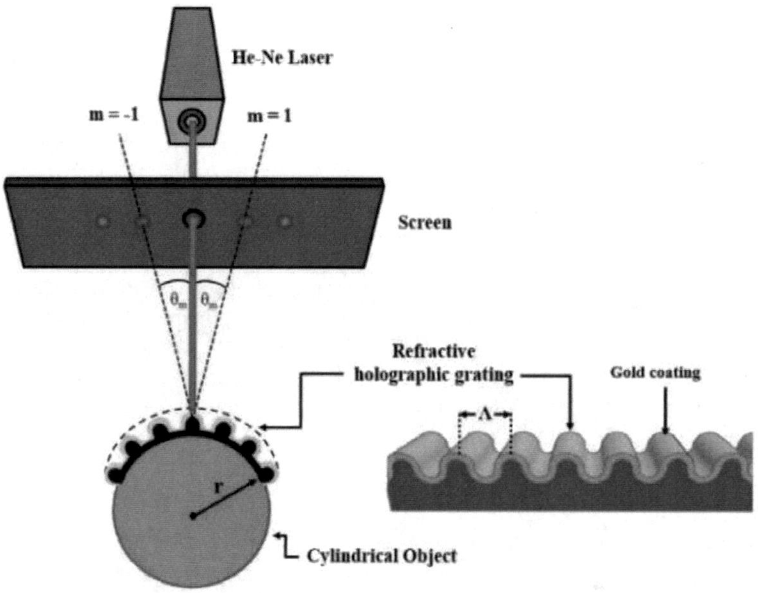

Figure 1. The schematic diagram for measuring the radius of the cylindrical object.

2.2. Experiment method

The experiment setup for measuring the radius of cylindrical objects is shown in figure 1. First, the reflective holographic grating was attached to the various radius of cylindrical object surfaces, where the grating is bent along the cylindrical object surface. The radius of objects are 0.0129 m, 0.0195 m, 0.0260 m, and 0.0319 m. Second, the He-Ne laser with a wavelength of 632.8 nm was used as a light source. By illuminating the laser beam normal to the bent grating surface with a different radius, a diffraction pattern was observed on the screen. Finally, the grating constant is calculated from the observed diffraction pattern. In addition, the grating constant was observed by Scanning Electron Microscope is shown in figure 2. The relationship between the grating constant and the reciprocal radius of the cylindrical objects is shown in figure 3.

3. Results and discussion

3.1. Measurement of grating constant by Scanning Electron Microscope

The grating constant was observed by Scanning Electron Microscope (SEM). Figure 2 shows the image of the reflective holographic grating. The grating constants are shown in table 1.

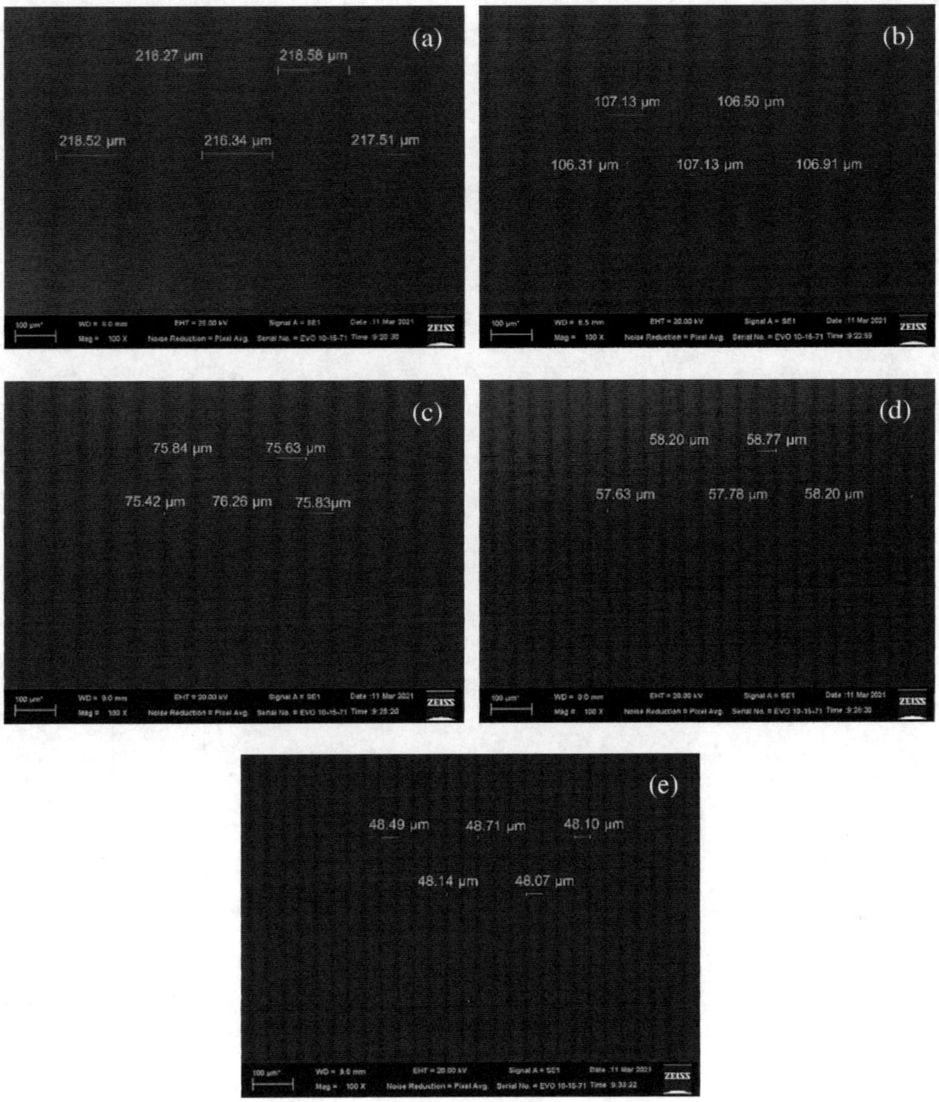

Figure 2. Image of reflective holographic gratings obtained by SEM, where (a) Sample 1, (b) Sample 2, (c) Sample 3, (d) Sample 4, and (e) Sample 5.

3.2. Measurement of grating constant by illumination of the laser beam

The laser beam is normally incident the bent grating surface. The diffraction pattern caused by the bent grating with different radius appears on a screen. The diffraction angle is measured, and the grating constant Λ is calculated from the following equation,

$$\Lambda = \frac{\lambda}{2\sin\theta_m} \tag{1}$$

where λ is the wavelength of the He-Ne laser and θ_m is the angle of the diffraction pattern. The relationship between the grating constant (Λ) and the reciprocal radius of a cylindrical object (1/r) is shown in figure 3. The y-intercept from the graph is shown in table 2.

Table 1. Comparison of the non-bend grating constants, measured by Scanning Electron Microscope (SEM) and measured by illumination of the laser beam.

Method	The grating constants (μm)				
	Sample 1	Sample 2	Sample 3	Sample 4	Sample 5
Measured by SEM	217.44	106.79	75.80	58.12	48.30
Measured by experiment	218.77	100.35	76.11	58.99	47.27
Percentage difference (%)	0.61	6.21	0.41	1.49	2.16

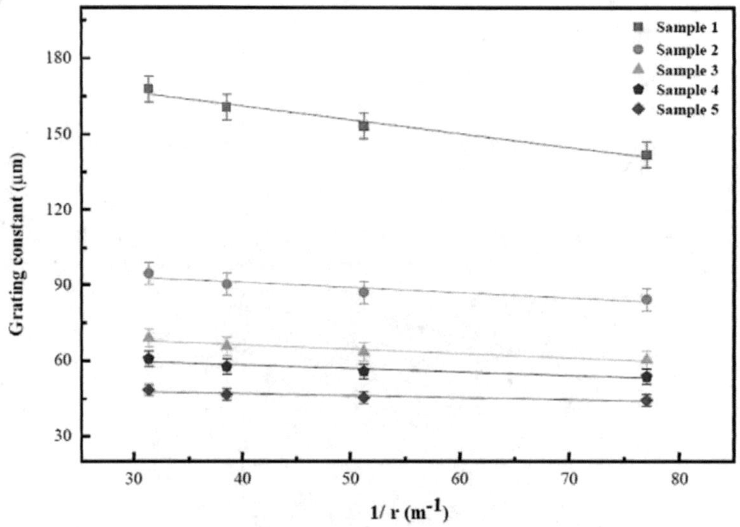

Figure 3. The relationship between the grating constant and the reciprocal radius of the cylindrical objects.

Table 2. The grating constants which measured by the illumination of the laser beam and y-intercept of the graph.

1/r (m^{-1})	The grating constants (μm)				
	Sample 1	Sample 2	Sample 3	Sample 4	Sample 5
31.37	167.84	94.65	68.98	60.80	48.48
38.54	160.62	90.32	65.67	57.70	46.72
51.23	153.23	87.08	63.41	55.77	45.43
77.04	141.98	84.25	60.41	53.77	44.46
Y-intercept	182.94	99.35	73.16	63.87	50.22
R^2	0.97	0.87	0.91	0.86	0.85

Figure 3 showed that the grating constant and the reciprocal radius of cylindrical objects is a linear relationship. The R-square is in the range of 0.85-0.97. Also, it was found that the y-intercept is consistent with the non-bend grating constant.

4. Conclusion

In this research, we presented the method and experimental results for measuring the radius of cylindrical objects. In the method, the grating is fixed to the surface of a cylindrical object, where the object radius is varied from 0.013 m to 0.032 m. Then the diffraction pattern was observed, and the grating constant of the bent grating of the various radius is calculated. The relationship between the grating constant and the reciprocal radius of the cylindrical object is linear, the R-square is in the range of 0.85 to 0.97, and the y-intercept corresponds to a non-bend grating constant, which is confirmed by two different methods. The non-bend grating constant measured by the illumination of the laser beam is in agreement with that obtained by the SEM imaging. From the experimental results, if the calibration graph between the grating constant and the reciprocal radius has been performed, this system will measure the curvature of an unknown radius, where the object radius is in the range 0.013 - 0.032 m with the error ± 1.02 %. Therefore, we could apply the method to measure the radius of a cylindrical object.

References

[1] Bang K, Jang C and Lee B 2019 *J. Inf. Disp.* **20** 9–23
[2] Li Y P, Zhang W G, Wang S, Chen L, Zhang Y X, Wang B, Yan T Y, Li X Y and Hu W 2017 *IEEE Photon. Technol. Lett.* **29** 224–7
[3] Yi Y *et al.* 2020 *Infrared Phys. Technol.* **111** 103520
[4] Zhu F, Zhang Y, Qu Y, Jiang W, Su H, Guo Y and Qi K 2020 *Opt. Fiber Technol.* **54** 102133
[5] Kadhim S A and Resen D A 2016 *J. Nanosci. Nanotechnol.* **2** 91–3
[6] Sun G, Hu Y, Dong M, He Y, Yu M and Zhu L 2019 *Optik* **176** 559–66
[7] Mao L, Lu P, Lao Z, Liu D and Zhang J 2014 *Opt. Laser Technol.* **57** 39–43
[8] Fu G, Li Y, Li Q, Yang J, Fu X and Bi W 2017 *IEEE Photon. J.* **9** 1–14
[9] Gong H, Song H, Li X, Wang J and Dong X 2013 *Sens. Actuator* A *Phys.* **195** 139–41
[10] Delgado G S, Newkirk A V, Lopez J E A, Rios M A, Schülzgen A and Correa R A 2015 *Opt. Lett.* **40** 1468
[11] Wang So, Zhang Y X, Zhang W G, Geng P C, Yan T Y, Chen L, Li Y P and Hu W 2017 *IEEE Photon. Technol. Lett.* **29** 822–5
[12] Meng Q Q, Dong X Y, Zhou Y, Ni K, Chen Z M, Zhao C L and Jin S Z 2012 *J. Electromagn. Waves Appl.* **26** 2438–44
[13] Wang R, Zhang J, Weng Y, Rong Q, Ma Y, Feng Z, Hu M and Qiao X 2013 *IEEE Sens. J.* **13** 1766–70
[14] Zhang S, Zhang W, Gao S, Geng P and Xue X 2012 *Opt. Lett.* **37** 4480
[15] Inneam C and Srinuanjan K 2021 *J. Phys.: Conf. Ser.* **1719** 012087

The overlapped latent fingerprint separation by using Fourier Optics (FO)

Nichchar Sommit* and Ratchapak Chitaree

Department of Physics, Faculty of Science, Mahidol University, Bangkok, 10400, Thailand

*Email: Nichchar.som@student.mahidol.edu

Abstract. Latent fingerprints are one of the most important pieces of evidence left at a crime scene and can be linked to all individuals involved. Each person's fingerprints are unique and permanent, becoming an ideal biometric trait for the personal identification by extracting and comparing characteristic points (minutiae) of ridges. The overlapping fingerprint cases are frequently encountered in a crime scene and causing a difficult interpretation for an investigator. The problem has been challenging for forensic scientists over a decade. The method proposed in this study for the separation of overlapped latent fingerprints is based on the well-known spatial filtering method in the Fourier Optics (FO). Instead of tackling the problem by experiment, an alternative and simple means of image processing was proposed and conducted. The working principles start form converting spatial domain patterns (an image of overlapped fingerprints) to spatial frequency domain patterns or power spectrum, filtering out unwanted components (unwanted fingerprint) by appropriate spatial filters, and finally converting the modified pattern back to spatial domain patterns (an image of suspect fingerprint). As a result, the final image is improved from its original state. The periodic pattern of ridges is the key that allows FO to be used in the separation of the overlapped fingerprints. In this work, the procedures described are simply performed by an open source software: ImageJ. The FO-based image processing technique satisfactorily demonstrated its ability to recover an individual fingerprint from overlapping fingerprints.

1. Introduction

A person's fingerprints are both unique and permanent, making them an ideal biometric trait for the person identification when characteristic points (minutiae) of the ridges are extracted and compared. In forensic work, the fingerprint verification is approved if the minutiae detection from a query fingerprint in comparison to a reference fingerprint is greater than ten points. If the minutiae detection score is less than ten points, the matching is considered to be inconclusive. Occasionally, overlapped fingerprints are found in a crime scene and the overlap makes the minutiae detection more difficult than usual for investigators. Therefore, it is essential to separate those overlapped fingerprints into individual ones. Several methods were proposed to deal with such a complex problem. In 2010, Ho-Wai Tang *et al.* [1]

proposed a method to separate overlapped latent fingerprints using gold nanoparticles during the latent developing stage. The separating technique was based on chemical reactions between gold nanoparticles and different fatty acids representing substances from overlapped fingerprints. In 2011, Fanglin Chen *et al.* [2] proposed a computational procedure to separate overlapped fingerprint images. The procedure used the Fourier analysis to detect ridge orientations of overlapped fingerprints. In 2018, Brian Dalrymple investigated the potential of the Fast Fourier Transform (FFT) to separate superimposed fingerprints. The Image Pro Premier, a commercial image analysis software, was used as a platform to apply the FFT in this study [3].

From the reviews, the method of Fourier Transform (FT) is of interest to us. In fact, the investigation based on the FT to separate overlapped fingerprints can be conducted either by experiment or computational tool. It is obvious that more tools and time are needed to recover a suspect fingerprint by the experiment than the image processing method. Although a lot of image analysis software offer the FT tools to manipulate images, we chose to explore an open-source imaging software: ImageJ developed by the National Institutes of Health and Laboratory for Optical and Computational Instrument [4]. This is because the software is freely available, easy to use and especially equipped with the FFT submenu supporting necessary functions for the spatial filtering operation. We believe the proposed idea and procedures in this study will technologically demonstrate capability of an open-source platform for image analysis in addition to applying the well-known FO to solve the imaging problem.

In the next section, theoretical background regarding the FT will be briefly described.

2. Theory

2.1. Fourier Transform [5]

The operating principle for editing an image either based on the experiment or computational tool follows a similar procedure. The procedure starts form conducting the Fourier Transform to convert spatial domain patterns (an image to be edited) to spatial frequency domain patterns and then apply a filtering process to remove noise (unwanted patterns) by appropriate spatial filters. Finally, the inverse Fourier Transform is used to convert the modified pattern back to spatial domain patterns (an edited image). As a result, the final image is improved from its original state.

Theoretically, Fourier Transform is the transformation of a function in the spatial domain into the spatial frequency domain and the inverse is also valid. The **Fourier-transform pair** take the form,

$$f(x,y)=\frac{1}{(2\pi)^2}\iint_{-\infty}^{+\infty} F\left(k_x,k_y\right) e^{-i\left(k_x x+k_y y\right)}\, dk_x dk_y \tag{1}$$

and

$$F(k_x,k_y)=\iint_{-\infty}^{+\infty} f\,(x,y)\, e^{i\left(k_x x+k_y y\right)}\, dxdy \tag{2}$$

where the function $F(k_x,k_y$ is the **Fourier transform** of $f(x,y)$, which is symbolically denoted by

$$F(k_x,\, k_y) = F\{f(x,y)\} \tag{3}$$

on the other hand, $f(x,y)$ was called the **inverse Fourier transform** of $F(k_x,k_y)$, or symbolically designated as

$$f(x,y)=F^{-1}\{F(k_x,k_y)\}=F^{-1}\{F\{f\,(x,y)\}\} \tag{4}$$

where k_x and k_y are the spatial frequencies along the two axes.

Note that $F(k_x,k_y)$ is the Fourier Transform and is, in general, a complex quantity with the real part $R(k_x,k_x)$ and the imaginary part $I(k_x,k_x)$ [6],

$$F(k_x,k_x) = R(k_x,k_x) + iI(k_x,k_x) = \left|F(k_x,k_x)\right|e^{i\phi} \tag{5}$$

where the amplitude in the frequency domain

Journal of Physics: Conference Series **2145** (2022) 012055 doi:10.1088/1742-6596/2145/1/012055

$$|F(k_x,k_x)| = \sqrt{(R(k_x,k_x))^2 + (I(k_x,k_x))^2} \tag{6}$$

and phase in the frequency domain

$$\phi = arctan(\frac{I(k_x,k_y)}{R(k_x,k_y)}) \tag{7}$$

Amplitude represents the distribution of magnitude of various spatial frequency components and phase contains the information of position of these spatial frequency components. In most cases, the display in the Fourier plane is based on only the magnitude, ignoring the phase.

The square of the amplitude is used and this term is the power spectrum: $P(k_x,k_y)$,

$$P(k_x,k_y) = |F(k_x,k_y)|^2 \tag{8}$$

The power at each location shows how much of that frequency and orientation is present in the image.

2.2. Spatial Filtering [7]

The objective of the spatial filtering in this study is to retrieve or remove specified frequencies in the frequency domain by setting the amplitude of those unwanted terms to zero. This will leave only the desired part of the information to be inverse transformed back to the spatial domain to produce a noise-free image. The FFT submenu in ImageJ provides several commands to design appropriate spatial filters to remove periodic points, lines, or regions in the power spectrum image. This suggests that the filter created in this program deals only with the amplitude modification.

3. Methodology for FFT spatial filtering

The following steps are applied to the implementation of the spatial filtering in the ImageJ.

(1) A digital image with noise is imported to the ImageJ via the open file command. The imported image appears in a spatial domain window. Some suggested image file formats include JPEG, BMP, TIFF and GIF.

(2) The Fourier Transform of the imported image file can simply be executed by choosing the FFT command in the FFT submenu.

(3) After the FFT computation is complete, the corresponding power spectrum is displayed in a new frequency domain window.

(4) Within the power spectrum window, filters or masks created from the ImageJ's selection tools can be applied to remove noise.

(5) Finally, the modified power spectrum in the frequency domain is transformed back into the spatial domain by the Inverse FFT command in the FFT submenu. As a result of these procedures, the noise-free image is obtained and illustrated in a modified spatial domain window.

4. Results

To show the FFT commands in ImageJ are applicable in separating the overlapped fingerprints, a number of overlapping fingerprint images and filtering schemes were investigated. The simplest fingerprint pattern selected for the first test was the plain arch. As illustrated in figure 1, the simple arch patterns overlap and create lines of intersection.

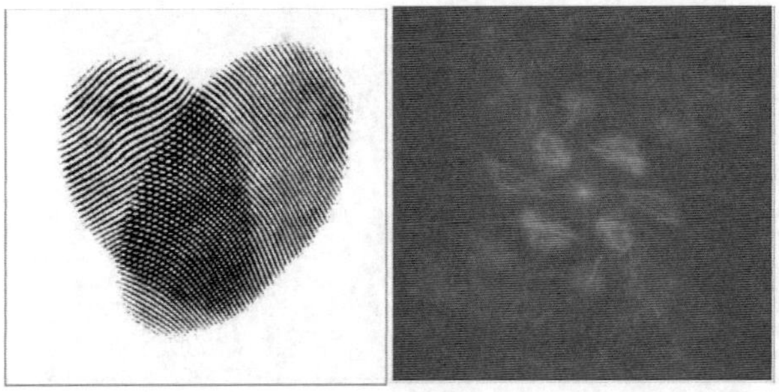

Figure 1. The overlapping of plain arch pattern and its power spectrum.

According to the theory of Fraunhofer diffraction, the size of the diffraction pattern is inversely proportional to the object's scale. If the fingerprint under interest is on the right, the left fingerprint must be deleted. As illustrated in figure 2(a), the spatial frequency components of the left fingerprint can be eliminated by applying appropriate area filters onto the power spectrum. Similarly, the spatial frequency components of the right fingerprints can be eliminated as well (figure 2(b)).

(a)

(b)

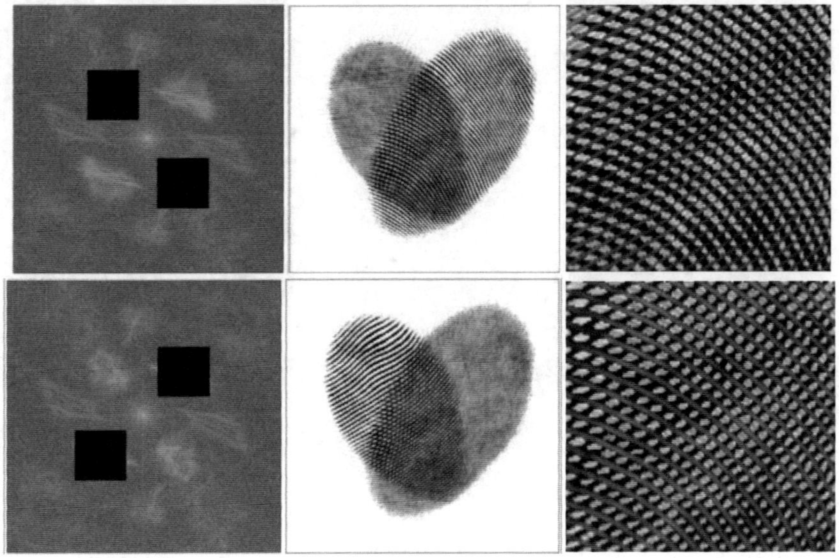

Figure 2. Removal of unwanted minutiae of plain arch pattern of (a) the left fingerprint (b) the right fingerprint.

Secondly, the investigation was conducted on the loop pattern (which is a more complicated fingerprint pattern than the plain arch pattern) as shown in figure 3. The fingerprint separation was processed by using the same procedures as conducted in the earlier study for removing the unwanted minutiae as shown in figure 4.

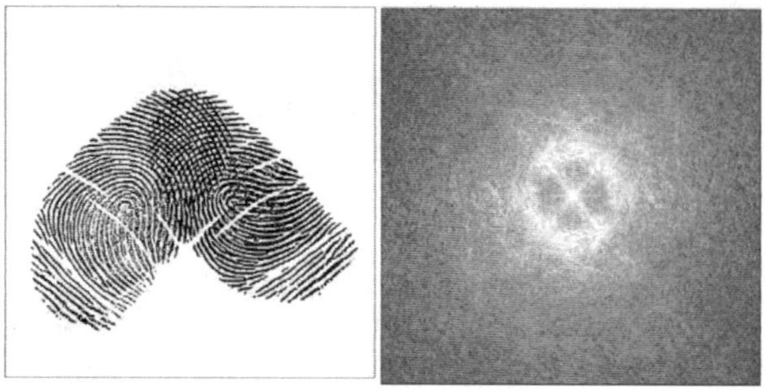

Figure 3. The overlapping of loop pattern and its power spectrum.

(a)

(b)

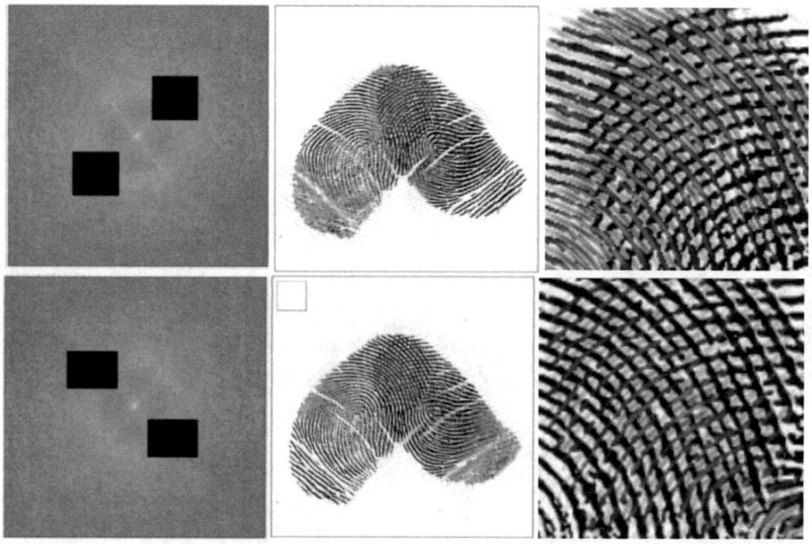

Figure 4. Removal of the unwanted minutiae of loop pattern of (a) the left fingerprint (b) the right fingerprint.

However, some wanted parts of the latent fingerprint also disappeared with the unwanted ridges. This is because the missing parts have the same spatial frequency components as the parts that being eliminated. Furthermore, some portions of the unwanted fingerprint were not entirely removed because they were not in the coverage of the filters. The effect of varying the filter location in the high spatial frequencies of each fingerprint was investigated and the results are shown in figure 5.

(a)
(b)
(c)

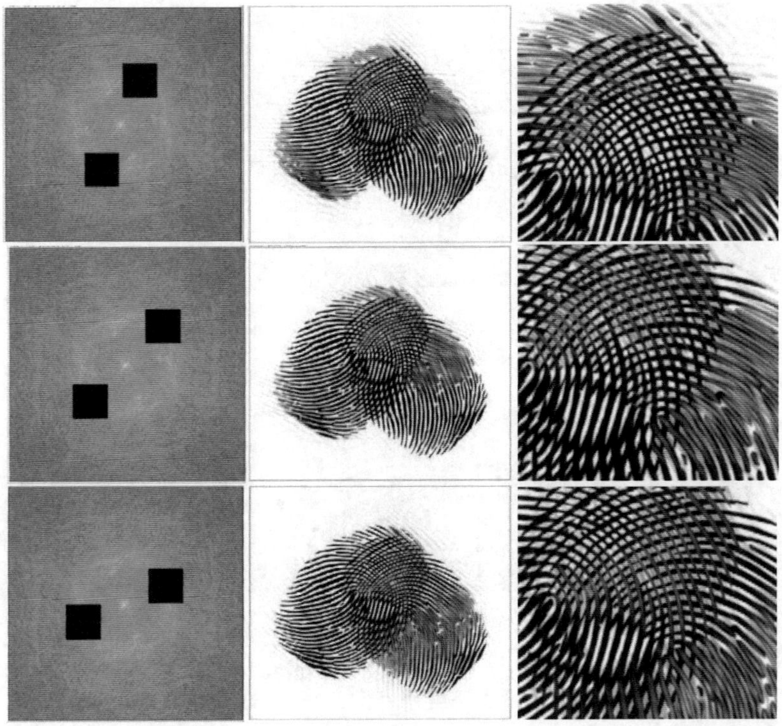

Figure 5. The elimination of the high spatial frequencies in the left fingerprint with filter positioned (a) close to the vertical low spatial frequency axis (b) at 45° with respect to the vertical low spatial frequency axis (c) close to the horizontal low spatial frequency axis. Modified images are also illustrated in each case.

The results show that, apart from the relative positions of the spatial filters with respect to the center of the image in the FT plane, the angular positions of the filter relative to a reference axis of the image also affect the modification. As shown in figures 5(a), (b) and (c), the fine detail of ridges that are eliminated depending on the angular position of the spatial filter relative to reference axis. For example, if the filters that blocked the high spatial frequencies was placed close to the vertical axis (figure 5(a)), the eliminated ridges were that close to the horizontal axis of the modified image.

As mentioned in the introduction, the fine fingerprint ridge patterns or minutiae are very important in identifying an individual's fingerprints. In reality, latent fingerprints are often ambiguous and difficult to check in crime scenes. This digital technique allows us to remove the unwanted minutiae from the wanted ones. Figure 6 shows an unclear image of overlapped fingerprints and its corresponding power spectrum. As seen in figure 7, the findings show that the unwanted minutiae from one finger print can partly be removed and vice versa.

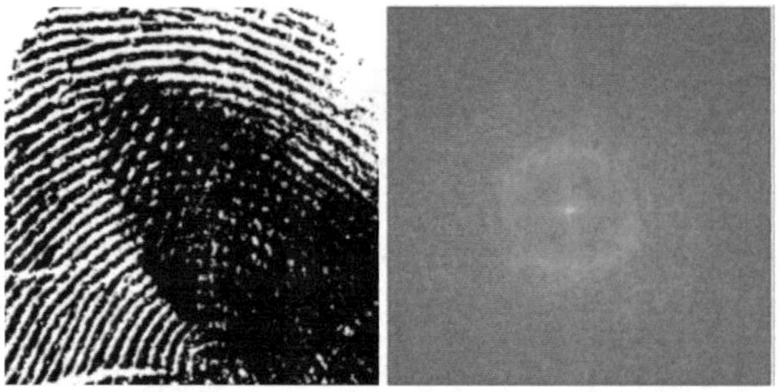

Figure 6. Unclear overlapped fingerprints and its power spectrum.

(a)

(b)

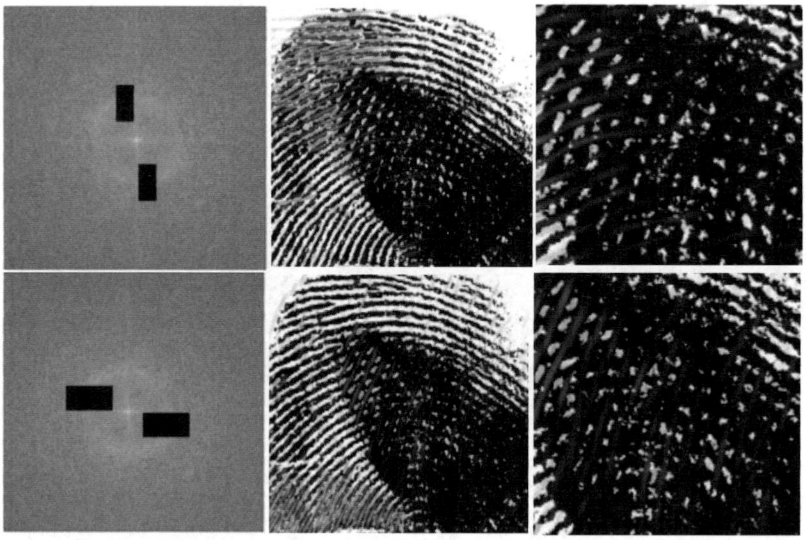

Figure 7. Removal of unwanted minutiae from (a) the left fingerprint and (b) the right fingerprint of unclear overlapped fingerprints.

5. Conclusions

The main goal of this study is to distinguish overlapped latent fingerprints using the Fourier Optics (FO) technique. ImageJ was used as a platform to apply the procedure of Fourier Transform to solve the problem. By employing the software, the modification of the overlapping fingerprint image could easily be done. However, there is a drawback which can hardly be avoided with the current FT. It is the missing of wanted parts while trying to eliminate some unwanted parts of fingerprints. This may overcome by also including the phase information (encoded in the imaginary part) of the image in the Fourier transform process. This point is actually worth further studying because usually only the power spectrum corresponding to the real part of the image FT involves in the reconstruction of the original image by applying the inverse FT. Yaov Shechtman *et al.* [8] pointed out the importance of the Fourier phase and also showed the significance of the phase term in the image recovery.

References

[1] Tang H W, Lu W, Che C M and Ng K M 2010 Gold nanoparticles and imaging mass spectrometry: Double imaging of latent fingerprints *Anal. Chem.* **82** 1589–93

[2] Chen F, Feng J, Jain A K, Zhou J and Zhang J 2011 Separating overlapped fingerprints *IEEE Trans. Inf. Forensics Secur.* **6** 346–59

[3] Dalrymple B 2018 Fast fourier transform: A Niche, but critical strategy in optimizing impression evidence *J. Forensic Sci. Med.* **4**(4) 192

[4] Abràmoff M D, Magalhães P J and Ram S J 2004 Image processing with ImageJ *Biophot. Int.* **11**(7) 36–42

[5] Haus H 1984 Waves and fields in optoelectronics (Englewood Cliffs, NJ USA: Prentice-Hall, Inc.,)

[6] Pan J J 1989 Spectral analysis and filtering techniques in digital spatail data processing *Photogramm. Eng. Remote Sens.* **55**(8) 1203–7

[7] Eisenkraft A 1977 A closer look at diffraction: Experiments in spatial filtering *Phys. Teach.* **15**(4) 199–211

[8] Shechtman Y, Eldar Y C, Cohen O, Chapman H N, Miao J and Segev M 2015 Phase retrieval with application to optical imaging: A contemporary overview *IEEE Signal Process. Mag.* **32** 87–109

Journal of Physics: Conference Series **2145** (2022) 012056 doi:10.1088/1742-6596/2145/1/012056

Investigation on optical integration between LED Mid-IR light sources and Si-based waveguides for sensing applications

Pawaphat Jaturaphagorn[1], Papichaya Chaisakul[1,*], Nattaporn Chattham[1] and Pichet Limsuwan[2]

[1] Department of Physics, Faculty of Science, Kasetsart University, Bangkok, 10900 Thailand
[2] Department of Physics, Faculty of Science, King Mongkut's Institute of Technology Ladkrabang, Bangkok 10520, Thailand

*Email: fscipac@ku.ac.th

Abstract. Research on mid-IR silicon-based waveguides has recently received strong interest. Particularly, this paper focuses on one of the critical issues in micron-scale photonic integrated circuits, which is to efficiently couple a mid-IR LED (light emitting diode) light source to an external micron-scale waveguide. The optical coupling scheme is crucial for the exploitation of LED light sources in waveguide-based spectroscopic sensing applications. This paper reports optical coupling scheme between an LED mid-IR light source and a silicon rich silicon nitride (SiN) waveguide that could enable the use of LED-based light sources. Finally, the detection limit of the investigated device for carbon dioxide gas detection is calculated.

1. Introduction

The potential of silicon photonics was largely based on the use of silicon (Si)-based platforms toward commercialization of photonic devices [1], that are compatible with the existing CMOS fabrication techniques used in most of the well-known electronic devices [2]. For optical interconnect applications, significant progress has been done regarding interconnection between light sources and Si-based waveguides. Optical integration between micro-disk quantum well lasers and external waveguides was reviewed [3].

In spectroscopic sensing applications, the new prospect of using silicon nitride (SiN) strip waveguides was recently discussed [4]. SiN strip waveguides growth on silicon dioxide (SiO_2) exhibit great electromagnetic confinement for a broad range of mid-IR wavelengths. Moreover, wideband gas detection with practical detection limits was also possible based on the absorption of the evanescent field by the targeted gas molecules. The incorporation of a light source was not addressed in [4]. As typical taper structures can be effective only when considering optical coupling between different photonic structures based on the same materials [5], interconnection between LED light sources which will not be Si-based and SiN strip waveguides is worth investigation. Moreover, almost all of the previous works investigate the optical integration based on laser-based light sources, which might prevent the devices from affordable and low power consumption usage [6]. For example, LED mid-IR light sources could have significantly-lower power consumption [7] compared to the laser-based ones.

In this paper, we investigate the optical integration between an LED mid-IR light source and a SiN waveguide on silicon dioxide to identify the optical coupling scheme that could enable the use of LED-based light sources for spectroscopic sensing applications. The efficient optical coupling allows the SiN waveguides to be employed for gas molecule sensing applications with good detection limits. As the refractive indices (n) of SiN can be usefully adjusted over a wide range of values from 2 to 3, it is a very promising platform to be used for on-chip spectroscopic sensing applications considering its flexibility to be specifically optimized for different needs [8-9]. In this proceeding, we focus our report on Si-rich SiN of n ~ 3.

2. Methodology

We employ two well-known techniques of Finite difference time domain (FDTD) and finite difference eigenmode (FDE) to approximate and optimize several parameters of the optical integration scheme between an LED mid-IR light source and a Si-rich SiN waveguide. Figure 1 shows the 3D schematic view and the corresponding side view of the integration scheme under investigation.

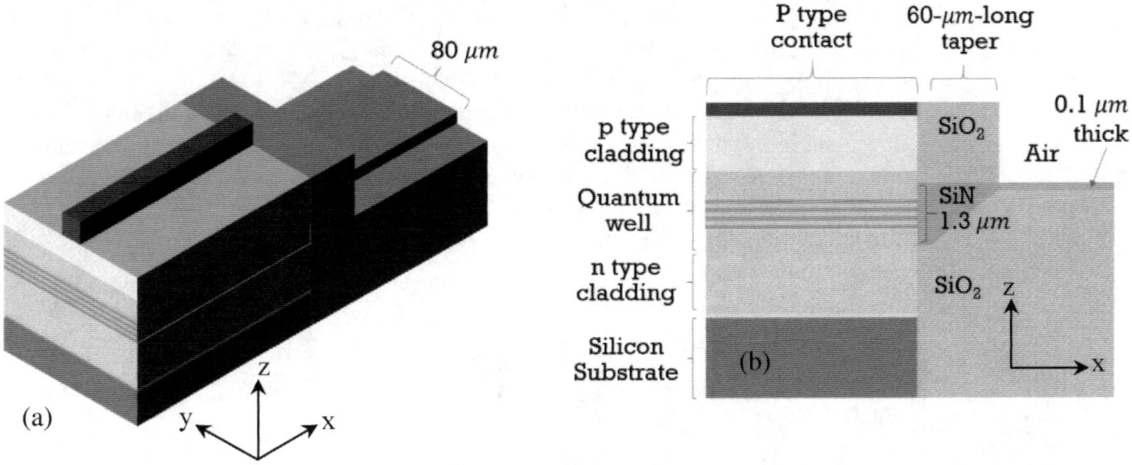

Figure 1. (a) The schematic view and (b) the side view of the structure under investigation.

In this paper, we adopt the LED mid-IR device from [7], in which all of the epitaxial details could be obtained. As in figure 2, from the FDE calculation, we confirm that the LED structure supports single mode propagation of transverse-electric (TE) light at the optical wavelength of ~ 2 μm, which is emitted from the quantum well layers embedded in the structure as indicated in figure 1.

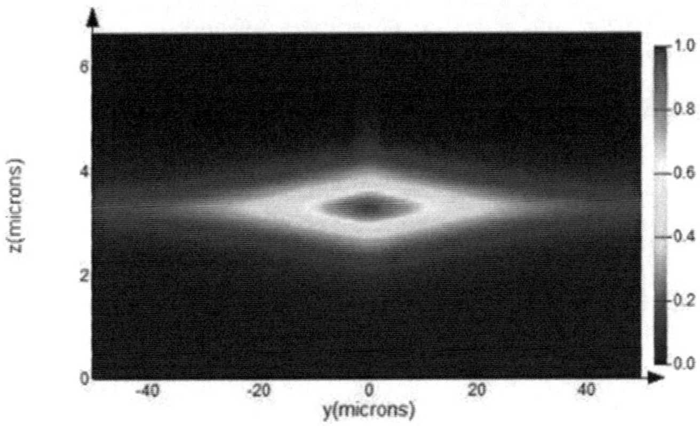

Figure 2. The optical mode profile of the LED mid-IR light source. The structure supports single mode propagation of light at the optical wavelength of ~ 2 μm.

3. Results and discussion

We used the FDTD method to simulate the optical integration between the LED mid-IR light source and the SiN waveguide at the optical wavelength of ~ 2 μm. After optimization, as in figure 1 we arrive at the coupling scheme that consists of a 60-μm-long taper structure which facilitates the optical coupling from LED mid-IR output, connected to the 1.3-μm-high and 80-μm-wide SiN waveguide (to enable an optimized optical mode matching between the LED and the SiN), via the taper to the 0.1-μm-high and 80-μm-wide SiN waveguide to be used for optical sensing based on the absorption of the evanescent field by the target gas molecules. Figure 3(a) shows the optical propagation pattern of the integration scheme from the LED mid-IR light source to the 0.1-μm-high and 80-μm-wide SiN waveguide with air top cladding that could be used for optical sensing. From figure 3(b-d), the coupling scheme can effectively transfer the relatively large optical mode from the LED-based mid-IR light source (as in figure 2) to the compact optical mode of the 0.1-μm-high and 80-μm-wide SiN waveguide. From the 3D FDTD simulation, 68% (as low as 1.7 dB loss, in which ~ 1.1 dB, ~ 0.04 dB, and ~ 0.56 dB are from the LED to figure 3(b), from figure 3(b) to figure 3(c), and from figure 3(c) to figure 3(d), respectively) of the optical power was impressively coupled into the 0.1-μm-high SiN waveguide with air top cladding. Moreover, at the 0.1-μm-high SiN waveguide of figure 3(d), we calculate that the optical mode has as much as 24% of the guided optical power in the air cladding which is comparable to the state-of-the-art report [10] making it promising to be used for spectroscopic sensing based on evanescent field absorption of the sensed molecules. Figure 3(e) reports optical coupling performance over the LED bandwidth of 200 nm from 1.90 μm to 2.10 μm [7]; the coupling performance decreases by less than 3% over the entire wavelength region.

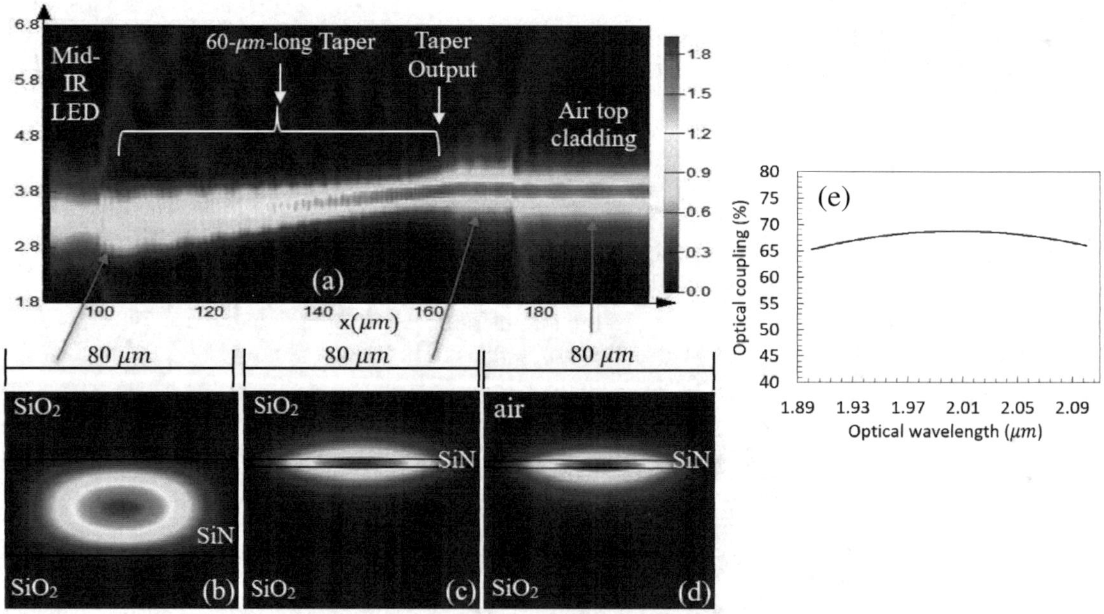

Figure 3. (a) The optical propagation pattern of the integration scheme from the LED mid-IR light source to the 0.1-μm-high and 80-μm-wide SiN waveguide with air top cladding. A 60-μm-long taper structure facilitates the optical coupling from LED mid-IR output, which is connected to (b) the 1.3-μm-high and 80-μm-wide SiN waveguide, to (c) the 0.1-μm-high and 80-μm-wide SiN waveguide and to (d) the 0.1-μm-high and 80-μm-wide SiN waveguide with air top cladding to be used for on-chip spectroscopic sensing based on the absorption of the evanescent field by the targeted gas molecules. (e) Wavelength-dependent optical coupling performance.

The detection resolution could be estimated from [11]

$$C(\text{mol/L}) = -\ln[1-\{(SNR \cdot NEP \cdot B^{1/2})/P_0\exp(-\alpha L_{opt})\}]/\varepsilon\eta L \quad (1)$$

in which *SNR*, *NEP*, and *B* represent the signal-to-noise ratio, the noise equivalent power, and the bandwidth of the employed detection scheme, respectively. *C* largely depends on the operating conditions employed in the measurement; hence, we use typical values of $SNR = 10$, $NEP = 5 \times 10^{-12}$ W/Hz$^{1/2}$, and $B = 0.5$ Hz (1s averaging time) [4, 11]. P_0(W) is the optical power from the LED which is conservatively assumed to be 1 mW, α(cm^{-1}) is intrinsic optical loss of the SiN waveguide which can be estimated to be 2dB/cm [4], η is the evanescent power factor representing the portion of optical power in the evanescent field as discussed previously ($\eta \approx 24$ %), and L(cm) is the waveguide length which is chosen to be only 2 cm for compactness. Considering the molar absorption of gas molecule ε of ≈ 1.1 Lmol^{-1}cm^{-1} for CO_2 around the optical wavelength of 2 μm [12], it is possible to estimate the detectable CO_2 concentration of around 6 ppm, which is lower than the occupational exposure limit of 5000 ppm for CO_2 recommended by the international environmental standard of the National Institute for Occupational Safety and Health (NIOSH), affirming the potential to employ the proposed integration scheme for gas trace detection in real-time environmental applications thanks to the fact that the coupling scheme can potentially transfer a relatively-large optical mode from the LED-based mid-IR light source to a much more compact optical mode of the SiN waveguide.

Acknowledgement
This project is funded by Kasetsart University Research and Development Institute (KURDI), and National Research Council of Thailand (NRCT): NRCT5-RSA63002-05.

References

[1] Boeuf F *et al* 2015 Recent progress in silicon photonics R&D and manufacturing on 300 mm wafer platform *Proc. Int. Conf. on Optical Fiber Communication (OFC)*(Los Angeles) **W3A.1**
[2] Zhang H *et al* 2012 CMOS-compatible fabrication of silicon-based sub-100-nm slot waveguide with efficient channel-slot coupler *IEEE Photon. Technol. Lett.* **24**(1) 10–2
[3] Roelkens G, Liu L, Liang D, Jones R, Fang A, Koch B and Bowers J 2010 III-V/silicon photonics for on-chip and inter-chip optical interconnects *Laser Photonics Rev.* **4**(6) 751–79
[4] Koompai N *et al* 2020 Analysis of Si_3N_4 waveguides for on-chip gas sensing by optical absorption within the mid-infrared region between 2.7 and 3.4 μm *Results Phys.* **16** 102957
[5] Fu Y, Ye T, Tang W and Chu T 2014 Efficient adiabatic silicon-on-insulator waveguide taper *Photonics Res.* **2**(3) A41–4
[6] Tarabrin M K *et al* 2015 Tunable CW solid-state Mid-IR Cr2+:CdSe single crystal laser with diode laser array pumping *Advanced Solid State Lasers, OSA Technical Digest* **AM5A.34**
[7] Vizbaras K *et al* 2015 High power continuous-wave GaSb-based superluminescent diodes as gain chips for widely tunable laser spectroscopy in the 1.95–2.45 μm wavelength range *Appl. Phys. Lett.* **107** 011103
[8] Ng D K *et al* 2015 Exploring high refractive index silicon-rich nitride films by low-temperature inductively coupled plasma chemical vapor deposition and applications for integrated waveguides *ACS Appl. Mater. Interfaces* **7**(39) 21884–9
[9] Lim K, Krishnamurthy V, Ying J, Pu J and Wang Q 2017 Ultrahigh index and low-loss silicon rich nitride thin film for NIR HAMR optics *IEEE Trans. Magn.* **53**(5) 1–7
[10] Ottonello-Briano F *et al* 2020 Carbon dioxide absorption spectroscopy with a mid-infrared silicon photonic waveguide *Opt. Lett.* **45**(1) 109–12
[11] Liu Q *et al* 2018 Mid-infrared sensing between 5.2 and 6.6 μm wavelengths using Ge-rich SiGe waveguides *Opt. Mater. Express* **8**(5) 1305–12
[12] Rothman L S *et al* 2013 The HITRAN2012 molecular spectroscopic database *J. Quant. Spectrosc. Radiat. Transf.* **130** 4–50

The simulation of the microwave shielding properties of the dual band pass frequency selective surface

Asma Samoh* **and Ratchapak Chitaree**

Department of Physics, Faculty of Science, Mahidol University, Bangkok, 10400, Thailand

* E-mail: asma.sam@student.mahidol.ac.th

Abstract. Without proper caution, the microwave leakage from a microwave oven door can be harmful to users' health. In practice, the leaked radiation has to be blocked while the visible light is allowed to pass for a visual inspection of the cooking progress inside the oven. To fulfil the requirements, the door design based on the principle of the frequency selective surface (FSS) was proposed and the gridded square loop pattern was chosen. In the simulation conducted by COMSOL Multiphysics software, the size of the proposed FSS was given as 40.7×40.7 mm with a dielectric thickness of 2.8 mm. Two important characteristics in terms of shielding effectiveness (SE) and optical transparency (OT) of the proposed FSS configuration at normal incidence were simulated and found to be 62.7 dB and 57.5%, respectively. The simulation results indicate that the proposed FSS is applicable to a safety design of a microwave oven door in suppressing the microwave leakage. Parametric studies on the characteristics due to geometrical dimensions and glass substrate thickness were also investigated. These parameters were found to affect the shielding and transmitting performances of the proposed FSS.

1. Introduction

Microwave radiations are waves of electric and magnetic energy propagating together through space. Microwaves are a form of non-ionizing radiation meaning that they do not have enough energy to alter atoms and molecules in substances. Some familiar applications of the microwaves in daily life are telecommunications, navigations, medical treatments and especially food cooking by microwave ovens. The last activity is perhaps the most common consumer use of the microwave ovens nowadays.

Principally, the microwave radiations in microwave ovens are confined in the chamber with solid metal plates on five sides and mesh-like glass screen on the sixth side or the front door. A typical microwave oven front door equipped with a circular perforated metal sheet gives low shielding effectiveness (<27 dB) and poor visibility (<50%) [1]. In other words, the radiation is likely to leak through the door risking users' health and the visually checking progress of cooking becomes difficult. To improve the shielding effectiveness and visibility, an alternative configuration based on the principle of the frequency selective surface (FSS) for a microwave oven door was numerically studied using COMSOL Multiphysics software.

A FSS is a periodic surface with two-dimensional array of element printed on a dielectric layer. It acts like a free-space filter, exhibiting total transmission or reflection around a resonance frequency. In terms of the frequency responses, FSSs are divided into four filtering types which are low pass, high pass, band pass and band stop [2]. The geometric configurations of the mentioned spatial filters can be constructed in the pattern of patch-type array or aperture-type array. Effective FSSs are required to exhibit stable resonant frequencies, at both normal and oblique angles of incidence [3]. It is interesting to note that some FSSs, especially operating at GHz range, can simply be fabricated by the etching method which is a subtractive method used for the production of printed circuit boards.

According to the investigation of the performances for FSS in various element shapes, the square loop (SL) pattern shows the best performance characteristics in terms of having larger bandwidth and stability under various angles of incidence and polarizations in comparison to all other element shapes [3]. However, the wide bandwidth of the SL pattern becomes redundant in our required shielding property. The grid structure has to be added to the SL pattern so as to restrict the filtering response over a much narrow bandwidth. Thus, the SL pattern has to be modified to the gridded square loop (GSL) pattern. A number of previous researches on the GSL pattern were mostly studied on general characteristics in which the narrow bandwidth responses were numerically [4-6] and experimentally [7] confirmed. Besides, none of the previous works were focused on any specific applications. This, therefore, prompted us to numerically investigate the use of the GSL pattern for FSS in blocking the microwaves and passing the visible spectrum for a microwave oven door. With these two targets, we therefore employed two corresponding parameters; namely, the shielding effectiveness (*SE*) and optical transparency (*OT*), to describe properties of the proposed FSS.

In all, the aim of this paper is to design the microwave oven's door of the gridded square loop (GSL) pattern to have a high pass transmission of the visible light and block the operating frequency of the microwave oven. The investigation in terms of the shielding and transmitting properties of the proposed FSS are conducted by the COMSOL Multiphysics software. The effects of the proposed geometry on *SE* and *OT* are explored. The simulation results in terms of *SE* and *OT* from the proposed FSS, finally, will be compared to other FSSs.

2. Methodology

In this proposed work, the front door structure was designed and simulated by using the COMSOL program to investigate the events of the microwave and visible light incident on the FSS. We started with the creation of a unit sample FSS to represent the whole periodic structures and using boundary conditions to solve the problem and finally obtained the required characteristic values.

The proposed configuration for the FSS is the gridded square loop (GSL) arrays of copper. The metallic thin layer has the conductivity (σ) of 5.998×10^7 S/m (represented as grey color in figure 1) and it is regularly arranged on the dielectric substrate layer (represented by white color in figure 1) with the relative permittivity (ε_r) of 6.5 (Highly UV-transmitting soft glass). Table 1 shows the optimal geometric parameters of the proposed FSS unit.

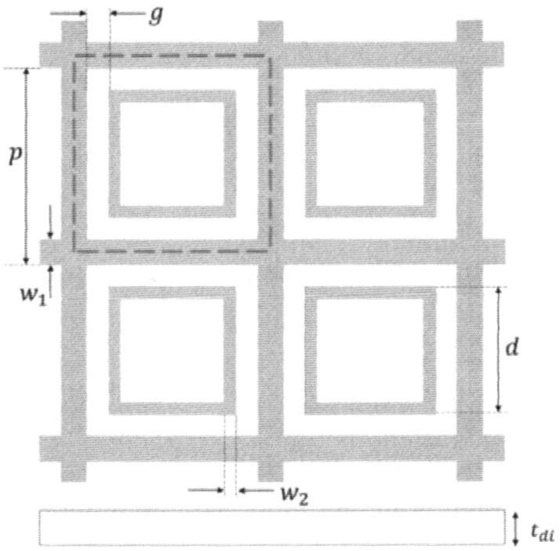

Figure 1. Gridded square loop (GSL) FSS and design parameters: top – top view with the FSS unit (red dash square) and bottom – side view. The grey and white colours represent the copper and glass areas, respectively.

Table 1. Geometric parameters of the proposed FSS unit.

t_{di} (mm)	p (mm)	d (mm)	g (mm)	w_1 (mm)	w_2 (mm)
2.80	40.7	26.4	4.65	5.00	2.50

The simulated model using COMSOL is shown in figure 2. Two ports are used at both sides of the proposed FSS unit cell located on the z-axis. The port 1 designated as the "source port" is excited by a normally incident electromagnetic (em) wave in the direction of the negative z-axis (red arrow) whereas the port 2 assigned as the "detector port" captures the transmitted em wave. Perfectly matched layers (PMLs) were applied on the top and bottom of the unit cell to absorb the excited mode from the source port and any higher order modes generated by the periodic structure. Then, the periodic boundary conditions on four sides of the unit cell were introduced to simulate the infinite 2D array. A frequency domain ranging from 0.4 to 4.3 GHz and parametric sweep solver were used for the simulation because the chosen range covered the dual band pass response of the proposed FSS.

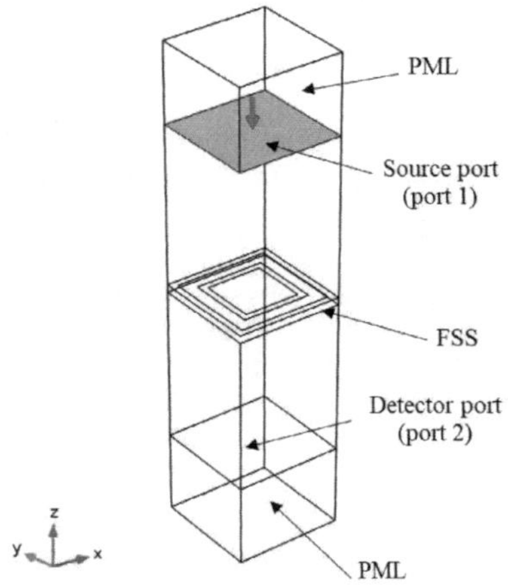

Figure 2. A model of unit cell for simulation of the proposed FSS using COMSOL.

3. Shielding and optical properties

3.1. Shielding effectiveness

The shielding effectiveness (SE) of any configuration is defined as a ratio, usually expressed in decibels, between transmitted and incident EM powers, i.e., electric field or magnetic field values. The SE is computed from the modulus of the total transmission scattering parameter s_{21} between two transmission lines separated by the device under test as follows [8]:

$$SE = 20\log_{10}\left(|s_{21}|\right). \tag{1}$$

3.2. Optical transparency

The theoretical optical transparency (OT) is the percentage of light transmitted through the proposed FSS determined by the ratio between transparent area and total area. The shape of the periodic array is shown in figure 1. p is the length of the grid, and each square aperture has sides of length $p - w_1$ with the square loop of length d and width w_2. So, the transmission percentage OT_{GSL} of the configuration can be calculated from

$$OT_{GSL} = \frac{(p-w_1)^2 - 4(dw_2 - w_2^2)}{p^2}. \tag{2}$$

As a metal layer is deposited on a dielectric substrate, the Fresnel losses due to air/glass interface have to be considered. The optical reflectance R is given by

$$R = \frac{(n_{glass} - n_{air})^2}{(n_{glass} + n_{air})^2}. \tag{3}$$

In this study, the value of $n_{glass} = 1.505$ for the highly UV-transmitting soft glass, then $R = 0.0406$. Therefore, the theoretical transparency OT of GSL unit is given by the formula:

$$OT(\%) = (1-R)^2 OT_{GSL} \times 100. \tag{4}$$

4. Results and discussions

When the plane waves were incident on the proposed FSS, the reflection and transmission were investigated. The frequency and parametric studies were thought to affect the shielding and optical performance of the proposed FSS.

4.1. Transmission and Reflection

This is obvious that some parts of the proposed FSS structure are metallic and the remaining parts are dielectric. They react to the electromagnetic waves differently. Generally, when the electromagnetic waves are incident on this surface, they are partly be transmitted in the forward direction and partly reflected in the specular direction. The quantity of the reflectance or transmittance depends on the areas of these materials. This should be noted that, in this study, the thickness of the metallic surface is assumed to be negligible as compared to the wavelength used. Consequently, such a structure can be estimated as a perfect conducting element. The effect of the penetration from the microwave below the conducting surface leading to the absorption can be ignored.

If the waves propagate in the z-axis, normal incident to the FSS, the electric field is laid on the plane of the structure. Then, the metals can induce the electric fields to cancel out the incident fields in the forward direction. As a result, the waves cannot penetrate through the structure. In addition, as the electric field interacts with a dielectric part, the material responds with a shift in charge distribution with the positive charges aligning with the electric field and the negative charges aligning against it. The event leads to the induction of the internal field and the overall field is decreased. If the field is high enough, these waves can penetrate the dielectric layer. Under the resonant condition, the amplitude of the transmitted waves may be equal to the incident waves and the reflected waves becomes zero. This situation is described as the total transmission.

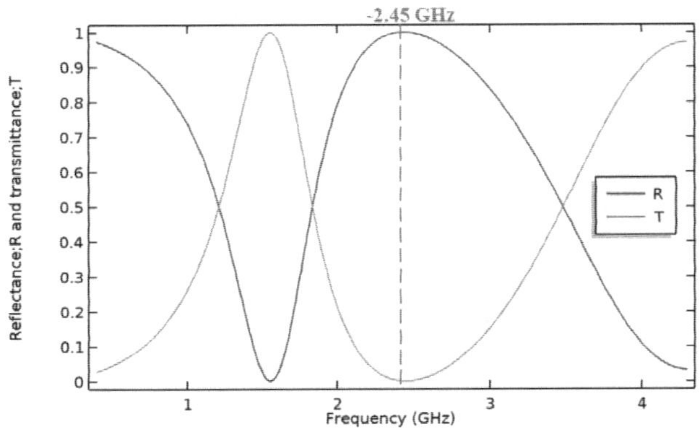

Figure 3. Reflection and transmission of the proposed FSS.

The simulation results of transmission and reflection responses for the proposed FSS are presented in figure 3. The optimization process was carried out step by step using the fine tuning. In the transmission pattern, there are two peaks around $f_{c,1}$ = 1.5 GHz and $f_{c,2}$ = 4.3 GHz of the total transmission. These peaks result from the matching incident waves to the size of the grid and square shapes respectively. In other words, the gridded square loop pattern works as a band pass filter and responds to two distinct frequencies. In the reflection mode, the structure shows a total reflection at 2.45 GHz which is the operating frequency of a microwave oven. This clearly indicates that the proposed FSS has a high potential to shield the microwave radiation.

4.2. Shielding performance

The simulated *SE* dependence on the frequency response is shown in figure 4 at the normal incidence. The proposed FSS provides a maximum shielding ability of around 62.7 dB at the resonance frequency of 2.45 GHz indicating that the proposed FSS is capable to provide the microwave shielding property.

Figure 5 shows the electric field distribution on the side view of the simulated FSS model. At the resonance frequency around 2.45 GHz, the weak field appears at the detector port while the strong field is seen at the source port. In other words, the incident waves just slightly went through the structure and most portions were reflected. These results again confirm that the proposed FSS can effectively shield the microwave at the frequency of 2.45 GHz.

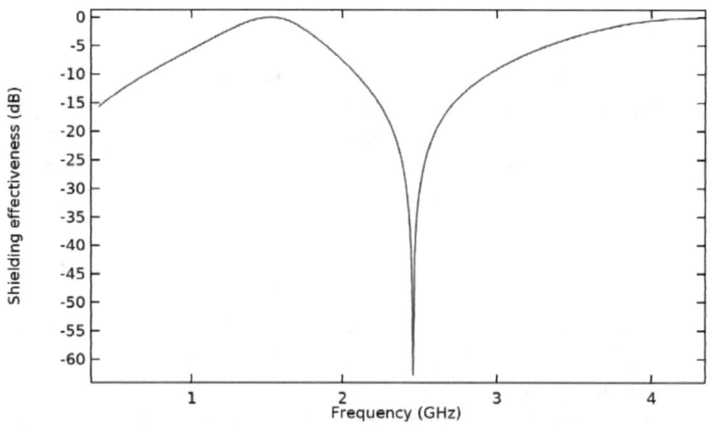

Figure 4. The simulated *SE* dependence on frequency at the normal incident angle.

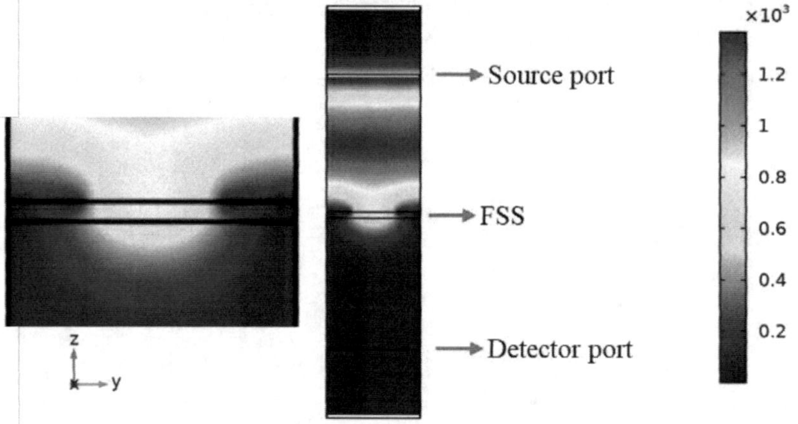

Figure 5. The electric fields distribution at 2.45 GHz.

The grid width (w_1) of the GSL pattern was varied from 4.0 – 6.0 mm, in 1 mm steps, at normal incident angle. The influence of w_1 on the *SE* of the FSS is shown in figure 6(a). The *SE* slightly changes with the increase in the width. On the contrary, the resonance points seem unchanged. This shows that the small increasement of the metal area cannot cause any significant effect to the reflection. We can also observe the effect from the square loop width (w_2) variations in figure 6(b).

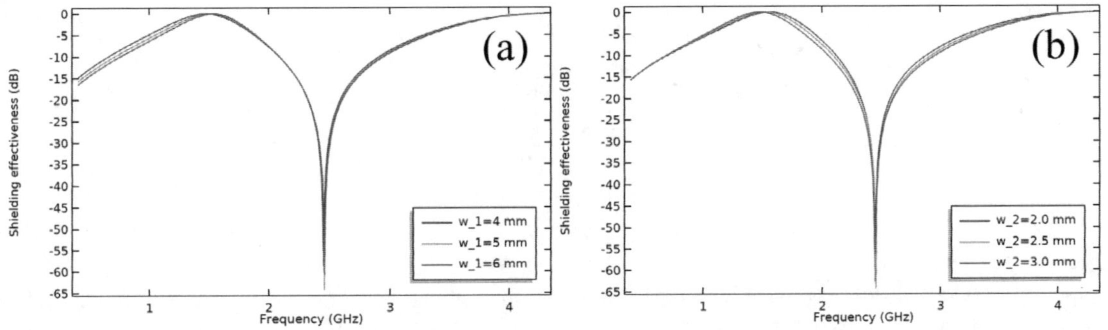

Figure 6. Influence on *SE* from (a) grid width and (b) square loop width of FSS.

The grid length (p) of the GSL pattern was studied and varied from 39.7 – 41.7 mm, in 1 mm steps, at normal incident angle. The influence of p on the *SE* of the FSS is shown in figure 7. The *SE* as well as the resonance point are changed due to the variation of *p*. This suggests that the shielding ability to the microwave at the specified frequency can be ineffective as the grid length varies.

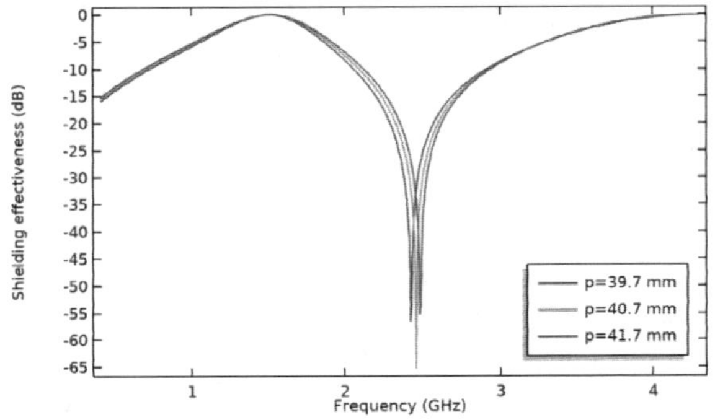

Figure 7. Influence of grid length of FSS on *SE*

.

The square loop length (d) of the GSL pattern was studied from 25.4 – 27.4 mm, in 1 mm steps, at normal incident angle. The influence of d on the *SE* of the FSS is shown in figure 8. The *SE* and resonant frequency decrease with the increase in d . This can alternatively be explained in terms of the equivalent circuit of the gridded square loop [9,10]. The parametric change affected the magnitude of the electrical characteristics of the structure and subsequently, the resonant point was changed. The frequency response of the structure to the incident waves occurs at lower resonance frequencies. Then, the shielding ability is accordingly altered.

We can conclude that the dimension of the structure directly affects to the shielding effectiveness *SE* . It was found that the length of square loop has more influence on the *SE* with a large shifting resonance frequency than any other parameters.

The thickness of the dielectric substrate was also found to affect the resonant frequencies in the transmission and reflection responses of the FSS. The resonance frequency exponentially decreases and subsequently converges to a value of $\left(0.5(\varepsilon_r+1)\right)^{-1/2} f_0$ for the thickness greater than $\lambda/4$ where f_0 is the resonance frequency of the FSS without the substrate and λ is the wavelength in the substrate at the resonance frequency [11].

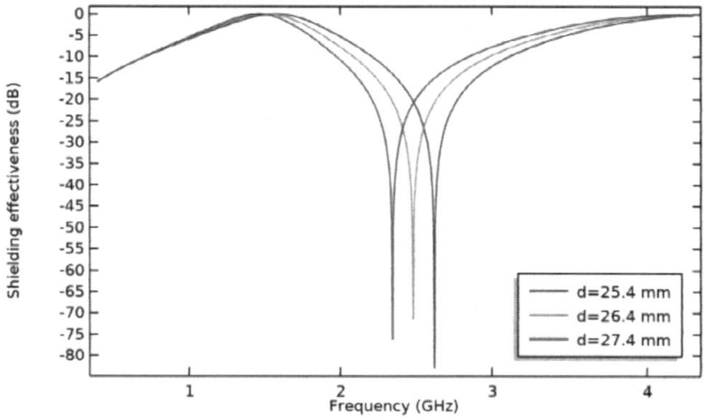

Figure 8. Influence of square loop length of FSS on *SE* .

The substrate thickness (t_{di}) was studied from 2.4 – 3.2 mm, in 0.4 mm steps, at normal incident angle. The influence of t_{di} on the *SE* of the FSS is shown in figure 9. The *SE* decreases with the increase in the substrate thickness. Meanwhile, the resonance points clearly moved backward. The reason of the shifting is that the increment of the dielectric layer with the value lower than $\lambda/4$ makes a rapid decrease of the resonance frequency [11]. The low-frequency transmission of the incoming waves is revealed.

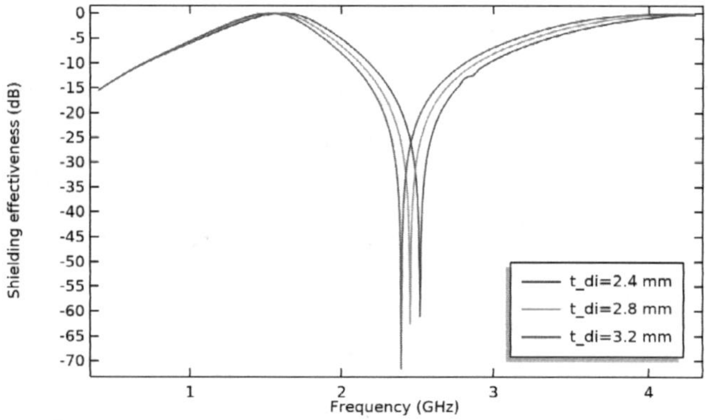

Figure 9. Influence of glass thickness substrate on *SE* .

4.3. Optical performance

According to the theoretical *OT* in equation (4), definition of the quantity is the ratio of the transparent area and the total area of the FSS. Then, we will define the parameters to readily describe the effect on the *OT* .

As shown in figure 10, the theoretical *OT* according to the grid aperture ($s_1 = p - w_1$), square loop aperture ($s_2 = d - 2w_2$), and square loop width (w_2) are plotted versus grid width (w_1). The *OT* decreases as w_1 and s_2 increase because the transparent area decreases. The increasement of s_1 makes the area of transparent increases, then the *OT* increases. In case of increasing w_2, the *OT* decreases and so do w_1 and s_2. However, the latter case is more significant than the former because of more metal area involved.

The low *OT* leads to a low quantity of the visible light transmission. For the optimal dimension, the theoretical *OT* of the proposed FSS is around 57.5%.

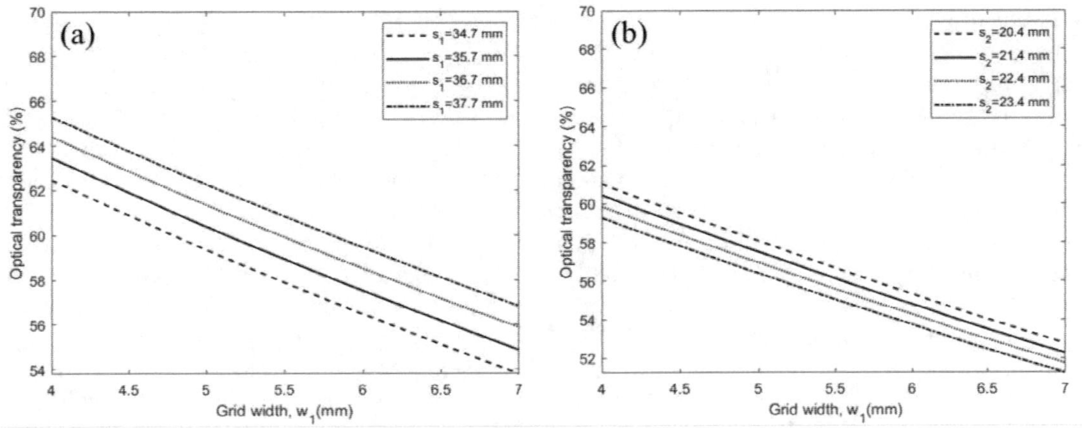

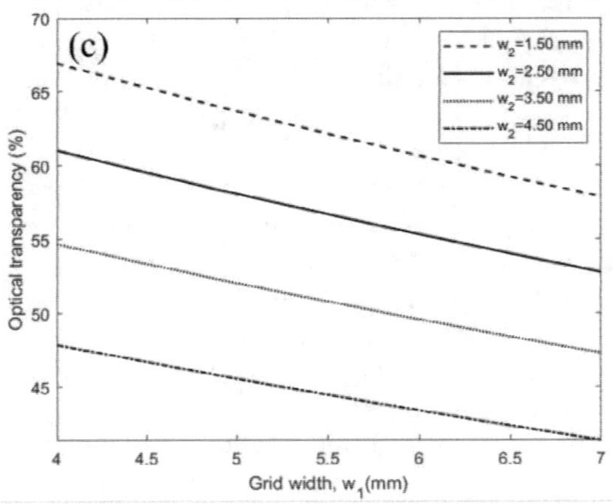

Figure 10. Influence on theoretical OT due to (a) grid aperture, (b) square loop aperture, and (c) loop width versus grid width.

In conclusion, improved shielding effectiveness can be obtained by altering the dimensions of the proposed FSS unit and glass substrate thickness. The simulation results of these parametric studies can be used to optimize the performances of the proposed FSS to achieve the expected goals of the design.

4.4. Comparison to other FSSs

The shielding effectiveness and optical transparency of this work, a metal mesh film (MMF), and a circular perforated metal sheet (PCS) at the generated frequency of 2.45 GHz from a microwave are listed in table 2.

Table 2. *SE* and *OT* of the proposed FSS, MMF, and PCS at operated frequency of microwave oven of 2.45 GHz

FSSs	*SE* (dB)	*OT* (%)
This work	62.7	57.5
MMF, [1]	32.4	73.0
PCS, [1]	26.6	48.5

These results show that the proposed FSS has the highest *SE* although the *OT* is lower than the MMF. This result indicates that the GSL pattern is a promising candidate for the microwave shielding designed for protecting users from an over-exposure of microwaves.

5. Conclusions

We have designed a FSS which is used for shielding the microwave and permitting the light transmission. The configuration of the proposed FSS is the gridded square loop pattern with the transmitting properties being in line with the band pass filters on the glass substrate. The design has been optimized according to the parametric studies by using the COMSOL software. The characteristics of the designed FSS fit the requirements which are the strong reflection of the microwave and ample transmission of the visible light at normal incidence.

To complete a whole spectrum of this research, some interesting points worth further investigating include (1) the study of some realistic effects of the interaction between the microwave and the proposed FSS such as the penetration depth [12] which is relevant to the absorbing properties of the FSS and the thermal insulation [13] which relates to the heat loss through the FSS while working, (2) the response of the proposed FSS to the oblique incident angles because, in practice, the incident waves can come from any angular directions, (3) the performance of the proposed FSS in terms of a well-known parameter, the Figure of Merit (FoM): a quantity describing the efficiency of the FSS [14], which is a parameter for a practical comparison and (4) the experimental study of the designed FSS structure in which the potential practical use of the proposed structure can be verified.

References

[1] Tung P D and Jung C W 2019 High optical visibility and shielding effectiveness metal mesh film for microwave oven application *IEEE Trans. Electromagn. Compat.* **62** 1070–81

[2] Munk B A 2000 *Frequency Selective Surfaces: Theory and Design* (New York: Wiley–Interscience)

[3] Ullah I 2012 *Measuring and Filtering Microwave Radiations using Frequency Selective Surface Through Energy Saving Glass* (Perth: Edith Cowan University)

[4] Majumdar P, Zhao Z and Liu R 2014 Parametric analysis of different configurations of loop elements in frequency selective surfaces *Adv. Electr. Electron. Eng.* **4** 161–8

[5] Filgueira J D B, Campos A L P S, Neto A G and Manicoba R H C 2016 Calculation of optimized parameters for the frequency selective surfaces by means of Taguchi's method *Microw. Opt. Technol. Lett.* **58** 1984–9

[6] Yilmaza A E and Kuzuoglu M 2009 Design of the square loop frequency selective surfaces with particle swarm optimization via the equivalent circuit model *Radioengineering* **18** 95–102

[7] Liu X, Li Z, Tang C, Yu J, Zhou J and Chen X 2018 Double-square and gridded-square loop frequency-selective surface in the K-band *Microw. Opt. Technol. Lett.* **60** 1136–42

[8] Corredores Y, Castel X, Besnier P, Dupeyrat C and Foutrel P 2018 Shielding effectiveness adjustment of planar mesh screen by fine-tuning metal thickness *J. Eng.* **2018**(4) 239–41

[9] Yang X, Wang L, Shen X, Liu X, Qi T and Zhou Y 2020 Single-substrate double-side high selectivity frequency selective surface *Prog. Electromagn. Res. Lett.* **92** 85–92

[10] Keyrouz S, Perotto G and Visser H J 2014 Lumped-elements tunable frequency selective surfaces *The 8th European Conf. on Antennas and Propagation (EuCAP 2014)* (Hague: Institute of Electrical and Electronics Engineers) pp 475–8

[11] Raspopoulos M 2007 *Radio Propagation in Frequency Selective Buildings* (Guildford: University of Surrey)

[12] Munk B A, Pryor J B and Marhefka R J 2003 Ohmic loss in frequency-selective surfaces *J. Appl. Phys.* **93** 5346–58

[13] Kiani G I, Karlsson A, Olsson L and Esselle K P 2007 Glass characterization for designing frequency selective surfaces to improve transmission through energy saving glass windows *Proc. of Asia-Pacific Microwave Conf.* (Bangkok: Institute of Electrical and Electronics Engineers)

[14] Haacke G 1976 New figure of merit for transparent conductors *J. Appl. Phys.* **47**(9) 4086–9

Low-cost laser diode pulse generator for quantum information applications

S Buathong[1,3], J Janpoon[1,3], N Suksawat[1,3] and S Deachapunya[1,2,3,*]

[1] Department of Physics, Faculty of Science, Burapha University, Chonburi Province, 20131, Thailand
[2] Thailand Center of Excellence in Physics, Ministry of Higher Education, Science, Research and Innovation, 328 Si Ayutthaya Road, Bangkok, 10400, Thailand
[3] Quantum and Nano Optics Research Unit, Burapha University, Chonburi Province, 20131, Thailand

*E-mail: sarayut@buu.ac.th

Abstract. A simple short-pulse generator circuit based on electronic gates is designed for short electric pulse of about 12 ns at Full Width at Half Maximum (FWHM) and 3.6 Volt amplitude for driving a laser diode. Using our circuit with a 780 nm laser diode designed and fabricated for producing short light pulses. The circuit utilizes an AND gate, a XOR gate, and a common function generator, provides a repetition rate from DC up to 1 MHz. The laser pulses were generated and then detected via an avalanche photodiodes (APD). This finding can benefit the field of light-based quantum information including single photon experiments.

1. Introduction

A pulsed laser has been at the center of interests due to its wide range of applications. The ultrashort light pulses are deployed for manufacturing methods which progress materials and product processing. The high-tech devices are fabricated requiring pulsed lasers for patterning and shaping of minuscule features at fast rate [1]. In comparison with a continuous-wave (CW) laser, the collateral damage caused by using a pulsed laser with high intensity for fabrication processes can be precisely controlled. The light pulses are also necessary for time-resolved spectroscopy to investigate the behavior of dynamics of various molecules [2]. In optical communication, the fast laser pulses are needed for test and measurement field [3]. Moreover, the pulsed lasers are employed in medical field, mostly ophthalmology, and shown to carry out clinical successes in the treatment of retinal diseases [4].

There are several techniques for creating laser pulses. Q-switching approach can produce the energetic short light pulses with time duration of nanosecond range and possibly extremely high peak laser power via manipulating intracavity losses [5]. The mode-locking laser involves mode lockers such as active and passive elements to generate the ultrashort light pulse width ranging from tens of femtoseconds to picoseconds [6]. Cavity damping was developed from the Q-switching technique to get rid of some Q-switching limitations [7]. For gain-switched laser pulses, the gain medium is modulated with the pump power at high-speed rate [8]. The gain-switching is often utilized in semiconductor laser diodes and achieves a pulse width of nanoseconds or picoseconds. Other approaches were also demonstrated, such as opto-electronic oscillator [9] and a technique associated with a beat signal provided via two diode lasers with different output frequencies [10].

Currently, quantum information has been intensively studied for its potential applications including quantum key distribution, quantum communication, and quantum computing. Light-based quantum information system has taken single photons as quantum bits and the regular way to produce single photons is to decrease the intensity of a laser beam or in other words the number of photons. This laser beam has to be pulsed at sufficiently narrow pulse width. Here, a low-cost laser diode pulse generator simply based on logic gates is constructed to create short laser pulses for future single-photon quantum experiment. The resulting light pulses are tested to be measured via the avalanche photodiode (APD).

2. Laser pulse generator circuit

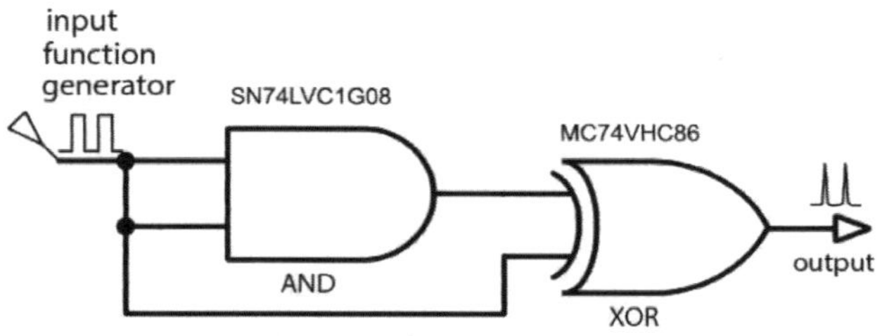

Figure 1. Electrical schematic of the laser pulse generator circuit.

The electrical schematic of the pulse generator circuit is seen in figure 1. The circuit simply utilizes two types of logic gates, the AND gate (SN74LVC1G08) and the XOR gate (MC74VHC86), to achieve short laser pulses. A voltage of 5 V is applied to the Vcc to run both gates. A function generator (GW Instek AFG-2012) basically provides a square wave signal with 500 kHz to the circuit input. Our design uses the AND gate to add a small delay to the signal at the top input leg of the XOR gate. The input signals therefore arrive at the XOR gate at different times (Ch1-yellow and Ch2-blue in figure 2). The output of the XOR gate is high if only the inputs are different (high or low). The brief period of time when both input signals are not identical leads to a pulse on the output of the XOR gate. The repetition rate of the output is therefore double that of the input.

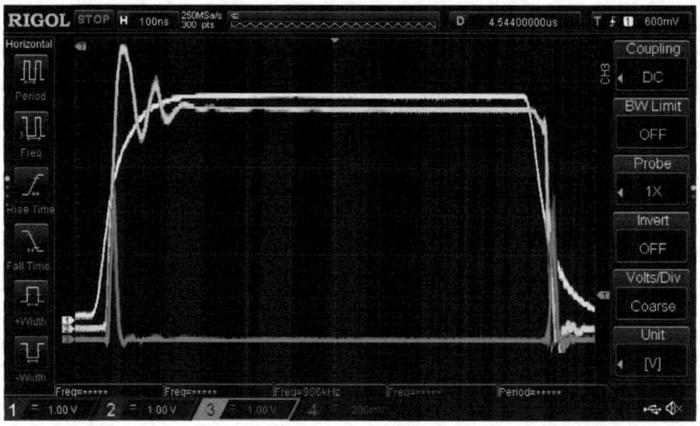

Figure 2. Waveforms of the measured electrical pulse signals (details see text).

As shown in figure 2, our circuit generates an electrical pulse with the duration and amplitude measured to be 12 ns and 3.6 V, respectively (Ch3- violet). The repetition rate was detected to be 1 MHz as expected. It was also found that the difference in the length of the input wires connected to the XOR gate also caused the delay and thus affected the pulse width.

3. Laser operation

The electrical pulses from the circuit were then exploited to pump the 780-nm laser diode (Thorlabs: L780P010) to obtain the laser pulses. The electrical amplitude was well above the operating voltage of the laser diode (1.8 V). The intensity of the pulsed laser had to be reduced for producing the single photons and this can be fulfilled via using a combination of optical elements such as a neutral density (ND) filter, linear polarizer (LP), half-wave plate (HWP), and a polarization beam splitter cube (PBS) (figure 3). The intensity of the laser can be adjusted to the single photon source by rotating the HWP. We used a homemade single photon counter module to measure single photons originated from our pulsed laser system. The counter module comprises the avalanche photodiode (APD) and its power supply circuit and the cooling circuit used for decreasing the temperature of the APD to -14 °C to eliminate the thermal noises. An example of the output of the APD is shown in figure 4. The photon signal pulses appear as the narrow peaks when the pulsed laser is operated. The repetition rate of photon signal pulses, few MHz, is far different from that of the fundamental dark counts arising from the counting module, which is in the scale of kHz.

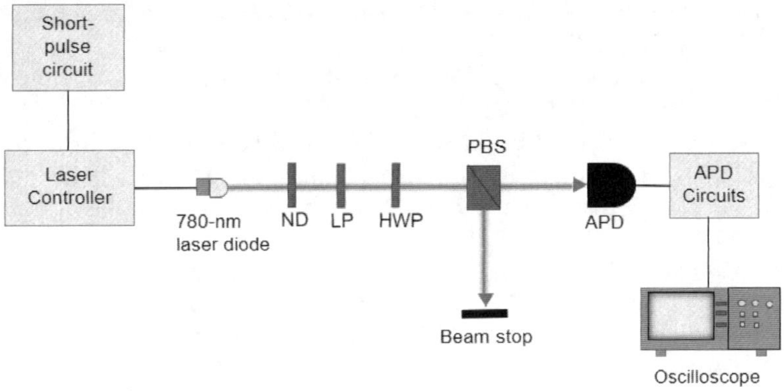

Figure 3. Experimental setup for single photon generation and detection. Short-pulse circuit as shown in figure 1, home-made laser controller for driving the laser, 780-nm laser diode as a single photon source, ND is a neutral density filter, LP is a linear polarizer, HWP is a half-wave plate, PBS is a polarization beam splitter cube, and APD is an avalanche photodiode. Details please see text.

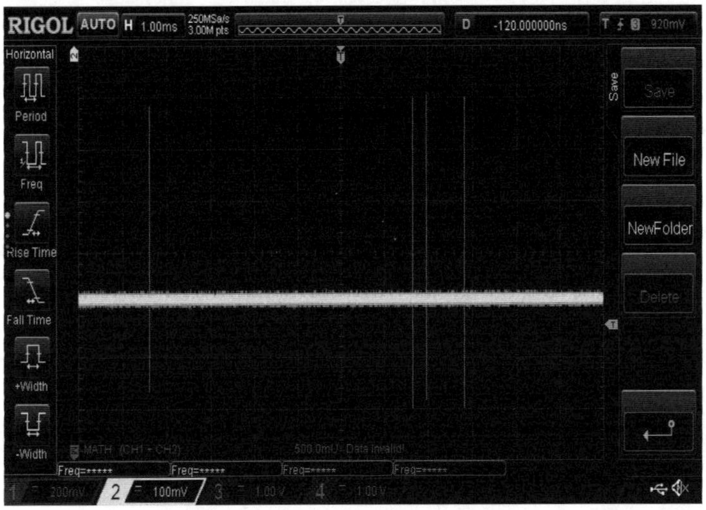

Figure 4. Output signal detected by the APD in the time histogram.

4. Conclusion

The simple short-pulse circuit has been designed and constructed. It comprises of two logic gates, i.e. AND gate and XOR gate. The circuit was used as a pulsed laser diode driver. The laser was then emitted as a single-photon source using weak coherent pulse configuration. These single photons were checked with a homemade single-photon counting module. This short-pulse circuit is aimed to utilize for single photon experiments in the near future.

Acknowledgments

We gratefully acknowledge the support from the research unit grant from Faculty of Science and Burapha University (RU01/2563). SB acknowledges the support from Faculty of Science, Burapha University (SC-NO2/2563).

References

[1] Booth H J 2004 Recent applications of pulsed lasers in advanced materials processing *Thin Solid Films* **453–4** 450–7
[2] Dantus M, Rosker M J and Zewail A H 1987 Real-time femtosecond probing of 'transition-states' in chemical reactions *J. Chem. Phys.* **87**(4) 2395–7
[3] Uhring W, Zint C V and Bartringer J 2004 A low-cost high-repetition-rate picosecond laser diode pulse generator *Proc. SPIE* **5452** 583–90
[4] Johnson T E, Keeler N, Dennis J E, Figueroa C L, Press H A, Rockwell R J, Stuck B E, Roach W P and Wartick A L 2003 Investigation of laser injuries *Proc. SPIE* **4953** 61–9
[5] Kovacs M A, Flynn G W and Javan A 1966 Q switching of molecular laser transitions *Appl. Phys. Lett.* **8**(3) 62
[6] Paschotta R and Keller U 2007 *Solid-State Lasers and Applications* ed A Sennaroglu (Boca Raton, FL: CRC Press) chapter 7 pp 259–318
[7] Myers J, Kokoczka C, Cook G and Bedford R 2017 High peak power cavity dumping semiconductor lasers *Opt. Lett.* **42**(1) 113
[8] Wang Y, Huang L, Cong M, Zhang H, Lei M and He F 2007 1 MHz repetition rate single-frequency gain-switched Nd:YAG microchip laser *Laser Phys. Lett.* **4**(8) 580
[9] Zhou P, Zhang F, Gao B and Pan S 2016 Optical pulse generation by an optoelectronic oscillator with optically injected semiconductor laser *IEEE Photon. Technol. Lett.* **28**(17) 1827–30
[10] Schimpf D N, Olgun H T, Kalaydzhyan A, Hua Y, Matlis N H and Kärtner F X 2019 Frequency-comb-based laser system producing stable optical beat pulses with picosecond durations suitable for high-precision multi-cycle terahertz-wave generation and rapid detection *Opt. Express* **27** 11037–56

Development of non-invasive alcohol analyzer using Photoplethysmographytle

Pornnapa Sanguansri, Nattapat Apiwong-ngam, Athipong Ngamjarurojana and Supab Choopun*

Applied Physics Research Laboratory, Department of Physics and Materials Science Faculty of Science, Chiang Mai University, Chiang Mai, 50200, Thailand

* Corresponding author's e-mail address: supab.c@cmu.ac.th

Abstract. Photoplethysmography (PPG) is one of the optical signals commonly used in clinical research for measuring the vital signs. Previously, PPG is often used as an indicator for detecting blood volume changes in the micro-vascular. The advantages of PPG signal mentioned in studies are non-invasive, lower operation cost, and the simplicity of the procedure. Although some the components of the PPG signal are not fully understood, it is generally accepted that it can provide valuable information in clinical study. Thus, it is interesting for finding a relation between PPG signal and blood alcohol concentration. The objective of this study is to classify two groups of ten-volunteer: (1) group of people who consumed alcohol and (2) non-consumed alcohol, by using the difference of PPG signals in these two groups and compared the results with fuel-cell breath alcohol analyzer. A set of PPG reflection data is recorded from optical sensors including the wavelength light of the red light and the infrared light from the fingertip of individuals. In additional, the changes of each signals for distinguishing two groups of volunteers are examined. The set of data is computed and analysed to find the correlation coefficient between significant variables in statistic domain. The analysis techniques are included (1) slope of the signals over time, (2) peak to peak of the heart rate, and (3) deep of waveform valley after rotation for training generalized linear (GLM) classifiers to create classification models. The accuracy of GLM classification can be obtained up to 87.50%. This suggests that PPG technique with our lab prototype has a potential for screening test to classify people who consumed alcohol and non-consumed alcohol.

1. Introduction

Driving under the influence (DUI) of alcohol is a major traffic accident for Thailand in the past decade. It is a leading cause of injury and death, especially, during the major festival such as Songkran festival or New Year festival. In 2021, according to The Road Safety Operation Center (RSOC) during the "Seven Dangerous Days" of the New Year, drunk driving cases accounted for nearly 33% of all traffic accidents. One solution for reducing the death rate of DUI in Thailand is limited to a minimum blood alcohol concentration (BAC) per legal at 0.05 percent [1]. Typically, two methods can be used as indicators for a BAC, breath test and blood test [2]. The blood test can be directly evaluated and more accurate. There are some drawbacks such as invasive methods, too many steps, and the need for experts for measuring. The breath test is commonly used for estimation BAC because of non-invasive and immediate results. Due to the epidemic COVID-19, these two methods are inconvenient, increase workload and procedure in medical, expand risk to contaminate COVID-19, and concern hygiene. Therefore, the PPG technique is an alternative way to investigate BAC. In brief, PPG is an optical technique that measures at the surface of the skin to detect volumetric changes in blood through the change in light absorption or reflection. PPG can be used to detect vital signs. Typically, it is used to measure blood oxygen levels and heart rate [3] with the benefits of non-invasive, inexpensive, convenient, and easy to merge with portable devices. It has been reported [4,5] that PPG has been used to measure BAC of the driver. So, this study aims to classify the person who consumed and non-consumed alcohol using the difference of PPG signals in these two groups and compared the results with a fuel-cell breath alcohol analyzer. The PPG signals are obtained from the reflection light of red and infrared light onto a finger.

2. Experimental

2.1. Device design and data collection

The primary PPG signal is collected by our lab prototype as shown in figure 1. The volunteers put the finger on the finger clip during the test. Each volunteer is controlled to consume the same amount of a bottle of 650 ml. beer with a 5% of alcohol. However, the amount of BAC of each volunteer does not expected to be the same due to the difference in gender, weight, and body metabolism. The PPG signals are obtained from the LED of red and infrared light. The PPG signal is processed by the Arduino zero SAND21, and the PPG is analysed by the spreadsheet software. The collected signal has band pass filter for noise reduction. The process of collecting data considered in this research is shown in figure 2.

Figure 1. Our lab prototype that designed for this experiment.

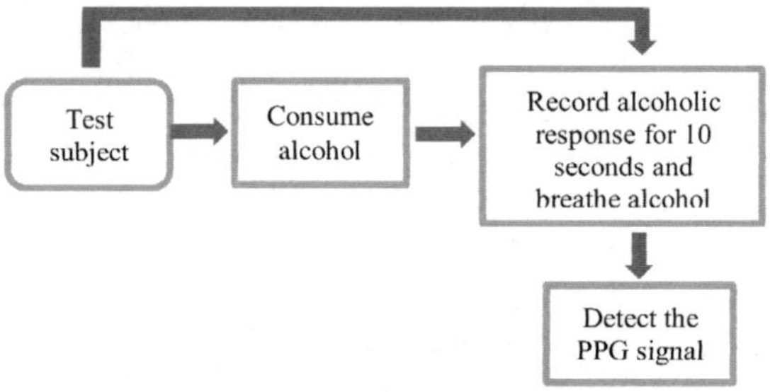

Figure 2. Flowchart of the data collection.

2.2. Processing a datum line rotation

Analytic data is measured for observing different signals from before and after consuming alcohol of the volunteers by using the Datum line rotation process. The primary part from PPG signals for the analysis is the maximum and minimum value of the waveform and is selected using a spreadsheet software.

In the datum line rotation [6] for this research, there will be a total of four procedures. The first step will be to estimate the maximum and minimum points of the peak as shown in figure 3 (a). Then the second step, a datum line is created as shown in figure 3 (b). A linear extracted from the maximum to the minimum can be indicated as in the equation (1).

$$y = mx + c \tag{1}$$

Next, the waveform signal is rotated with an angle until the datum line becomes a horizontal line as seen in figure 3 (c) and as shown in equation (2).

$$\theta = \tan^{-1} m \tag{2}$$

Finally, the new points are created by rotating the primary points using the equation (3)

$$\begin{aligned} X' &= X\cos\theta + Y\sin\theta \\ Y' &= -X\sin\theta + Y\cos\theta \end{aligned} \tag{3}$$

Thus, the datum line rotation can be defined as a rotation of the datum line created from the maximum and minimum points of the PPG waveform until the datum line becomes a horizontal line.

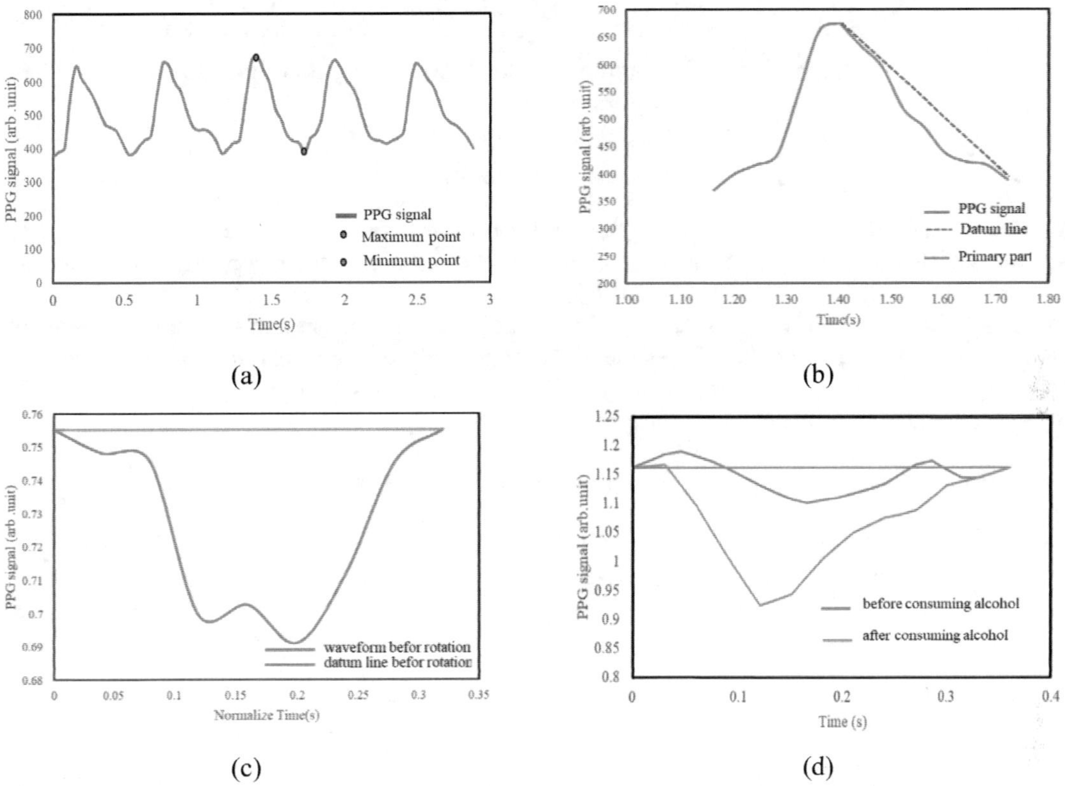

Figure 3. Datum line rotation procedure. (a) PPG signal indicating maximum and minimum point (b) datum line is created from the maximum to the minimum point (c) waveform signal is rotated with an angle until the datum line becomes a horizontal line (d) PPG signal after rotation for before and after consuming alcohol.

2.3. Prediction and classification

The statistical is analyzed by Package Program (R studio) to create a predictive model and classification. The data has been divided into two groups of PPG signals; before and after alcohol consumption. GLM classifiers is used for classification.

3. Results

3.1. Device design and data collection

The volunteers placed a finger on the finger clip for a stable measurement of PPG signals. PPG signals are plotted as a function of time. After band-pass filter, the PPG signals looks close to the waveform that is easier for analysis.

3.2. *Processing a datum line rotation*

The features of the datum line rotation generated for analyzing data are shown in table 1.

Table 1. The feature of datum line rotation generation

Feature	Description of feature
m	The slop of datum line linear extract
d	The maximum depth of waveform after rotation
sumpos	The sum of the depth under the datum line the waveform after rotation.
IR	PPG signal from IR LED
Red	PPG signal from red LED

The consistency of datum line rotation can be demonstrated by the standard deviation (SD) of the parameter, which is relatively low for each set of data. For example, the SD of dIR parameter of one volunteer is about 0.0158 with the mean of 0.176.

In addition, the results after the datum line rotation of 4 volunteers are shown in table 2. Most of the waveform after consuming alcohol shows the higher depth of waveform than that of before consuming alcohol. Figure 3 (d) shows an example of the different waveforms of the PPG signal of the volunteers (27-year-old female) before and after consuming alcohol.

Table 2. The result of datum line rotation generation.

Data		Subject 1		Subject 2		Subject 3		Subject 4	
		before	after	before	after	before	after	before	after
IR	sumpos	0.666	1.208	1.77	0.724	1.45	1.523	0.519	1.045
	d	0.071	0.122	0.149	0.063	0.15	0.156	0.069	0.122
	m	-644.7	-562.73	-2242.3	-3357.1	-605.85	-1154.7	-291.04	-487.21
RED	sumpos	1.068	1.165	1.424	0.724	1.238	1.61	0.618	1.294
	d	0.134	0.137	0.134	0.066	0.169	0.163	0.111	0.181
	m	-89.71	-200.92	-720.38	-981.35	-237.39	-438.27	-169.61	-203.27

3.3. *Prediction and classification*

After collecting the PPG signals of 10 volunteers before and after consuming alcohol for two times, the PPG signals of 40 set are obtained. The data are forwarded to the R studio for creating a predictive model and classify before and after consuming alcohol. It is found that 80 % of the collected data are splitted into a training set and 20 % into a testing set by using three features from analysis; for training GLM classifiers to create classification models. The main features for an alcohol classification are three parameters, including dIR, sumposRed and mIR. These parameters are selected from the statistical analysis using R studio with the p value less than 0.10. The p value of dIR, sumposRed and mIR is 0.0451, 0.0871, and 0. 0.0997, respectively. Before consuming alcohol, data is marked as the class "0" while after consuming alcohol, data is marked as the class of "1". Moreover, the best accuracy of classification is observed at 87.50%. The accuracy is calculated from the number of the volunteers that GLM classification can correctly predicted and divided by the total number of volunteers. For the best accuracy, the correct prediction is 7 out of 8 volunteers. This suggests that the PPG technique has a potential for detecting alcohol in blood.

4. Conclusion

The non-invasive alcohol analyzer by using PPG is successfully developed. The classification of the person who consumed and non-consumed alcohol using the difference of PPG signals in these two groups and compared the results with a fuel-cell breath alcohol analyzer can be achieved. The key features of PPG signals are dIR, sumposRed, and mIR for training GLM classifiers to create classification models. The accuracy of GLM classification can be obtained up to 87.50%. This suggests that PPG technique with our lab prototype has a potential for screening test to classify people who consumed alcohol and non-consumed alcohol.

References

[1] Suriyawongpaisal P, Plitapolkarnpim A and Tawonwanchai A 2002 *J. Med. Assoc. Thail.* **85** 496
[2] Savola O, Niemela O and Hillbom M 2004 *Alcohol Alcohol.* **39** 340
[3] Bagha S and Shaw L 2011 *Int. J. Comput. Appl.* **36** 45
[4] Wang W F, Yang C Y and Wu Y F 2018 *Pers. Ubiquitous Comput.* **22** 275
[5] Fukuda K, Shimizu Y and Omura Y 2010 *Proc. IEEE Sens.* 498
[6] Chen Y Y, Lin C L, Lin Y C and Zhao C 2018 *IET Image Process.* **12** 188

Determination of probe non-linearity and error due to measurement position for direct measurement type of gauge block comparator and its measurement uncertainty

A Pringkasemchai*, J Wongsaroj and K Mongkolsuttirat

Primary Length Standard Group, Department of Dimensional Metrology, National Institute of Metrology (Thailand), 3/4-5 Moo 3, Klong 5, Klong Luang, Pathumthani 12120, Thailand

*Corresponding author's e-mail address: angkoonnap@nimt.or.th

Abstract. The non-linearity of probes is one of the important components in gauge block calibration by the mechanical comparative method of two gauges blocks at the same nominal length. However, an advanced method for gauge block calibration is a mechanical direct measurement method of two gauge blocks showing the greatest difference in nominal length of 25 mm. This method uses a special probe based on the interferential scanning principle to produce the signals to measure the displacement. In this paper, non-linearity and error due to measurement position were investigated as they related to the accuracy of measurement results. The differences in central length of pairs of standard gauge blocks made of steel were measured by optical interferometry with the measurement uncertainty (k=2) 23 nm. Length in the range of 5 μm to 25 mm was used in the experiment. Non-linearity of the probe was evaluated by the simple linear regression model. Various factors such as origin setting point, temperature, and vibration have been analysed. In the preliminary experiment, the non-linearity, position error, repeatability and retrace error over the measuring range 25 mm are 13 nm, -18 nm, 15 nm, and 10 nm respectively. The standard uncertainty of direct measurement type caused by non-linearity is 4 nm.

1. Introduction

A gauge block comparator is a device used to calibrate the end standard of length or gauge blocks according to ISO3650:1998 [1]. Traditionally, a gauge block was measured using a comparative measurement method with the standard gauge block and the unknown gauge block having the same nominal length. However, this method had limitations on the non-linearity of the type of linear variable differential transformer (LVDT) probe. In other words, non-linearity is less than 25 nm in a different length up to 80 μm [2] caused the amount number of standard gauge blocks required for calibration. Differently, gauge blocks can be calibrated using a direct measurement method of two gauge blocks showing the greatest difference in nominal length 25 mm causing the number of the standard gauge blocks, calibration cost, and maintenance cost to significantly decrease. A special probe was used in this method based on the interferential scanning principle to produce the signal to measure the displacement [3].

In this paper, we investigated the non-linearity of a special probe and the error due to measurement position for a direct measurement type in the gauge block comparator and the investigated results were used to evaluate the measurement uncertainty and validate the measurement system to provide calibration services.

2. Measurement principle

2.1. Experiment and method

An experimental system was shown in figure 1. Direct measurement type of gauge block comparator (DGBC) manufactured by TESA, model UPD was used. This device can measure two different types of measurements: comparative type with 10 μm measurement range and direct type with 25 mm measurement range using a special probe manufactured by Heidenhain, model 2501 with display unit ND287 resolution 10 nm. The temperature correction system consisted of two platinum resistance thermometers (PRT) with a display unit manufactured by Fluke, model 1529 with a resolution of 0.001 °C.

Figure 1. Direct measurement type of gauge block comparator system.

Non-linearity of a probe and the error due to measurement position were determined by modifying the measurement method according to Euramet cg-2/v.2. The differences in central length of 13 pairs of gauge blocks made of steel, grade K were measured by optical interferometry with measurement uncertainty ($k = 2$) 23 nm. The 13 pairs of steel gauge blocks were presented in table 1. The non-linearity of the probe and its uncertainty were evaluated by the simple linear regression model [4].

Table 1. Measurement position of 13 pairs of gauge blocks made of steel.

Pair No.	Nominal length (mm)		Pair No.	Nominal length (mm)	
	A	B		A	B
1	25	50	8	25	26
2	25	45	9	25	25.5
3	25	35	10	25	25.1
4	25	30	11	25	25.01
5	25	29	12	25	25.005
6	25	28	13	25	25
7	25	27	-	-	-

2.2. Measurement setup

Zero position of a probe relative to a 25 mm measurement position on precision glass scale inside probe unit was determined using gauge block of 50 mm nominal length at A point. 25 measurement position on precision glass scale was defined at displacement 1.0645 mm ± 0.0002 mm of a display unit when origin point was the top of a probe unit.

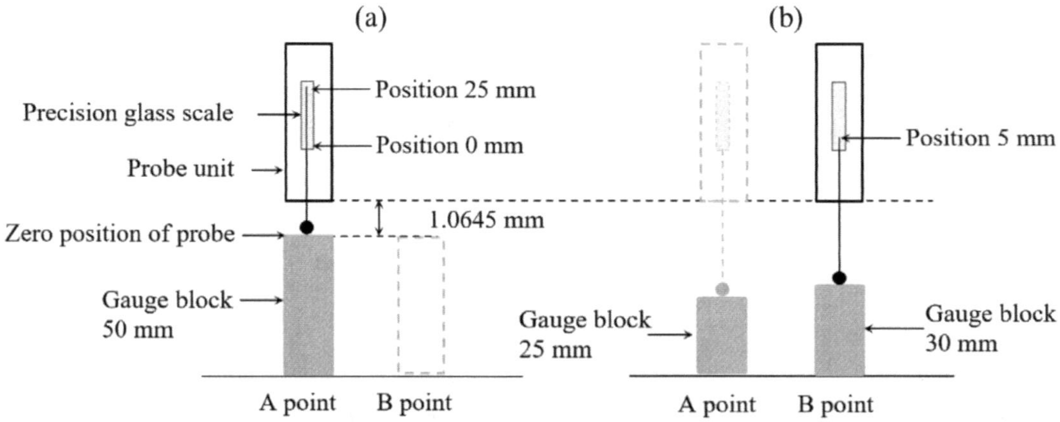

Figure 2. Zero position of a probe setup (a) and example of a direct measurement setup at 5 mm measurement position (b).

Next, a 0 measurement position was started using gauge block, nominal length 25 mm placed at point A and nominal length 25 mm placed at point B. Therefore, a 5 measurement position was shown as an example in figure 2. The measurement results of the measurement position on the precision glass scale were shown in table 2. Measurements at A and B points were performed using a special sliding system to control pitch and yaw motion and both gauge blocks were placed on a measuring table. The producibilities of the sliding system and variation of a measuring table were less than 10 nm and 20 nm respectively.

3. Results and discussions

The experiment was carried out under a laboratory environment of room temperature at 20.0 °C ± 0.3 °C, and relative humidity at 50 %RH + 10 %RH. The temperature difference between A and B points was less than ± 0.02 °C. All measurement results were compensated for the effect of temperature due to thermal expansion coefficient to reference temperature 20 °C refer to the ISO 1-1975. Since the special probe was considered as the high accuracy measuring system, the effect of vibration in this experiment was prevented using 20 mm rubber pads. The probe non-linearity of two comparative types over ± 10 µm and the direct measurement type over ± 25 mm determined by the linear interpolation of the measurement results was 2 nm and 13 nm sequentially as shown in figure 3.

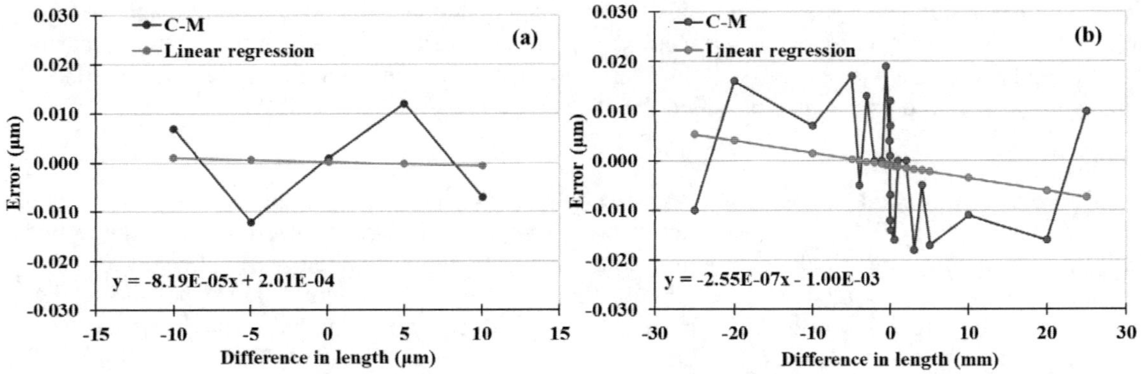

Figure 3. Non-linearity of the probe between comparative type (a) and direct measurement type (b).

The error due to a measurement position of the precision glass scale over 25 mm was shown in figure 4. A maximum error was -18 nm at a 3 mm measurement position. The results of the retrace errors as the difference between probe moving up and down were presented in table 2. A maximum retrace error was 10 nm at 0.1 mm. The experiment was repeated 10 times with repeatability less than 15 nm.

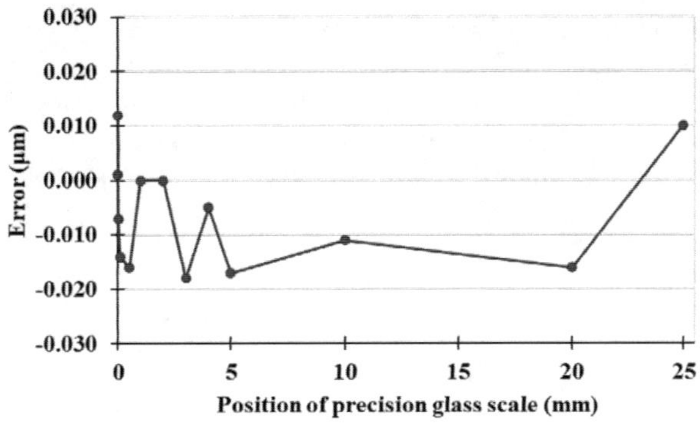

Figure 4. Relation between error and measurement position of precision glass scale inside probe unit.

Table 2. Errors of measurement position and retrace errors.

Position (mm)	Error (nm)		Retrace error (nm)	Position (mm)	Error (nm)		Retrace error (nm)
	Moved up	Moved down			Moved up	Moved down	
0	1	-	-	3	-18	-13	5
0.005	12	12	0	4	-5	-5	0
0.01	-7	-7	0	5	-17	-17	0
0.1	-14	-4	10	10	-11	-7	4
0.5	-16	-19	-3	20	16	16	0
1	0	0	0	25	10	10	0
2	0	0	0	-	-	-	-

Measurement uncertainty of non-linearity [2] and the error due to measurement position were evaluated by the EA publication EA-4/02 "Expression of Uncertainty of Measurement in Calibration"

[5]. The standard uncertainty of comparative type and direct measurement type caused by non-linearity was estimated as results of 1 nm and 4 nm severally. The standard uncertainty of measurement position error was also estimated given the result at 10 nm. Both standard uncertainties were considered as the rectangular distribution since they provided the effect of systematic uncertainty.

4. Conclusion

In this work, we investigated the parameters of a special probe of the direct measurement type gauge block comparator (DGBC). These parameters affected the measurement uncertainty on gauge block calibration using a gauge block comparator. Also, the measurement system was validated to provide calibration services. we found that the probe non-linearity of a comparative type and a direct measurement type were 2 nm and 13 nm, respectively. Moreover, we found a maximum error of a measurement position and a maximum retrace error were equal to -18 nm at 3 mm, and 10 nm at 0.1 mm, respectively. Finally, we evaluated the standard uncertainty of measurement position error with the result of 10 nm.

Therefore, it can be said that the accuracy of a DGBC probe was sufficient for providing calibration services on the gauge block by both of comparative type and direct measurement type.

References

[1] International Organization for Standardization (ISO) 1998 *ISO3650: 1998 Geometrical Product Specifications (GPS) – Length Standards – Gauge Blocks 2nd edition* (Geneva: ISO) pp 11–2
[2] Peingbangyang S and Wongsaroj J 2017 *J. Phys.: Conf. Ser.* **901** 012057
[3] Heidenhain 2021 *Length Gauges* (Traunreut: Heidenhain) pp 1–43
[4] Kreysig E, Kreysig H and Norminton E J 2011 *Advance Engineering Mathematics 10th edition* (New Jersey: John Wiley & Sons, Inc.,) pp 1077–109
[5] European Accreditation (EA) 2013 *Evaluation of the Uncertainty of Measurement in Calibration (EA-4/02: 2013)* (Paris: EA) pp 32–5

Journal of Physics: Conference Series **2145** (2022) 012061 doi:10.1088/1742-6596/2145/1/012061

Towards accurate non-contact moisture inspection using THz imaging and thickness information

N Cota[1], R Jintamethasawat[1,*], K Prasertsuk[1], P Rattanawan[1], N Cota[1], P Phukphan[1], C Jia-Yi[1], W Kusolthossakul[1], P Poomvised[2] and T Chulapakorn[1]

[1] National Science and Technology Development Agency, Thailand Science Park, Khlong Neung, Khlong Luang, Pathum Thani, 12120, Thailand

[2] King Mongkut's University of Technology North Bangkok (KMUTNB), 1518 Pracharat 1 Road, Wongsawang Subdistrict, Bang Sue District, Bangkok, 10800, Thailand

*E-mail: rungroj.jintamethasawat@nectec.or.th

Abstract. This study proposes an approach for a non-contact moisture inspection in dried food products, which is crucial to maintain optimal quality and shelf-life, using terahertz (THz) signal. To achieve this, a sample-specific calibration curve needs to be determined first. HAITAI crackers were chosen in this work for demonstration purposes. Fifteen stacks of crackers with different heights were prepared and moisturized by covering with a wet tissue paper for different time periods, resulting in moisture levels between 3 and 40% R.H.. Then, each sample was placed on a conveyor belt system between a THz source and THz a detector, and transmitted signal was measured 5 times. After that, moisture percentage of the sample was determined based on a gravimetric method, whose results served as a ground-truth measurement. A thickness of the sample was also measured using a vernier. All signal measurements, together with their corresponding known thicknesses and moisture percentages, were used to calculate necessary coefficients that define a sample-specific calibration curve. Once a calibration curve for the cracker was obtained, it was used to estimate the moisture percentages in samples with different thicknesses. Mean absolute error (MAE) of the moisture percentage is found to be less than 12% when the sample thickness is modelled as part of the calibration curve, which is over 50 times less compared to the case when the sample thickness is not modelled. Therefore, the utilization of an automatic thickness determination would be promising for real-time and accurate non-contact moisture inspection. This approach can be also integrated into a production line to improve quality control in the food industry without interrupting existing processes.

1. Introduction

Terahertz (THz) is an electromagnetic wave whose frequency lies between those from microwave and infrared, thus exhibiting several unique properties [1,2]. THz wave can penetrate non-polar materials such as papers, woods, clothes, and can be reflected by metallic materials. In addition, THz is strongly sensitive to liquids, such as water, which constitute polar molecules that absorb wave energy. Considering these advantages, it is possible to use THz to scan for concealed items in most of the packages as well as identify their moisture distributions and contaminants.

In food industry, moisture is a crucial factor affecting food quality [3–5]. Moisture in food products can promote spoilage that can lead to serious health impacts in consumers. Therefore, appropriate moisture level needs to be well maintained to allow long shelf life and also ensure that products are safe to consumers. Since THz wave can penetrate through most packages and is sensitive to moisture, it then has a potential in in-line quality-control (QC) systems, where the transmitted THz wave can be processed to obtain an image of internal moisture distribution as well as a moisture percentage of a given sample. There are already approaches that determine moisture level in food samples, such as ones based on gravimetric and infrared methods. Despite they are reasonably accurate, the gravimetric-based method is slow, and the infrared-based method can only detect moisture around surface region. Such limitations have led several researchers to attempt to utilize THz to detect moisture, which offers both good penetration and rapid scanning time.

Nonetheless, measuring the sample moisture by using THz signal intensity can be challenging. Unless acquiring THz signals in ideal settings [6], calculating moisture by only analyzing the THz signal intensity after passing through the sample does not always work well. This is because the remaining signal intensity is also affected by the sample thickness. Their have been attempts to calculate moisture non-destructively by utilizing THz spectrum [7–9], which contains phase shift information and thus can account for sample thickness. However, spectroscopic methods are not cost-effective and productive solutions in industry. Fast and cost-effective THz moisture measurement method is therefore necessary since otherwise it would delay the progress in translating THz in-line QC systems to food industry.

As such, we propose an approach to measure the sample moisture from both THz signal intensity and sample thickness. Our main contribution is to include effects due to sample thickness, which allows more accurate moisture calculation. The sample thickness can be measured easily and quickly by various tools or depth sensors. We also propose a protocol to collect datasets needed for constructing a calibration curve, which describes relationship between the remaining signal intensity in terms of sample moisture and thickness. Later, we will show that the proposed moisture model could yield a significant improvement over a conventional model. This is useful because it allows us to calculate moisture for various sample's sizes and shapes that are present in realistic situations.

This paper is organized as follows. Section 2 provides understanding about how transmitted signal intensity is affected by sample moisture and thickness, and proposes an absorption model that describes their relationship. Section 3 describes hardware components necessary for data collection, and also procedure to construct a sample-specific calibration curve. Section 4 presents results to confirm our hypothesis, and their discussion and conclusions can be found in section 5.

2. Background

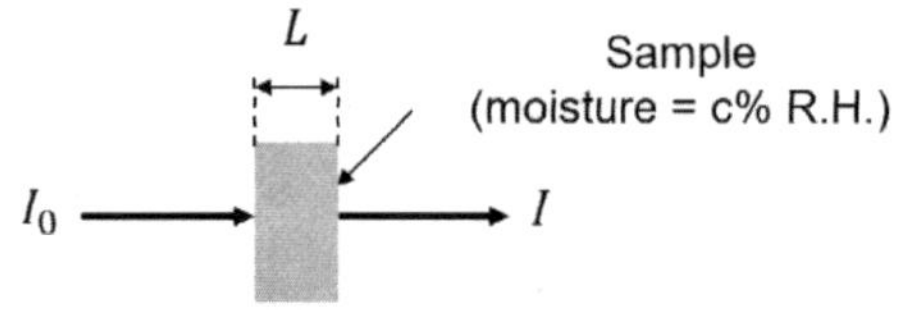

Figure 1. A graphical representation of how transmitted signal is absorbed by sample moisture (c) and thickness (L), resulting in a change of signal intensity from I_0 to I.

The interaction between THz wave and water molecules results in molecular vibrations and signal energy loss. Figure 1 demonstrates how the THz intensity is changed as it passes through the moist sample. In the figure,

- I_0 denotes THz signal intensity before passing through the sample (treated as constant)
- I denotes THz signal intensity after passing through the sample
- L denotes sample thickness

- c denotes sample relative moisture percentage

According to Beer-Lambert law of absorption, the remaining THz signal intensity can be modelled as equation (1):

$$I = I_0 \exp^{-\alpha c} \tag{1}$$

where α is a coefficient of absorption due to moisture c, which is constant and specific to the sample type. According to this model, it implies that the remaining THz signal intensity decreases exponentially as the sample moisture increases. By simply rearranging equation (1), one can calculate the moisture percentage from the known remaining THz signal intensity as shown in equation (2):

$$c = \frac{1}{\alpha} \ln\left(\frac{I_0}{I}\right) \tag{2}$$

Unfortunately, the model in equation (1) assumes constant sample thickness, so it does not account for effects due to sample thickness that also affect the remaining intensity. It turns out that the remaining THz signal intensity is also related to the sample thickness according to Beer-Lambert law, that is, the signal intensity decreases exponentially as thickness increases. Therefore, one could combine both moisture percentage and thickness into a single absorption model as equation (3):

$$I = I_0 \exp^{-\beta c L} \tag{3}$$

where β is a coefficient of combined absorption due to moisture c and thickness L, which is constant and specific to the sample type. As opposed to equation (1), this modified model includes the term L, which implies that the remaining THz signal intensity is dependent on both sample moisture and thickness. Also, note that β is modified from α in equation (1) in order to account for absorption due to both sample moisture and thickness. From this model, one can derive the moisture percentage from the known remaining THz signal intensity and sample thickness as:

$$c = \frac{1}{\beta L} \ln\left(\frac{I_0}{I}\right) \tag{4}$$

Note that, in order to calculate moisture percentage from equation (2) or (4), constants I_0, as well as α or β, must be determined. Since these constants are specific to sample type, they can be estimated from the set of samples with known thickness, moisture, and remaining THz signal intensity. Details about estimating these constants will be outlined in section 3.2.

3. Materials and methods

3.1. Imaging components

Figure 2 (a) shows a developed THz in-line imaging system that imitates production lines in food industry. Components related to imaging are displayed in figure 2 (b), which consist of 1) a lens-coupled THz source that generates continuous signal at frequency of 0.1 THz with maximum power up to 180 mW 2) a THz detector, placed under a moving belt, that possesses 1 x 256 pixels for receiving THz signals in 0.05 – 0.7 THz frequency range and generating THz images 3) an RGB-D camera that takes color (visible) and depth images. Image processing and registration algorithms developed in-house allow this in-line system to generate visible, THz, and overlay images in real-time, as demonstrated in figure 2 (c) by images of moist wafer. While it can be seen that the THz image can reveal internal moisture distribution, signals constituting

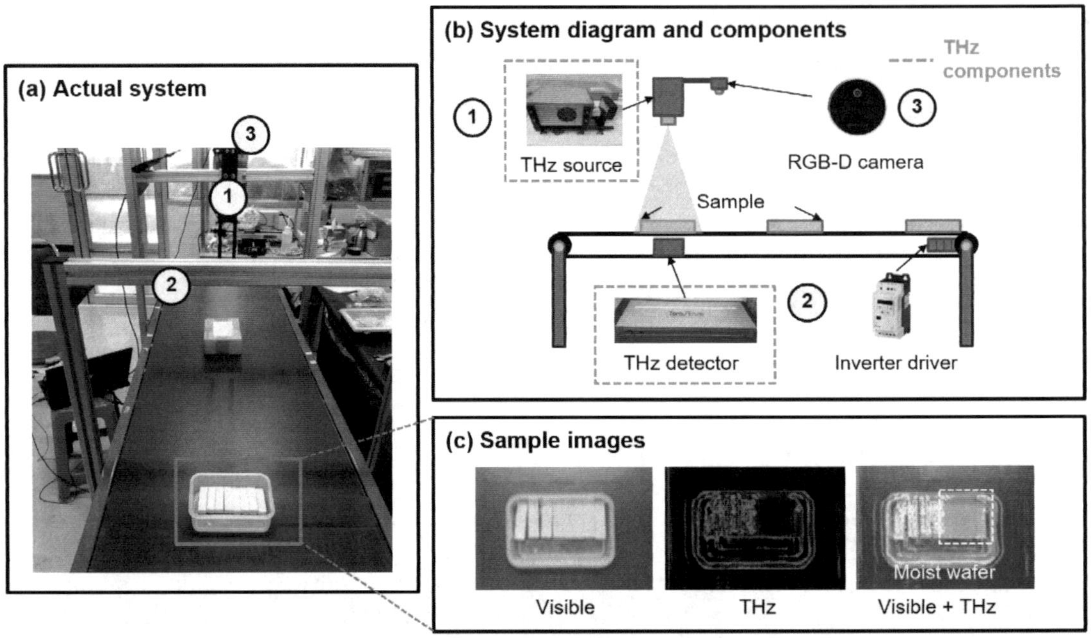

Figure 2. (a) A prototype THz in-line imaging system, whose components are diagrammed in (b). (c) Example of visible, THz, and overlay images of moist wafer.

THz image can also be used to analyze moisture percentage quantitatively. Section 3.2 will describe how this can be accomplished.

3.2. Obtaining calibration curves
For a given sample, calculation of its moisture percentage is possible if absorption model for that sample type, given as either equation (1) or (3), is known. This requires approximation of constants I_0 and α, or β according to procedure below.

(i) Prepare N samples with different thicknesses and moisture levels. Note that samples can be moisturized by slowly steaming them to maximize uniformity in moisture distribution.
(ii) For each sample, record remaining THz signals for m times, each at different poses.
(iii) After recording signals, measure sample moisture level using a gravimetric-based moisture analyzer, and sample thickness using a vernier.
(iv) After collecting all $N \times m$ datapoints of signal intensity, moisture level, and thickness, then apply a least-square method to estimate the constants I_0, and α or β in equation (1) or (3), respectively. Note that sample thickness is required only when estimating β.

The procedure above was used for collecting HAITAI cracker dataset, whose results and calibration curve are presented in section 4. Figure 3 shows an experimental setup used for collecting data necessary for constructing calibration curves. The resulting constants can describe the absorption model that serves as a calibration curve for a specific type of sample, and one can use it to calculate the sample moisture from the remaining THz signal intensity.

4. Results
Figure 4 (a) shows distribution of all datapoints consisting signal intensity, moisture level, and thickness. Datapoints were collected from 15 samples of stacked HAITAI crackers ($N = 15$) and

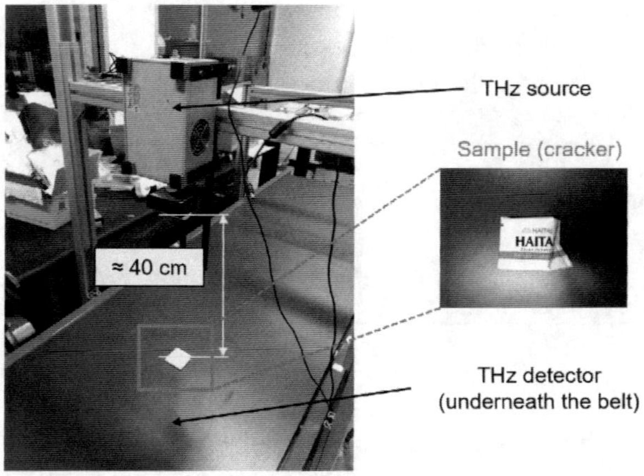

Figure 3. Experimental setup used for measuring THz signal intensity after passing through a given sample, i.e., remaining THz signal intensity.

signal measurement was repeated 5 times ($m = 5$) for each sample. Thickness of sample was varied from 1 to 7 pieces of cracker, where one piece corresponded to the thickness of 4.3 mm, and moisture of sample was varied between 3 and 40% R.H.. Observe that the remaining THz signal intensity, represented as normalized amplitude, tends to decrease as sample moisture or thickness increases. Therefore, this confirms validity of the collected datapoints. Figure 4 (b) shows the resulting calibration curve that is estimated from all 75 datapoints, which can be used for predicting sample moisture percentage. Note that a calibration curve corresponding to equation (1) was also obtained by the same approach for later performance comparison, except the sample thickness was not used.

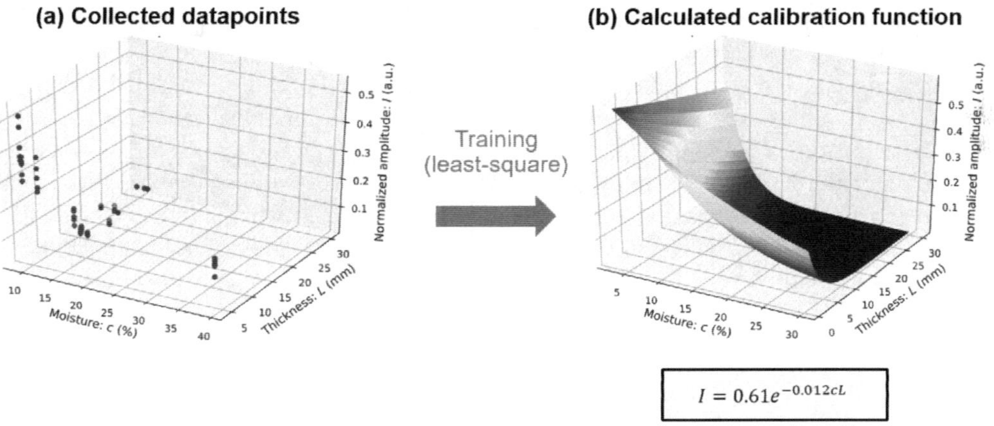

Figure 4. (a) Distribution of all collected datapoints. (b) Corresponding calibration curve, estimated by training the collected datapoints using least-square method to determine the required constants.

Table 1 shows the accuracy and fluctuation of the prediction results, calculated from two different calibration curves. In this comparison, datapoints were collected from 5 additional samples of stacked HAITAI crackers ($N = 5$) and signal measurement was repeated 5 times ($m = 5$) for each sample. This results in 25 test datapoints in total. Thickness of sample was varied from 2 to 7 pieces of cracker, and moisture of sample was varied between 3 and 40% R.H.. In calculation of moisture percentages, it is found that using the calibration curve that accounts

Table 1. Mean absolute error (MAE) of moisture prediction and standard deviation of absolute prediction errors, obtained as results of using different calibration curves.

Type of calibration curve	Mean absolute error (MAE) of prediction	Standard deviation of absolute prediction errors
With thickness information	11.5% R.H.	2.20% R.H.
Without thickness information	667% R.H.	350% R.H.

for the sample thickness leads to the mean absolute error (MAE) of prediction of 11.5% R.H.. This is significantly less than 667% R.H., obtained when using the calibration curve that does not account for the sample thickness. In addition, using the proposed calibration curve results in less fluctuation in the absolute prediction errors, as observed in the standard deviation of absolute prediction errors that is reduced from from 350% R.H. to 2.20% R.H..

5. Discussion and conclusions

This paper has demonstrated that THz technology shows promise in detecting moisture in crackers and potentially other food products. Furthermore, moisture calculation could be greatly improved by taking into account the sample thickness. It is important to note that even though the proposed calibration curve greatly improves the accuracy, the mean absolute error of 11.5 %R.H. is still too high to be used practical applications. Sources of error could be attributed to poor signal-to-noise ratio, vibrations, temperature fluctuations, or non-uniform moisture distribution in the sample. Nonetheless, this improvement is promising because, despite great challenge in THz moisture imaging, there are still lots of room for improvements, and this work has brought it one-step closer to practical applications.

In the future, it is suggested that the experiment should be conducted in well-controlled environments. Samples should be prepared in a closed chamber to ensure uniformity of the moisture distribution. The in-line imaging system could be refined to minimize vibrations. The THz source and detector locations could be optimized to maximize the signal-to-noise ratio. Most importantly, the proposed absorption model used in this work assumes linear absorption coefficients and independent effects of moisture and sample thickness. This might not be valid for all sample types. Therefore, to achieve more accurate moisture calculation, one could attempt to model more realistic relationship between intensity, moisture, and sample thickness.

References

[1] Dhillon S S *et al* 2017 *J. Phys.* D *Appl. Phys.* **50** 043001
[2] Mittleman D M 2017 *Opt. Express* **26** 9417–31
[3] Nielsen S S 2017 *Food Analysis* (New York: Springer)
[4] Hailu G and Derbew B 2015 *J. Biol. Agric. Healthc.* **5** 49–64
[5] Mathlouthi M 2001 *Food Control* **12** 409–17
[6] Oyama Y, Zhen L, Tanabe T and Kagaya M 2009 *NDT E Int.* **42**(1) 28–33
[7] Federici J F 2012 *J. Infrared Millim. Terahertz Waves* **33** 97–126
[8] Parasoglou P, Parrott E P J, Zeitler J A, Rasburn J, Powell H, Gladden L F and Johns M L 2010 *Terahertz Sci. Technol.* **3**(4) 172–82
[9] Banerjee D, von Spiegel W, Thomson M D, Schabel S and Roskos H G 2008 *Opt. Express* **16**(12) 9060–6

Color measurement of yellow sapphire by UV-Vis reflectance spectroscopy

N Tipkanon[1], N Monarumit[2], T Lhuaamporn[3] and W Wongkokua[1,*]

[1] Department of Physics, Faculty of Science, Kasetsart University, 50 Ngam Wong Wan Rd., Chatuchak, Bangkok, 10900, Thailand
[2] Department of Earth Sciences, Faculty of Science, Kasetsart University, 50 Ngam Wong Wan Rd., Chatuchak, Bangkok, 10900, Thailand
[3] The Gem and Jewelry Institute of Thailand (Public Organization), 140 ITF-Tower, Silom Rd., Bangkok, 10500, Thailand

*E-mail: wiwat.w@ku.ac.th

Abstract. Yellow sapphire is a gemstone variety of corundum (Al_2O_3). The yellow color is commonly caused by Fe^{3+} impurity substituting for Al^{3+} in the Al_2O_3 structure. The depth of yellow color is linked to Fe^{3+} concentration. The higher content of Fe^{3+} produces a deeper yellow color. However, there is low content of the impurity such as Be^{2+} in cooperating with the low content of Fe^{3+} to produce the yellow color in yellow sapphires. It is difficult to differentiate between the natural yellow sapphire caused by high Fe^{3+} content and the beryllium-treated yellow sapphire by naked eyes. Some advanced techniques to identify the Be heat-treated yellow sapphire have been practiced such as laser ablation inductively coupled plasma mass spectrometry (LA-ICP-MS), secondary ion mass spectrometry (SIMS) and laser-induced breakdown spectroscopy (LIBS). We measured the colors of naturals and beryllium-treated yellow sapphires by UV-Vis reflectance spectroscopy. The concentration of impurities in the sapphire samples was measured by the LA-ICP-MS technique. To compare the yellow colors of CIE $L^*a^*b^*$ and CIE L^*C^*h, we found that the lightness and hue values of the Be-treated yellow sapphires were higher than the natural yellow sapphires with high Fe^{3+} content.

1. Introduction

Ruby and sapphire are gems variety of corundum (Al_2O_3). The colors in corundum relate to impurities in the crystal structure, such as Cr^{3+} for ruby [1] and Fe^{3+}-Ti^{4+} for blue sapphire [2]. The yellow sapphire is caused by either Fe^{3+} replacing Al^{3+} or by color centers [3]. The intensity of the yellow color caused by Fe^{3+} depends on Fe^{3+} concentration. The quantitative relationship between color and impurities of natural corundum has been comprehensively studied [4]. However, there is a beryllium heat treatment technology to enhance the yellow color of yellow sapphire. The technique requires low content of Be^{2+} in cooperating with low content of Fe^{3+} [5].

It is difficult to differentiate between natural yellow sapphires with high Fe^{3+} content and beryllium-treated yellow sapphires by naked eyes. Hence, some advanced techniques have been developed to characterize the beryllium-treated sapphire, such as laser ablation inductively coupled plasma mass spectrometry (LA-ICP-MS), secondary ion mass spectrometry (SIMS) and laser-induced breakdown spectroscopy (LIBS) [6].

The spectra of the UV-Vis spectrophotometer could be converted to color scales such as ADMI (American Dye Manufacturers' Institute) for color measurement of wastewater [7] and CIE L^*c^*h for color measurement of pigmented rice grain [8]. In this work, we carried out UV-Vis reflectance spectroscopy to compare the CIE $L^*a^*b^*$ and CIE L^*C^*h values between natural yellow sapphires and Be-treated yellow sapphires.

2. Experiment

Two non-polished natural yellow sapphires with high Fe^{3+} content labeled as NY 3 and NY 5, and two polished Be-treated yellow sapphires labeled as BeY 8 and BeY 9, were collected as shown in figure 1. The trace elements of the samples have been measured by the LA-ICP-MS technique. The UV laser ablation system, a New Wave Research 266 nm Nd: YAG, is coupled to an Agilent 7500 ICP-MS. The LA-ICP-MS analyses for all samples were done under a pulse rate of 5 Hz and beam energy of approximately 0.5 mJ per pulse, producing a spatial resolution of 30 μm to 50 μm in diameter on the samples.

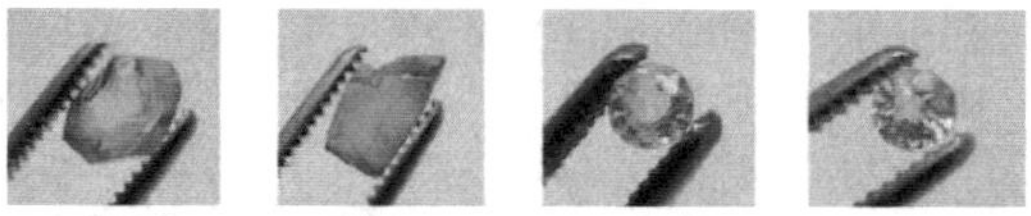

NY 3 NY 5 BeY 8 BeY 9

Figure 1. Natural yellow (NY) and Be-treated yellow (BeY) sapphires.

To measure the color of yellow sapphires, we carried out the UV-Vis reflectance spectroscopy using a portable spectrophotometer with an integrating sphere. The spectrometer was AvaSpec™ fiber optic type based on the Czerny-Turner design with a 2048 pixel CCD detector array. The spectrum range was about 275 nm to 1100 nm. The spectra were exported from raw data to 360 nm to 830 nm with the wavelength steps of 5 nm for color calculation. The CIE XYZ values can be calculated by [9]

$$X = k\sum_{\lambda} S(\lambda)R(\lambda)\bar{x}(\lambda)\Delta\lambda \quad (1)$$

$$Y = k\sum_{\lambda} S(\lambda)R(\lambda)\bar{y}(\lambda)\Delta\lambda \quad (2)$$

$$Z = k\sum_{\lambda} S(\lambda)R(\lambda)\bar{z}(\lambda)\Delta\lambda \quad (3)$$

where $S(\lambda)$ is the relative spectral power, $\bar{x}(\lambda)$, $\bar{y}(\lambda)$, $\bar{z}(\lambda)$ are the color matching function, $R(\lambda)$ is the spectral reflectance of an object surface and k is a normalizing factor,

$$k = \frac{100}{\sum_{\lambda} S(\lambda)\bar{y}(\lambda)\Delta\lambda} \quad (4)$$

The CIE 1931 standard observer matching function (2°) and the spectral power distribution illuminant source of D65 were selected. In practice, the CIE XYZ is transformed to a more uniform color space CIE $L^*a^*b^*$ in rectangular coordinate. L^* is represented as lightness with values from 0 (black) to 100 (white). The a^* value indicates red and green components of a color, where $+a^*$ and $-a^*$ indicate red and green values, respectively. The yellow and blue components are represented on the b^* axis as $+b^*$ and $-b^*$ values respectively.

The color space is defined by

$$L^* = 116f(Y/Y_n)-16 \quad (5)$$

$$a^* = 500[f(X/X_n)-f(Y/Y_n)] \quad (6)$$

$$b^* = 200[f(Y/Y_n)-f(Z/Z_n)] \quad (7)$$

where being $t = \frac{X}{X_n}, \frac{Y}{Y_n}$ or $\frac{Z}{Z_n}$:

$$f(t) = t^{1/3} \quad \text{if} \quad t > (24/116)^3 \quad (8)$$

$$f(t) = (841/108)t + 16/116 \quad \text{if} \quad t \leq (24/116)^3 \quad (9)$$

where X_n, Y_n, Z_n are the tristimulus values of a reference white object color stimulus.

However, a more natural representation of the quantities is CIE L^*C^*h color in cylindrical coordinate that is

$$C^* = \left(\left(a^*\right)^2 + \left(b^*\right)^2\right)^{1/2} \tag{10}$$

$$h = \tan^{-1}\left(\frac{b^*}{a^*}\right) \tag{11}$$

where C^* (chromaticity) gives vividness and h (hue) gives shades of color.

3. Results and discussion

Table 1 shows the quantitative chemical elements of the yellow sapphires analyzed by LA-ICP-MS to measure some trace elements (Be, Fe, Ti and Cr) involving in the causes of sapphire coloration. The Fe content of NY 3 and NY 5 were 12 478.5 mg kg^{-1} and 16 797.9 mg kg^{-1} respectively which is more than twice the Fe content of BeY 8 and BeY 9 that were 7 444.3 mg kg^{-1} and 6 182.5 mg kg^{-1} respectively. On the other hand, the Be content of NY 3 and NY 5 were 2.1 mg kg^{-1} and 2.2 mg kg^{-1} respectively which is less than half of the Be content of BeY 8 and BeY 9 that were 5.0 mg kg^{-1} and 10.5 mg kg^{-1} respectively. The higher content of Be in Be-treated yellow sapphire indicated the undergone Be heat treatment of the Be-treated yellow sapphires [10]. Ti and Cr content of all samples were less than about 50 mg kg^{-1} that was insufficient to affect the coloration of yellow sapphires.

Table 1. Trace element (Be, Fe, Ti, Cr) content in yellow sapphires measured by LA- ICP-MS.

Sample	Concentration / (mg kg^{-1})			
	Be	Fe	Ti	Cr
NY 3	2.1	12 478.5	15.2	6.3
NY 5	2.2	16 797.9	45.2	6.9
BeY 8	5.0	7 444.3	48.8	23.4
BeY 9	10.5	6 182.5	54.8	27.0

Figure 2 shows the UV-Vis reflectance spectra of natural yellow sapphires (NY 3 and NY 5) and Be-treated yellow sapphires (BeY 8 and BeY 9). The spectra of BeY 8 and BeY 9 are higher reflectance compared to NY 3 and NY 5. The fluorescence peak at 694 nm is the typical Cr^{3+} R line [11]. The peak was low intensity due to the low content of Cr. The fluorescence peaks at 630 nm and 840 nm are caused by Fe^{3+}-Be^{2+} mixed donor states [5]. The fluorescence peaks at 630 nm, 694 nm and 840 nm were observed due to the CCD type of the spectrophotometer. However, the surface polishing qualities of these two groups of samples might also influence the reflectance. The non-polished natural sapphire surface would result in more scattering and extinction of light.

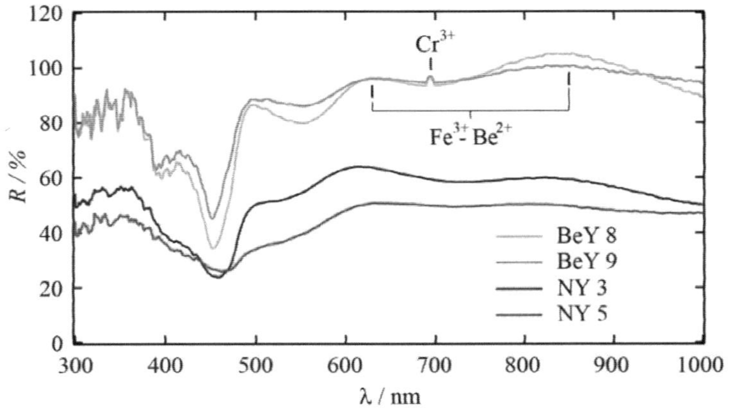

Figure 2. UV-Vis reflectance spectra of natural yellow sapphires and Be-treated yellow sapphires.

For quantitative analyses, we calculated CIE *XYZ*, CIE $L^*a^*b^*$ and CIE L^*C^*h as shown in table 2 to compare the natural yellow sapphires and Be-treated yellow sapphires.

Table 2. The CIE *XYZ*, CIE $L^*a^*b^*$ and CIE L^*C^*h of natural and Be-treated yellow sapphires.

	X	*Y*	*Z*	L^*	a^*	b^*	C^*	*h*
NY 3	51.78	55.33	33.45	79.23	-2.11	29.24	29.32	94.13
NY 5	40.62	40.86	31.42	70.08	5.59	16.25	17.18	71.02
BeY 8	77.24	83.67	57.29	93.31	-4.54	27.00	27.37	99.54
BeY 9	81.23	88.08	67.65	95.19	-4.78	21.05	21.58	102.80

The CIE *XYZ* values of Be-treated yellow sapphire (BeY 8 and BeY 9) were higher than natural yellow sapphire (NY 3 and NY 5). The values correspond to the higher reflectance of UV-Vis spectra shown in figure 2. For CIE $L^*a^*b^*$ values, L^* of BeY 8 and BeY 9 were higher than NY 3 and NY 5 indicated more lightness of Be-treated yellow sapphires than natural yellow sapphires while a^* values of BeY 8 and BeY 9 were lower than NY 3 and NY 5 indicated more greenness of Be-treated yellow sapphires than natural yellow sapphires and b^* values could not differentiate between Be-treated yellow sapphires and natural yellow sapphires. For CIE L^*C^*h values, C^* values could not differentiate between Be-treated yellow sapphires and natural yellow sapphires while *h* values of BeY 8 and BeY 9 were higher than NY 3 and NY 5 indicated difference shades of yellow color between Be-treated yellow sapphires and natural yellow sapphires.

4. Conclusion

The natural yellow sapphires and Be-treated yellow sapphires have been prepared for color measurement by UV-Vis reflectance spectroscopy. Some trace elements (Be, Fe, Ti and Cr) involving in the causes of sapphire coloration have been measured by the LA-ICP-MS technique. CIE *XYZ*, CIE $L^*a^*b^*$ and CIE L^*C^*h were calculated to compare the color of yellow sapphires. The L^* and *h* values of Be-treated yellow sapphires were higher than natural yellow sapphires. Hence, the color measurement by UV-Vis reflectance spectroscopy could be a promising technique to differentiate the natural yellow sapphires with high Fe^{3+} content and Be-treated yellow sapphires.

Acknowledgement

The authors would like to thank the Department of Physics and Department of Earth Sciences, Kasetsart University for sample collection and experimental instruments, and the Gem and Jewelry Institute of Thailand (Public Organization) for the LA-ICP-MS facility.

References

[1] Monarumit N, Lhuaamporn T, Satitkune S and Wongkokua W 2019 *J. Appl. Spectrosc.* **86** 486–92
[2] Wongrawang P, Monarumit N, Thammajak N, Wathanakul P and Wongkokua W 2016 *Mater. Res. Express* **3** 026201
[3] Nassau K and Valente G K 1987 *Gems Gemol.* **23** 222–31
[4] Dubinsky E V, Stone-Sundberg J and Emmett J L 2020 *Gems Gemol.* **56** 2–28
[5] Monarumit N, Lhuaamporn T, Sakkaravej S, Wathanakul P and Wongkokua W 2020 *J. Phys. Commun.* **4** 105018
[6] Abduriyim A and Kitawaki H 2006 *Gems Gemol.* **42** 98–118
[7] Suwanboriboon J, Meesiri W and Wongkokua W 2018 *J. Phys. Conf. Ser.* **1144** 012064
[8] Kaisaat K, Keawdonree N, Chomkokard S, Jinuntuya N and Pattanasiri B 2017 *J. Phys. Conf. Ser.* **901** 012069
[9] Schanda J 2007 *Colorimetry Understanding the CIE System* (New Jersey: John Wiley & Sons)
[10] Pisutha-Arnond V, Häger T, Wathanakul P and Atichat W 2004 *J. Gemm.* **29** 77–103
[11] Wongkokua W, Pongkrapan S, Dararutana P, T-Thienprasert J and Wathanakul P 2009 *J. Phys. Conf. Ser.* **185** 012054

Journal of Physics: Conference Series **2145** (2022) 012063 doi:10.1088/1742-6596/2145/1/012063

Luster measurement of pearl by UV-Vis reflectance spectroscopy

C Salyacheewin[1], N Monarumit[2] and W Wongkokua[1,*]

[1] Department of Physics, Faculty of Science, Kasetsart University, Bangkok, 10900, Thailand
[2] Department of Earth Sciences, Faculty of Science, Kasetsart University, Bangkok, 10900, Thailand

*E-mail: wiwat.w@ku.ac.th

Abstract. Pearl is an organic gem with its unique color and luster. One of the factors determining the quality and price of pearl is the luster. The different luster of pearls could be related to the different phase structures such as aragonite, calcite and vaterite. Previously, a gloss meter has been developed to measure light specularly reflected at 45° to the surface normal. Six different visual criteria for measuring gloss have been determined. Luster, a type of gloss, was defined as the ratio of specularly reflected light and that diffusely reflected normal to the surface. In practice, luster may be interpreted as relative brightness of specularly and diffusely reflecting areas. Due to the roundness of pearl, we measured the luster of pearl samples by a portable UV-Vis spectrophotometer with an integrating sphere. The luster could be calculated from the difference of CIELAB lightness measured by SCI and SCE geometries.

1. Introduction

Pearl is an organic gemstone widely popular due to its unique color and luster difference from other gems. Pearls originate from bivalves in both freshwater and seawater. When a foreign object such as the grain of sand or encapsulated parasite is caught into the shell and causes irritations, the bivalve releases a substance to coat the foreign object [1]. Its chemical composition is mainly composed of calcium carbonate ($CaCO_3$) more than 80 %, conchiolin 10 % to 14 % and water 2 % to 4 %. The outer layer of pearl is called nacre. If the nacre layer is thick, the pearl is lustrous. This surface structure gives rise to interference and diffraction of light which leads to the phenomenon of iridescence that produces the luster of a pearl [2].

The qualities that determine the overall value of a natural or cultured pearl are size, shape, color, surface quality, nacre quality and luster. The different luster of pearls could be related to the different $CaCO_3$ phases in their structures, i.e. aragonite, calcite and vaterite [3]. Luster is one of the identifiable features of a pearl because it can be judged qualitatively by the naked eye.

In 1934, Hunter developed a gloss meter to measure light specularly reflected at 45°. He determined six different visual criteria for measuring gloss. Luster, a type of gloss, was defined as the ratio of specularly reflected light and diffusely reflected normal to the surface. In practice, luster may be interpreted as relative brightness of specularly and diffusely reflecting areas [4].

However, the luster measurement method is related to the flatness of the measurand. Due to the roundness of pearl, we measured the luster of pearl samples by a portable UV-Vis spectrophotometer with an integrating sphere. The spectra of UV-Vis spectrophotometer could be converted to color scales such as CIE L^*C^*h for color measurements of rice grain [5] and ADMI (American Dye Manufacturers' Institute) for color measurement of wastewater [6]. The luster was calculated from the difference of CIELAB lightness measured by specular included and specular excluded geometries.

2. Experiment

Two sets of pearl samples were prepared shown in figure 1. All of the samples were freshwater pearl. The first set, on the top row, consists of four samples that were PK-3, PK-9, PKC3 and PKC8. The sample sizes varied in diameter with irregular shapes. The second set, on the bottom row, consists of five samples that were A1, A5, A6, E3 and F1. The sample size was about 8 mm in diameter with a regular round shape.

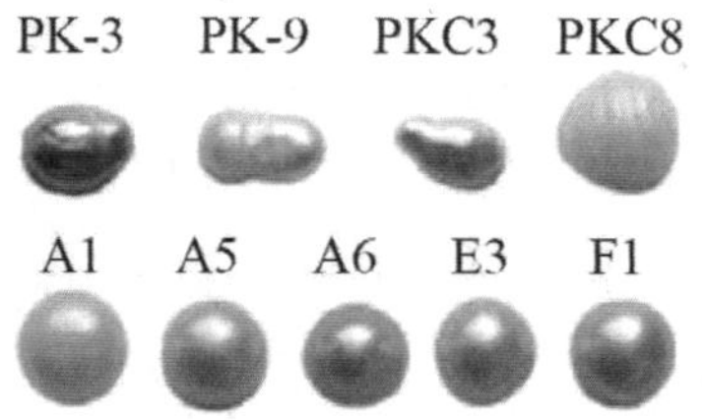

Figure 1. Freshwater pearl samples with different shapes, colors and luster.

We carried out UV-Vis reflectance spectroscopy using a portable spectrophotometer with an integrating sphere. The spectrometer was AvaSpec™ fiber optic type based on the Czerny-Turner design with a 2048 pixel CCD detector array. The spectrum range was about 275 nm to 1100 nm. The spectrophotometer was initialized by saving dark and white using specularly included (SCI) mode. Each sample was placed into the sample holder of the integrating sphere for UV-Vis reflectance measurements to obtain nine SCI spectra. Then, the specular port was opened for the specularly excluded (SCE) mode of measurements to obtain another nine spectra.

The spectra were exported from the raw data to 360 nm to 830 nm with the wavelength steps of 5 nm for color calculation. The CIE *XYZ* (tristimulus) values can be calculated by [7]

$$X = k\sum_\lambda S(\lambda)R(\lambda)\bar{x}(\lambda)\Delta\lambda \quad (1)$$

$$Y = k\sum_\lambda S(\lambda)R(\lambda)\bar{y}(\lambda)\Delta\lambda \quad (2)$$

$$Z = k\sum_\lambda S(\lambda)R(\lambda)\bar{z}(\lambda)\Delta\lambda \quad (3)$$

where $S(\lambda)$ is the relative spectral power, $\bar{x}(\lambda)$, $\bar{y}(\lambda)$, $\bar{z}(\lambda)$ are the color matching function, $R(\lambda)$ is the spectral reflectance of an object surface and k is a normalizing factor,

$$k = \frac{100}{\sum_\lambda S(\lambda)\bar{y}(\lambda)\Delta\lambda} \quad (4)$$

The CIE 1931 standard observer matching function (2°) and the spectral power distribution illuminant source of D65 were selected.

Then, the tristimulus values of *X*, *Y* and *Z* are converted to the CIE $L^*a^*b^*$ color space for both SCI and SCE. Each color indices can be interpreted separately. The L^* is the value of the lightness which ranged from 0 (black) to 100 (white). The a^* and b^* have no specific numerical limit. Positive a^* is red while negative is green, and positive b^* is yellow while negative is blue. The colors will deepen as the absolute values of a^* and b^* increase.

The color space is defined by

$$L^* = 116f(Y/Y_n) - 16 \quad (5)$$

$$a^* = 500[f(X/X_n) - f(Y/Y_n)] \quad (6)$$

$$b^* = 200[f(Y/Y_n) - f(Z/Z_n)] \tag{7}$$

where

$$f(X/X_n) = (X/X_n)^{1/3} \quad \text{if} \quad (X/X_n) > (24/116)^3 \tag{8}$$
$$f(X/X_n) = (841/108)(X/X_n) + 16/116 \quad \text{if} \quad (X/X_n) \leq (24/116)^3 \tag{9}$$

and

$$f(Y/Y_n) = (Y/Y_n)^{1/3} \quad \text{if} \quad (Y/Y_n) > (24/116)^3 \tag{10}$$
$$f(Y/Y_n) = (841/108)(Y/Y_n) + 16/116 \quad \text{if} \quad (Y/Y_n) \leq (24/116)^3 \tag{11}$$

and

$$f(Z/Z_n) = (Z/Z_n)^{1/3} \quad \text{if} \quad (Z/Z_n) > (24/116)^3 \tag{12}$$
$$f(Z/Z_n) = (841/108)(Z/Z_n) + 16/116 \quad \text{if} \quad (Z/Z_n) \leq (24/116)^3 \tag{13}$$

where X_n, Y_n, Z_n are the tristimulus values of a reference white object color stimulus.

For luster measurement, we calculated the difference lightness (ΔL^*) between SCI mode (L_i^*) and SCE mode (L_e^*) as

$$\Delta L^* = L_i^* - L_e^* \tag{14}$$

and difference color indices (Δa^* and Δb^*) between SCI mode (a_i^* and b_i^*) and SCE mode (a_e^* and b_e^*) as

$$\Delta a^* = a_i^* - a_e^* \tag{15}$$
$$\Delta b^* = b_i^* - b_e^* \tag{16}$$

3. Results and discussion

The UV-Vis reflectance spectra of the pearl samples measured by SCI and SCE mode are shown in figure 2 for the first set as the representation. The spectra measured by SCE mode were found to lower reflectance comparing to SCI mode in all samples. The spectra line shape were similar for each sample measured with the different modes. Different line shapes of the spectra are due to the different colors of pearl samples. The reflectance of PKC3 and PKC8 measured by SCI mode was higher than 100 % at the wavelength of 410 nm and 520 nm. These could be influent by the high specular reflectance of the samples.

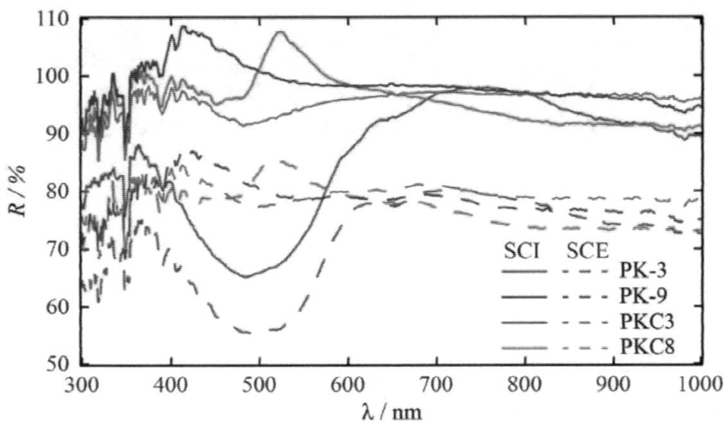

Figure 2. UV-Vis reflectance spectra of the represented pearl samples.

The pearl samples with different luster were inspected qualitatively by naked eyes. For the first set of samples, the luster could be classified from high to low as PK-3, PKC3, PK-9 and PKC8 respectively. Table 1 shows the color indices of L_i^*, L_e^*, a_i^*, a_e^*, b_i^*, b_e^*, ΔL^*, Δa^* and Δb^*. The values of L_i^* were significantly higher than L_e^* in all samples shown as ΔL^*, correspond to the UV-Vis reflectance spectra while the values of a_i^* and a_e^* were not significant difference shown as Δa^* as well as the values of b_i^*

and b_e^* shown as Δb^*. To classify the luster by the ΔL^* results, we found the luster from high to low as PKC3, PKC8, PK-9 and PK-3 respectively. The luster order of the results did not correspond to the visual inspection. Hence, the shape and size of the samples might influent the luster values.

Table 1. The values of color indices measured by SCI and SCE modes and the difference color indices of the first set of pearl samples.

Sample	L_i^*	L_e^*	a_i^*	a_e^*	b_i^*	b_e^*	ΔL^*	Δa^*	Δb^*
PK-3	92.37	85.71	11.85	11.59	5.35	4.86	6.65	0.26	0.49
PK-9	98.92	91.41	1.59	1.95	-0.86	-1.30	7.51	-0.36	0.45
PKC3	102.32	94.52	-5.47	-4.84	2.71	2.56	7.80	-0.64	0.15
PKC8	99.91	92.35	1.29	1.70	-4.57	-4.78	7.57	-0.42	0.21

Therefore, we measured the second set of pearl samples. The lusters of the samples were classified qualitatively by naked-eye inspection. We found that the luster from high to low was A6, A5, F1, E3 and A1 respectively. The measurement results of L_i^*, L_e^*, a_i^*, a_e^*, b_i^*, b_e^*, ΔL^*, Δa^* and Δb^* are tabulated in table 2. The values of L_i^* and L_e^* were higher than 100 for A1, E3 and F1. These results indicate high reflectance of the samples compared to the reference standard. To classify the luster by the ΔL^* results, we found the luster from high to low as A6, A5, F1, E3 and A1 respectively. The luster order of the results corresponded to the visual inspection. Hence, the size and shape of the pearl could influence the luster value.

Table 2. The values of color indices measured by SCI and SCE modes and the difference color indices of the second set of pearl samples.

Sample	L_i^*	L_e^*	a_i^*	a_e^*	b_i^*	b_e^*	ΔL^*	Δa^*	Δb^*
A1	116.30	110.05	-2.42	1.86	-15.28	-12.24	6.25	-4.29	-3.04
A5	104.95	97.28	1.26	1.83	-13.86	-13.29	7.67	-0.57	-0.57
A6	102.79	93.83	-1.41	-0.62	-14.61	-13.17	8.96	-0.79	-1.44
E3	112.80	106.08	2.10	5.59	-11.99	-9.47	6.73	-3.49	-2.52
F1	113.43	106.62	1.00	4.71	-13.57	-11.05	6.81	-3.71	-2.53

4. Conclusion

The UV-Vis reflectance spectroscopy by a portable spectrophotometer with an integrating sphere was used for luster measurement of pearl samples to calculate the difference of CIELAB lightness for the spectra measured by SCI and SCE geometries. The measurement results have well corresponded to the luster of pearl samples. Therefore, this is a promising method for luster measurement of pearl.

Acknowledgements

The authors would like to thank the Department of Physics and Department of Earth Sciences, Kasetsart University, for the sample collection and experimental instrument.

References

[1] Farn A E 1986 *Pearls: Natural, Cultured and Imitation* (London: Butterworths)
[2] Southgate P C and Lucas J S 2008 *The Pearl Oyster* (Amsterdam: Elsevier)
[3] Monarumit N, Noirawee N, Phlayrahan A, Promdee K, Won-in K and Satitkune S 2016 *J. Appl. Spectrosc.* **83** 298–301
[4] Hanson A R 2006 *Measurement Good Practice Guide No.94: Good Practice Guide for the Measurement of Gloss* (Middlesex: NPL)
[5] Kaisaat K, Keawdonree N, Chomkokard S, Jinuntuya N and Pattanasiri B 2017 *J. Phys. Conf. Ser.* **901** 012069
[6] Suwanboriboon J, Meesiri W and Wongkokua W 2018 *J. Phys. Conf. Ser.* **1144** 012064
[7] Schanda J 2007 *Colorimetry Understanding the CIE System* (New Jersey: John Wiley & Sons)

Improvement of simulated nuclear quadrupole resonance signals from explosive detection via a Red-Pitaya board

Puchong Kijamnajsuk[1,3], Piyachat Wungmool[2,3,a] and Chaiya Luengviriya[1,3,b]

[1] Department of Physics, Kasetsart University, 50 Phaholyothin Road, Jatujak, Bangkok 10900, Thailand
[2] College of Industrial Technology, King Mongkut's University of Technology North Bangkok, 1518 Pibulsongkram Road, Bangkok 10800, Thailand
[3] Research Center for Quantum Technology, Faculty of Science, Chiang Mai University, 239 Huay Kaew Road, Muang, Chiang Mai, 50200, Thailand

Email: [a]piyachat.w@cit.kmutnb.ac.th, [b]fscicyl@ku.ac.th

Abstract. In Nuclear quadrupole resonance (NQR), the interaction of the nuclear magnetic moments of quadrupolar nuclei (spin greater than 1/2) with the electric field gradient of the surrounding molecular orbitals produces an energy splitting. Because the resonant frequency is very specific to the molecular structure, the NQR can be used to detect explosive materials very accurately and it is extremely useful for detecting modern bombs whose containers made from plastics and wood instead of metals. However, NQR signals are generally very weak so they are difficult to be detected. Recently, Red-Pitaya boards, a Field Programmable Gate Array (FPGA) on Single Board Computers, have been being utilized in many electronic applications due to their small size and low cost. Since the boards can generate and acquire radio frequency signals, they can be taken as the console of portable bomb detectors. In this work, we study an improvement of the NQR signals of an explosive, ammonium nitrate with a resonant frequency of 423.6 kHz, acquired by using a Red-Pitaya board (STEMlab 125-14). To construct the NQR signals, we simulate free induction decay (FID) signals (exponential decay of sinusoidal functions) and add real measured noises from an input port of the Red-Pitaya board. To mimic real situations, the FID amplitude is varied, frequency fluctuations and phase shifts are added. The results show that averaging of signals from repeat measurements can improve the signals in all cases. To distinguish the signals from the noises, a minimal number of measurements is required. This necessary number of repeat measurements increases with frequency fluctuations and phase shifts but decreases when the FID amplitude grows.

1. Introduction

In Nuclear Quadrupole Resonance (NQR), energy gaps are generated by the interaction of the nuclear magnetic moments of quadrupolar nuclei (spin greater than 1/2) with the electric field gradient of the surrounding molecular orbitals. NQR works by sending radiofrequency (RF) waves with resonant frequency to excite the sample and detecting the respond signal with the same frequency [1-3]. This resonant frequency is very specific to the molecular structure of the sample. Therefore, the NQR is very useful for detecting explosives, especially modern bomb s whose containers made from plastics and wood instead of metals [4]. However, NQR signals are very weak and difficult to be detected. The signal-to-noise ratio (SNR) of NQR is affected by several factors, including molecular dynamics of relaxation, interference signals, electronic noise, explosive amount, distance, temperature, and the design of the detecting system [5,6].

Recently, Red-Pitaya STEMlab 125-14, single board computers containing a field programmable gate array (FPGA), have been used as control units in many electronic applications [7-9] due to their small size and low cost. Since the boards can generate and acquire signals within a frequency range of NQR, they are a suitable console of portable bomb detectors.

In this work, we study an improvement of NQR signals when acquired by a Red-Pitaya board (STEMlab 125-14). Thereby, the amplitude, frequency fluctuation and phase shift of the signals together with the electronic noises from the acquisition board are considered. We search for a necessary number of repeat measurements until the average signals has a sufficiently high SNR of 5 - 10.

2. Methods

To mimic the NQR signal obtained after an RF pulse is applied to the sample, we calculate an FID signal, an exponential decay of sinusoidal function of time t, as

$$FID(t) = A\sin[2\pi(f_0 + \Delta f \times rand)t + \phi \times rand]\exp\left(-t/T_2^*\right) \tag{1}$$

where A: amplitude of the FID signal, f_0: the resonant frequency of the sample at a given temperature and T_2^*: the spin-spin relaxation time. For ammonium nitrate, f_0 and T_2^* are 423.6 kHz (at 27°C) and 6 ms, respectively [5, 6]. A frequency fluctuation and a phase shift are added to mimic real situations where a temperature fluctuation may present. Δf and ϕ are the amplitude of the frequency fluctuation and the phase shift, respectively and *rand* is a uniformly distributed random number between -1 and 1.

Electrical noises from a fast analog input of a Red-Pitaya board (STEMlab 125-14) are acquired when the port is connected to a 50-ohm plug. The sampling rate is set to 1.953 MHz, about 4.6 times of the resonant frequency 423.6 kHz. Each set of acquired noises has 16384 samples (limited by the 16K-buffer size of the board) which results in an acquisition interval of 8.389 ms. 1000 sets of noises are collected as a noise bank for the simulations.

To simulate an NQR signal $S(t)$ acquired from the Red-Pitaya board, we add a set of noise randomly taken from the noise bank to the signal, $S(t) = FID(t) + noise$. The added noise together with the frequency fluctuation and the phase shift lower the signal-to-noise ratio (SNR) of a single acquired signal. The SNR is improved by averaging the signals from multiple measurements. The SNR is defined in the frequency domain as in [2]. We calculate the fast Fourier transform FFT of the average signal $S_{av}(t)$ obtain from n simulated measurements and the signal level is taken as the magnitude of FFT at the resonant frequency f_0, namely the NQR peak. Similarly, the noise level is estimated from the same n measurements (without FID signals) but it is defined as the maximal peak (the noise peak) in $f_0 \pm$ 2kHz of the resonant frequency.

3. Results

Examples of signals in time domain and the corresponding FFT are shown in figure 1. An electronic noise, acquired by a fast analog input of the Red-Pitaya board, fluctuate roughly between -0.4 and 1.0 mV during the interval of 8.389 ms and it is composed of many frequencies around the NQR frequency, figures 1(a) and 1(a′). A pure FID signal $S(t)$ with amplitude of 10 μV has a clear FFT, figures 1(b) and 1(b′). When the noise is added to the signal, its SNR is quite low (~1.7) so the NQR peak (at 423.6 kHz) and the noise peak are difficult to be distinguished, figures 1(c) and 1(c′). After the measurement is repeated 10 times, the average signal $S_{av}(t)$ has a better SNR (~5.7) so the NQR peak is much higher than the noise peak, figures 1(d) and 1(d′).

To illustrate the influence of the amplitude of FID signal, on the SNR, we perform simulations with different values of A in equation (1) while the frequency fluctuation and the phase shift are not added, i.e., both Δf and ϕ are set to zeros. As ones can expected, in the presence of given noises from the Red-Pitaya board the SNR increases with the FID amplitude A. We search for a sufficient repetition number N of measurements to improve the signal quality until the SNR reach to 5 - 10.

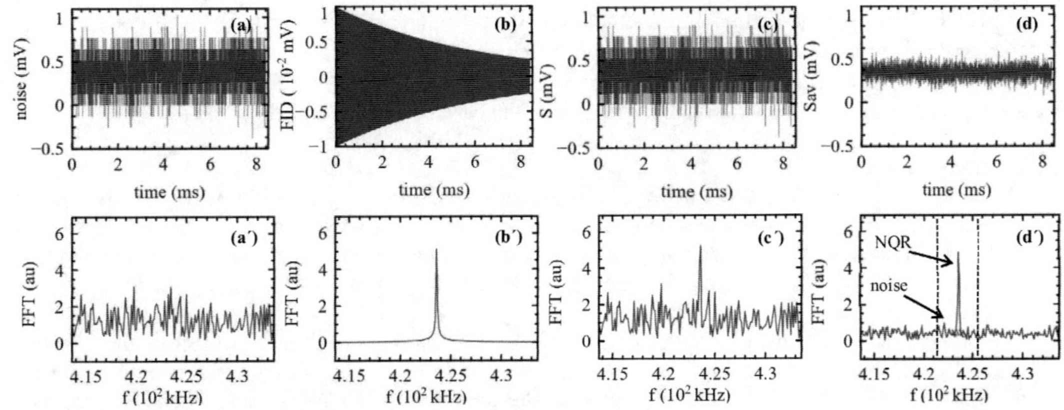

Figure 1. Examples of signals and their FFT. (a - a′) An electronic noise from a fast analog input of the Red-Pitaya board, (b - b′) an FID signal with amplitude of 10 μV, (c - c′) the signal with added noise *S*(*t*), and (d - d′) the average signal $S_{av}(t)$ from 10 measurements. In (d′), the dashed line corresponds to the average of 10 sets of noise (without FID signals) and the vertical dot lines, depict the ±2kHz zone around the resonant frequency.

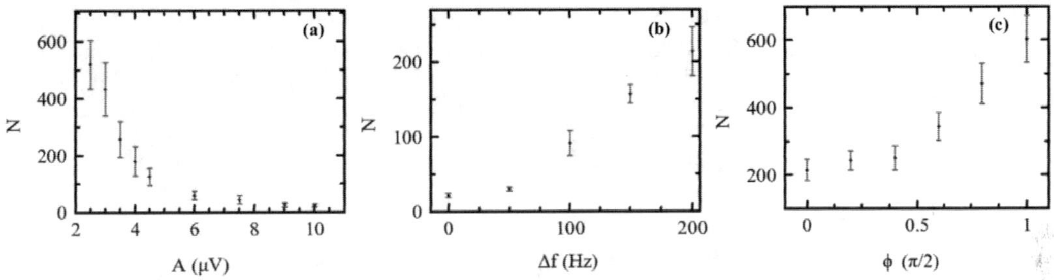

Figure 2. Repetition number *N* of measurements to reach the required SNR of 5 - 10 as functions of (a) amplitude A of FID, (b) amplitude Δf of frequency fluctuation, and (c) amplitude ϕ of phase shift. Each value of *N* is the average value obtained from 10 sets of experiments and the corresponding error bar shows the standard deviation.

To ensure that we get an appropriate approximation of *N* for each case of *A*, 10 sets of experiments are performed. For each experiment, a set of the repetition number *n*, that results in SNR 5 - 10, is collected. The average of 10 sets of *n* is taken as *N* and its standard deviation is also calculated. As shown in figure 2(a), *N* monotonously decreases when *A* increases. For very low *A* between 2.5 and 5.0 μV, *N* drops very fast. However, *A* gradually changes for 6 - 10 μV. In cases of high amplitude, $A > 50$ μV (data not shown), the SNR from a single measurement is sufficiently high (SNR > 10) so the repetition of measurements is not required.

Figure 2(b) shows the influence of the frequency fluctuation on the repetition number *N* of measurements. Note that the temperature coefficient for frequency of ammonium nitrate molecule is 91 Hz/K[6], thus the frequency fluctuation Δf is varied from 0 - 200 Hz to mimic a temperature fluctuation of about ±2°C during measurements. The amplitude *A* of FID is kept constant at 10 μV, the largest *A* in figure 2(a) and the phase shift is not added. These results shows that even for such high amplitude *A*, the repetition number *N* increases very fast as the amplitude Δf grows.

Figure 2(c) shows that one needs even larger N if phase shifts occur between the repeated measurements. This graph results from simulation with strong FID amplitude $A = 10\ \mu V$ and high frequency fluctuations $\Delta f = 200$ Hz. This graph shows that N grows with a very high rate for $\phi > \pi/4$. The maximal amplitude $\phi = \pi/2$ (the random phase shifts ranging between $-\pi/2$ and $\pi/2$) causes the number N increasing about 3 times in comparison to the case of no phase shifts $\phi = 0$.

4. Conclusion and discussion

We have demonstrated that averaging signals from repeat measurements could improve the SNR of the simulated NQR signals in a presence of electronic noise at Red-Pitaya board input. The method has been tested for different situations of signal amplitude, frequency fluctuation, and phase shift. In cases of low signal amplitudes or high frequency fluctuations as well as large phase shifts, one needs more repeat measurements to obtain a desired SNR.

We consider possible values of amplitude, frequency fluctuation and phase shift in real situations as follows. Typical NQR detection setups mainly consist of an NQR console, a transceiver unit, signal amplifiers, noise filters, a battery, and a probe. The amplitude of NQR signal depends on various parameters including the sample size, the sample-probe distance, and the gain of the electronic amplifiers. Based on an analysis in [10], the amplitude of amplified NQR signal reaching the console (e.g., a Red-Pitaya board) is in the order of μV. Concerning the frequency fluctuation, it has been shown that the NQR detection interval ranges from several seconds to minutes depending on the used pulse sequence type [6]. During such a detection interval, the temperature fluctuation of the sample under investigated is expected to be less than ±2°C so that the frequency fluctuation is possibly less than ±200 Hz in the case of ammonium nitrate. The phase shift depends on the timing of RF pulse excitations and NQR acquisitions which controlled by NQR console. Using the Red-Pitaya board as NQR console with the sampling frequency of 1.953 MHz, the resonant frequency of 423.6 kHz (at 27°C) and a given acquisition mismatch of ±0.5μs (i.e., the delay time between adjacent sampling points), the possible phase shift is about 0.43π radians. These considerations show that the ranges of amplitude, frequency fluctuation and phase shift of our study in figure 2 roughly correspond to those in real situations. Therefore, we expect that our simulation results can be utilized as a guideline for explosive detections.

Acknowledgments

This research was supported by the Program Management Unit for Human Resources & Institutional Development, Research and Innovation (grant no. B05F630113).

References

[1] Vij D R 2006 *Handbook of Applied Solid State Spectroscopy* (New York: Springer)
[2] Otagaki Y, Miyato Y, Barras J, Sato-Akaba H and Kosmas P 2021 *IEEE Sens. J.* **21**(5) 6922–9
[3] Sato-Akaba H 2014 *Solid State Nucl. Magn. Reson.* **63-64** 30–6
[4] Cardona L, Itozaki H, Jiménez J, Vanegas N and Sato-Akaba H 2019 *J. Magn. Reson.* **305** 104–11
[5] Cardona L, Jiménez J and Vanegas N 2015 *Ingeniare. Rev. chil. ing.* **23**(3) 452–72
[6] Cardona L, Miyato Y, Itozaki H, Jiménez J, Vanegas N and Sato-Akaba H 2015 *Appl. Magn. Reson.* **46** 295–307
[7] Hannig H, Mielke J, Fenske J A, Misera M, Beev N, Ospelkaus C and Schmidt P O 2018 *Rev. Sci. Instrum.* **89** 013106–12
[8] Preuschoff T, Schlosser M and Birkl G 2020 Rev. Sci. Instrum. **91** 083001–4
[9] Anand S 2018 OCRA: A Low-Cost, Open-Source FPGA-Based MRI Console Capable of Real-Time Control *MEng Thesis* Massachusetts Institute of Technology, Massachusetts
[10] Yinon J 2007 *Counterterrorist Detection Techniques of Explosives* (Amsterdam: Elsevier BV)

Simulation of wave propagation using graph-theoretical algorithm

K Ketthong[1], S Pulpirom[1], L Rianthakool[2], K Prasanai[2], C Na Takuathung[2], S Chomkokard[1], N Jinuntuya[1] and W Wongkokua[1]*

[1] Metrology Program, Department of Physics, Faculty of Science, Kasetsart University, Chatuchak, Bangkok, 10900, Thailand
[2] Department of Forest Engineering, Faculty of Forestry, Kasetsart University, Chatuchak, Bangkok, 10900, Thailand

*E-mail: wiwat.w@ku.ac.th

Abstract. We simulate the wave propagation through various mediums using a graph-theoretical path-finding algorithm. The mediums are discretized to the square lattices, where each node is connected up to its 4th nearest neighbours. The edge connecting any 2 nodes is weighted by the time of flight of the wave between the nodes, which is calculated from the Euclidean distance between the nodes divided by the average velocity at the positions of those nodes. According to Fermat's principle of least time, wave propagation between 2 nodes will follow the path with minimal weight. We thus use the path-finding algorithm to find such a path. We apply our method to simulate wave propagation from a point source through a homogeneous medium. By defining a wavefront as a contour of nodes with the same time of flight, we obtain a spherical wave as expected. We next investigate the wave propagation through a boundary of 2 mediums with different wave velocities. The result shows wave refraction that exactly follows Snell's law. Finally, we apply the algorithm to determine the velocity model in a wood sample, where the wave velocity is determined by the angle between the propagation direction and the radial direction from its pith. By comparing the time of flight from our simulation with the measurements, the parameters in the velocity model can be obtained. The advantage of our method is its simplicity and straightforwardness. In all the above simulations, the same simple path-finding code is used, regardless of the complexity of the wave velocity model of the mediums. We expect that our method can be useful in practice when an investigation of wave propagation in a complex medium is needed.

1. Introduction

Investigating wave propagation in a medium is important in many applications. To be specific, in our group we are mainly interested in wave propagation through the wood sample. We may inspect the defects inside a wood sample by analyzing the information about the time of flight of elastic waves passing through the wood. Measurements and numerical simulations are needed to do the task. However, wood is a complicated inhomogeneous medium. Simulation of wave propagation through a wood sample by solving the wave equation can be very complicated and computationally intensive [1].

In this work, we try to develop a different scheme to calculate the time of flight of a wave. Instead of solving the wave equation directly, we use Fermat's principle of least time [2] to find the path of propagation. It is very interesting that the goal of applying Fermat's principle, that is, to find the path of propagation that spends the least travelling time through a medium, is identical to that of the path-finding algorithm in the graph theory. In the path-finding problem, walking on a network from a source node to a destination node is performed to find the path that gives a minimum weight. We may apply this concept to our work by defining this weight as the time of flight of travelling waves between these nodes. The path with minimum weight will follow Fermat's principle, which we expect to be the actual path that happens in the real situation.

2. Dijkstra's algorithm

There are many path-finding algorithms in graph theory. We choose here Dijkstra's algorithm [3] due to its simplicity and straightforwardness. In this algorithm, the score and path history are given to a node when it is reached by the walker. The score is the cumulate weight along the path from the source, which is the starting point of the walker, to the node. Each node can be reached several times, depending on the number of branches that connect to the node. The score and path history of a node will be updated if the walker reaches the node with a smaller cumulate weight. Once all nodes are visited with all possible paths, the path with minimum weight connected the source point to each node in the network can be obtained, where the final score on the node represents the value of minimized weight.

Since the algorithm is discrete in nature, to apply it we have to discretize our space into a lattice. We choose a square lattice with node connection up to its 4th nearest neighbors, as shown in figure 1. The weight w_{ij} is defined as the time of flight of the wave travels from node i to node j, which can be calculated from $w_{ij} = 2d_{ij}/[v_{ij}(i)+ v_{ij}(j)]$. Here, d_{ij} is the Euclidean distance between nodes i and j, and $v_{ij}(i)$ and $v_{ij}(j)$ are the magnitudes of the wave velocity at node i and j, respectively, along the direction pointed from node i to j (see figure 2).

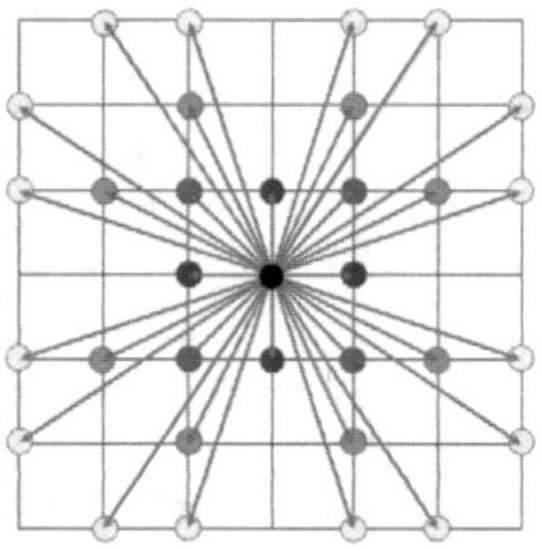

Figure 1. Connection of a node (black) to its 1st (purple), 2nd (blue), 3rd (green), and 4th (yellow) neighbors.

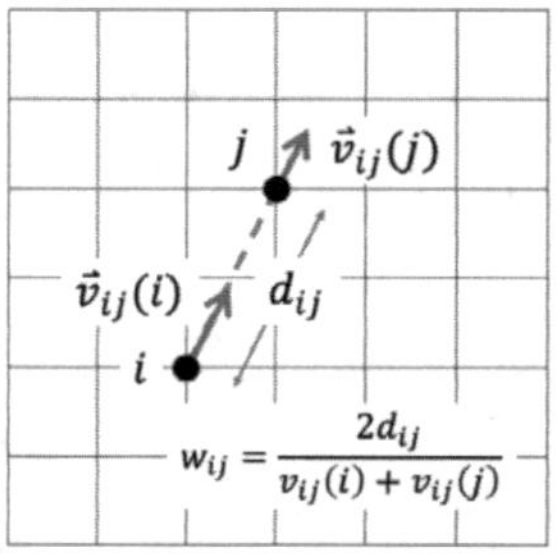

Figure 2. Definition of the weight w_{ij} as the time of flight of the wave travels from node i to j.

3. Testing the algorithm

As a test, a MATLAB code (https://www.mathworks.com/matlabcentral/fileexchange/12850-dijkstra-s-shortest-path-algorithm) is adapted to simulate wave propagation in 2 simple mediums. The first case is a homogeneous medium, where wave speed is a constant in all directions. We set a walker to walk from a point source to all nodes in a lattice. The least time of flights are recorded and assigned to each destination nodes as a color code. Since the wave speed is equal in all directions, the disturbances travel at the same time are on the same wavefront. As expect, we obtain a spherical wave propagates from the point source, as shown in figure 3.

The second case is the simulation of wave propagation through two adjacent homogeneous mediums with different wave speeds. A walker is set to find the least time path connected 2 nodes represented by 2 thick black dots in figure 4. Here, the horizontal dash line represents the border between two mediums. The connected red dots represent the path of the walker, and the green lines are the expected rays in each medium. We can clearly see the refraction at the medium boundary. The ratio of the wave speed in the upper medium to the lower is set to be 1.3, which resembles the refraction of light from water to air. We found that $\theta_i = 45.0°$ and $\theta_r = 66.3°$, which gives $\sin \theta_r/\sin \theta_i = 1.29$. The result agrees well with

Snell's law. From these two examples, we can see that our path-finding algorithm can simulate the behaviors of wave propagation quite well.

Figure 3. Simulation of a spherical wave propagates from the central point source.

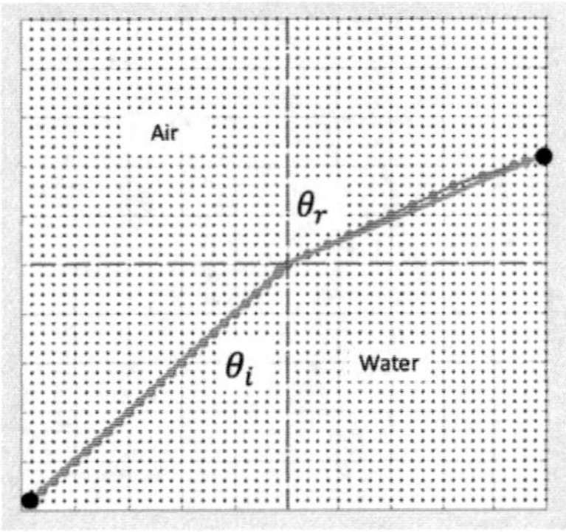

Figure 4. Simulation of light refraction from water to air. The connected red dots represent the path of the walker, and the green lines are the expected rays in each medium.

4. Determination of velocity model in a wood sample

As mentioned above, solving wave equations in an inhomogeneous medium like wood can be complicated. Despite computational complexity, lack of information on the velocity model in the wood can make the calculation even harder. Determination of this velocity model is thus an important task. We propose here the application of Dijkstra's algorithm to determine the velocity model in a wood sample. The time of flight of the travelling waves passing through a wood sample can be obtained by applying the path-finding algorithm with trial parameters of the velocity model. The results are compared with the measurements. The trial parameters are then updated. The iteration continues until the discrepancy of the time of flight is small enough to be excepted.

As an example, we use the algorithm to determine the velocity model in a teak wood block shown in figure 5. A pair of ultrasonic transducers are placed on the opposite sides of the wood block. The time of flight of the ultrasound travels between the transducers is measured at various positions. The results are compared with the time of flight obtained from Dijkstra's algorithm. We adapt the velocity model found in the literature [4, 5], in which the wave speed is depend on the angle between the direction of propagation and radial direction from the wood pith (see figure 6). We assume two linear functions of $v(\theta)$ in the intervals $\theta = 0°$ to $45°$ and $\theta = 45°$ to $90°$, as shown in figure 7. Trial values of the wave velocity $v(0°)$, $v(45°)$, and $v(90°)$ are put into the program. The calculated time of flight is compared with the measurement. The values of $v(0°)$, $v(45°)$, and $v(90°)$ are adjusted to minimize the root-mean-square error, or RMSE, of the time of flight between the measurements and calculations obtained from various position of the transducers. We find the optimized values of these parameters to be $v(0°) = 1.13 \times 10^3$ m/s, $v(45°) = 1.10 \times 10^3$ m/s, and $v(90°) = 0.79 \times 10^3$ m/s. The results of the measured and computed time of flight at various positions of the transducers are shown in figure 8. The RMSE of the time of flight is 0.5 μs, which is about 1% of the measured results. This small RMSE suggests that our velocity model should agree well with the actual wave velocity in the wood.

The velocity model obtained from our method can be very useful in the wood defect detection. Once we know the velocity model in a healthy wood, we may compare it with the velocity model of another wood sample. If some anomalies in the models are detected, it is reasonable to expect some defects or holes in that sample.

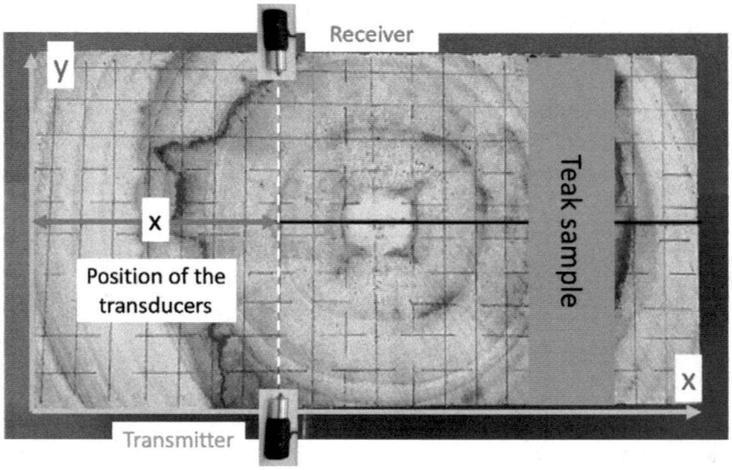

Figure 5. Positions of ultrasonic transmitter and receiver on the teak sample.

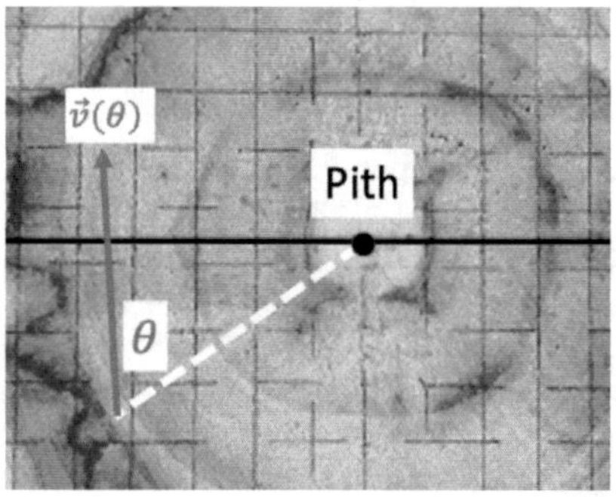

Figure 6. Velocity model in the teak sample.

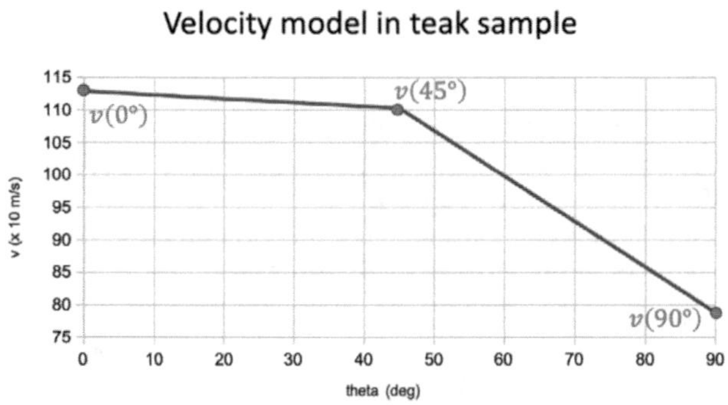

Figure 7. Piecewise linear approximation of the velocity model.

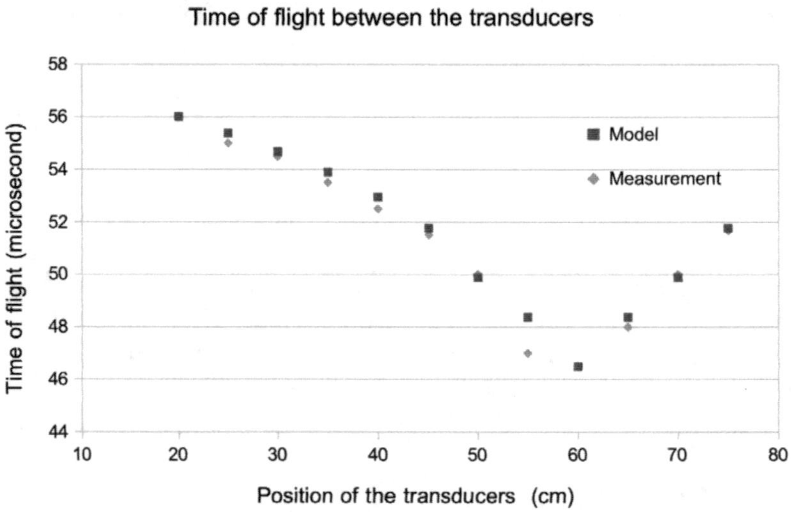

Figure 8. Comparison of the time of flight between the simulations and measurements.

5. Conclusion

We have shown the applications of the path-finding algorithm to simulate wave propagation in various mediums. In all the above simulations, the same path-finding code is used, regardless of the complexity of the wave velocity model. Of course, more technical details are needed for real-world application. However, we are quite certain that the core calculation of the algorithm will not change. From this simplicity, we expect that our method can be useful in wider applications when an investigation of wave propagation in a complex medium is needed.

Acknowledgments

This work is a part of the research administration program to support tree planting under the Valued-Tree Community Project. K. Ketthong would like to thank the Graduate Program Scholarship from the Graduate School, Kasetsart University.

References

[1] Espinosa L, Brancheriau L, Cortes Y, Prieto F and Lasaygues P 2020 *Ann. For. Sci.* **77** 68
[2] Lipson A, Lipson S G and Lipson H 2011 *Optical Physics* (4th Edition) (New York: Cambridge University Press)
[3] Dijkstra E W 1959 *Numer. Math.* **1** 269
[4] Dackermann U, Crews K, Kasal B, Li J, Riggio M, Rinn F and Tannert T 2013 *Mater. Struct.* **47(5)** 787
[5] Niemz P and Mannes D 2012 *J. Cult. Herit.* **13S** S26

Journal of Physics: Conference Series **2145** (2022) 012066 doi:10.1088/1742-6596/2145/1/012066

A simple improvement of an off-grid solar photovoltaic panel using an integrated reflector

Sawitree Wongrerkdee[1], Patcharawadee Kasemjit[2], Sasimonton Moungsrijun[2], Supphadate Sujinnapram[2], Sucheewan Krobthong[2] and Sutthipoj Wongrerkdee[2*]

[1] Department of Electrical Engineering, Faculty of Engineering, Rajamangala University of Technology Lanna Tak, Tak 63000, Thailand
[2] Department of Physics, Faculty of Liberal Arts and Science, Kasetsart University Kamphaeng Saen Campus, Kamphaengsaen, Nakhon Pathom 73140, Thailand

[*] Corresponding author's e-mail address: sutthipoj.s@ku.ac.th

Abstract. This work presents a simple demonstration of a solar photovoltaic (PV) panel integrated with mirror reflectors to increase electric energy generation. The reflector was integrated with a PV panel and its angle was adjusted to an incline to optimize sunlight collection. Current and voltage generated by PV panel was recorded by an Arduino data logger. The integrated-reflector PV panel at an appropriate incline angle of 70 degrees presented a 9.38% increased electric energy beyond that of a conventional PV panel. This result was because the reflector provided greater sunlight to the PV panel. Therefore, this method can be used to increase solar PV panel performance without the installation of additional panels.

1. Introduction

Each year, energy consumption rapidly increases due to population growth, industrial development, and economic growth [1,2], while conventional energy sources such as coal, oil, and natural gas are decreasing [3]. Alternative energy is thus proposed and considered for future development, especially solar energy [4] according to an energy policy that focuses on solar energy in Thailand [5]. Solar energy is a significant energy resource that is capable for meeting energy demands because it relies upon a free, limitless, clean, and available energy source [6]. To harvest solar energy, the solar photovoltaic system is a highly attractive technology to convert sunlight into electricity. Electricity also be stored for usability or sustainable-energy management. Moreover, energy in electric form is the most convenient to use since it can easily be converted to other energy-forms such as heat, light, and kinetic energy [7,8]. Photovoltaic systems have therefore been installed and developed in many areas and countries. Moreover, photovoltaic systems can be installed at different scales including the household, a community, and by industry [9-11]. Among these, the household and community scales are very difficult to develop due to capital limitations. Therefore, developing already-installed systems without additional installation should be considered as an alternative way to increase output.

To develop a PV system, increasing overall sunlight collection is an interesting and straightforward strategy. For example, Atif *et al.* [12] developed an open loop tracking PV system to improve electricity generation performance. The developed system showed 18% higher performance than a static PV system. However, moving component may be unsuitable for long-time stability since mechanical components require maintenance. Sornek *et al.* [13] investigated a simple method to improve building-integrated solar PV systems using Fresnel lenses to enhance sunlight collection.

They found that the developed PV system generated 7% more electricity than the conventional system. This study implies that a PV system can easily be improved using a concentrating strategy [14,15]. Similarly, Hadavinia and Singh [16] presented a simulation of a compact PV system with V-trough and compound parabolic concentrators. The simulated results indicated that both concentrators can potentially increase sunlight harvesting efficiency which resulted in improved solar PV performance compared to a flat PV system. These investigations suggest that the development of a stationary PV system using concentrating strategies can effectively and simply enhance PV performance. Concentrating strategies are an interesting method that can be promoted for rural or small-scale PV owners. This work therefore aimed to increase the electricity generation performance of a solar PV panel using a simple mirror reflector. The reflector was expected to reflect and collect more sunlight for the solar PV panel. This simple method could increase sunlight intensity on a solar PV panel and thereby improve electricity generation.

2. Experiment

A solar PV panel system integrated with two mirror reflectors (PV1) and a conventional PV (PV2, used as a reference system) were setup as shown in figure 1, at Kasetsart University Kamphaeng Saen Campus, Nakhon Pathom, Thailand (Latitude 14° 01′ 16.08″ N). The PV panels were turned southwards. The PV plate was inclined to 14 degrees to the south according to the experimental location. Both solar PV panels were silicon polycrystalline with identical characteristics of 12V30W and had an area of 35cm × 63cm. Current and voltage of each PV panel were recorded by a DIY Arduino data logger. Similarly, temperature at the PV surface was recorded by the data logger via a sensor. The DIY Arduino data logger system included an Arduino microcontroller, SD card, real time clock module, current sensor, voltage sensor, and temperature sensor. The Arduino microcontroller was programed to read the sensors and record the current, voltage, and temperature data from 9:00 to 15:00 local time. Meanwhile, sunlight intensity was measured by a handle solar power meter. Light emitting diodes (LEDs) were used to represent the external load. To improve PV panel performance, two mirror reflectors were attached at along with the PV1 length as presented in figure 1. To study the effect of the reflection of the mirror on PV panel efficiency, the reflectors were adjusted on the inclined angles of the plate from 40 degrees to 80 degrees.

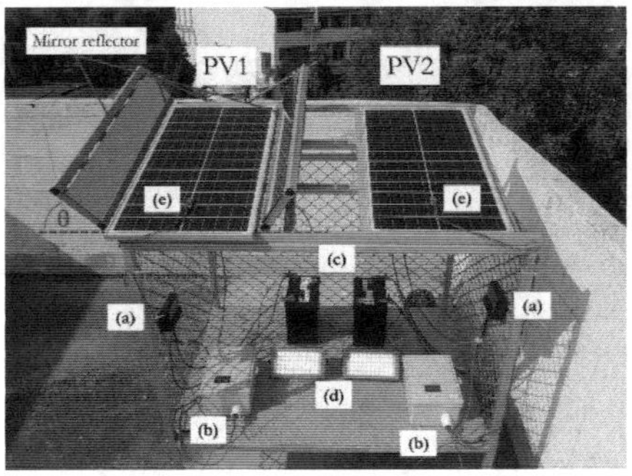

Figure 1. PV panel system setup with and without reflector: (a) charge controller, (b) Arduino data logger, (c) battery, (d) LED, and (e) temperature sensor.

3. Results and discussion

Figure 2 presents the power output of the solar PV panels versus duration period of dairy solar energy. It was found that PV1 exhibited higher power output than PV2 reference for reflector incline angles of 40, 50, 60, and 70 degrees. These results suggest that the reflector reflected more sunlight onto the PV panel. Thus, the PV panel integrated with the mirror reflectors is able to generate more electricity by collecting more sunlight. However, PV1 at an incline angle of 80 degrees showed lower power output than the PV2 reference. This effect was caused by shading of the reflector itself. This behavior implies the importance of considering shading. The power output trend shows that both PV1 and PV2 presented low power outputs at 9:00 and 15:00 because the sunlight rays diverged from the normal line of the PV panel. For quantitative evaluation of electricity generation of the PVs, the total generated electrical energy over 6 hours was calculated from the area under the power-time curve of figure 2, as listed in table 1. It is evident that the PV1 generated more electricity than PV2 for most angles with the exception of 80 degrees.

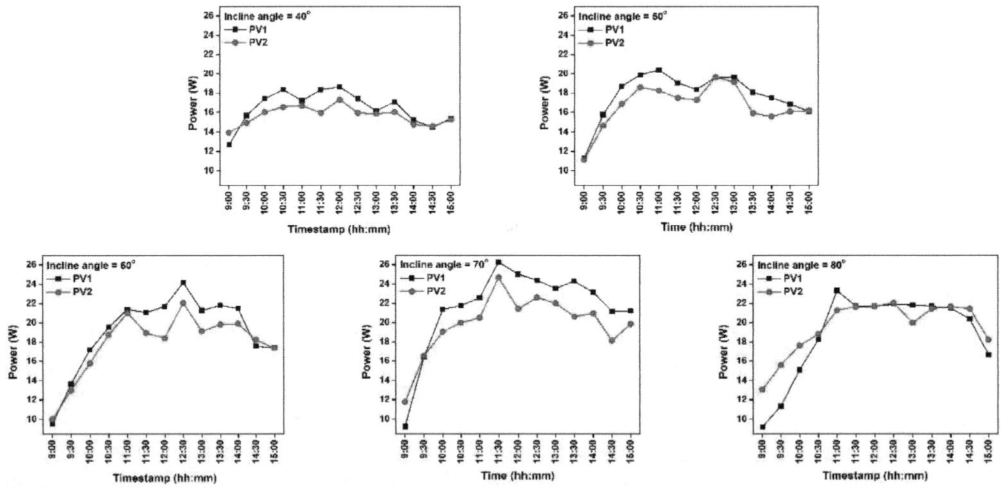

Figure 2. Power output of PV1 integrated with reflectors at different angles compared to PV2.

Table 1. Total generated electrical energy, sunlight energy, and average temperature for the PVs.

Incline angle (°)	Generated electric energy (kJ)		Total sunlight energy (kJ)		Temperature (°C)	
	PV1	PV2	PV1	PV2	PV1	PV2
40	359.9	340.6	3,814.5	3,782.5	46.3	46.5
50	391.6	365.7	3,321.7	3,258.2	49.4	48.4
60	421.6	393.8	3,195.3	3,164.5	47.5	46.9
70	477.2	436.3	3,914.3	3,872.9	48.9	48.3
80	417.0	429.8	3,808.8	3,767.6	47.5	47.6

Increased sunlight energy also resulted in higher temperatures for PV1. Higher temperatures can reduce electricity generation, resulting in lower efficiency. Yet the average temperature was insignificantly different between PV1 and PV2, indicating little temperature effect on electricity generation. Thus, the small increase in temperature might not be considered in this case, while sunlight energy was the significant factor for electrical energy generation. These results indicate that the PV panel integrated with reflectors can increase electricity generation. To evaluate the performance of PV integrated with reflectors, the total generated electricity was compared between PV1 and PV2 at the same angle, as shown in figure 3. It is found that PV1 exhibited a maximum performance around 9.38% higher than the conventional PV2 at an incline angle of 70 degrees, which is in agreement with another report [16]. This can be interpreted that solar PV panel integrated with reflectors at an appropriate incline angle can be used to increase electricity generation. This method can potentially produce more electricity without installation of additional panels, and should be considered for PV applications with low-cost management. However, the large scale of the solar farm must be considered on the economic analysis including to the maintenances cost during life time of PV panels.

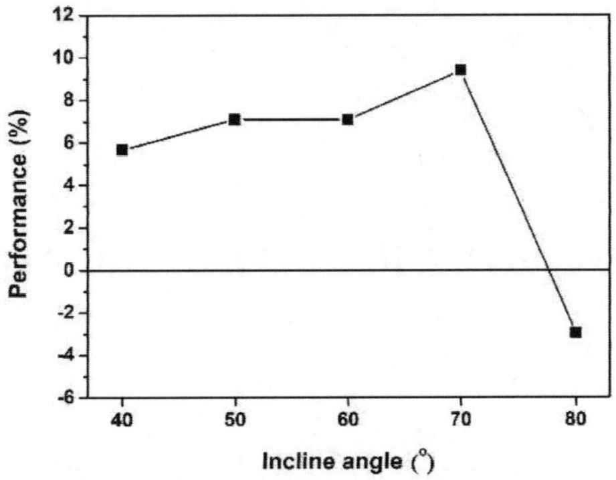

Figure 3. Performance evaluation of PV1 compared to PV2 at the same incline angle.

4. Conclusion

This work presented a simple method to improve electricity generation of solar PV panels using the integrated-mirror reflectors. The reflectors were integrated with the PV panel and its incline angle was adjusted to optimize sunlight collection. It was found that the PV panel integrated with the reflectors at an appropriate angle of 70 degrees presented the greatest improved electricity generation of up to 9.38% compared to a conventional-reference solar PV panel. The increased electricity generation occurred because more sunlight was reflected onto and collected by the PV panel. Therefore, this method can be used to increase solar PV panel performance without the installation of additional panels.

Acknowledgements

This work was supported by the Energy Conservation Promotion Fund (ENCON Fund), Energy Policy and Planning Office (EPPO), Ministry of Energy, Thailand. The authors would like to acknowledge the Department of Electrical Engineering, Faculty of Engineering, Rajamangala University of Technology Lanna Tak; and the Department of Physics, Faculty of Liberal Arts and Science, Kasetsart University Kamphaeng Saen Campus for facility support.

References

[1] Zhang S, Nuruzzaman M and Su B 2021 *Energy Policy* **156** 112420
[2] Li S, Meng J, Zheng H, Zhang N, Huo J, Li Y and Guan D 2021 *Environ. Res. Lett.* **16** 054002
[3] Madan D, Mallesham P, Sagadevan S and Veeramani C 2020 *Int. J. Ambient Energy* **41** 1110–7
[4] Akinbami O M, Oke S R and Bodunrin M O 2021 *Alex. Eng. J.* **60** 5077–93
[5] Chanchawee R and Usapein P 2018 *Int. J. Renew. Energy Res.* **8** 1553–62
[6] Kim S, Quy H V and Bark C K 2021 *Mater. Today Energy* **19** 100583
[7] Lenzen M 2010 *Energies* **3** 462–591
[8] Gottschamer L and Zhang Q 2016 *Renew. Sust. Energ. Rev.* **65** 164–74
[9] Alhammami H and An H 2021 *Renew. Energy* **167** 359–68
[10] Behura A K, Kumar A, Rajak D K, Pruncu C I and Lamberti L 2021 *Sol. Energy* **216** 518–29
[11] Shen W, He J and Yao S 2021 *Energy Policy* **150** 112129
[12] Atif E M, Dumrichob N, Jolivot R, Sterckx K and Mohammed W S 2020 *Eng. J.* **24** 65–76
[13] Sornek K, Filipowicz M and Jasek J 2018 *J. Sustain. Dev. Energy Water Environ. Syst.* **6** 415–26
[14] Sheikholeslami M, Farshad S A, Ebrahimpour Z and Said Z 2021 *J. Clean. Prod.* **293** 126119
[15] Lamnatou C, Vaillon R, Parola S and Chemisana D 2021 *Renew. Sust. Energ. Rev.* **137** 110625
[16] Hadavinia H and Singh H 2019 *Renew. Energy* **139** 815–29

Should we teach free-body diagrams before or after Newton's Laws?

N Nuchsirikulaphong*, N Emarat and K Arayathanitkul

Department of Physics, Faculty of Science, Mahidol University, Rama 6 Road, Ratchathewi, Bangkok, 10400, Thailand

*E-mail: nattaphong.nuh@student.mahidol.ac.th

Abstract. There are two interesting lesson sequences for teaching force and motion in high-school physics. These are teaching free-body diagrams before Newton's laws (FbN) and teaching Newton's laws before free-body diagrams (NbF). Both sequences were found in physics textbooks. Different authors adopted the sequence that they believe it would affect student understanding better. However, some physics experts did not agree with this. It is therefore interesting to know if we should teach with the FbN or NbF sequence. This motivates us to study the effect of such lesson sequences on student understanding of force and motion. The sample group was grade-10 students from two physics courses in 2020. One course was taught with the FbN sequence (29 students) and the other with the NbF sequence (34 students). Their understanding was evaluated by using an assessment test which consisted of three parts including (1) Newtonian concept, (2) problem solving, and (3) free-body diagrams. The result shows that for the Newtonian concept part, the average scores are 11% for the FbN and 13% for the NbF sequence. The average scores of the problem-solving part are 13% and 9% and those of the free-body diagram part are 41% and 48% for the FbN and NbF sequences, respectively. The scores of all parts between the two sequences were not significantly different. In addition, student difficulties found in all parts were similar. However, a larger number of students who could provide the equation of motion (F = ma) in the problem-solving part was found in the FbN sequence. We might conclude that teaching free-body diagrams before or after Newton's laws did not affect student understanding in the topic of force and motion. Detail of student difficulties in both sequences will be further discussed.

1. Introduction

Many teachers tried to develop their teaching approaches to help students understand what they learn. They also arranged the contents in the sequence that might help students. According to Driver (1994), a well-planned curriculum can help students understand the subject. However, a curriculum is normally constructed based on the logical structure of the subject and not on students' background knowledge [1]. Moreover, the logical structure depends on individual's ideas. An example is shown in force and motion chapter in several well-known textbooks. There are two interesting structures (or lesson sequences) which are introducing free-body diagrams before Newton's laws (FbN sequence) and introducing Newton's laws before free-body diagrams (NbF sequence). The first sequence can be found in textbooks [2] and [3] while the other is shown in textbooks [4] and [5]. Therefore, the objective of this work is to study the effect of lesson sequences on student learning on the topic of force and motion.

Journal of Physics: Conference Series **2145** (2022) 012067 doi:10.1088/1742-6596/2145/1/012067

2. Methodology

Two lesson sequences were developed and both included the same 4 subtopics which are Introduction of force, Newton's laws, Free-body diagrams and Problem solving. The only difference between the two sequences is the order of the 2nd and 3rd subtopics (see figure 1).

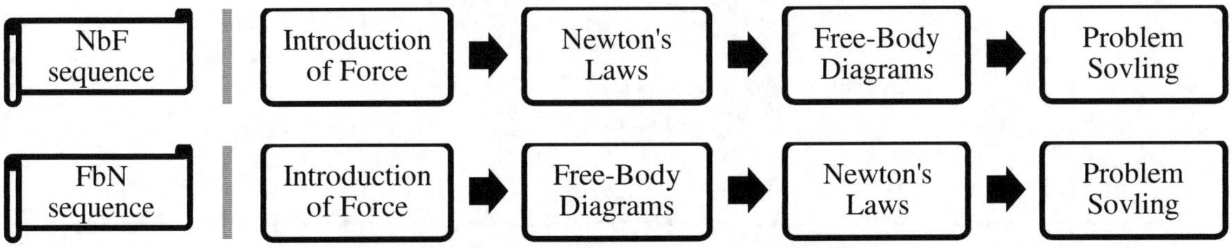

Figure 1. The NbF and FbN lesson sequences.

The lesson sequences were used in 2 high-school classes during the second semester of academic year 2020. One class used the FbN sequence (29 students) while the other used the NbF sequence (34 students). Although the two classes were instructed by different teachers, the researchers observed that both teachers instructed with a similar style. In addition, one teacher joined the other class to make sure that they would not make any difference other than the lesson sequences. Each subtopic was taught for 50 minutes except for the problem-solving subtopic when 100 minutes were spent.

The student knowledge was evaluated by using an assessment test. The test includes Newtonian concept, problem solving and free-body diagrams. The first part of the test was the force sled question set from the FMCE (Thai version) [6]. There are 7 situations in this part and the students must choose the condition that correctly explains each situation. The second part was a multiple-choice question translated from Rosengrant (2005). It asked for a magnitude of the friction force between a vertical pillar and a fireman who is moving down to the ground with a constant downward acceleration [7]. In this part, a blank area was provided in the test for students to write down their solving procedure. The last part consists of two questions to probe student understanding of free-body diagrams as shown in figure 2.

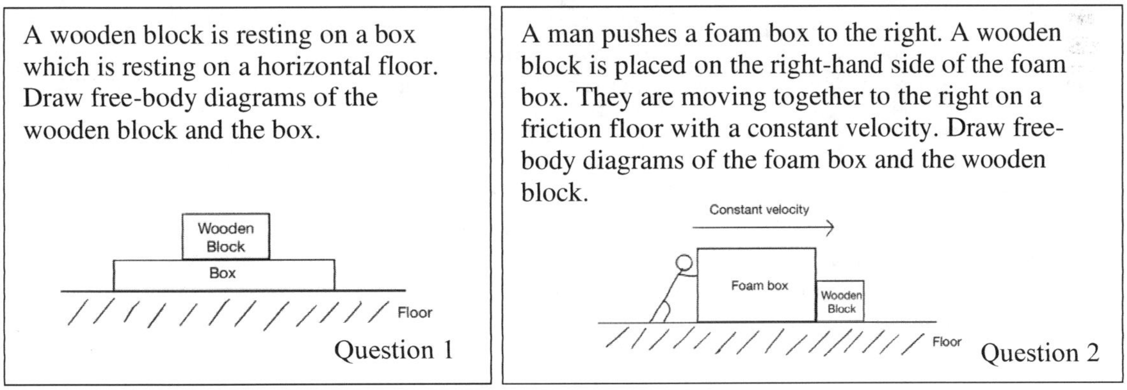

Figure 2. The questions in the free-body diagram part in the assessment test.

3. Results and discussion

The results from the assessment test were the same between the two groups. For the Newtonian concept part, the FbN students had the average score of 11% and the NbF of 13%. The data shows that the most popular answer for each situation is similar in both groups of students. They thought that the direction of motion defines the direction of force and a change of velocity magnitude is influenced by a change of the magnitude of force.

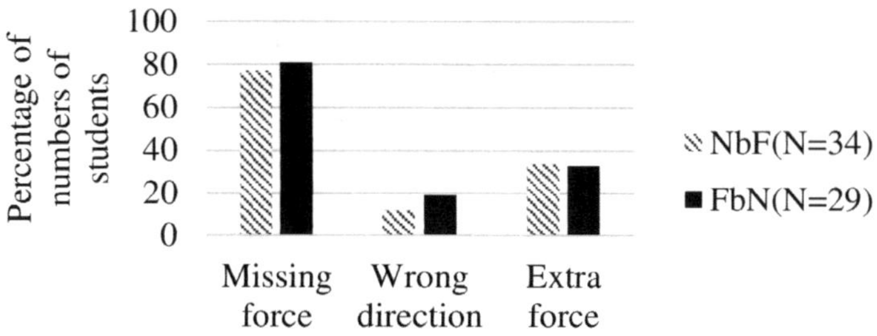

Figure 3. Numbers of students giving each category of mistakes in the free-body diagrams.

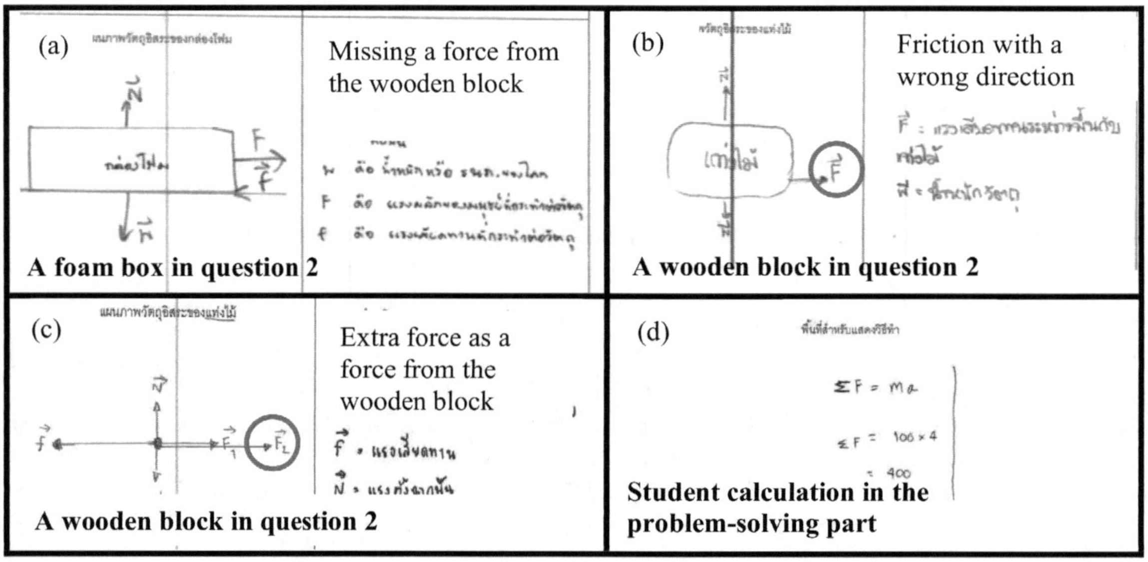

Figure 4. Examples of students' free-body diagrams (the second question): (a) missing a force from the wooden block (b) giving a wrong direction of friction force (c) giving an extra force in the free-body diagram of a wooden block, and (d) student calculation in the problem-solving part.

Next, for the problem-solving part, 13% and 9% of the students from the FbN and NbF sequences, respectively, could successfully find the correct answer. Finally, in the free-body diagram part, students from the FbN and NbF sequences provided average scores of 41% and 48%, respectively.

The mistakes in students' answers about free-body diagrams were categorized into three types. If some forces were missing in the free-body diagram, the data was collected as a "missing force" mistake. If the correct force was given but with a wrong direction, the data was collected as a "force with wrong direction" mistake. Finally, if there was an extra force which should not be involved in the free-body diagram, the data was collected as a "extra force" mistake. Figure 3 shows the percentage of students giving each mistake in their diagrams. The similar pattern of the results can be seen where the highest mistake found in both groups of students was the "missing force". Those missing forces are a force acted on the box by the wooden block in the first question and a force acted on the foam box by the wooden block in the second question (figure 4(a)). These mistakes of students from both sequences are the same. Figures 4(a) – 4(c) show examples of student drawing a free-body diagram in each category. For the "force with wrong direction" category, few students were confused about the direction of some forces such as a weight, a normal force and a friction force (figure 4(b)). For the "extra force" category, a few students thought that some forces should appear in the free-body diagram such as a friction force in question 1 and a force from the object itself (figure 4(c)).

The results from all three parts of the test seem to indicate that both groups of students gained similar knowledge after learning with different lesson sequences. However, an interesting point was found in the results from the problem-solving part. From the blank area provided in the assessment test for students to write down their calculations, it was discovered that students from the FbN sequence could realize that Newton's second law is the key to solve the problem. The evidence is that they wrote the equation $F = ma$ in the blank area as shown in figure 4(d) while most of the students from the other group did not write anything in the blank area.

4. Conclusion

The results from the assessment test which measure the knowledge of Newtonian concept, problem solving and free-body diagrams of the students from both sequences are significantly similar. This means that different lesson sequences did not affect student learning on force and motion. However, a larger number of students from the FbN sequence could realize about the use of the equation of motion in solving the problem. To improve the study, conducting the experiment with a larger sample group and improving the instruction for both sequences are suggested to confirm that lesson sequences cannot affect student learning.

Acknowledgments

I would like to thank every member in Physics Education Lab at Mahidol University, my family and myself.

References

[1] Driver R, Leach J, Scott P and Wood-Robinson C 1994 Young people's understanding of science concepts: Implication of cross-age studies for curriculum planing *Stud. Sci. Educ.* **24** 75–100
[2] Beichner R J and Serway R A 2015 *Physics for Scientists and Engineers with Modern Physics* 9th edn (Boston: Brooks/Cole Pub Co)
[3] Young H and Freedman R 2016 *University Physics with Modern Physics* 14th edn (New Jersey: Pearson Addison Wesley)
[4] The Institute for The Promotion of Teaching Science and Technology 2017 *Advanced Physics* vol 1 (Bangkok: Chulalongkorn University Press)
[5] Mazur E 2020 *Principle & Practice of Physics* 2nd edn (Boston: Pearson Education)
[6] Emarat N, Arayathanitlkul K, Soankwan C, Chitaree R and Johnston I D 2002 The effectiveness of the Thai traditional teaching in the introductory physics course: A comparison with the US and Australian *Proc. Scholarly Inquiry in Flexible Science Teaching and Learning Symp. (5 April 2002, Sydney)*
[7] Rosengrant D, Van Heuvelen A and Etkina E 2009 Do students use and understand free-body diagrams? *Phys. Rev. ST Phys. Educ. Res.* **5** 010108

Newton's law of cooling experiment set using Arduino temperature sensor

P Chanthamanee[1], P Jinda[1], M Mani[2] and S Prasitpong[1*]

[1]Department of Teaching Science and Mathematics, Faculty of Education, Thaksin University, Songkhla, 90000, Thailand
[2] Department of Basic Science and Mathematics, Faculty of Science, Thaksin University, Songkhla, 90000, Thailand

*Corresponding author's e-mail address: singha.p@tsu.ac.th

Abstract. The research aims to develop the experimental set of the temperature measurement in liquid by Arduino program displaying data on a smartphone via the Blynk application. The experimental set is composed of 1) 2 liquid temperature sensors (DS18B20 model), 2) Arduino program, and 3) LED screen for showing the temperature value in unit of °C and connect to a smartphone. The Arduino temperature sensor 1 and sensor 2 of the experimental set have 0.57% and 0.51% errors, respectively, compared with the temperature sensor of the B Smart Science Co., Ltd. company. The instrument is applied to the physics laboratory on Newton's law of cooling to find the cooling rate of water and coffee. This low-cost instrument revealed high accuracy results and easy to connect with other devices.

1. Introduction

Nowadays, there are several learning resources, easy to access and friendly to apply. Teaching and learning in educational institutions are required to promote the information and technology literacies as tools for learners' working. Moreover, learners must learn ways of thinking such as creativity, critical thinking, problem-solving, and learning ways of working such as communication and collaboration skills for success in today's world. It is the framework for 21st-century learning skills and knowledge students need to succeed in work and life [1]. Instructional tools and processes play a crucial role in activating learners to develop knowledge, skills, and attitudes, particularly in science and technology topics. This study aims to introduce a low-cost instrument developed by Arduino board illustrating results on smartphones via the Blynk application. This technology can be applied to encourage critical thinking, problem-solving, and collaboration skills of high school to university students in physics classrooms.

This study develops a temperature sensor based on Arduino and its technology to help students learn Newton's law of cooling concept. The Arduino temperature sensor is widely applied to classrooms because it has high accuracy and can measure extensive temperature range [2-4].

2. Instrument and method

2.1. NodeMCU V3 ESP8266

NodeMCU V3 ESP8266 is a microcontroller based on ESP8266 Arduino board and connected to a wifi signal. The main parts are nine digital input/output pins, one analog input pin, a USB connection, a power jack, and a reset button. The NodeMCU V3 ESP8266 can be programmed with the Arduino Software (IDE) and powered by an AC to DC adapter (or battery) to start.

2.2. DS18B20 temperature sensor

DS18B20 is a digital temperature sensor providing -55 °C to 125 °C measurable temperature ranges. It is one of the most popular temperature sensors on the market. It can be used as the temperature monitor tool in a smart farm, house, water station, and so on. It is similar to DHT11/DHT22 temperature and humidity sensor. The DS18B20 temperature sensor has many formats, such as IC and waterproof sensors. This study uses the ESPino32 board, reads data from a sensor, and displays results on a serial monitor. This Arduino board developed in this research was shown in figure 1.

2.3. Application blynk

Blynk is a free platform with iOS and Android applications to control Arduino. It is easy to program, real-time using, and simply to connect with Arduino, ESP8266, ESP32, NodeMCU, Rasberry pi.

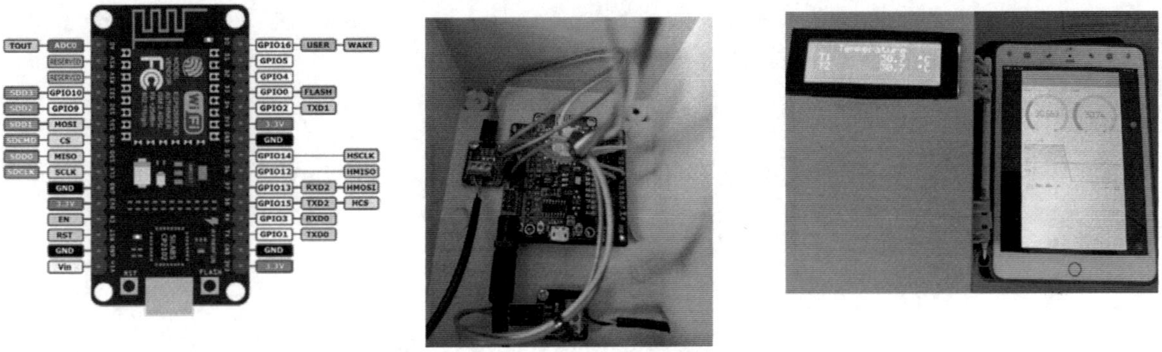

(a) (b) (c)

Figure 1. (a) diagram of NodeMCU V3 ESP8266, (b) Arduino board used in this study, (c) Blynk application.

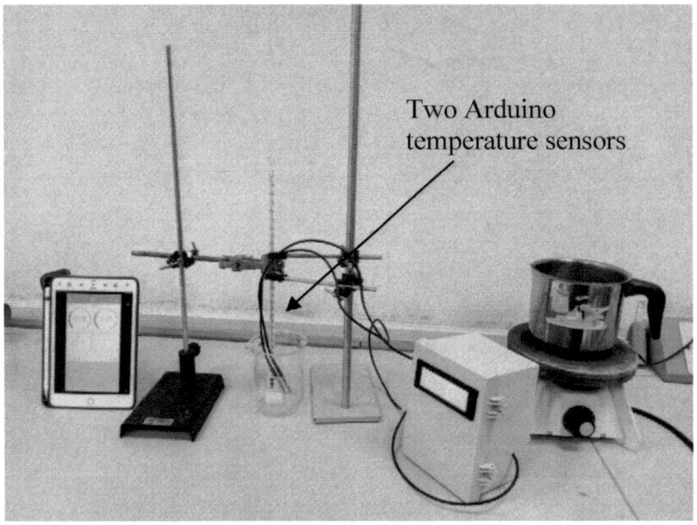

Figure 2. Newton's law of cooling experiment set using the Arduino temperature sensors.

2.4. Newton's law of cooling

Newton's law of cooling states that the rate of change of temperature of an object is proportional to the difference in temperature between the object and its surroundings [5], expressed by equation (1)

$$\frac{dT}{dt}(t) = -K(T(t) - A) \tag{1}$$

where $T(t)$ is the temperature of the object at time t, A is the temperature of its surrounding, and K is a constant of the cooling rate. A high K constant means a large change of the temperature over times, indicating a high cooling rate. The equation can be shown as

$$T(t) = [T(0) - A]e^{-Kt} + A \tag{2}$$

where $T(0)$is the initial temperature of the object. This study applied the Arduino temperature sensor to investigate the cooling rate of water, as shown in figure 2. As a preliminary work, we asked the 4-year physics students from the faculty of education to conduct Newton's law of cooling experiment as a group of 2-3 students. They were asked to design an experiment to observe the change of temperature of water and coffee, and to find the Kconstant using the developed Arduino temperature sensor.

3. Results and discussion

The two Arduino temperature sensors were used to measure water temperature during 31°C - 50 °C, and compared the result with the thermosensor of the B Smart Science Co., Ltd. company, which is commonly used in the physics laboratory. It revealed 0.57% and 0.51% errors for the temperature sensor 1 and sensor 2, respectively.

An example graph of water for one group was shown in figure 3. They measured the temperature of 200 mL water in a beaker at 90 °C initial temperature and put it at 27.5 °C surrounding temperature. The temperature was recorded every 30 seconds until it reaches the room temperature. The graph showed a decreasing trend of temperature versus times, which agrees with Newton's law of cooling concept. Moreover, equation (2) was reformatted as:

$$\ln[T(t) - A] = -Kt + \ln[T(0) - A] \tag{3}$$

The collected data were plotted and computed the K constant. They reported the K constant of water as $0.70 \times 10^{-3}\,s^{-1}$ with 2.3% error under the mentioned conditions. Furthermore, another group applied the Arduino sensor to measure the temperature of the 3-in-1 coffee contained in a 150 mL beaker at 90 °C initial temperature and put it at 27.5 °C room temperature. They found the K constant of coffee as $0.85 \times 10^{-3}\,s^{-1}$ with 3.2% error.

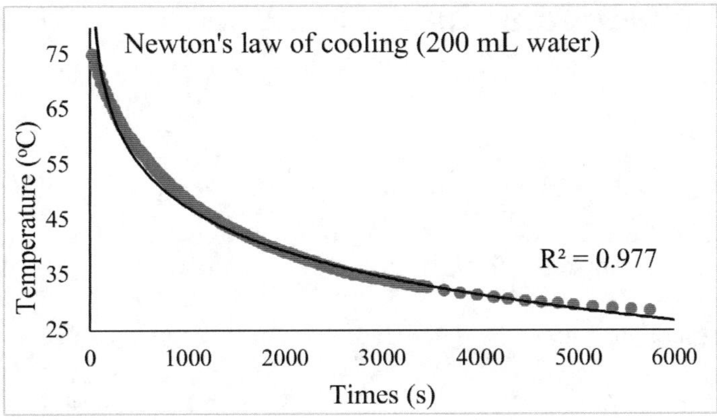

Figure 3. Newton's law of cooling for 200 mL water measured at 90 °C initial temperature and 27.5 °C room temperature.

Overall, the students were satisfied with this apparatus because it shows accurate values and easy to connect via wifi internet to other components. In addition, it can measure both solid and liquid materials within a large temperature range. Some students' groups planned and discussed using this instrument to measure the temperature of the coffee famous sells in the market, which has different ingredients to observe the cooling rate. It may guide sellers for decision-making. It reflects that working as a group of students in this study promotes both students' knowledge and higher-order thinking skills required for success in their life.

4. Conclusion

Newton's law of cooling experiment set based on Arduino temperature sensors developed in this study works well in finding the cooling rate constant of water and coffee. Advantages of the Arduino temperature sensor are low cost, high accuracy, and easy to link via wifi to free platform with iOS and Android applications or other devices. Thus, Arduino and its technology can be a highly effective instructional instrument in the active learning approach. Moreover, learners should have an experience in programming the Arduino and learn to extend the Arduino project.

Acknowledgments

The authors thank all participants, the Faculty of Education and the Faculty of Science at Thaksin University, Thailand.

References

[1] World Economic Forum 2015 *Ten 21st-century skills every student needs* retrieved from *https://www.weforum.org/agenda/2016/03/21st-century-skills-future-jobs-students/*
[2] Yesica D *et al.* 2021 *J. Phys.: Conf. Ser.* **1845** 012069
[3] Rahaman L M *et al.* 2016 *Procedia Comput. Sci.* **89** 320–3
[4] Vern S C *et al.* 2019 *TJEET* **3** 1–12
[5] Yunus C, Michael B and Mehmet K 2019 *Thermodynamics: An Engineering Approach (9th ed.)*(Singapore: McGraw-Hill Education (Asia))

Activities on reflection of light based on low-cost materials at eighth grade

A Kittiravechote* and N Thinarat

Program of General Science, Faculty of Education, Bansomdejchaopraya Rajabhat University, Bangkok, 10600, Thailand

*Corresponding author's e-mail address: aungtinee.ki@bsru.ac.th

Abstract. Studying science through hands-on activities within the context of teaching and learning using local environments and resources has been recognized as a fundamental prerequisite of school curriculum in many developing countries, including Thailand. Accordingly, we are interested in implementing such a concept in designing science teaching and learning for middle school students that emphasizes student participation in the provision of materials and equipment for scientific activities. The purposes of this study are to design hands-on activities for eighth grade students on the topic of light reflection using low-cost materials, and then to examine the results after they are implemented. The experimental group consisted of 30 eighth grade students who were chosen at random in one room. The instruments used in the experiment were two types of achievement tests that were administered as pretests and posttests on the topics of light reflection on smooth and curved surfaces. The paired sample t-test was performed after the collection of data to determine students' conceptual understanding. Our result displays a highly significant difference ($p < .01$) between the pretest score (39.7%) and posttest score (83.3%), indicating that students actively participated in hands-on activities and learned light reflection knowledge and skills through interaction and discussion with their classmates. We highlight the advantages of our study for science teachers in rural schools in ensuring students develop 21st Century Skills, which is an important part of STEM education.

1. Introduction

Participation and understanding of scientific processes are at the heart of science education [1]. The successful laboratory use is a requirement in science education, as science is an experimental, analytical, and laboratory-oriented discipline that provides a variety of benefits, including the acquisition of basic laboratory skills, the development of observational skills, the explanation of a specific concept, and the most authentic taste of what true science entails [2,3]. However, providing well-equipped labs appears to be one of the main roadblocks to the popularization of science education in low-income schools with limited resources. As a result, methods for improving the current situation in science laboratories include sponsoring various low-cost activities or training teachers to make the most of locally available tools for science teaching [4].

In this work, we are interested in designing and implementing hands-on activities on the topic of light reflection on smooth and curved surfaces using low-cost materials for 30 eighth-grade students. We used a randomized group: pretest and posttest design for educational experimentation. The effect on student achievement was examined using assessments that included a pretest, a posttest, and the

Journal of Physics: Conference Series **2145** (2022) 012069 doi:10.1088/1742-6596/2145/1/012069

difference between the two mean scores. A statistical test, the paired sample t-test, was used to evaluate the significance of the difference in average changes.

2. Research design

Our research was carried out in the following manner. (1) Preparation of the tests. We created two sets of multiple choice test items for light reflection on smooth and curved surfaces, which were adapted from Thai standardized achievement tests and represented Bloom's understanding, comprehension, and application, or related to arithmetic skills and data interpretation obtained through laboratory experiments [5]. Experts first qualified the desirable items in terms of content correctness and item-objective congruence index, and then a test-re-test technique was used to exclude incorrect questions in terms of difficulty index and discriminant index. Finally, five questions from each set were chosen as the pretest and posttest; note that the pretest and posttest were identical but alternate choices. (2) Designing activities with low-cost materials. We used activities based on light reflection, ranging from smooth to curved surfaces, as shown in figure 1. Figure 1 (A) depicted the transformation of a smartphone into a pseudo-hologram projector by demonstrating the concept of light reflection on a smooth surface using a plastic transparent frustum [6]. Figure 1 (B) displayed how a metal curved surface made from an unused soft drink can was used to demonstrate light reflection on the curved one [7]. (3) Running the pretest. We decided to perform an experiment in education using a randomized group: pretest and posttest design. In this study, 30 eighth grade students were randomly assigned to the experimental group in one room. (4) Implementation of the treatment. The experimental group was introduced to light reflection activities with low-cost materials using the modified 5E steps lesson plan shown in table 1. (5) Running the posttest. (6) Conducting the statistical inferences. The difference between the pretest and posttest was measured as an average. The paired sample t-test was used to assess the statistical significance of the difference in average changes.

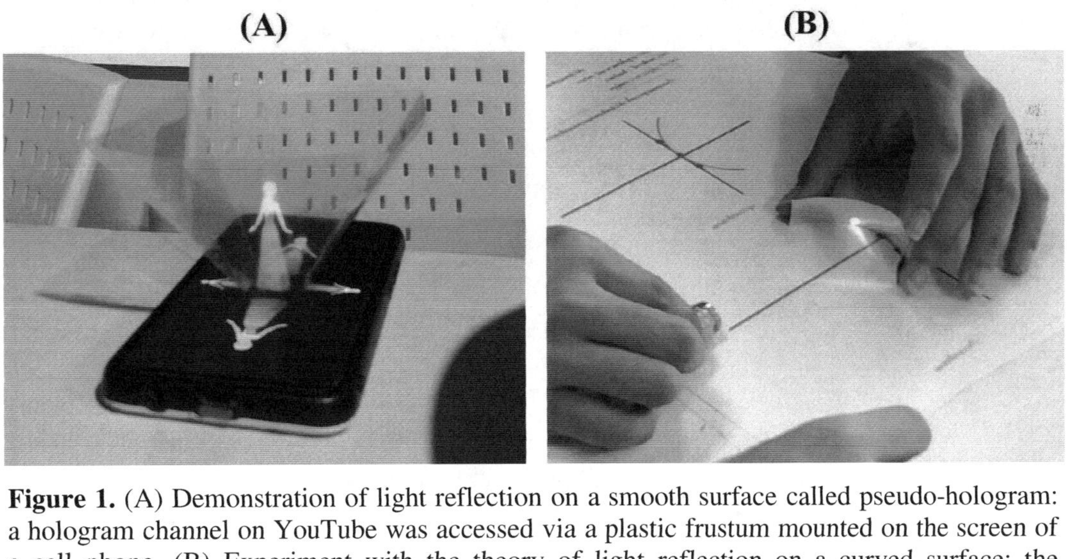

Figure 1. (A) Demonstration of light reflection on a smooth surface called pseudo-hologram: a hologram channel on YouTube was accessed via a plastic frustum mounted on the screen of a cell phone. (B) Experiment with the theory of light reflection on a curved surface: the remaining soft drink can was used to project a laser beam onto a curved metal plate.

Table 1. Lesson plans for light reflection using the modified 5E steps.

	Period 1 (50 min) Light reflection on a smooth surface	**Period 2 (50 min) Light reflection on a curved surface**
Warm up (3 min)	Students are introduced to the activity and lesson. Pairs of students are assigned to work together.	Students are introduced to the activity and lesson. Pairs of students are assigned to work together.
Introduction (10 min)	Students make the pseudo-hologram.	Students shine a laser light onto a curved surface and draw the incident ray and reflected ray.
Body (25 min)	Teacher demonstrates how to create a virtual image using light rays. Students discuss how to create image of the pseudo-hologram with their partner. Students give the presentations.	Teacher recalls the law of reflection. Students discuss how to find the focal point of a concave and convex mirror with a partner. Students give the presentations.
Closing (10 min)	The main idea of the lesson is summarized by all students. As light strikes a smooth surface, (i) the angle of incidence equals the angle of reflection (ii) the incident ray, reflected ray, and normal line are all in the same plane. Since the tails of light rays converge, the image obtained from a smooth surface is a virtual image. The visual image that appears within the pseudo-hologram can be seen from all lateral sides.	The main idea of the lesson is summarized by all students. An incident light is reflected according to the law of reflection at every point along the curved surface. As light that is parallel to the principal axis strikes the concave mirror, the reflected light gathers at a point known as the (real) focal point. As light that is parallel to the principal axis strikes the convex mirror, the reflected light diverges. However, if the tails of light rays are drawn indefinitely, a point of intersection known as the (virtual) focal point is discovered.
Exit ticket (2 min)	Assign students to write what they have learned today	Assign students to write what they have learned today

3. Result

Table 2 displays measures of central tendency by arithmetic mean from the pretest and posttest of light reflection on smooth and curved surfaces. The pretest and posttest mean scores are 3.97 (=39.7%) and 8.33 (=83.3%), respectively. The difference between the two mean scores is 4.36, indicating that activities conducted with low-cost materials had a positive effect on student achievement.

Table 2. Pretest and posttest arithmetic mean measures of central tendency.

No.	Pretest score	Posttest score	No.	Pretest score	Posttest score	No.	Pretest score	Posttest score	No.	Pretest score	Posttest score
1	2	7	**9**	3	7	**17**	9	10	**25**	3	8
2	6	10	**10**	2	7	**18**	8	10	**26**	4	10
3	3	7	**11**	2	8	**19**	4	8	**27**	6	10
4	4	7	**12**	2	7	**20**	2	9	**28**	4	9
5	5	10	**13**	2	8	**21**	3	9	**29**	4	10
6	5	8	**14**	4	7	**22**	4	7	**30**	4	9
7	4	8	**15**	5	9	**23**	3	7	**Sum**	**119**	**250**
8	6	9	**16**	2	7	**24**	4	8	**Mean**	**3.97**	**8.33**

Furthermore, as shown in table 3, the paired sample t-test score of students is used to determine the significance of the difference in average changes between the pretest and posttest. We begin by using the boxplot to examine the distribution of our differences. As long as the shape isn't too skewed and there aren't any outliers, the inference problem for paired data is simply a one-sample problem on the difference within each pair. Later, we established the null hypothesis: mean(Posttest) – mean(Pretest) = 0, and the alternative hypothesis: mean(Posttest) – mean(Pretest) > 0. Following the calculation, we discovered that our t-value was 18 with 29 degrees of freedom. Regarding the table with 29 degrees of freedom, our obtained value (18) exceeds the given value for both .05 level (1.699) and .01 level (2.462). As a result, the null hypothesis is rejected, and the significantly greater of the two means' differences than the test difference (0) is accepted. Therefore, the use of low-cost materials in science education was deemed a success overall.

Table 3. The significance of the difference in average changes between the pretest and posttest was indicated by the students' paired sample t-test score.

No.	*D*	d^2	No.	*D*	d^2	No.	*D*	d^2	No.	*D*	d^2
1	5	0.3969	**9**	4	0.1369	**17**	1	11.3569	**25**	5	0.3969
2	4	0.1369	**10**	5	0.3969	**18**	2	5.6169	**26**	6	2.6569
3	4	0.1369	**11**	6	2.6569	**19**	4	0.1369	**27**	4	0.1369
4	3	1.8769	**12**	5	0.3969	**20**	7	6.9169	**28**	5	0.3969
5	5	0.3969	**13**	6	2.6569	**21**	6	2.6569	**29**	6	2.6569
6	3	1.8769	**14**	3	1.8769	**22**	3	1.8769	**30**	5	0.3969
7	4	0.1369	**15**	4	0.1369	**23**	4	0.1369	**Sum**	**131**	**50.967**
8	3	1.8769	**16**	5	0.3969	**24**	4	0.1369	**Mean**	**4.37**	

Note: D = Posttest score - Pretest score

$d = D - \text{Mean of } D = D - 4.37$

$d^2 = (D - 4.37)^2$

$$S.D. = \sqrt{\frac{\sum d_i^2}{n-1}} = \sqrt{\frac{50.967}{30-1}} = 1.33$$

$$t = \frac{\overline{D}}{S.D./\sqrt{n}} = \frac{4.37}{1.33/\sqrt{30}} = 17.99 \approx 18$$

Degree of freedom = 30 – 1 = 29

Boxplot of the difference

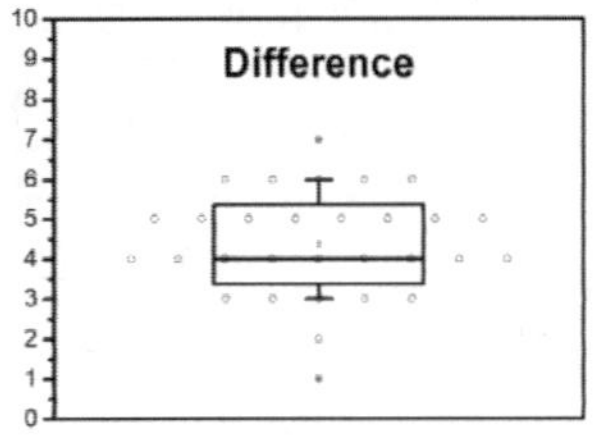

4. Discussion

We investigated whether activities with low-cost materials developed to teach concepts related to the reflection of light unit could contribute to high levels of science achievement for students, as well as students' understanding of science concepts and academic achievement. Multiple-choice tests were used as data collection tools to determine students' conceptual understanding. We found that there is a highly significant difference ($p < .01$) between the pretest and posttest results, according to the results of the question-based achievement score analysis of the pretest (39.7%) and posttest (83.3%).

After reviewing the student worksheet, we discovered that starting the class with a practice or an experiment can help the children visualize and understand further. For example, for the reflection of light on a smooth surface shown in figure 2, (i) students were instructed to work in pairs to construct the pseudo-hologram, (ii) the teacher taught students how to draw the rays of light on the smooth surface (figure 2 (A)), (iii) the teacher allowed students to work and discuss in pairs to draw light rays that created a virtual image on the pseudo-hologram (figure 2 (B)), (iv) the teacher then chosen students to present their findings in front of the class, and (v) the teacher employed leading questions to summarize the main idea of the science lesson. Likewise, such activity before concept was repeated in our class for the reflection of light on a curved surface, as shown in figure 3. Students were tasked with determining the focal point of the curved surfaces by incident and reflection of laser beams from the concave (figure 3 (A)) and convex mirrors (figure 3 (B)). As evidenced by worksheet samples, students can solve unfamiliar problems by applying abstract content knowledge to explain unknown phenomena, which is key science process skill that teachers seek to foster.

(A) **(B)**

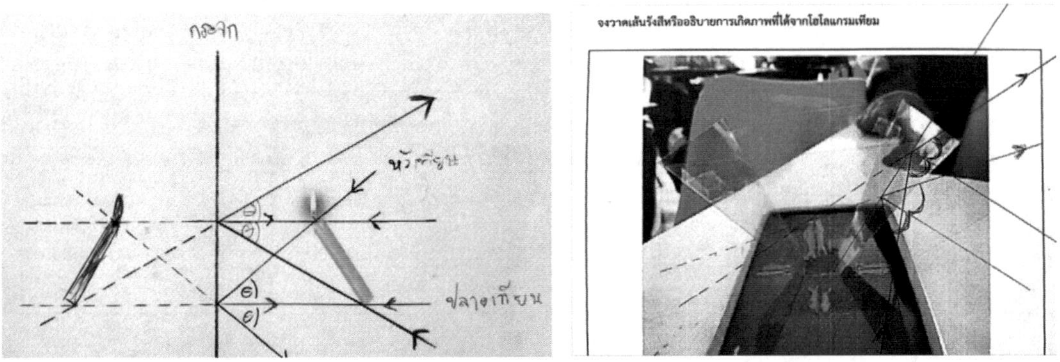

Figure 2. (A) Students practice drawing the light ray to create a virtual image. (B) Students discuss and report on how the image in the pseudo-hologram is formed with a partner.

(A) **(B)**

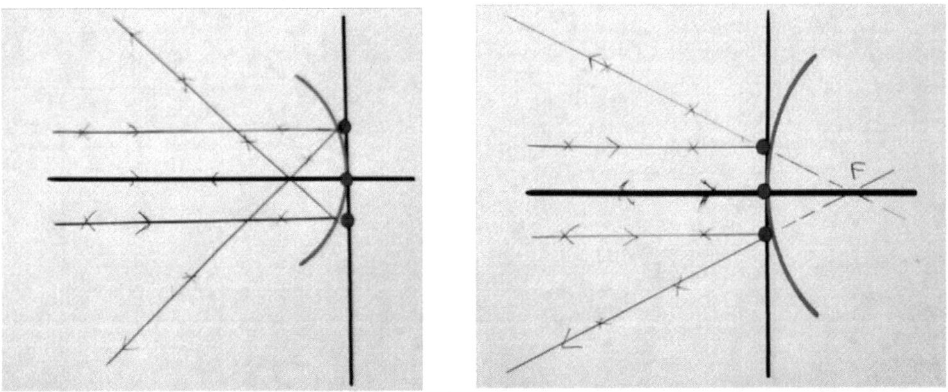

Figure 3. By shining a parallel laser light on a curved surface, the focal point of concave and convex mirrors can be determined. (A) A concave mirror serves as a focal point in front of the mirror, allowing light to converge. (B) A convex mirror provides a focal point behind the mirror to diverge the light.

In addition, educators strongly advise students to engage in hands-on activities in science classes. Many hands-on activities have the ability to greatly influence students' engagement in the activities because they are one situational factor that is often thought to elicit students' interest and inspire them to learn science [8], which are in good agreement when we perform the lesson, as shown in figure 4. The majority of observational findings support the hypothesis that engaging in hands-on tasks contributes to positive motivational results [9]. Moreover, it is not surprising that the students in our experiments have higher posttest scores than pretest scores, given that previous studies show that students who engage in hands-on studies have a higher GPA, complete more science courses, and have a greater willingness to pursue STEM degrees [10]. Our research backed up this conclusion, claiming that the effect is most likely due to the active design of the integrated instruction discussed earlier. We conclude that our activities, which were conducted with low-cost materials, had a positive impact on student achievement and could provide an opportunity for students in rural areas to experiment.

(A) (B)

Figure 4. In the beginning of science class, students participated in hands-on activities on light reflection on a smooth surface (A) and a curved surface (B), which drew their attention to the next step, the discussion and report stages.

5. Conclusion

We have demonstrated how low-cost materials can be used in science hands-on activities to provide purposeful and meaningful experiences for eighth graders. Activities based on low-cost materials, in particular, improve conceptual understanding of light reflection on smooth and curved surfaces, resulting in substantially higher mean posttest scores than pretest scores at the .01 level. The findings of this study show that enabling students to perform science experiments with low-cost materials has a positive effect on their interest in those subjects and related skills.

Acknowledgements

We are grateful for the financial support by the Research and Development Institute, Bansomdejchaopraya Rajabhat University.

References

[1] Taber K S 2017 *Science Education: An International Course Companion* (Rotterdam: Sense Publishers)
[2] Li R and Wong T 2018 *IAFOR J. Educ.* **6** 79–92
[3] Kittiravechote A and Sujarittham T 2020 *Int. J. Adv. Sci. Technol.* **29** 884–9
[4] Alves G H V S, Fragel-Madeira L, Varandas de Azeredo T, Castro H C, Pereira G R and Coutinho-Silva R 2020 *Creat. Educ.* **11** 760–82
[5] IPST 2020 *Teacher's Manual for Fundamentals of Science and Technology: Secondary Education 3 Vol. 1* Retrieved 20 Apr. 2020, from https://www.scimath.org/ebook-science/item/11459-3-1-11459
[6] Kittiravechote A, Pakdeethai T, Chandot A, Panaudom K, Dansai P, Kardkaew P, Hao T and Eiamsri T 2019 *Using Pseudo-Hologram to Demonstrate How to Make Pseudo-Hologram* Retrieved 29 Mar. 2019, from https://so06.tci-thaijo.org/index.php/hsudru/article/view/180959/128400
[7] IPST 2018 *How is the Reflection of the Curved Surface?* Retrieved 27 Nov. 2018, from https://www.scimath.org/video-science/item/9510-2018-11-27-02-14-25
[8] Shahali M, Hafizan E, Halim L, Rasul M S, Osman K and Zulkifeli M A 2017 *Eurasia J. Math. Sci. Technol. Educ.* **13** 1189–211
[9] Sadi Ö and Cakiroglu J 2011 *J. Balt. Sci. Educ.* **10** 87–97
[10] Vanmeter-Adams A, Frankenfeld C L, Bases J, Espina V and Liotta L A 2014 *CBE Life Sci. Educ.* **13** 687–97

Smartphones as detector the speed of sound: A classroom explanation and demonstration

T Thongsuk and A Intanin*

Department of General Science, Faculty of Education and Educational Innovation, Kalasin University, Kalasin, 46230, Thailand

* E-mail : Angkhan.in@ksu.ac.th

Abstract. This paper presents how smartphones determine the speed of sound (C) with a classroom explanation and demonstration to design a variety of lab instruments. Smartphone sensors such as mics and speakers were used as experimental tools by students for calculating the value of speed of sound. Mathematics is used to describe physics principles using only the mean of repetitive experimental results. After conducting an experiment with 43 students, majoring in general science, faculty of education and educational innovation, Kalasin University, the students report the value of the speed of sound nearly to theoretical values with a percentage difference of less than 0.1%, equipment used in everyday life in the classroom, equipment that is cheap, along with a simple calculation of speed of sound, is an advantage of this experiment.

1. Introduction

In the realm of teaching sound waves in general physics, learners are enquired to participate in the practical laboratory of a resonance tube apparatus. Learners will able to understand of wavelength λ, the frequency of the sound wave f, and proficiency of computation in the speed of sound C. Regarding the Traditional resonance labs, have normally basic equipment which is involving a scale-mounted closed-end tube with adjustable pistons or fluids, speakers, and a variable-frequency amplifier, a tuning forks replace a sound source, a thermometer, and a microphone to measure the sound intensity level in decibels. According to Lab-based learning, learners gained positive performances in both scientific reasoning skill and practical skills [1-2]. Nevertheless, laboratories require exclusive lab instruments which is costly to cooperate with a research laboratory. The worthy instruments can generate the accuracy of the operating parameters on the resonance tube device.

In the aspect of the experimental laboratory, the resonance tube apparatus device used basic tools such as plastic bottles and common measuring instruments in the classroom [3]. Also, the smartphones performed as the detector equipment instead of trial laboratory session [4]. We found that the resonance tube apparatus device proved precisely accurate when compared to the related study [5-7] In addition, the standardized device can be applied to general science classrooms to enhance learners' interest in using costless equipment. The purpose aims to enhance and encourage learners' scientific reasoning abilities and practical skills. The study tends to propose a suitable method for applying syringe barrel replacement to the substitution of a closed tube. The microphones' sensors within smartphones are used to measure sound-intensity levels by application of The Decibel X (ios) Sound Meter (Android). Likewise, the speaker of the smartphone is mainly used as the sound source that can adjust the sound frequency with the application f Generator (ios) Frequency Generator (Android). The results in a resonance tube revealed that it can be implemented in a general science classroom.

2. Experimental methods

2.1. Research objectives
To develop scientific reasoning skills and practical skills from the study of sound-speed detector.

2.2. Participants
The participants of the study were 43 undergraduate students, Bachelor of Education (General Science), Faculty of Education and Educational Innovation, Kalasin University. The applicants of the study who study general sciences course were being classified into 10 groups since the beginning of the semester.

2.3. Experimental setup
Each of the groups were prepared the following equipments:

- The replacement of truncated syringe and syringe plunger use to interchange the substitution of a closed tube and the adaptable pistons respectively.
- The initial smartphones (speaker) use for generating the variable-frequency from sound source application.
- The second smartphones (microphone) use for measuring sound-intensity levels in decibels from sound measure application. The quiet classroom is needed for the experiment.
- In this research, a ruler was used as the equipment for measuring the length.
- Thermometer which is used for measuring the temperature of the room.

2.4. Recording of different lengths of syringe barrel from two nearby positions that measured the highest sound-intensity level
The participants organized the equipment which was included a truncated syringe and two smartphones (or tablets). Regarding different functions, smartphone was used for generating the frequency from sound source. Whereas, the other smartphone was used for measuring sound-intensity levels respectively. Meanwhile, participants prepared a thermometer in order to measure the room's temperature. The position of the smartphone was placed close to the mouthpiece of the syringe. The students pulled down the syringe plunger to locate the highest sound intensity level at the first position l_1, as shown in figure 1 (a). Owning to the sound waves moved along to the syringe barrel to encounter the piston, it was reflected by the standing wave amplitude. Then became zero at the surface of the piston. This point was called node where the sound waves reflected from the piston surface. At the same time, the sound wave reflected from the piston was phased shift by 180°. The reflection of either incoming waves or reflect waves was caused to generate a standing wave at the tip of the syringe barrel. In accordance, resonance conditions are met at the open end of the tube, the maximum amplitude of the standing sound wave was called anti-node. Similarly, students pulled down the syringe plunger to locate the position with the highest sound intensity in the second position l_2, as shown in figure 1 (b).

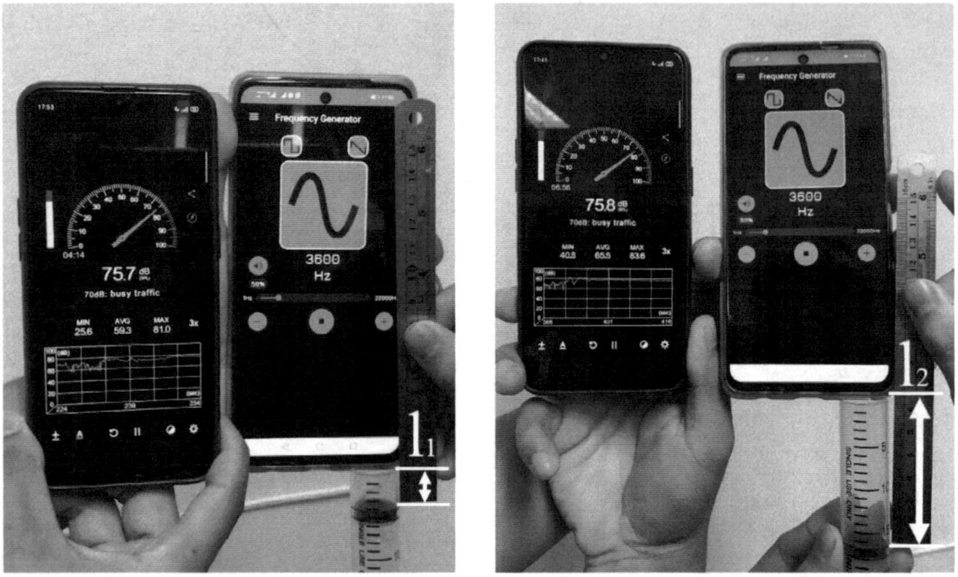

(a) (b)

Figure 1. (a) the first resonance position (b) the second resonance position.

2.5. *Data analysis*

From the study of general physics classroom, students have gained understandable precisely relationship of wavelength λ, the frequency of the sound wave f and speed of sound C according to the equation:

$$C = f\lambda \tag{1}$$

Regarding, the resonance condition, the anti-node at the open end of the tube has the maximum amplitude of a standing sound wave. The distance between the positions of the anti-node (Δl) is related to the sound wavelength of the sound source correspondingly, according to the equation:

$$\Delta l = (1/2)\lambda \tag{2}$$

Where Δl is the position between 1/4 wavelength position and 3/4 wavelength position. In other words, Δl is 1/2 wavelength. Consequently, equations (1) and (2), the speed of sound wave is defined by [3]

$$C = 2f\ (l_1 - l_2) \tag{3}$$

where l_1 and l_2 are the length of the syringe barrel for the first and second resonance respectively. When consider to the theory of the speed of sound upon the temperature, the speed of sound C is equal to the square root of the ratio of specific heats $\gamma = 1.4$ for air at STP, the gas constant $R = 8.314\ J/(mol \cdot K)$ for air, the molecular mass of gas $M = 29 \cdot 10^{-3}\ kg/mol$ and the temperature $T = 273.15 + {}^\circ C$ [8] according to the equation:

$$C = sqrt[\gamma RT / M] \tag{4}$$

Here, T must be specified on an absolute scale.

3. Results

3.1. *Student's scientific reasoning skills and student's practical skills*

Students in each group were requested to analyze and comment on the likely range of temperature in the science classroom in Thailand. The experiment demonstrated that the greatest temperatures were commonly ranged from approximately 24-38 °C based on the regional area and the personal facilities used of the air conditioner in a greater number of classrooms learning. From the scientific learning collaboration, all groups calculate the feasible speed of sound based on equation (4) theoretical. As presented in the illustration in table 1 includes the feasible speed of sound which was estimated range 344-353 m/s.

Table 1. The relationship of the speed of sound waves and range of temperature.

Temperatures (°C)	38	36	34	32	30	28	26	24
speed of sound (m/s)	352.9652	351.829	350.6891	349.5455	348.3981	347.247	346.092	344.9331

By the particular results, the explicit area of the temperature was ranged at approximately 24-38 °C. Most students were capable to calculate the applicable sound frequency for the experiment which was corresponded to the length of the syringe barrel. The results from students' opinions determined the frequency of the sound wave f in the range of 3-4 kHz. Because of the length of the largest syringe barrel, approximately 0.075 meters necessity be greater than the 3/4 wavelength that is the second resonant position. The reason is that the length of the largest syringe barrel, approximately 0.075 meters, must be greater than 3/4 wavelength where the second resonant position was taking place.

3.2. The result of the experiment of speed of sound

The result of the experiment of the speed of sound indicated that all students distributed the frequency of the sound waves to each group in the experiment session. The results of the experiment shown in table 2.

Table 2. The relationship of the velocity of sound, was calculated from the measurement results from frequency indicated 5 different frequency values from each group with 26 °C room temperature.

Group No.	Frequency (Hz)	Δl_{avg} (cm)	Speed of sound (m/s)
1 and 6	4,000	4.325	346.00
2 and 7	3,900	4.450	347.10
3 and 8	3,800	4.550	345.80
4 and 9	3,700	4.700	347.80
5 and 10	3,600	4.800	345.60
Speed of sound average			346.46

4. Discussion and conclusion

From the students' perspective of the relationship of parameters index of the equipment resonant experimental. Some positive pieces of evidence revealed that students gained the methodical study. To begin with the study, students examined the possible temperature values which were normally 24-38 °C for the general science classroom. Also, students analyzed and simulated the appropriate frequency within the Excel program that equivalent to the length of the syringe barrel, resulting in the frequency of the study in the range 3-4 kHz. In the following step, students were participated in calculating the speed of sound from the theoretical value according to equation (4) is 346.092 m/s. Later, students measured Δl of the syringe barrel in each frequency band and calculated the speed of sound value is 346.460 m/s. The computation indicated that the results differed from the theoretical value of 0.1%. Similar to development of scientific reasoning skills and student's practical skills [1-2] defined as those developed through the application of scientific principles, observation, manipulation and demonstration.

According to the results, the researcher and co-workers found that the scores of the Student's scientific reasoning skills and practical skills were 97% and 98% respectively. The consistency of the results was taken by consideration and students' experiment cooperation. Within the condition, if values of Δl accurate and measure temperature scale deeply on a more detailed scale. The results will be able to calculate the speed of sound exactly accurately. By the effective procedure and condition, students were classified into a group to find out and experiment with parameter index. In the aspects of enhancing students' performance both scientific reasoning skills and practical skills, the instructor performed as the facilitator in the class such as setting conditions for the use of ready devices, encourage students by asking, analyzing, and discussing during the experiment. Moreover, small-group discussion can develop students in practical skills, and cooperative working [9]. All approaches aim to facilitate learners to understand the content and be able to implement a common device with the smartphone. In this experiment, the devices can be used as additional tools for other experiments either or demonstrate tools for learning in the scientific internship classroom teaching as well.

Acknowledgements
The authors would like to thank all participating students and wish to acknowledge Faculty of Education and Educational Innovation, Kalasin University.

References

[1] Darling-Hammond L, Flook L, Cook-Harvey C, Barron B and Osher D 2020 *Appl. Dev. Sci.* **24** 97−140
[2] Wood B K and Blevins B K 2019 *Phys. Educ.* **54** 035004
[3] Wijitwongwan P and Wuttiprom S 2019 *J. Phys.: Conf. Ser.* **1380** 012096
[4] Neva J, Benjamin R, Heidi R, Claire M, Shannon G and Richard N 2020 *Appl. Acoust.* **166** 107408
[5] Çoban A and Çoban N 2020 *Phys. Educ.* **55** 043005
[6] Lutfiyah A, Adam A S, Suprapto N, Kholiq A and Putri N P 2018 *J. Phys.: Conf. Ser.* **997** 012024
[7] Amrani D 2013 *Phys. Educ.* **48** 65
[8] Kirtskhalia V G 2012 *Open J. Acoust.* **2** 80-85
[9] Slavin R E 2010 *Cooperative Learning (Teori, Riset, Praktik)* (Bandung: Nusa Media)

Students' understanding of the inverse square law in electrostatics

Trai Unyapoti[1]*, Thanida Sujarittham[2] and Siri Sirininlakul[3]

[1] Department of Curriculum and Instruction, Faculty of Education, Srinakharinwirot University, Bangkok, 10110, Thailand
[2] Department of General Science, Faculty of Education, Bansomdejchaopraya Rajabhat University, Bangkok, 10600, Thailand
[3] Department of Physics, Faculty of Science, Srinakharinwirot University, Bangkok, 10110, Thailand

*E-mail: trai@g.swu.ac.th

Abstract. One problem of learning Electrostatics is that students often learn from their common-sense beliefs about electric force and electric field. This study investigated students' conceptual understanding of finding electric force, electric field, and electric potential of point charge after learning an introductory physics course. We administered the Electrostatics Conceptual Evaluation Test to four lecture-based classes in high school. The first question was a comparison of the electric force from two-point charges at two different positions and the second question was a comparison of the electric field from a point charge at two different positions. The use of the inverse square law is required to find the electric force and the electric field at various positions. It was found that many students answered incorrectly. They described that the electric force and the electric field decrease whereas the distance increases by neglecting the inverse square law. This finding can be particularly used to suggest high school teachers to develop their effective strategy to support student learning.

1. Introduction

Many reports have been found that many students have misunderstanding in Electrostatics, especially in the concept of the inverse square law of electric field [1-3]. This misunderstanding causes student learning difficulties in the topics of electric force, electric field, and electric potential in introductory physics course. Our preliminary study is an investigation of 162 first-year university students' conceptual understanding about finding the electric field due to a point charge by using the Electrostatics Conceptual Evaluation Test as pre- and post-tests. There were two questions of interest. The first question was about determining the electric field due to a point charge at two different positions with one being twice the distance of the other. The use of the inverse square law was required to solve the problem. It was found that 69.8% and 25.9% of the students could not give the correct answer in the pre- and post-tests respectively. The second question was about determining the electric field due to a point charge enclosed by the different Gaussian surfaces. It was found that 77.8% and 31.5% of the students were not able to find the correct solution in the pre- and post-tests respectively. From this finding, most of students expressed electric field by neglecting the inverse square law. This initial study encouraged us to pursue students' understanding of the inverse square law in high school in order to be a guidance for teachers to develop their own teaching strategy.

2. Methodology
2.1. Samples
125 high school students from four lecture-based classes who enrolled in the Physics course in 2020 were the samples of this study.

2.2. Measurement and data collection
In this study, we administered the Thai version of the Electrostatics Conceptual Evaluation Test to the students after they learned Electrostatics. The test was developed from a standardized test and physics textbooks [4, 5]. There were three questions of interest and the students were required to use the inverse square law to solve the problems. The first question was a comparison of the electric force due to point charges at two different positions. The second question was a comparison of the electric field due to a point charge at two different positions. Subsequently, the last one was a comparison of the electric potential from a point charge at two different positions.

3. The questions
Question 1: The electric force between the point charges + Q and +2Q has a magnitude of F as shown in figure 1. If the charge + 2Q is shifted to positions A and B. What is the magnitude of the electric force acting between the two point charges for each position?

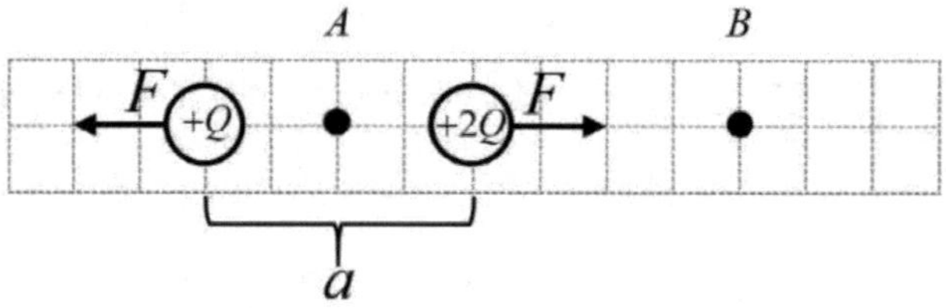

Figure 1. Positions A and B from the two point charges.

Question 2: The particle with the charge + Q is fixed. The electric field from this particle at position P has a magnitude of E_0 and its direction as shown in figure 2. What is the magnitude of the electric field at positions A and B?

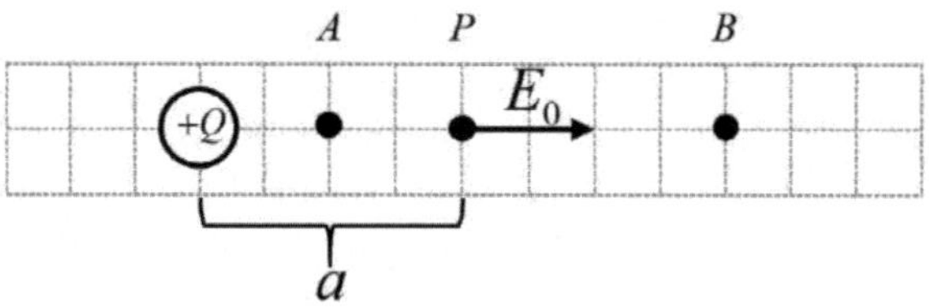

Figure 2. Positions A and B from the point charge +Q.

Question 3: The particle with the charge + Q is fixed. The electric potential from this particle at position P is V_0 as shown in figure 3. What is the electric potential at positions A and B?

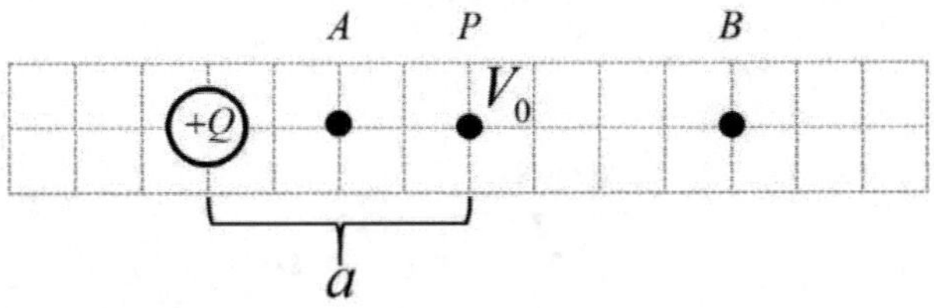

Figure 3. Positions A and B from the point charge +Q.

4. Results and discussion

From testing, only 8% provided the correct answers. This result confirmed that most students had misunderstandings in finding the magnitude of the electric force between the two-point charges and the electric field due to the point charge (see table 1).

It was found that most students answered incorrectly. They described that the electric force and the electric field decrease whereas the distance increases by neglecting the inverse square law (See figure 4). Only 7-9% of students were able to link the inverse square law to determine the electric force and the electric field.

In addition, the result shows that students continuously used the similar alternative ideas to solve the third problem—finding the electric potential at two different positions. Interestingly, 90% of them who could find the electric potential correctly were from the same group of the students determining the magnitude of electric force by neglecting the inverse square of the distance (See figures 4, 5, 6, and 7).

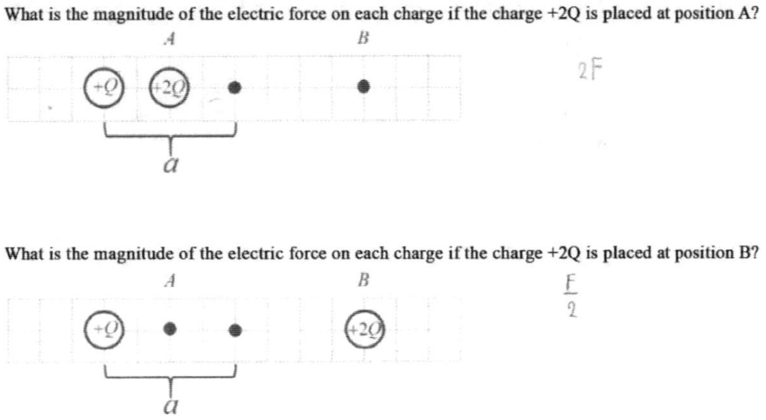

Figure 4. The example of student's answers who determined the electric force being inversely proportional to the distance in question 1.

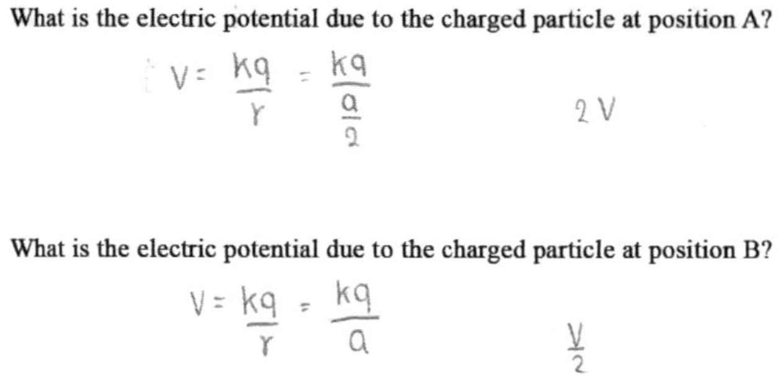

Figure 5. The example of student's answers who determined the electric potential being inversely proportional to the distance in question 3.

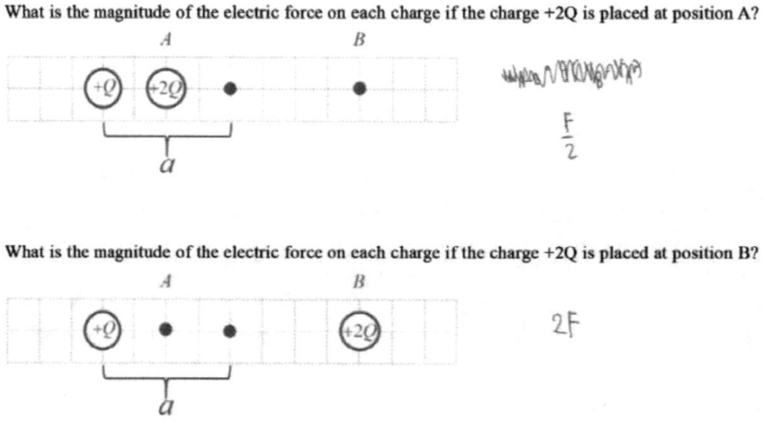

Figure 6. The example of student's answers who determined the electric force being directly proportional to the distance in question 1.

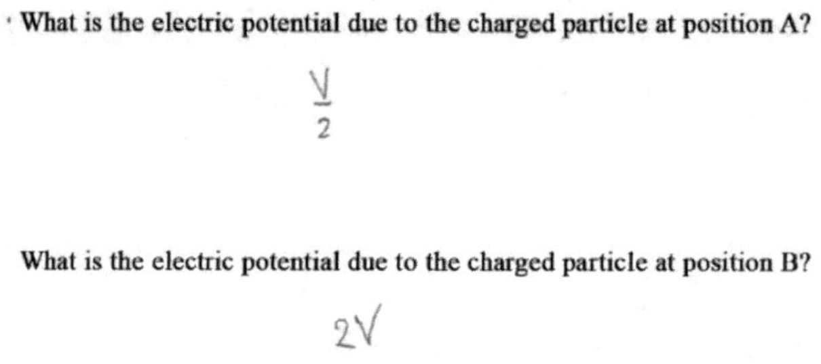

Figure 7. The example of student's answers who determined the electric potential being inversely proportional to the distance in question 3.

Table 1. The students' answers to the test for questions 1 and 2.

	Percentage of students
When the distance increases, the magnitude of the electric force decreases at the same rate. (figure 4)	67%
When the distance increases, the magnitude of the electric force increases at the same rate. (figure 6)	17%
The magnitude of the electric force is inversely proportional to the square of the distance between the point charges.	9%
When the distance increases, the magnitude of the electric field decreases at the same rate.	65%
When the distance increases, the magnitude of the electric field increases at the same rate.	10%
The magnitude of the electric field is inversely proportional to the square of the distance from the point charge.	7%

5. Conclusions

After completing the lesson, most students employed their common-sense beliefs to solve the electric force and electric field problems. They described that electric force and electric field by neglecting the inverse square law. According to the investigation deploying the three basic questions, it particularly suggests that high school teachers need to develop an effective teaching strategy to support their student learning in Electrostatics. In addition, this result clearly implies that physics teachers are required to devote their time for solidifying students' conceptual understanding together algebraic problem solving in the inverse square law as it relates not only Coulomb's law but also Newton's law of gravitation and the definition of intensity.

Acknowledgments

We would like to thank Physics Education Network of Thailand (PENThai) for their valuable suggestions. This project is supported by Faculty of Education, Srinakharinwirot University.

References

[1] Ding L, Chabay R, Sherwood B and Beichner R 2006 Evaluating an electricity and magnetism assessment tool: Brief electricity and magnetism assessment *Phys. Rev. ST Phys Educ. Res.* **2** 010105

[2] Maloney D P, O'Kuma T L, Hieggelke C J and Van Heuvelen A 2001 Surveying students' conceptual knowledge of electricity and magnetism *Am. J. Phy.* **69** S12–23

[3] Singh C 2006 Student understanding of symmetry and Gauss's law of electricity *Am. J. Phy.* **74** 923–36

[4] Knight R D and Pritchard D 2008 *Physics for Scientists and Engineers: A Strategic Approach with Modern Physics* (San Francisco: Pearson Addison Wesley)

[5] Young H D, Freedman R A and Ford A L 2016 *University Physics with Modern Physics* (San Francisco: Pearson Addison Wesley)

Measuring refractive index gradient of sugar solution

Pattarapon Tanalikhit*, Thanabodi Worakitthamrong, Nattanon Chaidet and Wittaya Kanchanapusakit

Department of Physics, King Mongkut's University of Technology Thonburi, Bangkok, 10140, Thailand

*E-mail: pattarapon.t@mail.kmutt.ac.th

Abstract. To measure refractive index at a particular altitude in a solution with vertical refractive index gradient, a transparent wedge-shaped container was constructed altogether with the development of mathematical formula derived from the Snell's law. The refractive index of the solution can be calculated by measuring the angles of incoming and outgoing laser beams relative to respective normal line. By varying height of the laser beam, the refractive index as a function of height of a sugar solution was obtained. This technique is applied to investigate Fata Morgana which is a kind of superior mirage resulting from bending of light in a medium with density gradient.

1. Introduction

Mirage is an optical phenomenon caused by formation of refractive index gradient in a medium in which a light ray follows a curved path. A well-known example of the mirage is the puddles on hot road. In this so-called inferior mirage, the projected image is below the actual altitude of the object due to positive refractive index gradient of air with height above the road. There is another type of mirage called superior mirage where the projected image is above the actual altitude of the object. This is commonly seen as a flying ship above water level, and the phenomenon is attributed to the negative change in the refractive index of air with height. The mirage phenomena can be simulated experimentally in a solution-filled container where the solution's concentration and hence the refractive index varies with height [1]. There are also qualitative investigations on Fata Morgana mirage at Lake Geneva, which is a more complex mirage resulting from increasing and decreasing refractive index with height [2].

Studies on mirage rely on measurement of the refractive index gradient. When a sugar solution is left awhile in a container, gravity-facilitated diffusion causes the solution to be more concentrated at the bottom. The negative concentration gradient and hence negative refractive index gradient is set up. This paper presents a simple method of finding refractive index of a sugar solution as a function of height together with a careful design of the shape of container. By using Snell's law, the refractive index of the solution can be calculated from the characteristics of the incoming laser beam on one side of the container and the outgoing beam from another side. Internal reflection must be prevented. The conventional rectangular container approves only a small range of permissible angles of incidence before the occurrence of the internal reflection; this limits the number of data required to produce solid results. The shape of the container was then redesigned to be a wedge in order to allow a wider range of angles of incidence without the internal reflection.

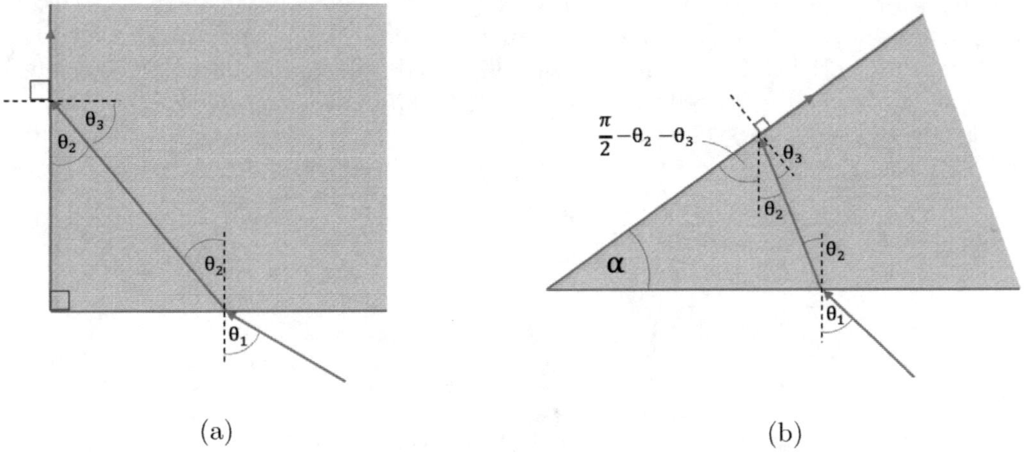

Figure 1. Light rays entering (a) a rectangular prism and (b) a triangular prism. The outgoing rays are parallel to the interface, being on the verge of internal reflection.

2. Experimental design

2.1. Theoretical motivations

Figure 1 shows light ray travelling from air into a medium with refractive index n. The outgoing ray parallel to the interface indicates the verge of internal reflection. Only a certain range of angles of incidence θ_1 can be used to avoid internal reflection and to allow accurate measurement of angles of refraction of the outgoing ray. In figure 1(a), the internal reflection in case of rectangular prism will not occur if

$$\sin\theta_3 < \frac{1}{n}. \tag{1}$$

From geometry, one finds that $\theta_2 + \theta_3 = 90°$. By assuming refractive index of air $n_{\text{air}} = 1.00$, Snell's law reads $\sin\theta_1 = n\sin\theta_2$ and leads to

$$\sin\theta_1 > \sqrt{n^2 - 1}. \tag{2}$$

For example, water with $n = 1.33$ gives $61° < \theta_1 < 90°$. This determines the experimentally permissible range of the angles of incidence.

However, for triangular prism as shown in figure 1(b), $\theta_2 + \theta_3 = \alpha$, where α is the apex angle. By using Snell's law, the angles of incidence θ_1 that ensure the nonoccurrence of the internal reflection must satisfy

$$\sin\theta_1 > \sqrt{n^2 - 1}\sin\alpha - \cos\alpha. \tag{3}$$

To have the greatest range of θ_1, the right hand side of above expression can be set to zero. Hence,

$$\alpha = \arctan\left(\frac{1}{\sqrt{n^2 - 1}}\right). \tag{4}$$

For water, $\alpha = 48.7°$ and the apex angle of the triangular container must be chosen to be less than or equal to this value in order to maximise the range of allowed values of θ_1.

2.2. Container design

From the analysis in the previous section, the shape of the container was therefore designed to be a wedge with the wedge angle α as in figure 2. The data to be recorded are θ_1 of the incoming beam and θ_4 of the outgoing beam. The refractive index of the solution in the container at a height is denoted by n. Because thickness of the container's wall is uniform across the region through which the laser beam is shone, it can be ignored in the calculation. The application of Snell's law at the interfaces reads $\sin\theta_1 = n\sin\theta_2$ and $\sin\theta_4 = n\sin\theta_3 = n\sin(\alpha-\theta_2)$ respectively. By eliminating θ_2, one obtains

$$n = \sqrt{\sin^2\theta_1 + \left(\frac{\sin\theta_4 + \sin\theta_1\cos\alpha}{\sin\alpha}\right)^2}. \tag{5}$$

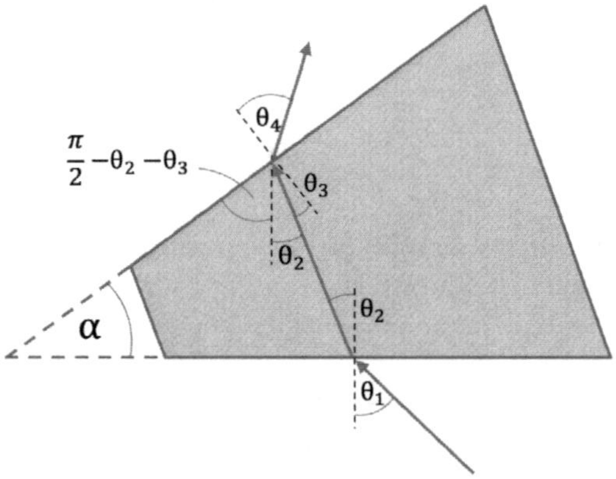

Figure 2. Schematic diagram for top view of a wedge-shaped container with the wedge angle of α

2.3. Experimental setup

The determination of n involves measurement of θ_1 and θ_4 without total reflection inside the container. To have a full range of the allowed θ_1, a triangular container with apex angle no greater than equation (4) should be used instead of the conventional rectangular container. The wedge-shaped container with the wedge angle of 10.0° was built and employed in the experiment. Figure 3 shows the experimental setup. Since the solution used to fill the container has a vertical concentration gradient, the light ray travelling through the solution will bend. The bending is minimised by keeping the container narrow. The liquid-filled wedge-shaped container is held in place. The movable laser holder allows a full range of angles of incidence. The height of the holder can be adjusted and locked. By measuring θ_1 and θ_4, the refractive index of the solution at a particular height can be obtained by using equation (5).

3. Results and discussion

3.1. Test of accuracy

The protractor scale attached to the base of the laser holder reads θ_1 with approximately 3% difference from the angles obtained from image analysis program, *Tracker*. To test the operation of the setup, distilled water at 25°C was used to fill the container to measure its refractive index. By shining a 532-nm laser beam and varying multiple angles of incidence θ_1 at different heights on a container's wall in a low-humidity room, the average refractive index of the distilled water was found to be 1.33 ± 0.01. To verify the obtained value, the Abbe refractometer at wavelength of 589 and 644 nm was used and it gives the refractive index of 1.33 and 1.34 respectively. This establishes the accuracy of the designed device in finding the refractive index to one hundredth. Moreover, the obtained value of refractive index of the distilled water at 25°C is in agreement with the results reported in [3].

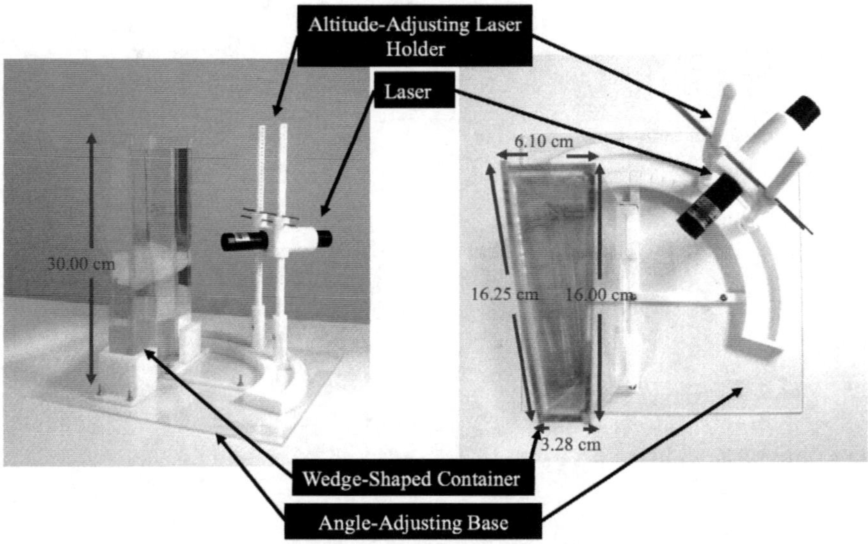

Figure 3. The experimental setup consists of a wedge-shaped container filled with liquid and a laser holder which can be moved vertically and azimuthally. The side view (left) and top view (right) are illustrated.

3.2. Refractive index measurement of sugar solution

There are several ways to prepare the gradient of refractive index. One can firstly fill the container with distilled water and then pass the sugar syrup to the bottom of the container by using a funnel and a pipe [1]. However, in this paper, the preparation can simply be done by filling half of the container with 80% w/w syrup and then pouring the distilled water gently on the inside of a container's wall. The water layer sits on the top of the syrup layer. The setup was left for four days to allow significant diffusion of the syrup to occur and the refractive index gradient to set. Of course, the gradient changes over time. The perturbation from pouring water on the syrup layer only has initial effect on the formation of the gradient but no impact after some period of the diffusion.

Experiments are performed by shining the laser at each height with angles θ_1 of 10°, 20° and 30°. The height was then varied from 8.5 cm to 20.0 cm with an increment of 0.5 cm. By measuring the angles of refraction θ_4 and using equation (5), the refractive indices can be calculated. Figure 4(a) shows a plot of measured refractive index against height whose data were collected on the fourth day after the syrup-water mixture preparation. Near the top of the container where the solution is so dilute that it is essentially water, the refractive index approaches 1.33. Near the bottom of the container, the solution concentration is 80% and the refractive index is close to 1.48 which is the value also reported in [4]. The data points in between clearly exhibit the sigmoid or logistic trend. The logistic trend results from solving differential equation of the form $dn/dz = r(n_0 - n)n$ [5]. By curve fitting, the equation for refractive index n as a function of height z is given by

$$n = n_0 + \frac{c}{1 + de^{-fz}}, \tag{6}$$

where n_0, c, d and z are the coefficients displayed in the inset of figure 4(a). The average refractive index gradient $\Delta n/\Delta z$ can also be calculated. As time goes on, the size of the average gradient gets weaker as illustrated in figure 4(b).

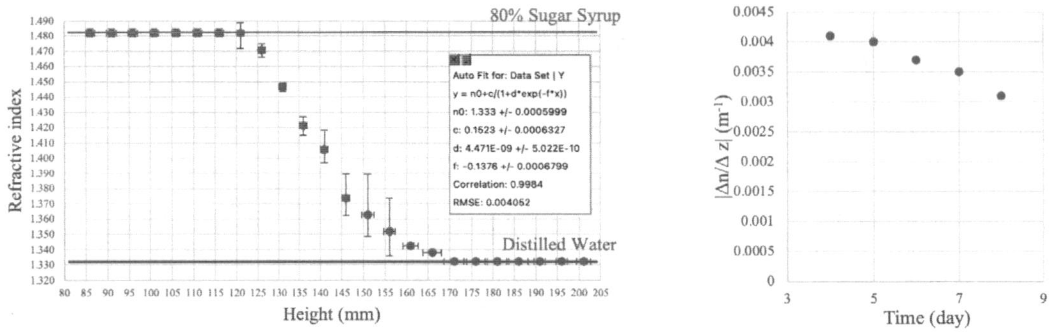

(a) (b)

Figure 4. Experimental results showing (a) the refractive index at different heights and (b) the change of the gradient strength as a function of time

4. Conclusion

In this work, a portable refractive-index measuring device, suitable for solution with concentration gradient, is designed, constructed and tested. A wedge-shaped container with an appropriate wedge angle is filled with a carefully-prepared sugar solution. Investigating refractions of laser beams, based on Snell's law, gives the refractive index of the sugar solution as a function of height, and hence the refractive index gradient. The relationship between the refractive index and height follows logistic function with coefficients experimentally determined. The device and technique that were developed in this research would benefit demonstrating Snell's law in classroom for teachers and facilitate physics projects for students.

Acknowledgments

This work is financially supported by VISTEC and the Department of Physics, KMUTT. The authors would like to thank Dr Ian Jacobs and Dr Supattanapong Dumrongrattana for ideas regarding experimental techniques. Many thanks go to Dr Wandee Onreabroy, Dr Tula Jutarosaga and Dr Monsit Tanasittikosol for emotional support, and to Mr Suphakorn Koknoi for necessary material.

References

[1] López-Arias T, Calzà G, Gratton L M and Oss S 2009 *Phys. Educ.* **44** 582–8
[2] Young A T and Frappa E 2017 *Appl. Opt.* **56** G59–68
[3] Segelstein D J 1981 *MSc Thesis* University of Missouri-Kansas City, Missouri pp 116–64
[4] Belay A and Assefa G 2018 *J. Laser Opt. Photonics* **5** 1000187
[5] Thomas G B, Weir M D, Hass J, Heil C and Behn A 2016 *Thomas's Calculus, Thirteenth Edition with SI Units* (Harlow: Pearson Education Limited) pp 563–5

Exploring the links between Physics and Astronomy education teaching and learning in the Philippines

R M Guido[1,2*] and A Orleans[2]

[1] Center for Astronomy Research and Development, Rizal Technological University, Mandaluyong 1550, Philippines
[2] College of Graduate Studies and Teacher Education Research, Philippine Normal University, Manila 1000, Philippines

* E-mail: rmdguido@rtu.edu.ph

Abstract. Students consider physics to be one of the most prevalent and troublesome topics since they find physics to be a challenging course in high school and become more deceptive when they have it as an undergraduate course. The Rizal Technological University (RTU) in the Philippines offers undergraduate and graduate academic programs in astronomy. The university utilizes physics diagnostic tests to classify students' suitability towards the course as RTU houses the Philippines' most advanced optical telescopes and the Center for Astronomy Research and Development. The latter was established through the Department of Science and Technology to promote a research hub in the field of astronomy and space science, as well as providing support for research funding and scholarships. The study performed causal-comparative research design and correlational analytics toward students' attitude and motivation, study habits, learning styles, professor factors, social factors, and academic grades in physics and astronomy. The study explored the links between physics and astronomy and revealed their faith in their study habits in both astronomy and physics. The findings of this study provide evidence of the link of physics and astronomy education in students' academic programs and its value from the previous learnings of students and teaching of teachers as a critical role in predicting student performance. The increased enrolment in the academic program demonstrates the validity of the above-mentioned statements, as it enticed students to pursue astronomy as their scientific career. Other efforts are discussed in detail in this paper.

1. Introduction

This study delves into the associations for both physics and astronomy education teaching and learning in the Philippines. Many students avoid studying physics since it is often viewed as a difficult subject. For most students, physics was seen as the most challenging course in high school, and this perception has never changed since then [1,2]. The learning of physics in the Philippines presents a significant puzzling effect specifically for the undergraduate course.

Gifted and bright students in both the sciences and mathematics have a true understanding of physics in their daily life. Physics lecturers will often use humor and entertainment to help the students understand the lesson very well. When teaching physics, several teaching techniques can be employed to achieve that goal [3].

In the Philippines, astronomy has been always taught as a component of the general education subject at the basic education level. The University of the Philippines – National Institute of Physics (NIP) elevated the teaching of astronomy through the offering of "Physics and Astronomy for Pedestrians" in 2002 [4,5], which explored the introductory of physics and astronomy, from inception to their contemporary advances.

In 2005, the RTU made a significant breakthrough in the history of Philippine education by pioneering the offering of the Master of Science in Astronomy program [4,5]. Furthermore, RTU has made another step forth into its unwavering lead in astronomy education by offering the Bachelor of Science in Astronomy Technology in 2007. RTU has the most advanced observational astronomy equipment in the Philippines [5]. It is equipped with the most sophisticated optical astronomy equipment, allowing it to provide high competitive astronomy academics and research [6]. Despite its urban location, the university has embraced the issues of light pollution in the metropolis and launched the country's first light pollution research project.

With these assertions, the researchers wanted to seek the links between physics and astronomy education teaching and learning in the Philippines. The following are the study's measurements that were used to establish the connection between physics and astronomy: Attitudes and Motivation in Astronomy (AMA), and Attitudes and Motivation in Physics (AMP); Study Habits in Astronomy (SHA), and Study Habits in Physics (SHP); Learning Styles in Astronomy (LSA), and Learning Styles in Physics (LSP); Professor Factor in Astronomy (PFA), and Professor Factor in Physics (PFP); Social Factor in Astronomy (SFA), and Social Factor in Physics (SFP); Academic Grade in Astronomy (AGA), Academic Grade in Physics (AGP).

2. Methodology

The study purports the links between physics and astronomy through a validated questionnaire. The questionnaire has been utilized in the previous studies [1-3] and have had undergone the validation process. A survey technique causal-comparative research design was used in this study. The samples consisted of 177 BS Astronomy students from RTU who took co-requisite physics and astronomy courses. The instrument used was a researcher-made questionnaire to measure the responses of students on the factors such as study habits, learning styles, professor factors, social factors. This was validated by the different licensed professionals and experts in the field. Moreover, the Attitude Inventory Test and Achievement Motivation Scale was used to determine the links between physics and astronomy attitudes, and physics and astronomy achievement motivation. Also, the links between physics and astronomy grades were explored using their final grades in the courses. Data were processed through regression analysis and correlation to determine the apparent links between physics and astronomy based on the students' responses.

3. Results

Figure 1 shows the significant responses of the students specifically about SHP which falls among the other criteria. This signifies that students have the least median response in their study habits in physics. Moreover, a higher median response was evident in AMA rather than in AMP which reveals that students are well motivated to study astronomy rather than physics. Also, it reflects that students received mean higher grades in astronomy rather than in physics, and have shown desirable results in PFA rather than in PFP.

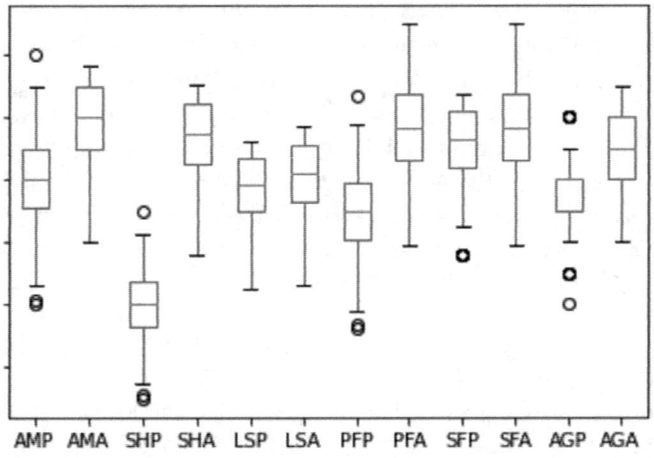

Figure 1. Mean boxplot of student responses between physics and astronomy.

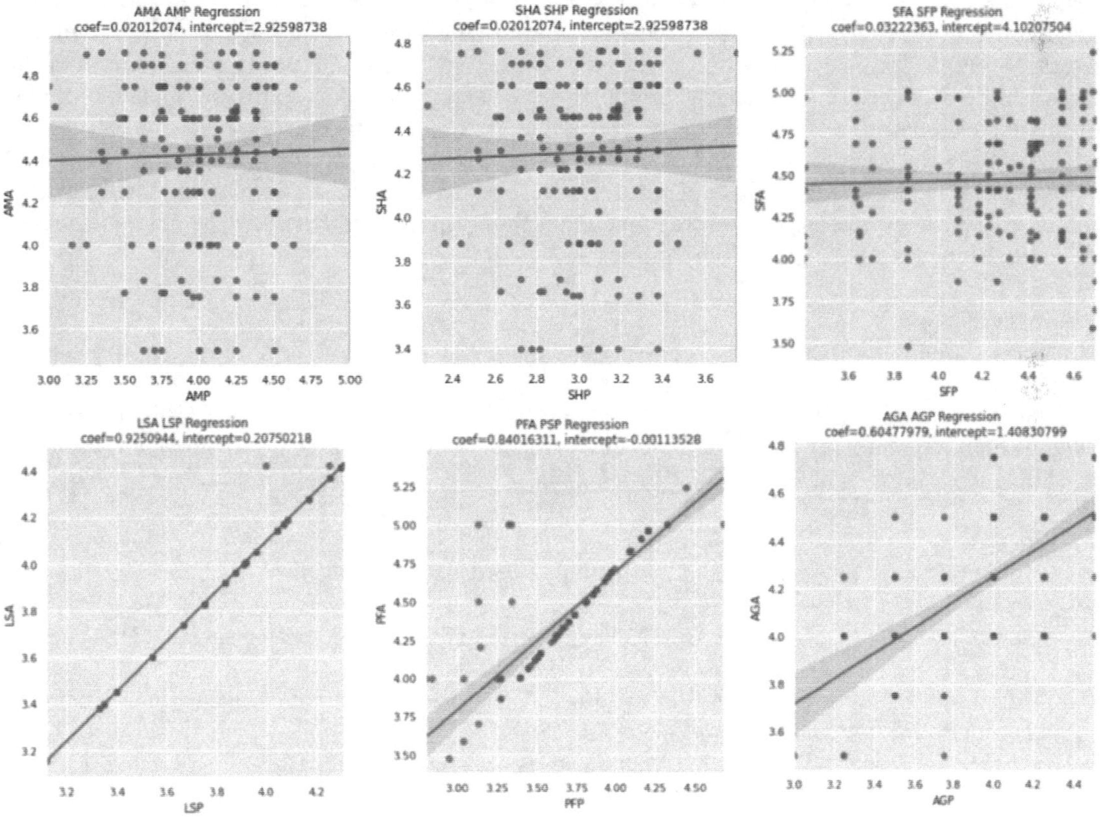

Figure 2. Regression analysis of factors between physics and astronomy (a) AMA and AMP, (b) SHA and SHP, (c) SFA and SFP, (d) LSA and LSP, (e) PFA and PFP, (f) AGA and AGP.

Figure 2 (a) shows the regression analysis of the factors which shows that AMA and AMP have a correlation value of r = 0.025700 which implies a negligible correlation. Also, (b) SHA and SHP show a negligible correlation of r = 0.028621, and (c) SFA and SFP have a correlation of r = 0.031167 which all reflects negligible correlation. The students may have different measures of association for physics and astronomy with these factors as they have perceived astronomy and physics as different entities. It reflected different perspectives on their attitudes and motivation, study habits, and social factors.

A significant relationship was established in (d) LSA and LSP, (e) PFA and PFP, and (f) AGA and AGP with correlation values r = 0.99740 very high correlation, r = 0.870502 high correlation, and r = 0.568871moderate correlation, respectively. With these factors, it shows that students establish connections between physics and astronomy as they see it to as a single criterion to which they treat it as one. Students have the same learning styles in physics and astronomy as it poses the same context of learning, computation, principles, and concepts. Astronomy and physics have different topics and discussion contexts. The similarity between the two discloses problem-solving/computational reinforcement, scientific critical thinking, and both are under the realm of physical science. Additionally, it demonstrates that professor factors in physics and astronomy are identically treated by students. The reflection of responses might have come from the teaching styles of the teachers, as most of the astronomy professors were also physics graduates or with physics backgrounds which reflect having the same teaching styles, and classroom management strategies.

4. Conclusion

The findings of the study appear that students are confident in their study habits in astronomy, their attitudes and motivation in astronomy, and their professors in astronomy. This also confirms that a relationship exists between students' learning styles, professor factors, and their academic grades in physics and astronomy.

This study shows that physics and astronomy have links and differences. While the Philippines is starting to elevate astronomy, it is imperative the fact that physics and astronomy are interrelated with each other. However, researchers foresee that delineation can be seen between physics and astronomy education teaching and learning in the Philippines.

Acknowledgment

The authors would like to express utmost gratitude for the financial support given by the Rizal Technological University - Center for Astronomy Research and Development (RTU-CARD) of the Niche Centers in the Region (NICER) under the Science for Change Program (S4CP) of the Department of Science and Technology - Philippine Council for Innovation, Energy, and Emerging Technology Research and Development (DOST-PCIEERD).

References

[1] Guido RM D 2013 Attitudes and motivation towards learning physics *Int. J. Eng. Res. Technol.* **11** 2087

[2] Guido RM D and Dela Cruz R B 2013 Factors affecting academic performance of BS Astronomy technology students *Int. J. Eng. Res. Technol.* **12** 84

[3] Guido RM D and Dela Cruz R B 2015 Physics diagnostic test: An approach to qualify astronomy students *Int. Lett. Soc. Humanit. Sci.* **63** 129

[4] Celebre C P 2008 Astronomy in the Philippines: A hundred years after *Phil. J. Astro.* **1** 11

[5] Torres J R F 2007 Bachelor of science in Astronomy technology: A model *Astron. Educ. Rev.* **6** 163

[6] Guido RM D 2020 The progress of Astronomy program in the Rizal Technological University *Preprint* arXiv:1911.06656

Journal of Physics: Conference Series **2145** (2022) 012074 doi:10.1088/1742-6596/2145/1/012074

Demonstration of light reflection concepts for rendering realistic 3D tree images

Tiantada Hiranyachattada*, Kampanat Kusirirat, Kasem Kamolchaipisit and Panadda Jaiboonlue

Program of Animation and Digital Media, Faculty of Science and Technology, Bansomdejchaopraya Rajabhat University, Bangkok, 10600, Thailand

*Corresponding author's e-mail: tiantada.hi@bsru.ac.th

Abstract. With advancements in computer graphics, creating natural images has always been the main purpose, image rendering is all based on principles of physics. So, understanding the physics of image rendering will enable us to create the most realistic images. A ray of light hit a surface with different orientation and reflects as per the rules of physics. It is difficult to calculate the light reflection of complex foliage, such as trees, so, the reflection of this natural complexity needs to be adapted to rendering situations. In this research, the researchers provide demonstrations to enable students to understand the light reflection in nature, light calculation in computer graphics and methods to apply them to render realistic tree images. The researchers assign students to render 3D realistic tree images to assess the students' understanding by applying the diffuse reflection value, specular reflection value and surface normal direction to render realistic tree images. The researchers find that most students understand of diffuse reflection, specular reflection, and surface normal direction causes the rendering results to be most realistic.

1. Introduction

With advancement in computer graphics for rendering realistic image nowadays, creating natural images has always been the main purpose of computer graphics. With the geometric complexity of plants, the complex light interactions, especially inner foliage is leaves, make a realistic and efficient display of plants extremely challenging [1], especially in various visualization applications, such as virtual environment and computer games that occur in nature with trees. These image rendering is all based on principles of physics, therefore, understanding the physics of image rendering will enable to create the most realistic images. In the development of digital art, by combining science and technology, it can bring unprecedented development through fully applying the physics concepts and technology in computer graphics, which can create high-quality 3D computer graphics works, the importance and necessity of learning method and technology in 3D computer graphics design are fully reflected [2]. A polygon is a collection of vertices, edges and faces that defines the shape of a polyhedral object with material and texture. In 3D image rendering, these values need to be adjusted correctly based on an understanding of the physics concepts of light reflection which is the imitation of natural phenomenon. However, physics of light in computer graphics has limitations in calculating the reflection of rays of light because, light flows continuously in nature but does not in the computer graphics.

Over the years, with the rapid development of digital technology, teaching physics theory, the fundaments of rendering concepts, has drawn much attention. Lecturers have innovated teaching methods to improve students understanding of physics concepts. Physics is a principle that any natural phenomenon may be explained, which can also encourage students' curiosity. However, students often lack conceptual understanding of basic concepts when studying concepts of light. Students are mostly focused on the examination and performance outcomes rather than learning. They only memorize the important points get high grades in examinations rather than inquiring for scientific facts [3]. Therefore, lecturers should focus on teaching students to understand the whole process and all contents linked together.

With the rapid development of computer technology, modern computer-based software tools in classrooms are often used to assist in the teaching, helping students to understand physics concepts and equations [4]. Therefore, with the difficulty of rendering 3D realistic tree images, the researchers propose learning activities to understand the physics concepts of light reflection. To create 3D realistic rendering, students need to understand light reflection both in nature and computer graphics. In order to assess students' understanding, the assessment are given to students with the learning activities present in the research so that it can help students that lack understanding of the concepts. Also, it allows students to focus on what to do and helps with their thinking process to create their work.

2. Methodology

The researchers represent the learning steps to students to render realistic 3D tree images. In table 1 shows the overall students' learning activities in our research, which we divided into 3 steps.

Table 1. Steps of overall learning activities.

Step	Learning activities
Step 1. Understanding the law of light reflection in nature	Lecturers use a mirror, paper, and a laser pointer to demonstrate law of reflection. Then, students do an experiment in group and observe the law of reflection results.
Step 2. Understanding the light reflection in computer graphics	Students watch a short animation video about light reflection in computer graphics. They also learn the difference between diffuse reflection and specular reflection. Then, students discuss and calculate the light reflection of diffuse reflection value and specular reflection value using the equation.
Step 3. Applying light reflection concepts to render 3D realistic tree images by themselves	Each student creates a 3D tree model, renders the 3D realistic tree images, and adjusts the value that is referenced from the light reflection concepts from step 1 and step 2.

2.1. Light reflection demonstration

For the first step, students must understand how light reflects in nature. A ray of light hits a surface with different orientation and reflects as per the rules of physics. If the ray of light can be observed by approaching and reflecting of a flat mirror, then the behavior of the light as it reflects would follow a predictable law known as the law of reflection [5]. According to the law of reflection, the ray that hits the surface is called the incident ray, or L, and the ray of light that leaves the mirror is known as the reflected ray, or R. At the point of incidence where the ray strikes the mirror, a line that is perpendicular to the surface of the mirror is known as a normal line or N. The normal line divides the angle between the incident ray and the reflected ray into two equal angles.

For the learning activities of the law of reflection, the researchers use a plane mirror and a laser pointer. The experiment in figure 1 shows that when emitting the laser beam to the plane mirror, the laser reflected in an equal angle in accordance with the law of reflection. Students are able to observe the activities results. Students also do the experiment in groups and observe the law of reflection results. After the learning activities, the researchers assess the students' understanding giving a quiz to calculate the light reflection angle.

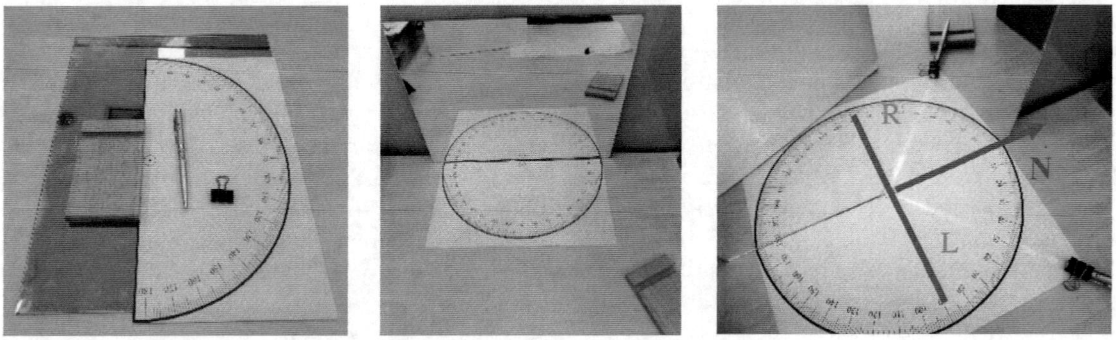

(a) (b) (c)

Figure 1. Light reflection demonstration (a) Experimental instruments consist of a laser pointer, plane mirror, paper, and a paper clip (b) Experiment set up (c) Shooting a laser beam to the plane mirror to observe incidence ray (L) and reflected ray (R).

2.2. Light reflection calculation in computer graphics

The second step is understanding the light reflection in computer graphics. The main difference is to simulate the real-world reflection, light will reflect into two types - specular reflection and diffuse reflection. Once a normal surface at the point of incidence is drawn, the angle of incidence can then be determined. The light ray will then reflect in such a manner that the angle of incidence is equal to the angle of reflection. Reflection of smooth surfaces, such as a plane mirror, is a type of reflection known as specular reflection. Reflection of rough surfaces, such as clothing and paper, is a type of reflection known as diffuse reflection [6]. In figure 2 shows the overall learning activities in this step.

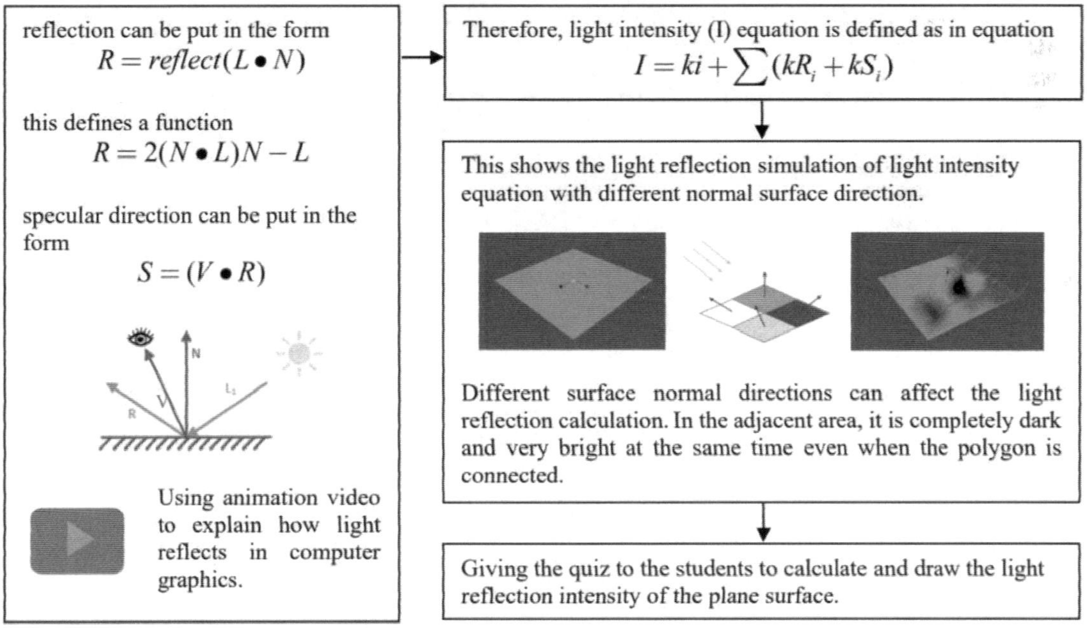

Figure 2. Overall learning for light reflection in computer graphics.

The learning activities consist of three parts as follows: first, the researchers use an animation video to demonstrate how light rays reflect the surface and bounce [7] and show the light reflection equation. It is important to understand that specular reflection of the eyes only sees the image of the object that reflection direction coincidences with the view direction. Second, the researchers explain the equation to calculate light reflection with Phong lighting model [8] with the reflection value from the first part. In this part, students must see the relationship and nuances of light reflection in nature and in computer graphics. We use the computer rendering software to demonstrate how light calculation works in 3D rendering. It represents that different normal direction will affect the light reflection calculation. When the normal line is at zero degree to the direction of light, it will make the surface as bright as possible. On the other hand, if the normal line is at an angle opposite to the direction of light, it will make the surface as dark as possible. Last, the researchers assess the students' understanding using a quiz to calculate the intensity of the light that will occur when the direction of the normal on the surface is different.

2.3. Applying light reflection concepts to rendering realistic tree images

The last step is to apply the light reflection concepts from step 1 and step 2 to render 3D realistic tree images. The sample is third-year bachelor's degree students from Animation and Digital Media Program at Bansomdejchaopraya Rajabhat University that are enrolled in Shading Lighting and Rendering subject. The assignment is given to the students to rendering 3D realistic tree images with the condition that students have to setup the scene and assign the appropriate variables. In figure 3 shows the example of 3D tree model setup with surface normal direction.

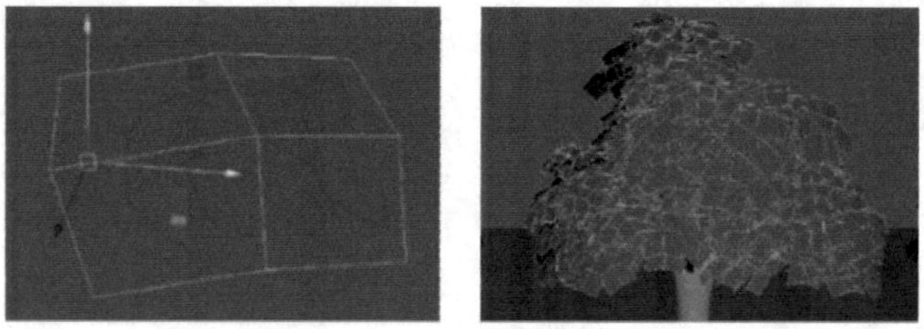

(a) (b)

Figure 3. 3D tree model setup (a) A single tree leaf. (b) Leaves join together to form a tree.

The main variables for realistic 3D tree rendering are lighting, surface normal direction, diffuse value, and specular value. We define each variable as follows: 1) Lighting is the light setup in the scene which consist of light from the sun and sky to be the main light source. 2) Surface normal direction is the direction of the leaf surface which is important to calculate the light intensity. 3) Diffuse reflection value is the color value that light reflected to viewer's eyes in every direction which is green color. 4) Specular reflection value controls the reflection of light that reflects into the viewer's eyes. Specular reflection value depends on the angle of viewer. The students' assessment criteria of understanding are shown in table 2.

Table 2. Assessment criteria.

Level of understanding	Criteria
Clearly understand	Lighting, edit surface normal direction, diffuse reflection value and specular reflection value
Partially understand	Lighting, diffuse reflection value and specular reflection value
Not understand	Lighting and diffuse reflection value

3. Results

The researchers analyze the understanding from the assignment of 42 students to assess the students' understanding. It is found that 83 percent of the sample understands clearly, 10 percent understands partially, and 7 percent of the sample does not understand.

Students who understand can correct the surface normal direction which make the tree light intensity in bush look clearly separated and look more diorama to give the most realistic rendering results. They also adjust the diffuse reflection value and specular reflection value correctly which can make the rendering results look more realistic with the same appearance at the bottom of the bush, which is typically dark. Students who partially understand can adjust the diffuse reflection value and specular reflection value but they cannot correct the surface normal direction to be most realistic. For example, it can be seen that all bush area has almost the same brightness, which makes the rendering image tree look flat. Students who do not understand cannot adjust the surface normal direction and cannot adjusted the surface normal direction and specular reflection value. The random surface normal direction can cause wrong light reflection calculation. They also do not know the relationship between the values. In figure 4 compares the students' rendering results in each level of understanding.

(a) (b) (c)

Figure 4. Comparing 3D tree image rendering results for (a) clearly understand, (b) partially understand and (c) not understand.

4. Discussion and conclusion

In conclusion, the learning activities show that the students are interested in and pay attention to the demonstration in all steps although the physics concepts of light reflection is difficult to understand. However, visual activities will make students understand more, develop more skills, become more engaged in doing activities [9] and they also increase students' interest in the subject. Dividing the learning activities into 3 steps will give students a good understanding of the physics concepts of light reflection in both nature and computer graphics, and also the relationship between each variable. The researchers are aware that some students do not understand some of the light properties presented in depth, but they discover some aspects of light reflection demonstration that can build up their interest [10] in learning. Explaining the overall concepts allows them to better understand the purpose of the learning activities. Finally, the result shows that most students can understand the physics concepts of light reflection to render 3D realistic tree images.

For further research, it is expected to be able to use more diverse learning activities to demonstrate the light reflection concepts and make the learning activities in each step more connected. Also, creating 3D image rendering may not require realism in some cases of rendering because the beauty of the rendering results depends on personal judgement, which should have more specified criteria to assess the image rendering results.

Acknowledgements

The author would like to express sincere thanks for financial support from Program of Animation and Digital Media, Research and Development Institute, Bansomdejchaopraya Rajabhat University and Faculty of Science and Technology, Bansomdejchaopraya Rajabhat University, Bangkok, Thailand.

Journal of Physics: Conference Series **2145** (2022) 012074 doi:10.1088/1742-6596/2145/1/012074

References

[1] Habel R, Kusternig A and Wimmer M 2007 *Proc. of the 18th Eurographics Conf. on Rendering Techhniques (Grenoble)* (Goslar: Germany/Eurographics Association) p 253
[2] Haiyan C 2020 *J. Phys.: Conf. Ser.* **1607** 012016
[3] Chanserm T, Tupsai J and Yuenyong C 2019 *J. Phys.: Conf. Ser.* **1340** 012086
[4] Li A, Yin C, Gan Y, Zhang Q and Wu W 2021 *Eur. J. Phys.* **42** 045203
[5] The Physics Classroom 2021 *The Law of Reflection* Retrieved 20 Jan. 2021, from https://www.physicsclassroom.com/class/refln/Lesson-1/The-Law-of-Reflection
[6] The Physics Classroom 2021 *Specular vs. Diffuse Reflection* Retrieved 22 Jan. 2021, from https://www.physicsclassroom.com/class/refln/Lesson-1/Specular-vs-Diffuse-Reflection
[7] Pharr M, Jakob W and Humphreys G 2016 *Physically Based Rendering: From Theory to Implementation 3rd Edition* (Massachusetts: Morgan Kaufmann Publishers) p 870
[8] Buck J 2019 *The Ray Tracer Challenge: A Test-Driven Guide to Your First 3D Renderer* (Raleigh: Pragmatic Bookshelf)
[9] Dearamae R, Safkolam R and Yacob M 2021 *J. Phys.: Conf. Ser.* **1835** 012065
[10] Gargallo A, Gómez-Varela A I, Gónzalez-Nuñez H, Delgado T, Almaguer C, Cambronero F, García-Sánchez Á, Pallarés D, Aymerich M, Aragón Á L and Flores-Arias M T 2015 *J. Phys.: Conf. Ser.* **605** 012040

Journal of Physics: Conference Series **2145** (2022) 012075 doi:10.1088/1742-6596/2145/1/012075

The interactive computer simulation and learning activity for facilitating students' conceptual understanding on the buoyant force through the CoSci learning platform

Wachirawut Wongsuwan[1], Jiradawan Huntula[1,*] and Chen-Chung Liu[2]

[1] Science Education Program, Faculty of Education, Khon Kaen University, Khon Kaen, 40002, Thailand
[2] Graduate Institute of Network Learning Technology, National Central University, Taoyuan, 32001, Taiwan

*E-mail address: jirahu@kku.ac.th

Abstract. Learning science, especially in the physics field, there are many varieties of invisible and abstract phenomena that are hard and difficult for students to observe and learn. One of the tools that can help students to understand those phenomena in a better way is computer simulations. The computer simulations are usually used in both on-site classroom and on-line learning platforms. Learning in the COVID-19 pandemic era at present, the computer simulations are very important for helping students to understand the physics concept. Interactive computer simulation can be considered as one of the effective methods of facilitating inquiry learning in science, as it allows students to experience the scientific inquiry process and facilitates students to understand an abstract conception and to understand the relationship between variables of invisible phenomena more clearly in reasonable ways. This study aimed to develop the interactive computer simulation and learning activity for enhancing students' conceptual understanding of the buoyant force on the CoSci learning platform. Totally eighteen participants were studied in the twelfth grade in science classrooms of a university-affiliated school project (SCiUS), Khon Kaen University, Thailand, in 2019. The learning activity was developed based on students' alternative concepts and used to facilitate students' conceptual understanding of the buoyant force. There were six basic concepts related to the buoyant force constructed based on the predict-observe-explain strategy (POE) with the interactive computer simulation (i.e., the CoSci learning platform) in the learning activity. The learning activity on the CoSci learning platform consisted of eight pie charts such as 1) main question pie chart, 2) density pie chart, 3) water level pie chart, 4) volume pie chart, 5) mass pie chart, 6) weight pie chart, 7) submerged depth pie chart, and 8) answer pie chart. There were six interactive computer simulations used in this research including 1) density simulation, 2) water level simulation, 3) volume simulation, 4) mass simulation, 5) submerged depth simulation, and 6) weight simulation. All of these simulations were developed on the CoSci learning platform (https://cosci.tw/). The findings showed that 72% of students performed better in the post-test scores than in the pre-test score in all six basic concepts related to the buoyant force after learning buoyant force on the CoSci platform. Furthermore, the most difficulty in changing misconception in learning of the buoyant force was the concept related to the mass of the object.

1. Introduction

Since late 2019, the COVID-19 pandemic has created the largest disruption of education systems in history, affecting learners in more than 190 countries around the world [1]. As the COVID-19 transmits by person-to-person, many countries launch the policy to close the school, university, and other learning spaces such as a library, and a museum for preventing the COVID-19 pandemic. Responsibility of the distancing policy, online learning has been forced and brought to take action into the educational field. But it does not seem like online learning is only used for preventing the pandemic, because it has its own advantages [2]. However, there are many varieties of online learning platforms available to use. It becomes a challenge for teachers to try to find online learning platforms that suitable for their field. One of the online learning platforms that suitable for learning science is CoSci learning platform (https://www.cosci.tw/, hold a copyright since 2013 by LTLab, NCU, Taiwan). On the CoSci platform, students can learn independently after exploring them on the activities provided during their learning. Learning science, especially in the physics field, there are many varieties of invisible and abstract phenomena that are difficult for students to observe and learn. One of the tools that can help students to understand those phenomena in a better way is computer simulations. The interactive computer simulations, provided in the CoSci platform, may change students' alternative conceptions directly. Students can vary the variables, examine how the variables are changed, and investigate the problem by using interactive computer simulation. There are many benefits for students to learn in the interactive computer simulation, for example, replace such a kind of risk, visualizing an abstract concept, and saving time for conducting experiments [3]. The interactive computer simulation can be considered as one of the effective methods of facilitating inquiry learning in science education, as it allows students to experience the scientific inquiry process [4]. The conception of the buoyant force in previous research was considered that students' conceptual understanding of the buoyant force in Thailand is very low (O-net, 2015). Most students still have alternative conceptions on the buoyant force even they have been taught before [5]. Therefore, this research needed to develop the interactive computer simulation and learning activity for facilitating students' conceptual understanding on the buoyant force concept.

2. Methodology

The purposes of this study were to develop the interactive computer simulation and learning activity for facilitating students' conceptual understanding on the buoyant force concept. There were eighteen students of twelfth-grade from a university-affiliated school project (SCiUS), Khon Kaen University participated in this study. They were selected because they have already learned the buoyant force concept since they were in ninth-grade. None of the participants used the CoSci learning platform before this study. To examine students' conceptual understanding before and after they learned the buoyant force through the interactive computer simulation and learning activity, the conceptual test was revised from the prior study was used as pre-test and post-test in this study [5]. The conceptual test is a two-tier multiple choice, consisted of eight items which are related to the concept of buoyant force including density (ρ), mass (m), weight (W), submerged depth (h), volume (V), water level (L). After students finished the pre-test, they learned the activity provided on CoSci learning platform which is the problem related to the buoyant force in 100 minutes or until he or she finished learning. An activity consisted of eight pie charts included 1) main question pie chart, to ask students to solve the problem of buoyant force., 2) density pie chart, 3) water level pie chart, 4) volume pie chart, 5) mass pie chart, 6) weight pie chart, 7) depth pie chart, and 8) answers pie chart for answering the question posted on a main question pie chart. Pie chart number 2) to 7) were created from students' alternative conception of buoyant force. The POE teaching strategy was adapted in pie chart number 2) to 7) for also changing students' alternative conceptions of buoyant force. Interactive computer simulations used in this study were also developed from students' misconceptions on the buoyant force. There were six interactive computer simulations used in this research including 1) density simulation, 2) water level simulation, 3) volume simulation, 4) mass simulation, 5) submerged depth simulation, and 6) weight simulation. All of these simulations were developed on the CoSci learning platform (https://cosci.tw/). We created new interactive computer simulation which are 1) density simulation, 2) water level simulation, 3) volume simulation, and 4) mass simulation. The simulations of submerged depth and weight were revised from the previous simulations, created by Ming-Hua Chang, which were available on the CoSci learning platform. The details of each simulation are described as followed in table 1.

Table 1. Description of interactive computer simulation.

Simulations and Interface	Goals	Descriptions
Density simulation 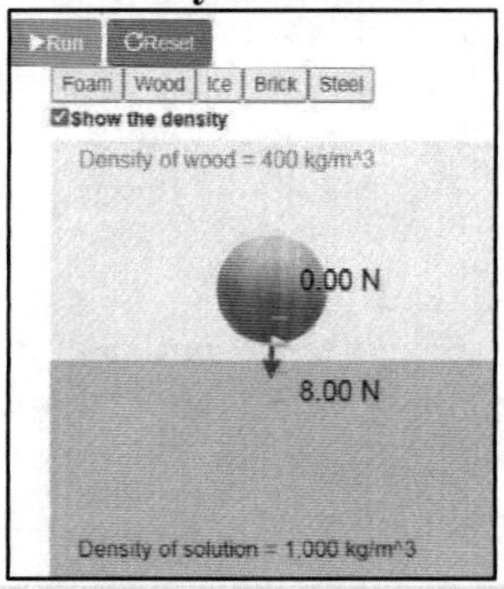	To help students to understand the effect of density on floating and sinking in the solution.	This simulation simulates the floating and sinking of different materials of an object which has a different density. There are five kinds of object available for students to choose by clicking consists of foam, wood, ice, brick, and steel. The simulation's interface comprises an object and a fixed density solution.
Water level simulation 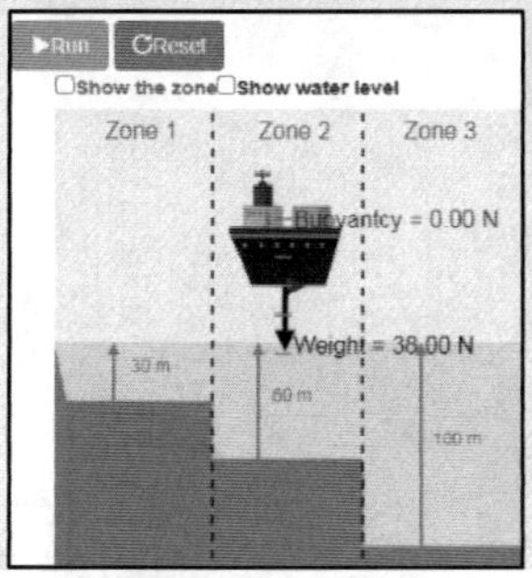	To enhance students' understanding on water level effect to the buoyant force.	This simulation simulates a cargo ship that floating in different level of water. The simulation's interface consists of a cargo ship and a zone of water. There are three zones of different levels of water including zone1: 30 meters depth, zone2: 60 meters depth, and zone3: 100 meters depth.
Volume simulation 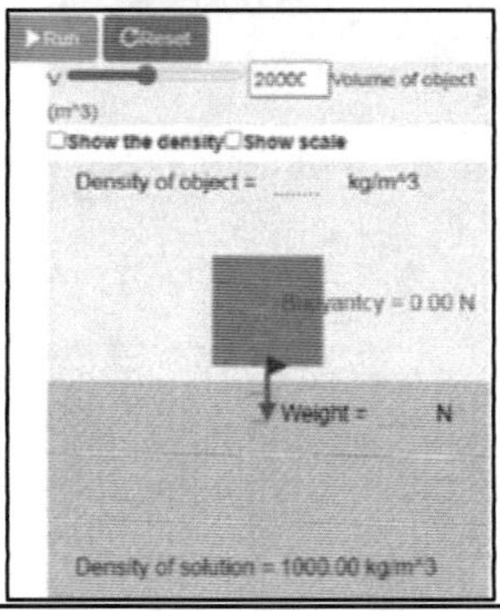	To enhance students' understanding about the effect of volume of object on buoyant force.	This simulation simulates dropping an object into the solution when it was fixed mass. In this simulation, students can vary the volume of an object from 12,000 m^3 up to 30,000 m^3. The colour and size of an object will be changed while the volume is varied for representing a fixed mass object when the volume is changing.

Simulations and Interface	Goals	Descriptions
Mass simulation 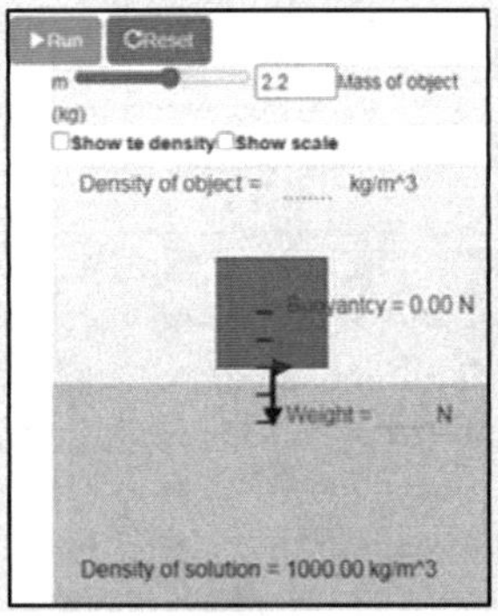	To enhance students' understanding about the effect of mass of object on buoyant force	This simulation simulates dropping an object into the solution, but it was fixed volume of an object. In this simulation, students can vary the mass of an object from 1.2 kg up to 3 kg. Only the colour of an object will be changed while the mass is varied for representing a fixed volume object when the mass is changing.
Weight simulation 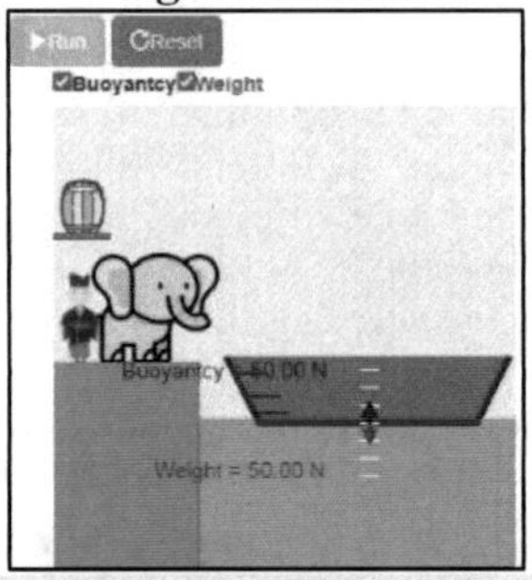	For enhancing the understanding of the impact of weight on the buoyant force.	This simulation simulates the sinking of a boat when adding more weight into it. To run the simulation, students need to drag the elephant, the monk, or tanks into the boat.
Submerged depth simulation 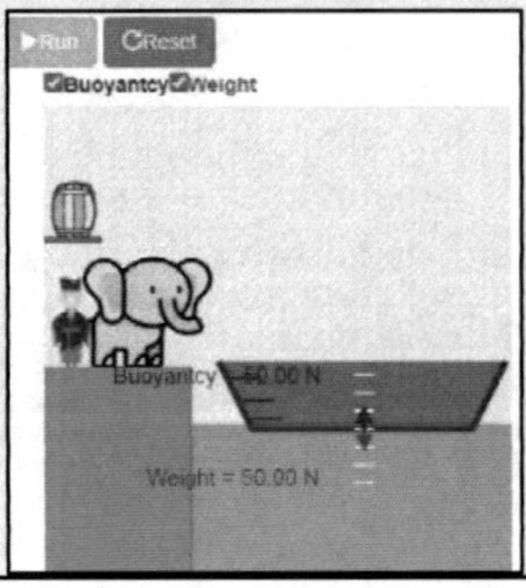	To enhance student to understand the effect of weight on the buoyant force after an object fully submerged in the water.	To run the simulation, students need to drag the elephant, the monk, or tanks into the boat. The weight simulation will stop when the boat completely submerged under the water while the boat of submerged depth simulation will not stop.

The students' responses, behavior, and using time would be recorded in CoSci system while students were using the CoSci learning platform. The post-test was provided to students after they finished learning immediately. To achieve the aim of this study, as shown in table 2, students' responses of the pre-test and the post-test would be scored according to the following criteria by Costu [6]. The conceptual test score of the pre-test and the post-test were analyzed by using the Wilcoxon paired test ranks between pre-test and post-test scores.

Table 2. The criteria for evaluation of student responses to the conceptual test.

Understanding level	Score	Tier 1: Answer (multiple choice)	Tier 2: Reason (multiple choice)
Scientific understanding (SU)	3	True	True
Partial understanding (PU)	2	True False	No response True
Specific misconception (SM)	1	True	False
No understanding (NU)	0	False False No response	False No response No response

3. Research findings

This study examined the impact of the interactive computer simulation toward students' concept on the buoyant force to see students' conceptual understanding after they learned the buoyant force concept on CoSci learning platform which was provided in this study. The finding shows that the post-test score was significantly higher than the pre-test score ($p < 0.01$) as shown in table 3. There was 72% of students had a progression after they learned buoyant force concept on CoSci learning platform.

Table 3. Shows the Wilcoxon test ranks between pre-test and post-test score.

Posttest - Pretest	N	Mean Rank	Sum of Ranks	Z	Asymp. Sig. (2-tailed)
Negative Ranks	0	0.00	0.00	-3.266	0.001 *
Positive Ranks	13	7.00	91.00		
Ties	5				
Total	18				

Note: The pre-test and post-test score were graded following the criteria shown in table 2.

As mentioned above, the conceptual test consisted of eight items of basic concepts related to the buoyant force. The scores of each item in pre-test and post-test were analyzed by using the Wilcoxon paired test ranks to examine students' basic concepts related to the buoyant force more deeply. The finding shows that the percentages of students who were successful in learning the buoyant force (SU level) in each item were as follows: 1) item 1 related to submerged depth concept which there was 100% of students who were in SU level, 2) item 2 related to water level concept which there was 100% of students who were in SU level, 3) item 3 related to density concept which there was 94% of students who were in SU level, 4) item 4 related to volume concept which there was 100% of students who were in SU level, 5) item 5 related to mass concept which there was 72% of students who were in SU level, 6) item 6 related to weight concept which there was 89% of students who were in SU level, 7) item 7 related to submerged depth concept which there was 94% of students who were in SU level, and 8) item 8 related to density concept which there was 100% of students who were in SU level. Most of the students had difficulty with item 5, which was related to the mass concept. Significant difference ($p < 0.01$) was found between the pre-test and the post-test score of this item, while the others had no difference. The number of students who performed unsuccess in the pre-test and post-test of each concept is shown in figure 1.

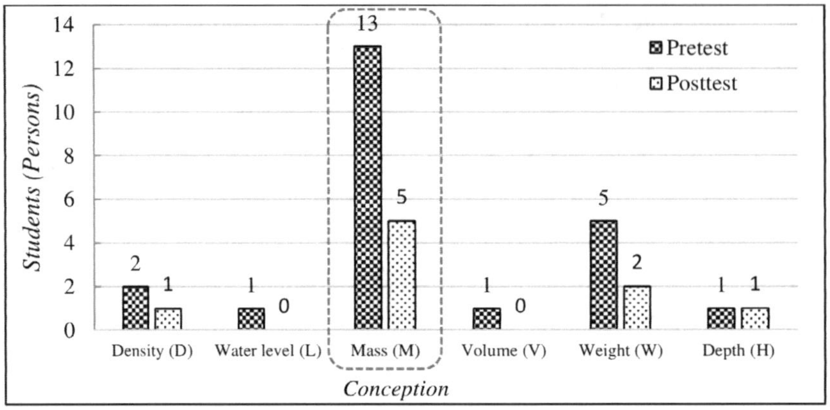

Figure 1. Students who performed unsuccess in each concept.

4. Conclusion and discussion

The interactive computer simulation can use to facilitate students' learning process and to enhance students' conceptual understanding on the concept of buoyant force. Because while students were learning on the CoSci, they were enhanced to observe phenomenon by using the computer simulation carefully. We found that most students predicted the result incorrectly. However, after students used the computer simulation with the guided question carefully and clearly, they could correct the target conception of the activity and they could explain the situation correctly.

References

[1] Tomasik M J, Helbling L A and Moser U 2021 Educational gains of in-person vs. distance learning in primary and secondary schools: A natural experiment during the COVID-19 pandemic school closures in Switzerland *Int. J. Psychol.* **56** 566–76

[2] Adedoyin O B and Soykan E 2020 Covid-19 pandemic and online learning: the challenges and opportunities *Interactive Interact. Learn. Environ.* doi:10.1080/10494820.2020.1813180

[3] Chang C J *et al* 2017 An analysis of student collaborative problems solving activities mediated by collaborative simulation *Comput. Educ.* **114** 222–35

[4] Chang C J, Liu C C, Wen C T, Tseng L W, Chang H Y, Chang M H, Chiang S H F, Hwang F K and Yang C W 2020 The impact of light-weight inquiry with computer simulations on science learning in Classrooms *Comput. Educ.* **146** 103770

[5] Wongsuwan W and Hantula J 2019 The students' basic conceptions of Buoyant force *J. Phys.: Conf. Ser.* **1380** 012139

[6] Costu B, Ayas A, Niaz M, Unal S and Calik M 2007 Facilitating conceptual change in students' understanding of boiling concept *J. Sci. Educ. Technol.* **16** 524–36

Development creativity of grade 5 students on electricity through STEM education

Supaluk Sasangbong and Jiradawan Huntula*

Science and Technology Education Program, Faculty of Education, Khon Kaen University, Khon Kaen, 40002, Thailand

*E-mail: jirahu@kku.ac.th

Abstract. The research aim was to develop creativity of grade 5 students through STEM education in electricity. There were 15 students of Grade 5 in Khon Kaen Province in 2020 participated in this study. The action research was implemented to this study with three loops of action research to improve creativity of students. The physics contents consisted of simple circuit, conductor, insulator, switch, connecting battery in a series circuit and a parallel circuit, series and parallel circuit and the motor connection. The fifteen lessons were separated into three loops of actions. Each loop of lessons consisted of three Predict-Observe-Explain (POE) lessons, and a lesson which is an Engineering design process lesson for designing product to improve student's creativity in electricity. The students were fluent in designing various tasks to solve problems arising from a given classroom situation and had flexibility and elaboration in given reasons in selecting the materials to create the works. At the last lesson of the each action loops the data was collected to show how students improved their creativity. The student's work and their presentation were evaluated by the creativity's rubric scores in four dimensions of Originality, Fluency, Flexibility and Elaboration. The results show that student's creativity were improve by STEM education in electricity. All dimension of creativity which are originality, fluency, flexibility and elaboration were improved from loop 1 to loop 3 of action research.

1. Introduction

In the 21st century, the organizing of learning activities needs to change to develop the learner. We are in the 21st century, Learning is not just for understanding but for improving process skills of learning. Therefore, the ability of a person to live in the present and in the future consists of three skills, Learning and Innovation Skills, Media and Technology Skills, and Life and Career Skills. In order to develop people in the 21st century, the development of knowledge and creativity is very important to provide the development of innovations [1]. A psychologist, Guilford, said in 1967 that creativity is divergent thinking which is multi-directional, multifaceted, and broad. This kind of thinking leads to the invention of novelty as well as the idea of successful solutions to problems. It includes 1) Fluency, 2) Flexibility, 3) Originality, and 4) Elaboration [2]. In medically, our brain is divided into two hemispheres that work together but exhibit different characteristics depending on the individual's aptitude. Its left hemisphere controls thinking, reasoning, language, logic, critical thinking, and number, and the right hemisphere controls creative thinking, imagination, synthesis, appreciation for music and art, and making use of forms and geometries. STEM education is a teaching and learning approach in which science, technology, engineering, and mathematics (STEM) are purposely targeted to connect the real-life that provided the learners developed or created the new processes or new products along with the development of 21st-century skills. STEM education aims to motivate learners to work on real-world problems by generating ideas and designing relevant products based on integrated STEM knowledge and STEM skills [3]. Additionally, STEM education which is focused on the process of engineering design consisted of identifying problems, finding related concepts, designing solutions, planning and implementing the solutions, testing and evaluating, presenting results or solve the problem [1]. The electricity for elementary school students is important because in learner's daily life involved with using electronic devices. Therefore, the students are necessary to understand the electrical circuit components, including explaining the connecting of a simple circuit and writing circuit diagrams. Moreover, the electrical experiment must be designed with the appropriate way, whether the circuit designing of series, parallel, or mixing, which will provide the learners understanding of the electrical devices and the ability of using appliances efficiently. These are the indicator of learning in the core of sciences [4].

Therefore, to develop the creativity of student STEM education was used on electricity in order to encourage students to apply their understanding and to realize creativity of designing products.

2. Methodology

This is an action research consisted of three loops of action with fifteen lessons totally [5]. There were fifteen Students in grade 5 participate in this research. They were separated into five groups.

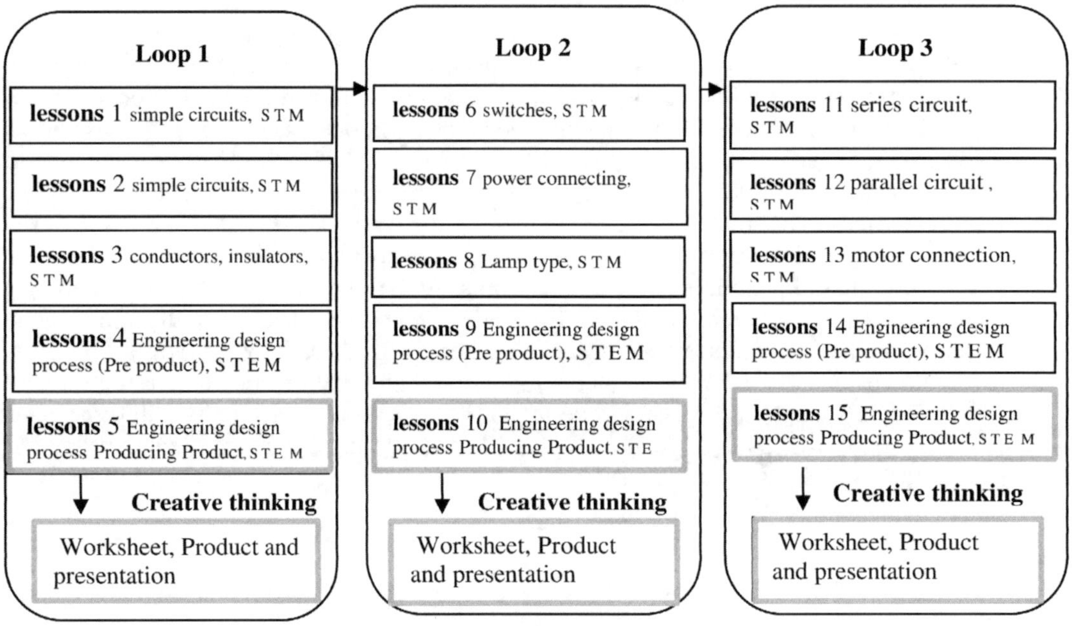

Figure 1. Research methodology design.

From the figure 1, in each loop of action, there are three POE lessons in lessons 1-3. More addition the fourth and the fifth lessons were constructed following engineering design in order to let students design and produce their products. Then in the last lessons of each loop which are lessons 5, 10, and 15 the data was collected from worksheets, products, and presentations of students. In table 1 shows conception of electricity in activity and example products in loop 1 to loop 3.

Table 1. Conception of electricity and example product in activity in loop 1 to loop 3.

Loop	Conception of electricity		Product	
1	simple electrical circuit Conductor and insulator	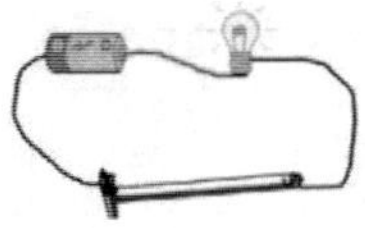	The invention of waste separator. conductor or insulator ?	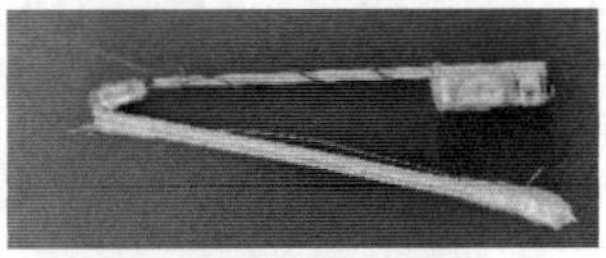
2	switch circuit		city road switch	
3	motor circuit	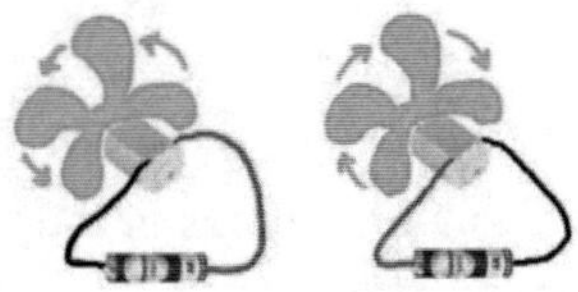	toys and appliances from garbage	

3. Data analysis

The student's worksheets, Products, and presentations in lessons 5, 10, and 15 were collected and interpreted by the rubric score to show the creativity of students following criteria used in table 2 [2].

Table 2. Rubric score for the creativity.

creativity	Description	Scores
Originality	Concept of designing is different from friends in class.	Student answers, more than 61% gets 0 points 21-60% gets 1 point, less than 20% gets 2 points
Fluency	Number of designing concept difference in the class in 10 minute for the activity.	Student answers, more than 61% gets 0 points 21-60% gets 1 point, less than 20% gets 2 points
Flexibility	Reasonable of choosing material in design product.	1 point for each alternative approach
Elaboration	The abilities to add details of designing product, for example; fill the gaps of products attaching, distance measurement, size indication, beautiful color rendering, and explanation for the material of producing.	1 point for each alternative approach

4. Results

This study developed creativity of Grade 5 students on electricity through STEM Education. The finding shows that the characteristics of creativity increased from loop 1 to loop 3.

Figure 2. Shows the designing concept of product in loop 1-3.

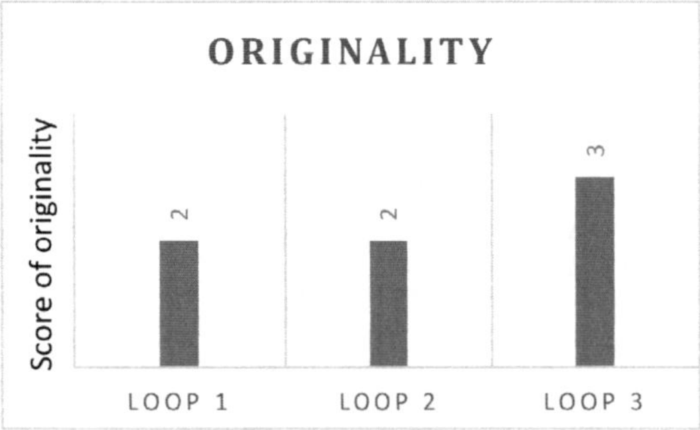

From the figure 2. The originality score is ability of designing product in difference way of friends. There are two scores of designs product in loop 1 and 2 and three scores in loop 3 which are increasing from loop 1 to loop 3.

Figure 3. Shows the idea of design fluency in 10 minutes of loop 1-3.

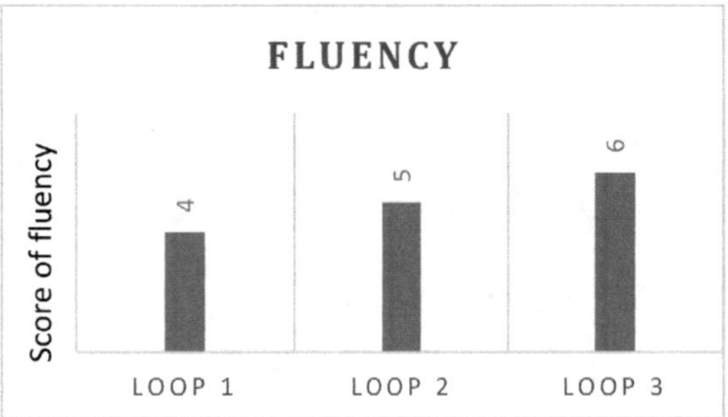

From the figure 3. The fluency score is the idea of design fluency in 10 minutes. There are four ideas of designing task in loop 1, five ideas of designing task in loop 2 and six ideas of designing task in loop 3 which are increased from loop 1 to loop 3.

Figure 4. Shows the guidelines for choosing and reasoning materials of flexibility of loop 1-3.

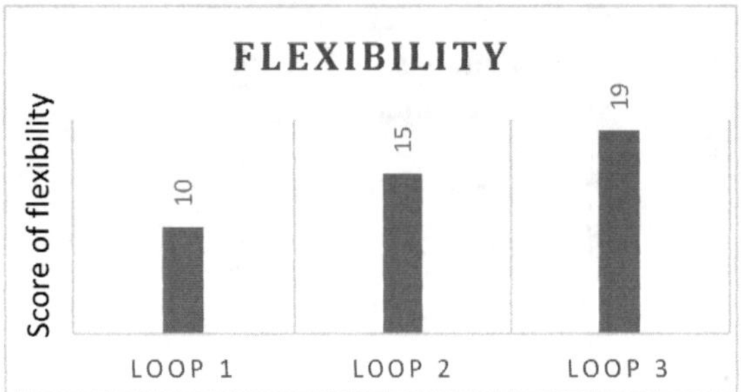

From the figure 4. The flexibility score is the guidelines for reasonable choosing materials of producing product. There are ten ideas in loop 1, fifteen ideas in loop 2 and nineteen ideas in loop 3 which are increased from loop 1 to loop 3.

Figure 5. Shows the elaboration of product in loop 1-3.

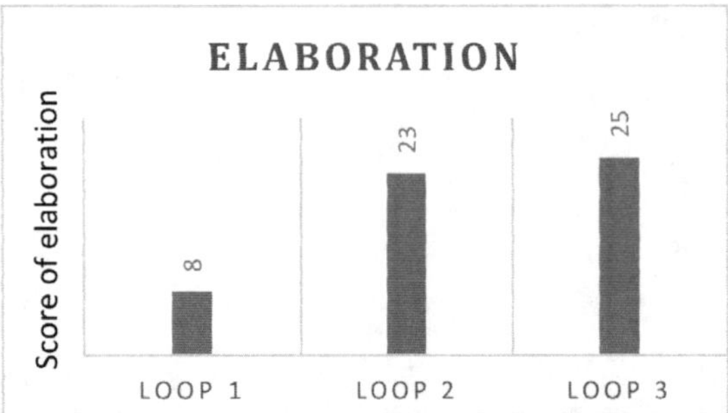

From the figure 5. The elaboration score is the abilities to add details of designing product. There are eight ideas in loop 1, twenty-three ideas in loop 2 and twenty-five ideas in loop 3 which are increased from loop 1 to loop 3.

5. Conclusion and discussion

The results showed that creativity was developed by STEM education which led learners to understand the content and integrate knowledge with the electrical engineering design process. Moreover it was found that; six step of engineering design process to developed creativity as following;

1. Problem Identification, the students share idea and use the content and integrate knowledge for identity the problems in electricity.
2. Related Information Search, the students share ideas of science, technology and mathematic for inventing product that can be accomplished with the limitations of equipment for each problem.
3. Solution Design, the student are fluency increasing idea in 10 minutes.
4. Planning and Development, students have reasonable in choosing material for creating products and can invent suitable product.
5. Testing, Evaluation and Design Improvement, the student can testing and edit product. They can choosing suitable material to create product.
6. Presentation, the student can present ideas of inventing product and explain reasonable in choosing material for product.

According to the results, STEM education focused on engineering design process encouraged students to able to design products in variety of way. In this research students worked together in group and share opinions with their classmates. These kind of activities help them to develop their creativity in creating products.

References

[1] Ministry of Education (Thailand) 2017 *Educationplan 2017-2036* vol.1 (Bangkok: Office of the Education Council) pp 15–6

[2] Genek S E and Küçük Z D 2020 Investigation of scientific creativity levels of elementary school students who enrolled in a STEM program *Elem. Educ. Online* **19**(3) 1715–28

[3] Chamrat S 2017 The definition of STEM and key features of STEM education learning activity *STOU Educ. J.* **10**(2) 13–34

[4] Ministry of Education (Thailand) 2017 *Indicators and subjects of knowledge in the science center (Revised version 2017) are follow up the core curriculum of basic education* vol.1 (Bangkok: The Agricultural Co-operative Federation of Thailand, Ltd.) pp 68–9

[5] Intharaksa P 2019 Active learning management for creative problem-solving *UDRU Educ. J.* **1**(1) 35–43

Journal of Physics: Conference Series **2145** (2022) 012077 doi:10.1088/1742-6596/2145/1/012077

Corrigendum: Efficiency of supercapacitor with $CaTiO_3$-filled polysulfone separators (2021 *J. Phys.: Conf. Ser.* **2145** 012041)

S Suwanwong*, C Yuenyao and A Hutem
Physics Division, Faculty of Science and Technology, Phetchabun Rajabhat University, Phetchabun 67000, Thailand

Page 3:
In the Results and discussion section, the following text appears:
"The efficiency of supercapacitor using $CaTiO_3$- filled PSF separators in the weight percentage of 0.5, 1.0 and 2.0 wt% to analyze specific energy (as shown in figure 5) and specific capacitance (as shown in figure 6) using Battery Testers system,"

This should read:
"The efficiency of supercapacitor using $CaTiO_3$- filled PSF separators in the weight percentage of 0.5, 1.0 and 2.0 wt% to analyze specific energy and specific capacitance (as shown in figure 6) using Battery Testers system,"

Page 3:
In the Results and discussion section, the following ***Figure*** appears:

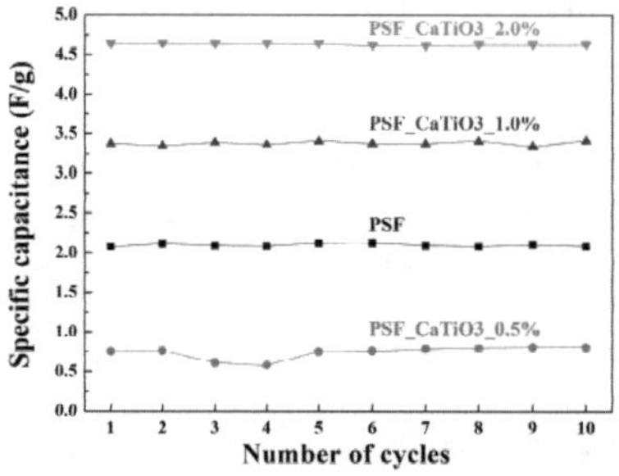

Figure 4. *FT-IR spectra of $CaTiO_3$-filled PSF separators.*

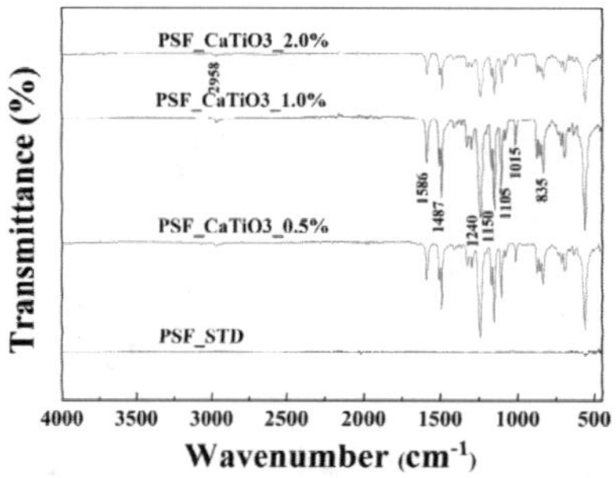

Figure 6. *Specific capacitance of supercapacitors with $CaTiO_3$-filled polysulfone separators.*

This should ***Figure***:

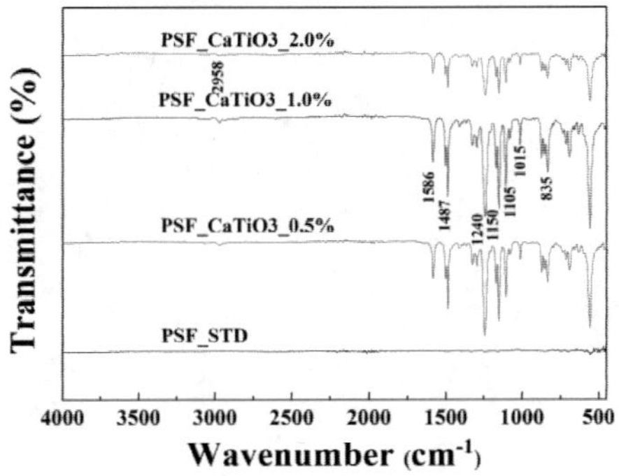

Figure 4. *FT-IR spectra of $CaTiO_3$-filled PSF separators.*

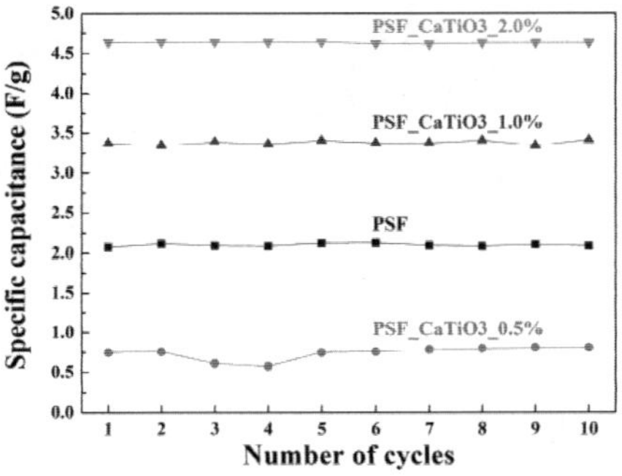

Figure 6. *Specific capacitance of supercapacitors with $CaTiO_3$-filled polysulfone separators.*

AUTHOR INDEX